P9-CQD-210

Statistics and Calculus

Statistics and Calculus

A FIRST COURSE

James Murtha
Department of Mathematics
Marietta College

Earl Willard
Department of Mathematics
Marietta College

PRENTICE-HALL, INC., ENGLEWOOD CLIFFS, NEW JERSEY

ISBN: 0-13-846089-2

Library of Congress Catalog Card No. 72-13441

10 9 8 7 6 5 4 3 2 1

PRINTED IN THE UNITED STATES OF AMERICA

PRENTICE-HALL INTERNATIONAL, INC., *LONDON*
PRENTICE-HALL OF AUSTRALIA, PTY. LTD., *SYDNEY*
PRENTICE-HALL OF CANADA, LTD., *TORONTO*
PRENTICE-HALL OF INDIA PRIVATE LIMITED, *NEW DELHI*
PRENTICE-HALL OF JAPAN, INC., *TOKYO*

Contents

chapter 3 Discrete Statistics *110*

chapter 4 An Introduction to Calculus *221*

chapter 5 Continuous Statistics *385*

Preface

This book grew out of an introductory course for students planning to major in the social, biological, or management sciences, where an increasing emphasis on quantitative methods has caused these students to seek some mathematical training at the college level. Our course was designed as a third alternative to the two most common responses to the needs of such an audience.

One response has been to offer these students a standard calculus course. While this approach may have some virtues, it hardly seems appropriate for students who are not primarily interested in either the level of rigor or the areas of application—principally physics and engineering—associated with traditional calculus.

The other response is to present a cookbook course in statistics and/or finite mathematics. While this is somewhat more acceptable to the intended audience, it does not entirely meet their needs because much of the contemporary literature in their fields of interest is inaccessible without some knowledge of calculus.

We have tried to provide the student with the elements of both statistics and calculus. Our approach, while hardly rigorous, could not be classified as cookbook. We do not expect the student to spend much time proving theorems, but we do try to discuss fundamental ideas and their influence on problem solving.

During the four years in which we have taught this material we have included some elementary computer programming. We feel that in a statistics course students should occasionally handle sets of data of a realistic size. This means some computational device is needed, and we opted for the computer. On the other hand, the textbook has been kept independent of the computer and the material has been taught elsewhere without its use. For the instructor who wants to make the computer a part of the course there are some representative problems scattered throughout the text as well as a section on flow charting in Chapter 1. A more extensive discussion of programming and a number of suggested problems are available in an accompanying instructor's manual.

We have tried to make the calculus arise naturally as we extend statistical procedures from discrete to continuous variables. The concepts of the integral and the derivative are important in understanding statistics. However, honesty compels us to admit that few of the density functions arising in statistics can be handled by elementary means. Much of the development of the calculus beyond the conceptual material in Sections 1–5 of Chapter 4 is there because it is useful and interesting in its own right.

One of the features of the book is the supplementary exercises at the end of the first five chapters. Some of these provide routine drill for the student who needs more practice. Others can be used to extend the course in various directions.

There are always several hard decisions as to whether a particular topic should be included or omitted. Several important topics which we omitted to keep the course of reasonable length appear as supplementary exercises so that the instructor who disagrees with a few of our decisions can to some extent reshape the course. In particular, in the calculus chapter there are several extended supplementary exercises on such subjects as integration by parts, implicit differentiation, trig functions, and so on, which an instructor could use to make the calculus a larger part of the course.

We have included summaries at the end of most sections to give the student a clearer picture of what he should have learned from that section. The Appendix contains the various tables needed in the course. Also included are answers to most odd-numbered exercises.

We have been able to teach essentially all of the material (exclusive of supplementary exercises) in a two-semester course. A two-quarter course could easily cover the first five chapters or could include Chapter 7 (Non-Parametric Techniques) by omitting the last few sections of the calculus chapter.

Any book represents a great deal of work, not only by the authors, but by many others. The three preliminary versions of this text which were used in the classroom and the final manuscript employed the labors of many typists, principally Pat Loreno, Cheryl Hohman, Cynthia Brown, and Jeri Bursae. We thank them all. We also want to thank our colleagues at Marietta, Roger Pitasky and Ray Zarling, who have taught this material several times and who have given us invaluable criticism. Ray Zarling also should be thanked for allowing us to take advantage of his expertise in computer programming. We are grateful to Bill Brown and Bill Topp who taught from a preliminary version of this material at the University of the Pacific. They spotted several weaknesses in our exposition, and they tested the material (successfully) without using a computer. We also want to thank the people at Prentice-Hall: the Mathematics Editor, Art Wester, who signed the book and encouraged us; Bruce Williams, who ably led us through the production stages; and, particularly, the principal reviewer, Ken Hoffman, of M.I.T., whose criticism, patience, and sense of humor were extremely helpful in the final preparation of the manuscript.

JAMES MURTHA / EARL WILLARD

Statistics and Calculus

CHAPTER

1 Descriptive Statistics

0 INTRODUCTION

The first objective of this chapter is to examine methods of analyzing and representing sets of numbers called data. We will see how to organize large sets of numbers into smaller groups, how to draw pictorial representations of data, and how to extract measurements of the data sets that help provide us with distinguishing features among various sets of numbers. We hasten to point out that data comes to us in a variety of forms in everyday life, and that much of what we will be doing can be just as meaningful to the layman as to the statistician. As often as not, we encounter the data in pictorial form (charts, pie diagrams, histograms, bar graphs, and the like) in newspapers and magazines, so that it is useful to understand the interplay between numerical tables or lists and their corresponding graphical representations.

A second objective of this chapter is to develop some working tools. One of these is a notational device, called summation notation, that offers us a shorthand technique for expressing formulas that involve sums of large sets of numbers. We shall also see how to use flow charts to help us break down a complex procedure into a sequence of simple steps. Flow charts have become popular in recent years in large part because of their applicability to computer programs, which are simply sequences of very basic instructions.

1 ORGANIZING AND REPRESENTING DATA

The old saw, "A picture is worth 1,000 words," should be amended by the phrase, "provided the picture is a good one and the reader knows how to interpret it." In this section we shall examine pictures called *histograms* and study the attendant terminology with the aim of learning how such pictures come about and how we should interpret them.

EXAMPLE 1 Student Enrollment

The information in the following table appears in a publication by the Ohio Board of Regents. It shows the distribution of student course loads in Fall, 1968, among the State supported schools.

Hours of Credit	Number of Students	
	4-Year Schools	2-Year Schools
1–3	16,598	10,830
4–6	17,908	8,273
7–9	10,969	4,487
10–12	22,821	6,016
13–15	60,617	10,335
16–18	56,586	6,932
19 or more	7,583	557
Total	193,082	47,430

Each of these categories

1–3, 4–6, 7–9, 10–12, 13–15, 16–18, 19 or more

is called a *class*. The number of items (in this case, students) in a particular class is called the *absolute frequency* of the class. Thus, for instance, the absolute frequency of the class 4–6 in two-year schools is 8,273.

The two pictures that follow are called *histograms*. Figure 1.1 is a pictorial representation of the first two columns of the above table.

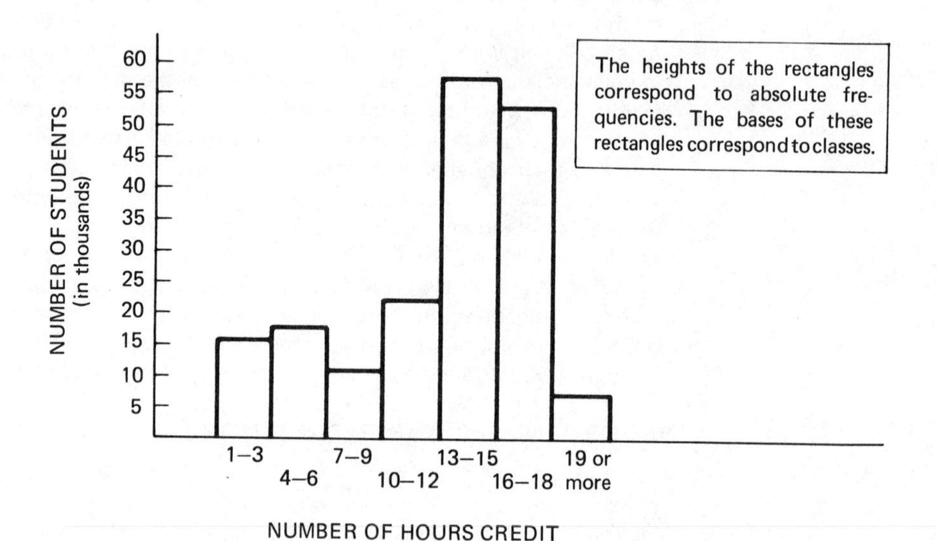

FIGURE 1.1 Histogram for student enrollment load—4-year schools.

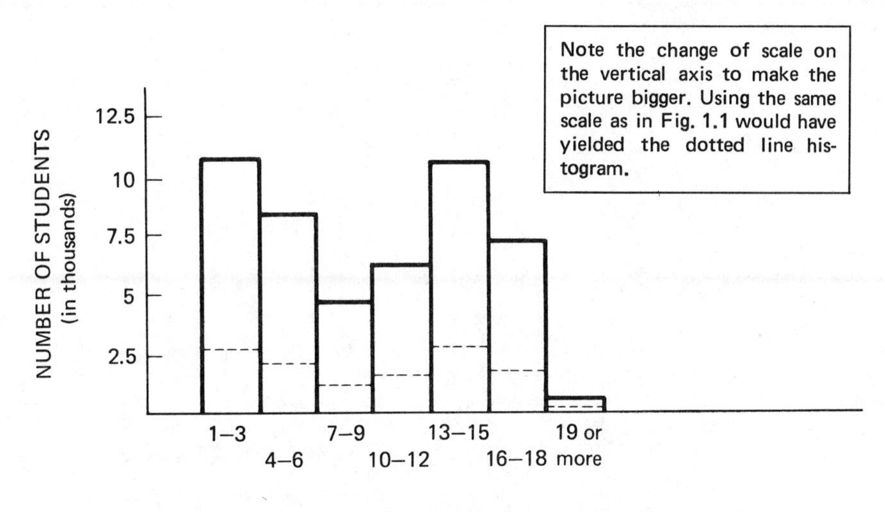

FIGURE 1.2 Histogram for student enrollment load—2-year schools.

QUESTION 1 Is the pattern of student enrollment in two-year schools different from that in four-year schools?
ANSWER Yes. From the histograms it is clear that a larger percentage of two-year college students were enrolled in light course loads (1 to 6 hours credit).

QUESTION 2 If 15 credit hours is considered "average," how many four-year college students had heavier than average loads?
ANSWER From the table we add the frequencies for the two classes (i.e., categories) "16–18" and "19 or more":

$$56{,}586 + 7{,}583 = 64{,}169$$

Thus 64,169 students had heavy loads.

QUESTION 3 How many two-year college students had lighter-than-average loads?
ANSWER If lighter-than-average means less than 15 hours, then *we cannot tell* from the information given, because the class 13–15 includes some students who should be counted as having lighter loads as well as some who are taking average loads. We would have to examine the *raw data*, that is, the information on the 47,430 students before it was grouped into classes.

EXAMPLE 2 Borrowers' Family Income

The distribution of family incomes for 140 students who borrowed money from the Ohio Student Loan Commission is shown below.

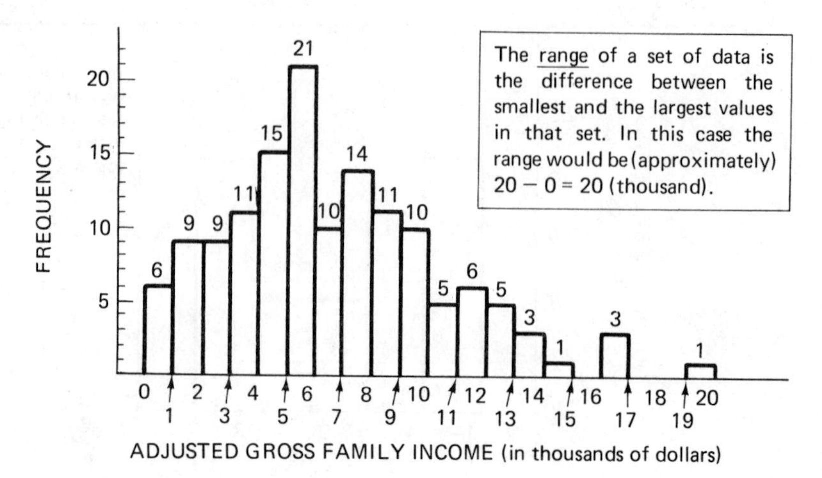

FIGURE 1.3 Histogram of family income for student borrowers.

QUESTION 1 What is a typical class in this case?
ANSWER Although it is not clear from the scale on the horizontal axis, it is customary to use the classes

0 to 999,	denoted by	─┼──┼─ (0, 1)
1000 to 1999,	denoted by	─┼──┼─ (1, 2)
2000 to 2999,	denoted by	─┼──┼─ (2, 3), and so on.

These are called *class boundaries*

QUESTION 2 What percentage of students came from families with income of at least 10,000 dollars?
ANSWER We add the frequencies for the classes 10–11, 11–12, . . . , 19–20

$$5 + 6 + 5 + 3 + 1 + 0 + 3 + 0 + 0 + 1 = 24$$

Since there are 140 students in all, we divide

$$\tfrac{24}{140} \approx .1714$$

$\approx$ means "approximately equal to."

Thus approximately 17.14% of the students came from families whose income was at least 10,000 dollars.

EXAMPLE 3 Gum Ball Diameters

This time we begin with raw data. Suppose we measured the diameter (in centimeters) of 48 large gum balls and got

3.63	3.56	3.44	3.47	3.75	3.64	3.46	3.47
3.70	3.64	3.39	3.73	3.40	3.36	3.48	3.57
3.76	3.65	3.46	3.54	3.47	3.55	3.52	3.56
3.50	3.57	3.49	3.53	3.54	3.46	3.61	3.72
3.46	3.56	3.55	3.54	3.51	3.58	3.62	3.39
3.47	3.56	3.57	3.54	3.55	3.60	3.60	3.68

We group these numbers into eight classes, arriving at the following table, and then draw the corresponding histogram.

Class	Class Mark	Absolute Frequency	
3.36–3.40	3.38	IIII	(4)
3.41–3.45	3.43	I	(1)
3.46–3.50	3.48	𝍸𝍸 I	(11)
3.51–3.55	3.53	𝍸𝍸	(10)
3.56–3.60	3.58	𝍸𝍸	(10)
3.61–3.65	3.63	𝍸 I	(6)
3.66–3.70	3.68	II	(2)
3.71–3.76	3.735	IIII	(4)

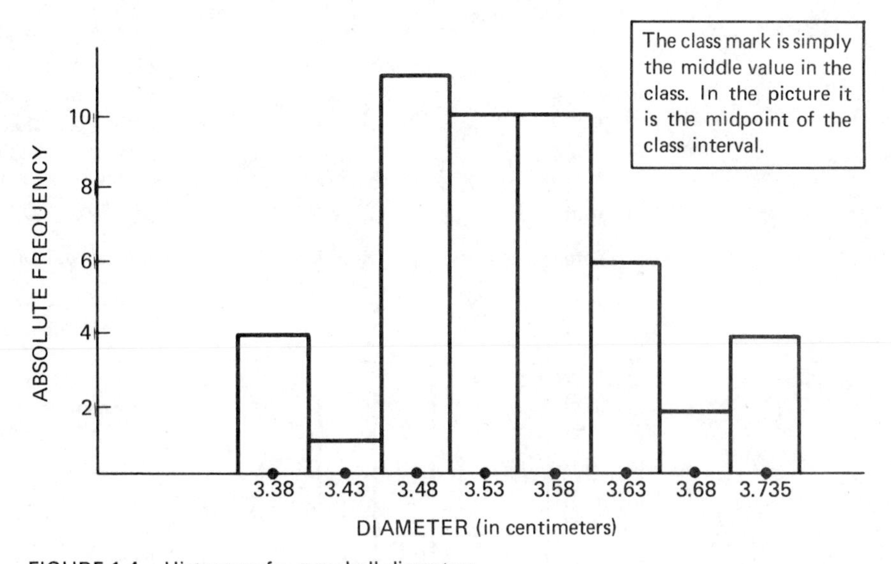

FIGURE 1.4 Histogram for gum ball diameters.

QUESTION 1 What is the range of the gum ball diameters?
ANSWER Since we were given the raw data, we can find the exact range by noting that 3.36 and 3.76 are, respectively, the smallest and the largest values in the data set. Thus the range is 3.76–3.36 = .40 (centimeters).

QUESTION 2 What percentage of the gum balls measured had diameters larger than 3.45 centimeters but smaller than 3.71 centimeters?
ANSWER We add the frequencies for the third through seventh classes:

$$11 + 10 + 10 + 6 + 2 = 39$$

Since there are a total of 48 measurements, we divide

$$\tfrac{39}{48} = .8125$$

Thus, 81.25% of all gum balls measured have diameters of at least 3.46 but not more than 3.70 centimeters.

QUESTION 3 Are all eight classes the same size?
ANSWER No. The last class is slightly larger than the other seven.

EXAMPLE 4 Persons Wanted for Crimes in the United States

According to the National Crime Information Center, in April, 1969, there were 28,404 persons wanted for crimes in the United States. Their ages were distributed as follows.

Class (Age)	Absolute Frequency (Number of Persons)	Relative Frequency (Per cent of Total)
under 20	1,230	4.3% or .043
20–24	5,795	20.4% or .204
25–29	6,032	21.2% or .212
30–34	4,393	15.5% or .155
35–39	3,481	12.3% or .123
40–44	2,974	10.5% or .105
45–49	2,027	7.1% or .071
50 or above	2,472	8.7% or .087
Total	28,404	100.0% or 1.00

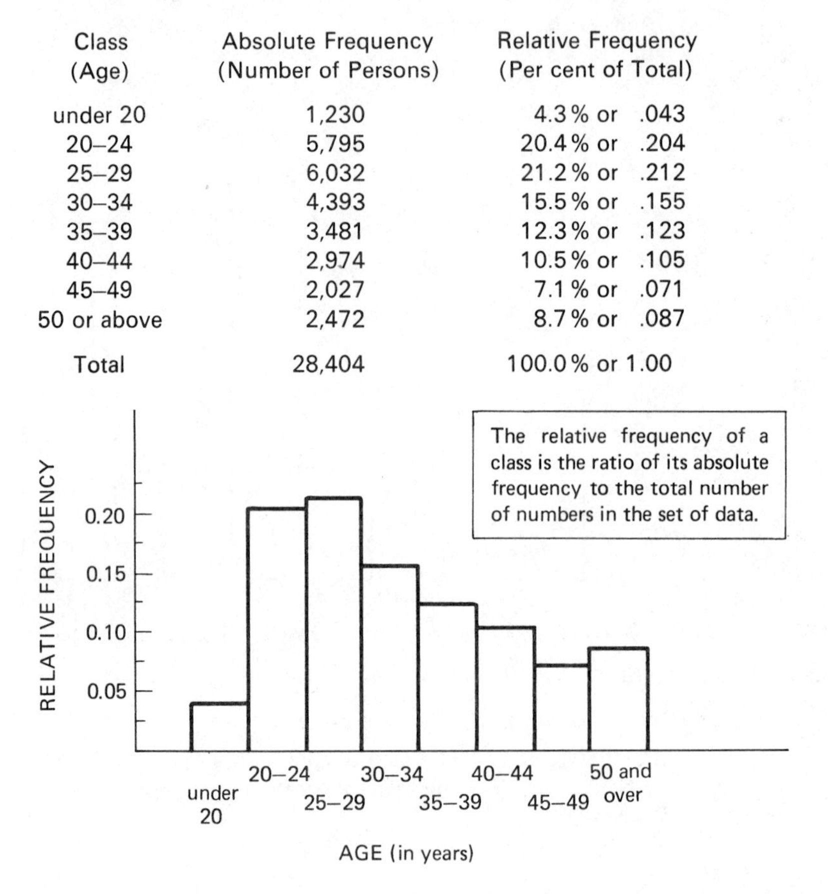

FIGURE 1.5 Relative frequency histogram for age of persons wanted for crimes in U.S.—April, 1969.

QUESTION 1 How does the shape of the relative frequency histogram compare to that of the absolute frequency histogram?
ANSWER They look alike; only the scale on the vertical axis is changed.

QUESTION 2 What is the range of this set of data?
ANSWER We can't tell exactly from the information given. The class "under 20" could conceivably contain values as low as 10 or 12; while the class "over 50" might contain values as high as 80 or more. A reasonable estimate might be 70 (years).

QUESTION 3 What per cent of criminals wanted in April, 1969, were under 30?
ANSWER We add the relative frequencies of the first three classes:

$$4.3 + 20.4 + 21.2 = 45.9$$

Thus approximately 45.9% of the criminals wanted were under 30 years old.

QUESTION 4 Why *approximately* 45.9%?
ANSWER The relative frequencies were calculated by long division and rounded off to the nearest decimal point. Indeed in some cases the error caused by this rounding off will cause the total relative frequency to be slightly different from 100%.

Summary

We have examined several sets of *data*. In three examples, we were presented with the data already grouped into *classes*; once we began with *raw data* and did the grouping ourselves. In each instance we constructed a *histogram* to summarize graphically our results. The horizontal axis of the histogram was marked off according to the various classes into intervals and labeled in an assortment of ways: by indicating the *class boundaries* (the endpoints of the intervals); by indicating the class itself (e.g., "10–14"); or by indicating the *class mark* (the midpoint of the interval). The vertical axis was labeled either according to *absolute frequency* (the number of values in each class) or *relative frequency* (the ratio of absolute frequency to total). We observed that once data has been grouped into classes, some detailed information is lost since we cannot tell how the various values within a given class are distributed. In particular, we found that the *range* (the difference between the smallest and the largest value in the raw data) must be estimated when working with grouped data.

When one does the grouping of raw data and the construction of its histogram, he relies on his judgment to a large degree in deciding upon the number of classes and their boundaries. Nevertheless, there are some rules of thumb that should be taken into account.

1. There shouldn't be too few or too many classes: generally at least five but no more than 20 classes should be used.
2. When possible, each of the classes should be approximately the same size; otherwise comparisons of frequencies among classes are apt to be misleading. Of course, we have seen situations where one or two classes were somewhat larger than the others (e.g., "50 or more," "under 20," etc.).

3. Classes must be *disjoint* (they may not overlap) and *exhaustive* (every available value in the data set must fall into one class).

Exercises Section 1

1 Construct a histogram for the following data.

Age (in years)	Absolute Frequency (nearest thousand)
under 5	66
5 to 9	59
10 to 14	50
15 to 19	35
20 to 24	33
25 to 29	38
30 to 34	39
35 to 39	40
40 to 44	35
45 to 49	32
50 to 54	27
55 to 59	23
60 to 64	17
65 to 69	13
70 to 74	9
75 to 79	5
80 to 84	3
85 and over	2

This table represents the male population in Dallas, Texas in the year 1959.

2 Regroup the data in Exercise 1 using the classes

0–9, 10–19, 20–29, and so on

and draw the corresponding histogram.

3 Some of the questions that follow cannot be answered on the basis of the census information in Exercise 1. Decide which questions can be answered, and answer them.

(a) What percentage of males were over 50?
(b) What percentage of males were over 49?
(c) What percentage of males were under 6?
(d) What percentage of males were under 10?

4 The following table represents an inventory of trees on commercial forest land in Ohio in 1968, where the number of trees is rounded off to the nearest million. Construct two histograms, one for softwoods and one for hardwoods, indicating the class marks rather than the endpoints in each but the last category. To get reasonable pictures, you will have to use different vertical scales for the two histograms.

Breast-height Diameter (in inches)	Number of Softwoods (in millions)	Number of Hardwoods (in millions)
1.0– 2.9	36	824
3.0– 4.9	23	312
5.0– 6.9	9	150
7.0– 8.9	5	95
9.0–10.9	2	60
11.0–12.9	1	39
13.0–14.9	.4	25
15.0–16.9	.1	15
17.0–18.9	.08	9
19.0–28.9	.02	9
29.0 or larger	.00	.7

5 Referring to the data in Exercise 4:

(a) What percentage of softwood trees had diameter at least 13 inches?

(b) What percentage of hardwood trees had diameter at least 13 inches?

(c) What percentage of hardwood trees had diameter less than three inches?

(d) What percentage of softwood trees had diameter less than three inches?

6 Regroup the raw data on gum ball diameters into classes and draw the corresponding histogram.

7 Rewrite the table in Example 1 by rounding off each of the class frequencies to the nearest thousand, and add each of the resulting columns (under "Number of Students") to compare the total with the rounded off total.

8 Give an example of a set of raw data that could yield the following histogram. (Interpret the class with boundaries 60 and 70 as being "60 through 69.")

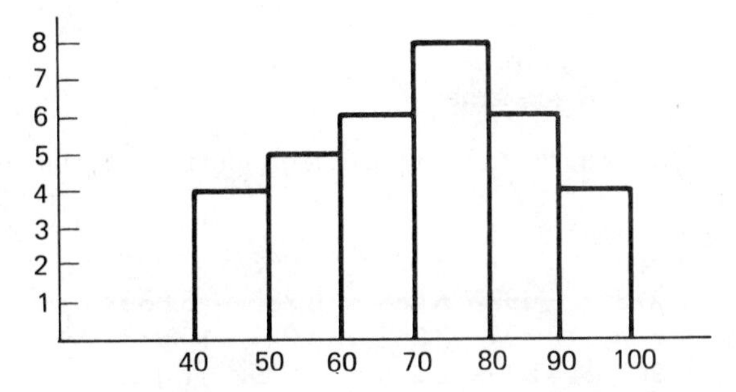

9 The following data sets were gathered by a group of students who (more or less) randomly selected two passages from each of three novels and recorded the number of words per sentence in 30 consecutive sentences from each passage. For each novel, combine the two data sets into one, group the resulting 60 data points into classes 1–5, 6–10, 11–15, . . . , 46–50, 51 or more, and draw the resulting histograms.

(a) From *Tom Jones* by Henry Fielding:

22, 44, 51, 79, 50, 54, 56, 27, 77, 55, 63, 12, 44, 58, 14, 28, 18, 25, 13, 15, 23, 14, 6, 17, 27, 38, 23, 10, 22, 43

33, 68, 50, 22, 28, 20, 25, 11, 37, 87, 24, 28, 36, 31, 24, 25, 23, 51, 48, 32, 57, 106, 57, 22, 15, 52, 47, 20, 27, 15

(b) From *The Sun Also Rises* by Ernest Hemingway:

21, 9, 4, 52, 8, 10, 2, 36, 7, 6, 4, 7, 5, 3, 3, 5, 4, 9, 6, 16, 5, 8, 6, 3, 5, 13, 8, 13, 18, 5

1, 16, 8, 6, 8, 5, 7, 4, 8, 11, 4, 11, 2, 5, 3, 5, 6, 4, 2, 17, 11, 31, 9, 15, 7, 7, 8, 15, 11, 11

(c) From *Far From the Madding Crowd* by Thomas Hardy:

42, 49, 18, 36, 48, 16, 31, 67, 73, 31, 35, 49, 31, 19, 7, 8, 49, 29, 9, 9, 17, 5, 7, 8, 1, 4, 1, 6, 3, 4

12, 17, 2, 9, 9, 5, 8, 19, 9, 8, 6, 3, 9, 17, 7, 4, 25, 35, 11, 16, 17, 13, 9, 4, 23, 15, 49, 24, 14, 12

10 Roll a pair of dice 72 times, record the sum of the faces showing on each roll, and draw the resulting frequency histogram using the singleton classes 2, 3, 4, 5, 6, 7, 8, 9, 10, 11, 12.

11 **(a)** Draw an absolute and a relative frequency histogram for the following data set.

−2	1	2	3
−2	1	2	3
0	2	2	4
0	2	3	4

(b) Now group the data and draw the grouped (absolute) frequency histogram. Use the groupings.

−2 to −1
0 to 1
2 to 3
4 and larger

12 The following set of numbers represents the populations (in thousands) of the fifty states as of 1967.

3,540	741	5,421	1,003	674
273	699	8,584	18,335	3,888
1,635	10,894	3,582	5,027	10,873
1,969	4,999	2,348	639	1,022
19,163	2,753	4,605	10,462	416
1,975	2,275	701	2,496	4,533
2,925	3,292	1,435	1,999	3,089
523	3,660	444	11,626	1,798
5,996	973	685	901	4,188
4,511	3,685	7,004	2,603	315

Draw the grouped data histogram using the groupings

0–1 million
1–3 million
3–5 million
5–7 million
over 7 million

Use the histogram you have drawn and/or the raw data to answer the following questions.

(a) How many states had a population larger than 5,000,000?
(b) What percentage of states had population larger than 1,000,000?
(c) How large does a state have to be to put it in the top 10%?
(d) How many states have between 1 and 6 million people?
(e) What is the range of the data presented in this problem?

13 Represent the following 21 grade point averages as a histogram using 0, 1, 2 and 3 as class marks.

1.035	2.096	1.172
2.111	1.411	1.121
.892	1.000	1.757
.705	1.647	1.000
1.400	2.406	2.527
1.533	.931	.760
1.371	1.322	1.633

2 MEANS AND STANDARD DEVIATIONS

The methods outlined in Section 1 enable us to begin interpreting data by organizing it and representing it pictorially. Usually, however, a satisfactory interpretation of data demands a more detailed analysis. In particular there are two common measurements one imposes on a data set: the *mean* (or *average*) and the *standard deviation*. The objective of this section is to study these two measurements through algorithms and examples.

Rule for Calculating the Mean of a Set of Numbers

Step 1: Add together all the numbers; call the sum SUM.
Step 2: Count how many numbers there are in the set; call the tally NUM.
Step 3: Divide SUM by NUM; call the result MEAN.

EXAMPLE 1 To minimize the computational difficulties, we find the mean of a small set of numbers: 1, 1, 2, 2, 3, 3, 3, 3, 4, 4, 5, 5.

Step 1: SUM $= 1 + 1 + 2 + 2 + 3 + 3 + 3 + 3 + 4 + 4 + 5 + 5 = 36$.
Step 2: NUM $= 12$.
Step 3: MEAN $= \frac{36}{12} = 3$; the mean is 3.

EXAMPLE 2 Find the mean of the set

1, 1, 1, 1, 2, 2, 4, 4, 5, 5, 5, 5

Step 1: SUM = 1 + 1 + 1 + 1 + 2 + 2 + 4 + 4 + 5 + 5 + 5 + 5 = 36.
Step 2: NUM = 12.
Step 3: MEAN = $\frac{36}{12}$ = 3; the mean is 3.

EXAMPLE 3 A student has taken three exams in a math course and received scores of

70, 65, 62

His final grade will be based on the average of four exams. In particular, if he needs an average of 70 or better to pass the course, what grade must he receive on his fourth exam? Let us call the score on the fourth exam x. Then the average of the four exam scores will be

$$\frac{70 + 65 + 62 + x}{4} = \frac{197 + x}{4}$$

and to get an average of 70, we must have

$$\frac{197 + x}{4} = 70$$

or

$$197 + x = 4 \cdot 70 = 280$$

Thus

$$x = 83$$

So, if the student gets an 83 on the fourth exam, he will pass the course. (It should be clear that any score greater than 83 would yield an average greater than 70.)

QUESTION Can we find the average family income of the students who borrowed money from the Ohio Student Loan Commission in 1969? This data was presented in Section 1.
ANSWER No, we cannot. In order to compute the mean of a set of data, we must have access to the *raw data*. It is possible (and indeed practical), however, to *estimate* the mean from a grouped data histogram. To do so, we make the assumption that each number in a given class has the value of the class mark.

That is, we interpret the class

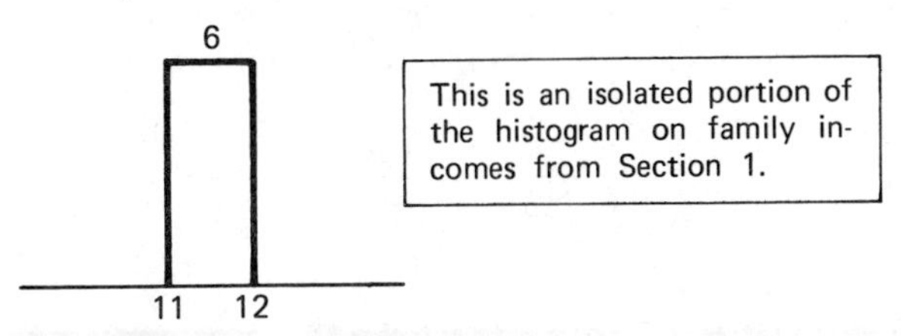

as representing the data

11.5, 11.5, 11.5, 11.5, 11.5, 11.5

Thus, we pretend that the 140 data points in the set are

.5, .5, .5, .5, .5, .5 (or 6 values of .5)
1.5, 1.5, 1.5, 1.5, 1.5, 1.5, 1.5, 1.5, 1.5 (9 values of 1.5)
2.5, 2.5, 2.5, 2.5, 2.5, 2.5, 2.5, 2.5, 2.5 (9 values of 2.5)

and so on. In general one does not write down all these numbers. Instead, the following kind of table is helpful.

Class	Class Mark	Frequency	Contribution to the Mean
0– .999	.5	6	$6 \cdot (.5) = 3.0$
1– 1.999	1.5	9	$9 \cdot (1.5) = 13.5$
2– 2.999	2.5	9	$9 \cdot (2.5) = 22.5$
3– 3.999	3.5	11	$11 \cdot (3.5) = 38.5$
4– 4.999	4.5	15	$15 \cdot (4.5) = 67.5$
5– 5.999	5.5	21	$21 \cdot (5.5) = 116.5$
6– 6.999	6.5	10	$10 \cdot (6.5) = 65.0$
7– 7.999	7.5	14	$14 \cdot (7.5) = 105.0$
8– 8.999	8.5	11	$11 \cdot (8.5) = 93.5$
9– 9.999	9.5	10	$10 \cdot (9.5) = 95.0$
10–10.999	10.5	5	$5 \cdot (10.5) = 52.5$
11–11.999	11.5	6	$6 \cdot (11.5) = 69.0$
12–12.999	12.5	5	$5 \cdot (12.5) = 62.5$
13–13.999	13.5	3	$3 \cdot (13.5) = 40.5$
14–14.999	14.5	1	$1 \cdot (14.5) = 14.5$
15–15.999	15.5	0	$0 \cdot (15.5) = 0$
16–16.999	16.5	3	$3 \cdot (16.5) = 49.5$
17–17.999	17.5	0	$0 \cdot (17.5) = 0$
18–18.999	18.5	0	$0 \cdot (18.5) = 0$
19–19.999	19.5	1	$1 \cdot (19.5) = 19.5$
Totals		140	921.0

$$\text{Estimated Mean} = \frac{921.0}{140} \approx 6.579 \text{ (thousand dollars)}$$

Thus, we estimate that the average family income of borrowers was $6,579.

The reader may have observed that the numbers in Example 1 had the same mean as those in Example 2. Using histograms, we can illustrate how data sets with obviously different ranges and distributions can share a mean value:

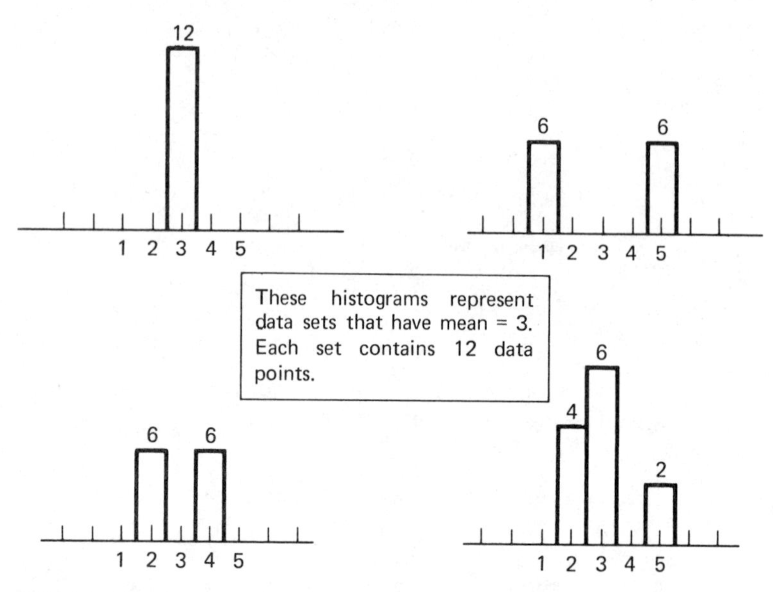

FIGURE 1.6 Histograms for sets of numbers having mean 3.

The *standard deviation* of a set of numbers is a measurement of the extent to which the numbers are spread out from their mean. Roughly speaking, the standard deviation is a kind of average distance away from the mean. There are two methods for calculating the standard deviation. One method allows us to see how the standard deviation measures the dispersion or spread of the data, but it is difficult to compute with. The second method is much more convenient for actual computation, but it obfuscates the underlying idea. In both methods, we first compute an intermediate value, called the *variance*; the standard deviation is then the square root of the variance.

Rule 1 for Computing the Standard Deviation

Step 1: Find the mean of the given set of numbers.

Step 2: Produce a new set of numbers by subtracting the mean from each number in the given set.

Step 3: Square each of the numbers in Step 2.

Step 4: Find the mean of the set of numbers produced in Step 3; this value is the variance, and is denoted VAR.

Step 5: Take the square root of the answer to Step 4; this value is the standard deviation, and is denoted STDEV.

EXAMPLE 4 Find the (mean and the) standard deviation of

1, 1, 2, 2, 3, 3

Step 1: $(1 + 1 + 2 + 2 + 3 + 3)/6 = 2$ is the *mean*.
Step 2: $1 - 2, 1 - 2, 2 - 2, 2 - 2, 3 - 2, 3 - 2$ is the new set.
Step 3: $(1 - 2)^2, (1 - 2)^2, (2 - 2)^2, (2 - 2)^2, (3 - 2)^2, (3 - 2)^2$ or

$$1, 1, 0, 0, 1, 1$$

is the new set.
Step 4: $(1 + 1 + 0 + 0 + 1 + 1)/6 = \frac{2}{3}$ is the *variance*.
Step 5: $\sqrt{\frac{2}{3}} \approx .816$ is the *standard deviation* (using Table A in the Appendix).

Remark: To justify our interpretation of the standard deviation as a sort of average distance of the data points from the mean, let us trace through the steps in Example 4.
Step 1: We found the mean to be 2.
Step 2: We computed the directed distances from the mean to each data point.
Step 3: We squared each of these directed distances from the mean.
Step 4: We found the average squared distance from the mean.
Step 5: We found the square root of the average squared distance from the mean to each data point.
Note that the true average distance from the mean to each data point would be the average of the absolute values of the differences between the mean and the various data points. In the above case,

$$\frac{(|1 - 2| + |1 - 2| + |2 - 2| + |2 - 2| + |3 - 2| + |3 - 2|)}{6} = \frac{2}{3}$$

While the standard deviation is not precisely this average, there are certain advantages, both computational and theoretical, to dealing with squares rather than absolute values of numbers.

EXAMPLE 5 Find the standard deviation of the set

1, 1, 1, 3, 3, 3

Step 1: $(1 + 1 + 1 + 3 + 3 + 3)/6 =$ ____ is the mean.
Step 2: $1 - 2, 1 - 2, 1 - 2, 3 - 2, 3 - 2, 3 - 2$ is the new set.
Step 3: $(1 - 2)^2, (1 - 2)^2, (1 - 2)^2, (3 - 2)^2, (3 - 2)^2, (3 - 2)^2$ or

$$1, 1, 1, 1, 1, 1$$

is the new set.

Step 4: $(1 + 1 + 1 + 1 + 1 + 1)/6 = 1$ is the variance.
Step 5: $\sqrt{1} = 1$ is the standard deviation.

EXAMPLE 6 Find the standard deviation of

0, 0, 0, 4, 4, 4

Step 1: $(0 + 0 + 0 + 4 + 4 + 4)/6 =$ ____ is the mean.

Step 2: ___, ___, ___, ___, ___, ___ is the new set.
Step 3: $4 + 4 + 4 + 4 + 4 + 4$ is the new set.
Step 4: $(4 + 4 + 4 + 4 + 4 + 4)/6 = 4$ is the variance.
Step 5: $\sqrt{\_\_\_} = \_\_\_$ is the standard deviation.

The following sequence of histograms illustrates several data sets with mean 2 and different standard deviations.

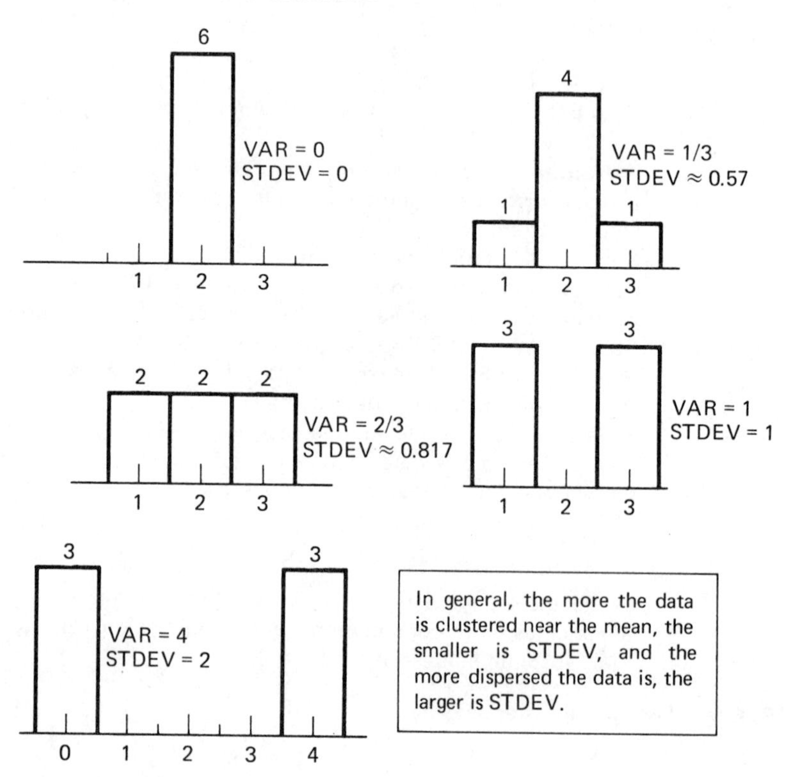

In general, the more the data is clustered near the mean, the smaller is STDEV, and the more dispersed the data is, the larger is STDEV.

Rule 2 for Computing the Standard Deviation

Step 1: Find the mean of the given set of numbers.
Step 2: Find the squares of each of the original data points.
Step 3: Find the mean of the set of numbers in Step 2.
Step 4: Subtract the square of the answer in Step 1 from the answer in Step 3; the result will be VAR.
Step 5: Take the square root of VAR; this will be STDEV.

EXAMPLE 7 Find the standard deviation once again for

1, 1, 2, 2, 3, 3

Step 1: $(1 + 1 + 2 + 2 + 3 + 3)/6 = 2$ is the mean.
Step 2: $1^2, 1^2, 2^2, 2^2, 3^2, 3^2$ is the new set of numbers.

Step 3: $(1^2 + 1^2 + 2^2 + 2^2 + 3^2 + 3^2)/6 =$ ____ is the average of the squares of the original data set.

Step 4: VAR = ____ $- (2^2) =$ ____

Step 5: STDEV $= \sqrt{\_\_\_\_} \approx .817$.

Remark: Rule 2 can be summarized verbally as "the standard deviation equals the square root of the mean of the squares minus the square of the mean." More symbolically, we have

$$\text{STDEV} = \sqrt{(\text{mean of squares}) - (\text{mean})^2}$$

EXAMPLE 8

Find an *estimated* standard deviation for the borrowers' family income whose histogram was given in Section 1. Once again, the best we can do is to *estimate* the variance and the standard deviation by assuming that each data point within a class has the same value as the class mark. The following table, which is quite similar to the one done earlier for estimating the mean, utilizes Rule 2 for finding the standard deviation. In the exercises the reader will be invited to try using Rule 1 for this purpose to see how much harder it would be.

Class	Class Mark	Frequency	Contribution to Variance
0– .999	.5	6	$6 \cdot (.5)^2 = 1.50$
1– 1.999	1.5	9	$9 \cdot (1.5)^2 = 20.25$
2– 2.999	2.5	9	$9 \cdot (2.5)^2 = 56.25$
3– 3.999	3.5	11	$11 \cdot (3.5)^2 = 134.75$
4– 4.999	4.5	15	$15 \cdot (4.5)^2 = 303.75$
5– 5.999	5.5	21	$21 \cdot (5.5)^2 = 635.25$
6– 6.999	6.5	10	$10 \cdot (6.5)^2 = 422.50$
7– 7.999	7.5	14	$14 \cdot (7.5)^2 = 787.50$
8– 8.999	8.5	11	$11 \cdot (8.5)^2 = 794.75$
9– 9.999	9.5	10	$10 \cdot (9.5)^2 = 902.50$
10–10.999	10.5	5	$5 \cdot (10.5)^2 = 551.25$
11–11.999	11.5	6	$6 \cdot (11.5)^2 = 793.50$
12–12.999	12.5	5	$5 \cdot (12.5)^2 = 781.25$
13–13.999	13.5	3	$3 \cdot (13.5)^2 = 546.75$
14–14.999	14.5	1	$1 \cdot (14.5)^2 = 210.25$
15–15.999	15.5	0	$0 \cdot (15.5)^2 = 0$
16–16.999	16.5	3	$3 \cdot (16.5)^2 = 816.75$
17–17.999	17.5	0	$0 \cdot (17.5)^2 = 0$
18–18.999	18.5	0	$0 \cdot (18.5)^2 = 0$
19–19.999	19.5	1	$1 \cdot (19.5)^2 = 380.25$
Totals		140	8,139.00

$$\text{Mean of the squares} = \frac{8139}{140} \approx 58.14$$

Mean (computed earlier) ≈ 6.58

Square of the mean ≈ 43.29

VAR $\approx 58.14 - 43.29 = 14.85$

STDEV $\approx \sqrt{14.85} \approx 3.85$

The mean of a data set is often referred to as a *measure of central tendency*. It helps us locate a sort of typical value. Two other measures of central tendency are the *mode* and the *median*. The mode is simply the data value that has the highest frequency; the median is the middle value when the data are arranged in ascending order. If there happen to be an odd number of data points, then there is a unique "middle value." If, however, there are an even number of data points, the median is defined to be the average of the two middle values.

EXAMPLE 9 We find the median and the mode of some data sets.

(a) Given the data set

$$-1, -1, 5, 5, 2, 3, 2, 2, 6$$

We rearrange to get

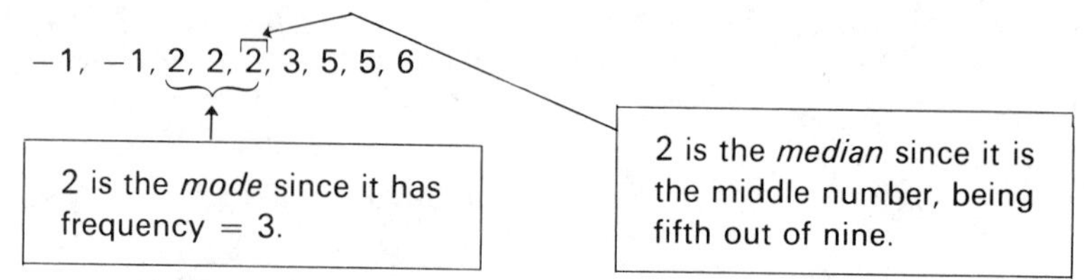

(b) Given the data set

$$100, 90, 1, 1, 1, 1, 4, 10$$

We rearrange to get

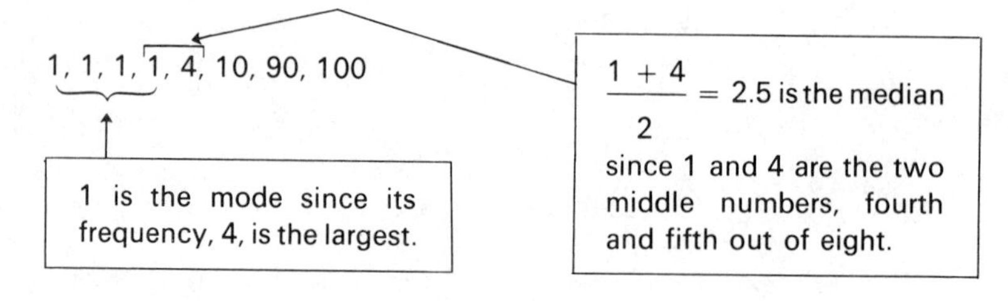

QUESTION If the mean, the mode, and the median are all measures of central tendency, how does one choose among them when analyzing data?

ANSWER Circumstances often dictate which measure is most appropriate. For instance, when dealing with income data, the median is often used rather than the mean to single out a "typical" value. Suppose that the data in Example 8(b) represented the yearly royalties (in thousands of dollars) for eight authors with textbooks in a certain area of mathematics. The *mean* yearly royalty would be $(1 + 1 + 1 + 1 + 4 + 10 + 90 + 100)/8 = 26$ thousand dollars. Yet six of the eight authors have royalties that fall far short of that figure. The *mode* of 1 thousand dollars or the *median* of 2.5 thousand dollars might be more realistic choices of typical values.

Summary

When we are presented with a set of data, it is frequently illuminating to determine its *mean* (or perhaps its *mode* or *median*). These values are called *measures of central tendency* and help us summarize the data set in terms of a "typical" value. In addition, we often determine the *variance* or the *standard deviation* of the data set. These values, along with its *range*, referred to in the literature as *measures of dispersion*, give us an indication of the extent to which the data is spread out from its mean. When the data is given in grouped form rather than as raw data, we can still find *estimates* of the mean and standard deviation by making the assumption that each data point in a given class takes on the value of the class mark.

Exercises Section 2

1 Find the mean, mode, median, variance, and standard deviation for each of the following sets of data.

(a) 1, 2, 3
(b) 100, 200, 300
(c) −1, 0, 1
(d) −2, 0, 2
(e) −200, 0, 200
(f) 1, 3, 5
(g) −1, −1, −1, −1, −1
(h) 5, 4, 5, 3, 5, 2, 5, 1, 5, 0

2 Find a number x such that the set of numbers

$$-2, 1, 4, 6, x$$

has mean 2. Is there more than one possible answer to this problem?

3 Find the values for x and y such that the *mean* and the *median* of

$$2, 6, 1, 0, x, y$$

are, respectively, 2 and $\frac{3}{2}$.

4 **(a)** Find the mean and the variance of

$$1, 2, 3, 4, 5$$

(b) Construct a new data set consisting of the means of all possible pairs of numbers chosen from the above list. (*Hint:* There will be 10 such pairs: (1, 2), (1, 3), etc.)

(c) Find the mean and the variance of the data set in (b).

5 Find a data set consisting of two numbers such that

(a) MEAN = 1 and STDEV = 0
(b) MEAN = 1 and STDEV = 1
(c) MEAN = 1 and STDEV = 2

6 In a certain forest there are three kinds of trees: lodgepole pines, white pines, and aspen. Their approximate average diameters are, respectively, 8 inches, 10 inches, and 4 inches. Suppose that there are twice as many aspens as lodgepoles and 3 times as many whites as aspens. What is the approximate average diameter of all trees in the forest? (*Hint:* Pretend you know how many lodgepoles there are in the forest.)

7 Verify that each of the sets of numbers represented in Fig. 1.1 has MEAN = 3.

8 Two golfers, Lee and Jack, play the same course ten times each getting scores of:

(Lee)	74	69	70	70	73	70	68	71	71	74
(Jack)	73	72	65	65	72	68	74	67	74	70

(a) Which golfer has the better average score?
(b) Which golfer is the more consistent? (*Hint:* Compute the standard deviations of each set of scores.)
(c) How often did each golfer shoot his average score?

9 **(a)** Find the mean of the gum ball diameters using the raw data given in Section 1.
(b) Estimate the mean of the gum ball diameters from the grouped data histogram.

10 **(a)** Find the mean and variance of the set of numbers

$$2, 3, 3, 5, 7$$

(b) Find the mean and variance of the set of numbers obtained by subtracting the mean [from part (a)] from each of the original numbers.

11 **(a)** Find a formula for MEAN and VAR of the set

$$x, y$$

(b) Construct a new data set by subtracting MEAN [from part (a)] from each of x and y.
(c) Find a formula for MEAN and VAR for this new set of numbers.

12 **(a)** Find the average word length of the 60 sentences given in Exercise 9(a) of Section 1. Also use the grouped data histogram to estimate this average.
(b) Repeat (a) using the data in Exercise 9(b) of Section 1.
(c) Repeat (a) using the data in Exercise 9(c) of Section 1.

13 The following basketball scores occurred on the same day.

Professional teams:
110–105, 113–93, 107–102, 113–111, 120–91, 136–112, 104–101, 110–98, 129–116, 122–98

Selected colleges:
115–65, 100–93, 107–89, 97–84, 93–84, 89–44, 74–65, 75–74, 115–70, 88–86, 74–68, 73–71, 103–54, 69–63, 73–72

Selected high schools:
74–72, 66–60, 71–60, 71–51, 68–66, 67–63, 81–58, 70–55, 74–65, 74–60, 64–42, 77–71, 82–74, 111–100, 65–56

(a) Find the mean winner and loser scores in each category.
(b) Find the mean winning margin in each category.
(c) What conclusions would you draw concerning comparisons between these three categories?

3 SUMMATION NOTATION

In Section 2 we learned how to find the mean and standard deviation of a set of numbers. The discussion was phrased in standard English and no appeal was made to a specialized "mathematical notation." While keeping the discussion in standard English has certain advantages, it also has a number of liabilities. In particular, we gave two algorithms for calculating the standard deviation. These two came out the same for each set of numbers we looked at but we have no means of seeing whether or not they are always the same. The algebraic notation which we shall introduce in this section gives a technique for comparing these and other quantities.

Let us begin thinking about the matter of formulae and algebraic expressions in the following way. Suppose we simply want a compact way to communicate the instructions telling how to find the mean and standard deviation of a set of numbers. For the sake of being concrete let us suppose we only want to find the mean and standard deviation of sets consisting of three numbers. The instructions could be written as follows.

Let a, b, and c denote three numbers. If we denote their mean by $\bar{x}$ and their standard deviation by s_x, then

$$\bar{x} = (\tfrac{1}{3})(a + b + c)$$

$$s_x = \sqrt{(\tfrac{1}{3})[(a - \bar{x})^2 + (b - \bar{x})^2 + (c - \bar{x})^2]}$$

With the aid of these formulae, you can calculate the mean and the standard deviation of a set of three numbers. Suppose you wish to consider the set of numbers 27, 13, 43. If you let $a = 27$, $b = 13$, $c = 43$ in the above formulae, you will have the mean and standard deviation (after a bit of arithmetic).

However the instructions (formulae) written above do not suffice to replace the instructions which we gave in Section 2. We will have to give instructions which will serve for any finite set of numbers, not merely for sets of three numbers. When we try to write instructions which will serve for any finite set of numbers, we begin to realize that the form in which the instructions were written in the previous paragraph has certain features which will be difficult to carry over directly to the general situation.

In particular there is a potential difficulty in thinking of enough letters to represent all the numbers. When we are only going to deal with three numbers, it is not hard to think of three letters. But what happens when we want 30 numbers, or 30,000, or 3,000,000,000? The problem is compounded by the fact that we not only need a lot of symbols to stand for numbers but we cannot be sure in advance how many will be needed. We know we will want one for each number in the set to be considered, but until someone submits a set of numbers to have the mean and standard deviation computed, we don't know how many there will be. Happily we do not have to expend time and energy on dreaming up millions of distinct symbols to represent numbers. Mathematicians have already thought of a clever and simple way to handle the problem.

Let n denote a positive integer. (A positive integer is a number of the form 1, 2, 3, 4, 5, etc. They are the numbers with which you count. Note that the number of numbers in any set for which we wish to compute the mean and standard deviation will be a positive integer.) For each integer i between 1 and n (inclusive) let x_i be a number. The integer i attached to the symbol x is called a *subscript*. The beauty of this device is that we no longer need to use distinct letters for the various numbers in our set; we use only one letter and as many different values of the subscript as are needed. For example, if we wanted to talk about a set of eight numbers we could say, let $x_1, x_2, x_3, x_4, x_5, x_6, x_7, x_8$ be numbers. Of course, if we wished to talk about a set with 7,156,859 numbers, writing out the list as above might become somewhat tedious. In cases such as this, mathematicians customarily abbreviate the list by writing

$$x_1, x_2, \ldots, x_{7,156,859}$$

In this case, even though only three numbers appear on the page, you are expected to realize that 7,156,859 are being described. The missing 7,156,856 numbers are hiding under the . . . .

We can now begin our algebraic instructions for finding means and standard deviations by saying, "Let n be a positive integer and $x_1, x_2, \ldots, x_n$ be numbers." Indeed, in this context it is usually understood that n is a positive integer and one simply says, "Let $x_1, x_2, \ldots, x_n$ be numbers."

The instruction to add the numbers is usually written as

$$x_1 + x_2 + \cdots + x_n$$

Once again we should note that

$$x_1 + x_2 + \cdots + x_{7,156,859}$$

tells you that you should add together all 7,156,859 numbers and not just the three which you see written down.

Finally, to obtain the average of $x_1, x_2, \ldots, x_n$ we should divide their sum by the number of numbers. Fortunately the notation tells us how many numbers there are; there are n of them. We can now write in algebraic notation the formula for the mean of any set of numbers.

Definition 1 The mean (or average) denoted $\bar{x}$, of the numbers $x_1, x_2, \ldots, x_n$, is defined by

$$\bar{x} = \left(\frac{1}{n}\right)(x_1 + x_2 + \cdots + x_n)$$

In a similar manner we can write a formula for finding the standard deviation of a set of numbers.

Definition 2 The standard deviation, denoted s_x, of a set of numbers $x_1, x_2, \ldots, x_n$ is defined by

$$s_x = \sqrt{\left(\frac{1}{n}\right)[(x_1 - \bar{x})^2 + (x_2 - \bar{x})^2 + \cdots + (x_n - \bar{x})^2]}$$

where $\bar{x}$ is the mean of the set of numbers $x_1, x_2, \ldots, x_n$.

The alternative definition of the standard deviation is

$$s_x = \sqrt{\left(\frac{1}{n}\right)(x_1^2 + x_2^2 + \cdots + x_n^2) - \bar{x}^2}$$

You should now go back to the description of the standard deviation in Section 2 and see that the algebraic formulation we have just given is the same.

There is an alternative to the "..." notation and, since we will have so many uses for this alternate notation, this is a good time to discuss it. Instead of

$$x_1 + x_1 + \cdots + x_n$$

one frequently writes

$$\sum_{i=1}^{n} x_i$$

Using this notation, we would replace

$$\bar{x} = \left(\frac{1}{n}\right)(x_1 + x_2 + \cdots + x_n)$$

with

$$\bar{x} = \left(\frac{1}{n}\right)\left(\sum_{i=1}^{n} x_i\right)$$

Similarly, the formula for standard deviation would become

$$s_x = \sqrt{\left(\frac{1}{n}\right)\sum_{i=1}^{n}(x_i - \bar{x})^2}$$

or

$$s_x = \sqrt{\left(\frac{1}{n}\right)\sum_{i=1}^{n} x_i^2 - \bar{x}^2}$$

This notation, using the Σ for sum, is often referred to as *summation notation*. Since we shall be using it so frequently in the future, it will be helpful if we consider some of its basic properties. The first properties we will consider relate to the *index of summation* (that is, the subscripts). Note first that

$$\sum_{i=1}^{n} e_i = e_1 + e_2 + \cdots + e_n$$

$$\sum_{k=1}^{n} e_k = e_1 + e_2 + \cdots + e_n$$

and so

$$\sum_{i=1}^{n} e_i = \sum_{k=1}^{n} e_k$$

This says that changing the index of summation, from i to k for example, does not alter the value of the expression.

Let us look at a few examples concerning summation notation.

EXAMPLE 1

$$\sum_{i=1}^{5} i^2 = 1^2 + 2^2 + 3^2 + 4^2 + 5^2$$
$$= 1 + 4 + 9 + 16 + 25$$
$$= 55$$

EXAMPLE 2

$$\sum_{i=1}^{4} (2i + 1) = (2 \cdot 1 + 1) + (2 \cdot 2 + 1) + (2 \cdot 3 + 1) + (2 \cdot 4 + 1)$$
$$= 3 + 5 + 7 + 9$$
$$= 24$$

EXAMPLE 3

$$\sum_{i=1}^{5} \left(\frac{2}{i}\right) = \left(\tfrac{2}{1}\right) + \left(\tfrac{2}{2}\right) + \_\_\_\_ + \_\_\_\_ + \_\_\_\_$$
$$= 2 + \_\_\_\_ + \_\_\_\_ + \_\_\_\_ + \_\_\_\_$$
$$= \_\_\_\_$$

Additional flexibility is lent to the notation by using it as in the following example.

EXAMPLE 4

$$\sum_{i=2}^{5} (i + 1) = (2 + 1) + (3 + 1) + (4 + 1) + (5 + 1)$$
$$= 3 + 4 + 5 + 6$$
$$= 18$$

EXAMPLE 5

$$\sum_{i=1}^{6} 7 = 7 + 7 + 7 + 7 + 7 + 7$$
$$= 42$$

EXAMPLE 6

$$\sum_{i=1}^{4} (i^2 + i) = (1^2 + 1) + (\_\_\_\_\_\_) + (\_\_\_\_\_\_) + (\_\_\_\_\_\_)$$
$$= (1^2 + 2^2 + 3^2 + 4^2) + (1 + \_\_\_\_ + \_\_\_\_ + \_\_\_\_)$$
$$= \sum_{i=1}^{4} i^2 + \sum_{i=1}^{4} i$$

EXAMPLE 7

$$\sum_{k=1}^{6} 2k = 2 \cdot 1 + 2 \cdot 2 + 2 \cdot 3 + 2 \cdot 4 + 2 \cdot 5 + 2 \cdot 6$$
$$= 2(1 + 2 + 3 + 4 + 5 + 6)$$
$$= 2 \sum_{k=1}^{6} k$$

These last two examples illustrate properties of the summation notation which are of considerable importance and should be stated in general.

$$\sum_{i=1}^{n} (x_i + y_i) = \sum_{i=1}^{n} x_i + \sum_{i=1}^{n} y_i$$

$$\sum_{i=1}^{n} cx_i = c \sum_{i=1}^{n} x_i$$

We have said several times that the two expressions we have for the standard deviation are equal. Let's see that this is so for $n = 2$.

If $n = 2$, then

$$\bar{x} = \tfrac{1}{2}(x_1 + x_2)$$

One expression for the standard deviation is

$$\tfrac{1}{2} \sum_{i=1}^{2} (x_i - \bar{x})^2$$

while the other is

$$\tfrac{1}{2} \sum_{i=1}^{2} (x_i)^2 - \bar{x}^2$$

The first of these gives

$$\begin{aligned}
\tfrac{1}{2} \sum_{i=1}^{2} (x_i - \bar{x})^2 &= \tfrac{1}{2}[(x_1 - \bar{x})^2 + (x_2 - \bar{x})^2] \\
&= \tfrac{1}{2}[x_1^2 - 2x_1\bar{x} + \bar{x}^2 + x_2^2 - 2x_2\bar{x} + \bar{x}^2] \\
&= \tfrac{1}{2}[(x_1^2 + x_2^2) - 2\bar{x}(x_1 + x_2) + 2\bar{x}^2] \\
&= \tfrac{1}{2}[x_1^2 + x_2^2 - 2\bar{x} \cdot 2\bar{x} + 2\bar{x}^2] \\
&= \tfrac{1}{2}[(x_1^2 + x_2^2) - 2\bar{x}^2] \\
&= \tfrac{1}{2}(x_1^2 + x_2^2) - \bar{x}^2 \\
&= \tfrac{1}{2} \sum_{i=1}^{2} x_i^2 - \bar{x}^2
\end{aligned}$$

Let us make one final remark concerning the use of these formulae for calculating the estimated mean and standard deviation of grouped data.

Suppose we have n data points grouped into r classes with class marks $x_1, x_2, \ldots, x_r$. Moreover, suppose that these classes have absolute frequency $f_1, f_2, \ldots, f_r$. Then it can be shown that

$$n = \sum_{i=1}^{r} f_i$$

$$\bar{x} = \left(\frac{1}{n}\right) \sum_{i=1}^{r} f_i x_i$$

and

$$s_x^2 = \left(\frac{1}{n}\right) \sum_{i=1}^{r} f_i x_i^2 - \bar{x}^2$$

You will find it helpful to compare these formulae for $\bar{x}$ and s_x with the calculations of estimated mean and standard deviation for grouped data on pages 13 and 17. These formulae can be modified to

$$\bar{x} \approx \sum_{i=1}^{r} \left(\frac{f_i}{n}\right) \cdot x_i$$

$$s_x^2 \approx \sum_{i=1}^{r} \left(\frac{f_i}{n}\right) x_i^2 - \bar{x}^2$$

and you should note that f_i/n is the relative frequency of the ith class.

Summary of Formulae

Mean

$$\bar{x} = \left(\frac{1}{n}\right) \sum_{i=1}^{n} x_i$$

Standard Deviation

Formula 1 $$s_x = \sqrt{\left(\frac{1}{n}\right) \sum_{i=1}^{n} (x_i - \bar{x})^2}$$

Formula 2 $$s_x = \sqrt{\left(\frac{1}{n}\right) \sum_{i=1}^{n} x_i^2 - \bar{x}^2}$$

Modified Formula 2 for use with computer

$$s_x = \sqrt{\left(\frac{1}{n}\right) \sum_{i=1}^{n} x_i^2 - \left(\frac{1}{n^2}\right)\left(\sum_{i=1}^{n} x_i\right)^2}$$

Exercises Section 3

1 Evaluate each of the following.

(a) $\sum_{i=1}^{3} i^3$

(b) $\sum_{i=1}^{7} (2i - 1)$

(c) $\sum_{k=1}^{8} k^3 - \sum_{i=1}^{7} i^3$

(d) $\sum_{i=1}^{100} (i^2 + 1) - \sum_{i=1}^{100} i^2$

(e) $\sum_{i=1}^{4} (-1)^i \frac{1}{i}$

(f) $\sum_{k=3}^{6} 4$

(g) $\sum_{k=2}^{5} \frac{1}{k}$

(h) $\sum_{i=1}^{4} (-1)^{i+1} i^2$

2 Write each of the following in summation notation.

(a) $2 + 4 + 6 + 8 + 10 + 12$ **(d)** $1 - 2 + 3 - 4 + 5$

(b) $1 + \sqrt{2} + \sqrt{3} + 2 + \sqrt{5}$ **(e)** $5 + 5 + 5 + 5$

(c) $\frac{1}{3} + \frac{1}{4} + \frac{1}{5}$ **(f)** $6 + 9 + 12 + 15 + 18 + 21$

3 What is $x_1 + x_2 + \cdots + x_n$ when $n = 2$?

4 Are $\sum_{i=1}^{n} x_i^2$ and $\left(\sum_{i=1}^{n} x_i\right)^2$ equal?

5 Verify the following identities for $n = 1, 2, 3, 4$.

$$\sum_{i=1}^{n} i = \frac{n(n+1)}{2}, \qquad \sum_{i=1}^{n} i^2 = \frac{n(n+1)(2n+1)}{6}$$

6 Which is bigger for positive numbers x_1 and x_2?

$$\frac{x_1 + x_2}{2} \quad \text{or} \quad \sqrt{x_1 x_2}$$

(*Hint:* Try some particular numbers for x_1 and x_2, then guess.)

7 Show that for any $x_1, x_2, \ldots, x_n$,

$$\sum_{i=1}^{n} (x_i - \bar{x}) = 0$$

(*Hint:* Write out the equation when $n = 2$ and $n = 3$.)

8 In the text the equality

$$\left(\frac{1}{n}\right) \sum_{i=1}^{n} x_i^2 - \bar{x}^2 = \sum_{i=1}^{n} \left(\frac{1}{n}\right)(x_i - \bar{x})^2$$

was shown for $n = 2$.

(a) Establish the equality for $n = 3$.

(b) Establish the equality for all n.

9 Consider data sets $A = \{1, 6, 12, 16\}$ and $B = \{4, 9, 15, 19\}$. Notice each number in B is precisely 3 more than the corresponding number in A.

(a) Without calculating, what can you predict about the relationship between the mean of A and that of B?

(b) Calculate the two means to check your prediction.

(c) Without calculating, what can you predict about the relationship between the standard deviation of A and that of B? (*Hint:* Consider what the standard deviation measures.)

(d) Calculate the two standard deviations to check your answer.

(e) Find the mean and standard deviation of the following data set: $\{99, 106, 98, 97, 100\}$. [*Hint:* Use what you learned in (a), (b), (c), and (d).]

4 FLOW CHARTS

In Section 2 we presented algorithms, or sets of rules, by which the mean and standard deviation of sets of numbers can be calculated. One of the advantages of the algorithms is that they can be performed by anyone with a mastery of grade school arithmetic. One of the disadvantages is that, particularly for large sets of numbers, the necessary work is apt to be very time consuming. In this section we will show you how to write algorithms in such a way that they can be used by a computer.

The discussion will be kept independent of any particular computer or any particular programming language by presenting the algorithms in the form of flow charts. The rules for forming the flow charts may, at times, seem a bit artificial. However we are going to make these rules so that there is a reasonably straightforward way to translate from the flow chart to any particular programming language now in common use.

Let us write a simple flow chart. It is a set of instructions for finding the mean of the set of numbers $\{1, 2, 3\}$.

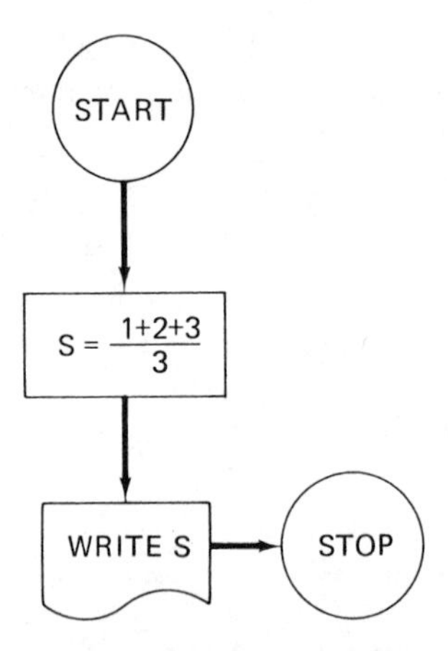

This flow chart is a fairly simple one and probably most of you could read it without any explanation. However there are a few points which we should make explicit before going on to more complicated flow charts.

The circles with the words START and STOP in them serve an obvious function. You start at the circle labeled START and you have completed the algorithm whenever you arrive at a circle containing the word STOP. These may seem superfluous now but are very necessary when we encounter more complicated flow charts. The rectangle computing the mean is self-explanatory. The grand piano shaped box containing the phrase WRITE S may seem foolish but one of the things you always have to keep in mind about computers is that, while the machine may have the solution to the problem somewhere inside it, that solution does you no good unless you tell the machine to write out the answer where you can see it. The lines going from one figure to the next tell you how to proceed through the flow chart and the arrows tell you which way to go along the line. Once again, this may seem unnecessary in a simple flow chart such as this but becomes more important as the flow chart becomes more complicated.

The flow chart we have drawn is fine in the sense that it serves the purpose for which it was developed, that is, it does find the mean of the set of numbers $\{1, 2, 3\}$. However this purpose is somewhat limited. Let us see how to modify the flow chart to calculate the mean of *any* set of three numbers. The flow chart (or more properly *a* flow chart) which will do the job is the following.

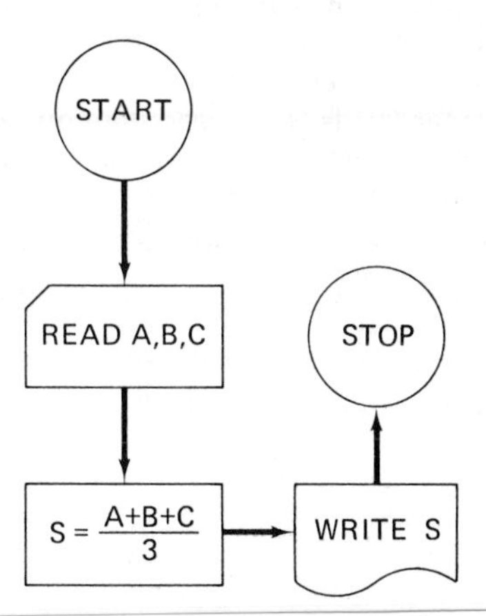

The START statement is the same; however the second statement we come to is a new one, READ A,B,C. Note that this statement is not enclosed in a rectangle but in a rectangle with the upper left-hand corner cut off to suggest a data card (such as the now familiar IBM card).

When the computer encounters a READ statement, it goes back to its card hopper and reads the next card there. (That card, once it goes through the reader, cannot be read again.) It expects to find on that card as many numbers as there are variables in the READ statement. For our flow chart it expects to find three numbers there.

What does the machine do with these numbers? To answer this question we need to understand a little bit about the machine. It consists of a number of memory storage locations. To any or all of these a program can assign a name and a number. In our flow chart, when the computer comes to the statement READ A,B,C it will note that it has not, up to that point, named any locations by the names A, B, or C so it will take three unused locations, name one of them A, a second one B, and a third one C. It then goes to the data card, puts the first number on it into the location it named A, the second number into the location it named B, and the third number into the location it named C.

Next we come to the *assignment statement*

$$S = \frac{A+B+C}{3}$$

The computer sees this statement, notes first that it has already got storage locations labeled A, B, and C, but none named S. It names an unused location S, looks at the numbers in A, B, and C, adds them, divides by 3, and puts the answer into the location it just named S. The numbers that were in A, B, and C are still there. They cannot be changed by appearing on the right hand side of the = in an assignment statement.

The rest of the flow chart consists of the familiar statements WRITE S and STOP.

One additional comment may be useful. In connection with the READ statement we talked of data cards. We shall continue to do so though this may be misleading. Most computers are capable of reading other things besides cards; indeed some cannot read cards. They read data off teletypes, paper tape, magnetic tape, disks, and drums. We shall continue to use the idea of cards because they emphasize certain aspects of the READ process which we will want to emphasize as we go along. These aspects are also important when the computer is reading from something else.

We now have a flow chart which will find the mean of any set of three numbers, but suppose we want the mean of 10 numbers or 50 or 17,482 numbers. The following flow chart will handle that.

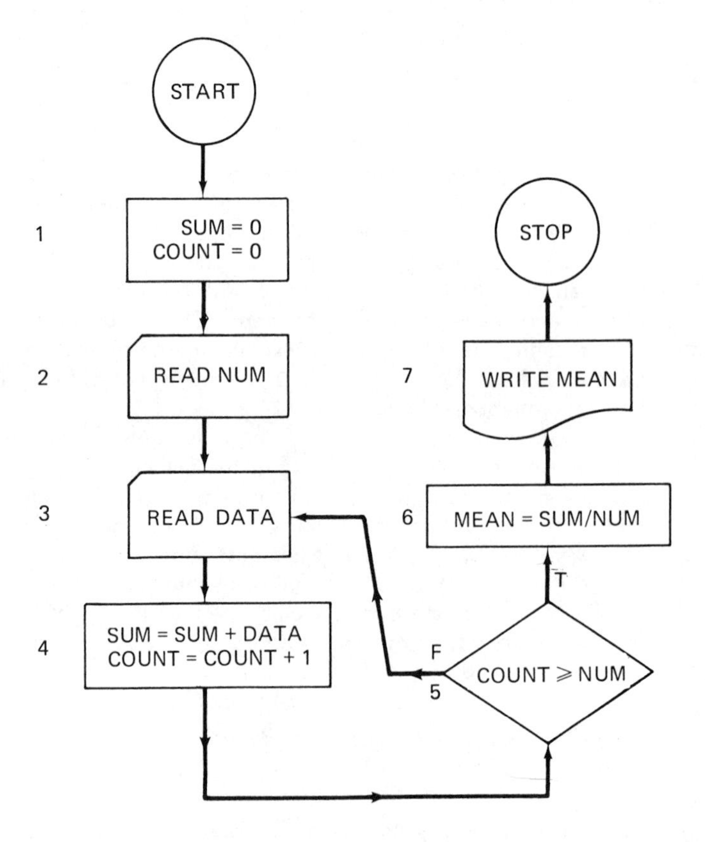

This flow chart contains a number of new techniques. Let's look closely and see what has been done. We might begin by noting that there is no statement such as the statement

$$S = \frac{A+B+C}{3}$$

which appeared in the last flow chart. The change is dictated by the fact that, as we write the flow chart, we do not know how many numbers are in the set whose mean we wish to compute. Before, when we knew that there were to be exactly three numbers in the set we could reserve three memory storage locations, call them A, B, and C, put the three numbers in those locations and compute the mean. But now we don't know how many numbers there will be so we don't know how many locations to reserve for them.

The new flow chart solves this problem by circumventing it. We note that we don't necessarily need all the numbers in the data set to be stored in the computer. What we do need for calculating the mean is their sum. What we do is set aside a storage location, which we name SUM, where we will put the sum of the numbers in the data set. In box 1 we put the initial value 0 in this location, and we put the numbers in our data set onto cards, one number per card. Each time we read a card we put the number on it into a location named DATA. We then add the number in DATA to the number in SUM and put the total in SUM. This may be a bit confusing now, but it will clarify itself in a moment when we perform a *trace* of the flow chart. The idea is that after all the data cards have been read in, the sum of all the numbers will be in the location named SUM.

What about the locations named NUM and COUNT and the diamond shaped box with the two arrows coming from it? The location NUM is where you tell the computer how many numbers are in the set to be averaged. This is the first number to be read in and must be on the first data card. The location COUNT is where we keep track of the number of numbers which have been read into the machine. We start it at 0 (notice the initialization in box 1) and each time we read a data card we add 1 to the number in COUNT (see box 4) so that the number in COUNT is always the number of data cards we have read into the machine. When the number in COUNT is as large as the number we have read into NUM, we have read in all of the numbers in our data set.

Box 5 is what is known as a *decision box*. A computer can make a choice between alternative procedures based on the relative size of the numbers in two locations. In particular we can tell the computer to do one thing (compute the average) if the number in COUNT is as large as the number in NUM, and do something else (go back and read another data card) if the number in COUNT is smaller than the number in NUM.

This discussion is apt to be confusing. The only way to understand what this flow chart does is to trace its steps. Let us use this flow chart to compute the mean of the set of numbers $\{1, 2, 3\}$. The first card that the machine will read has to be the number of numbers in the set, 3, so the four cards for the computer to read will contain (one to a card and in order)

3
1
2
3

Here is the trace.

STEP	BOX	SUM	COUNT	NUM	DATA	MEAN
1	START					
2	1	0	0			
3	2	'	'	3		
4	3	'	'	'	1	
5	4	1	1	'	'	
6	5	$1 \geq 3$ is false				
7	3	'	'	'	2	
8	4	3	2	'	'	
9	5	$2 \geq 3$ is false				
10	3	'	'	'	3	
11	4	6	3	'	'	
12	5	$3 \geq 3$ is true				
13	6	'	'	'	'	2
14	7	the computer writes 2				
15	STOP					

The symbol ' means that the value in that location is unchanged.

Let's follow this trace carefully and see what is happening. At step 1 we start the machine. At step 2 we get to box 1 and so we put the initial value 0 into locations named SUM and COUNT. At step 3 we encounter our first read statement so we read the number on the first card, 3, and put that in NUM. Now the machine knows how many numbers are in the set to be averaged.

At step 4, the machine gets to box 3 which is another read statement. It reads the next card, gets the value 1, and puts that number into a location named DATA.

At step 5, the computer gets to box 4. It looks at the numbers in SUM and DATA, 0 and 1 respectively, adds them, and puts the total, 1, into SUM. It also looks at COUNT, sees that there is a 0 in there, adds 1 to 0 and puts the result, 1, into COUNT. At step 6 the computer compares the numbers in COUNT and NUM, notes that $1 \geq 3$ is false and so, follows the arrow labeled F out of box 5.

This takes it, at step 7, back to box 3. It reads the next data card and puts the number on it, 2, into the location named DATA.

The machine continues to cycle in this "loop" until it gets to step 12. Now it has $3 \geq 3$, which is true, so all the cards have been read and it is now ready to compute the mean. It has in the location named SUM the sum of the numbers and in the location named NUM the number of numbers. In step 13 the computer goes to box 6, calculates the mean and puts it into a location named MEAN. In step 14, the computer is instructed to write out the value which it has in the location named MEAN. It does so and in step 15 it stops, with its task completed.

In order to understand better what this flow chart is doing, you may find it instructive to work out for yourself the trace which would occur if the four data cards got shuffled. For example you should convince yourself that the machine would come up with the value 3 if the cards were put into the machine in the order

1
3
2
3

i.e., if the first two cards were interchanged.

Now that we have seen a few flow charts, the next question is, how do you develop a flow chart to solve a problem. Let's consider an example to see how they are developed.

One thing we have done is the grouping of data. Can we have a computer do this for us? To make the problem concrete, let us suppose that we have several test scores. We want to assign an A to those who got 85 or better, a B to those whose score ranges from 70 through 84, a C to those from 55 through 69, a D to those from 40 through 54 and an F to those under 40. We want to count the number of people getting the various grades A, B, C, D, and F. How do we do it?

Since we want, as output, the number of people getting these various grades, we will need memory storage locations where the numbers can be found at the end of the flow chart. We might well (and will) decide to name these locations A, B, C, D, and F. How do we get the appropriate numbers into these locations? The idea is that we will count the number of people getting A's, B's, etc. Our experience counting in the last flow chart might suggest to us that we should begin by putting, initially, the number 0 into each of the locations. Our flow chart begins

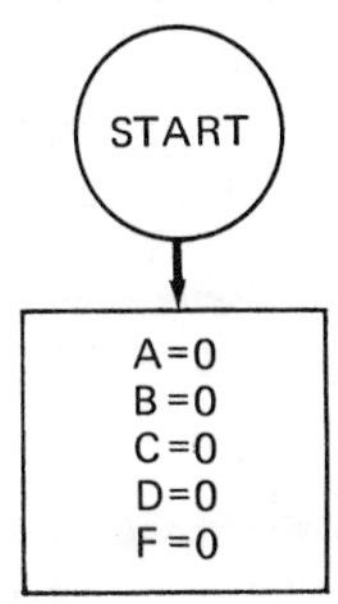

What next? We want to read a card with a test score on it, decide whether it is an A, B, C, D, or F, and add 1 to the number in the appropriate storage location.

Let's do it. We read in a test score and put it into some location, we might as well call it SCORE. Now we have the flow chart

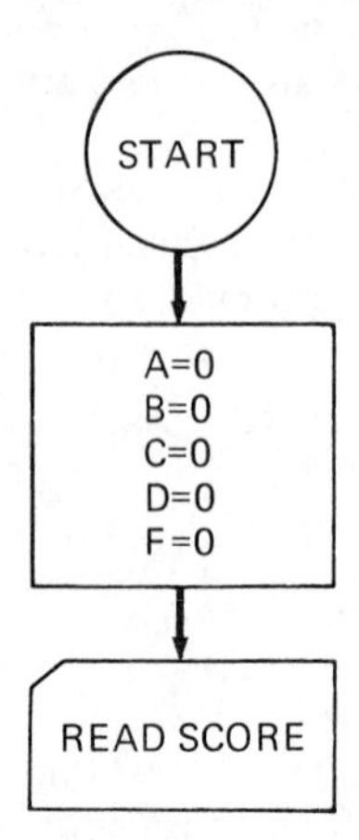

In order to decide whether or not this test score is an A, we have to see whether or not it is 85 or larger. But we know that the computer can make this kind of decision in a decision box. So we can see if SCORE is 85 or better. If so we want to add 1 to the number in A and read the next score. If not we want to go on and see what grade we should assign. At this point the flow chart is as follows:

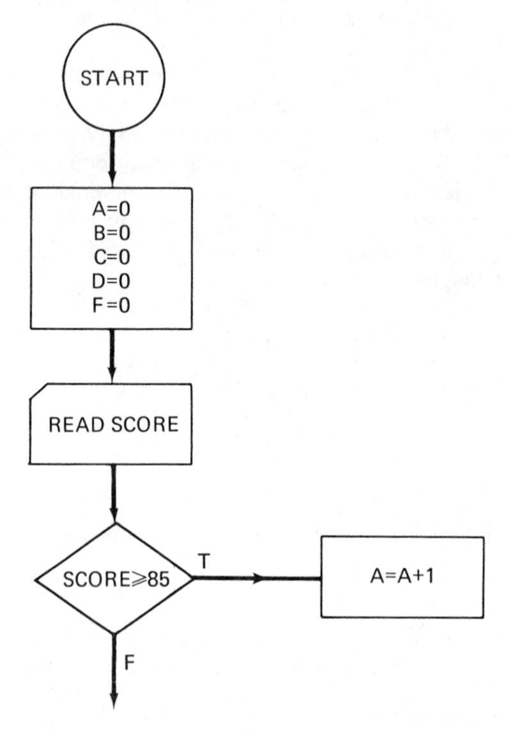

Now we have to decide what to do if SCORE $\geq$ 85 is false. One reasonable thing to do is to see if it is a B. If SCORE $\geq$ 70, then the test is a B (the possibility that it is an A has already been ruled out). Continuing this reasoning process, we obtain the following flow chart.

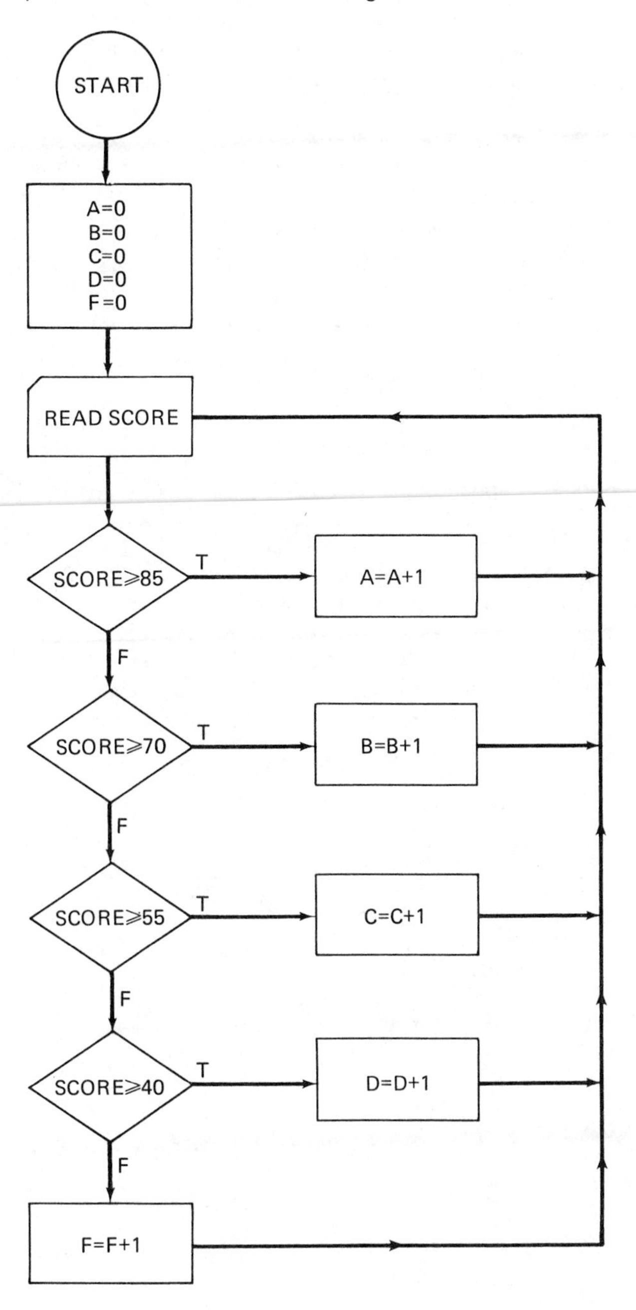

Our flow chart now does a nice job of deciding upon the letter grade corresponding to each numerical score and it counts nicely the number of A's, B's, etc. However it has no way to know when it has read all of the cards. We can deal with this in precisely the way we dealt with this problem in our flow chart for finding the mean. That is, we can read in the number of scores to be read and insert a counter to count the number of cards which have been read. We then add a decision box to get the flow chart out of its cycle of reading test scores and deciding on their letter grade whenever we have handled the predetermined number of scores.

Finally we have to print out the results. There is one additional convention concerning the WRITE statement which we did not mention earlier. Most commonly used programming languages have a method which allows you to print out labels as well as numbers. That is, if the number of A's is 16 you can make the computer print

(†) The number of students receiving an A is 16

rather than simply the number 16. The most common way to do this is to enclose the label

The number of students receiving an A is

in quotes or apostrophes. We shall, in our flow charts, write

WRITE 'THE NUMBER OF STUDENTS RECEIVING AN A IS' A

to get (†) as a printout. Note that there are delicacies concerning writing expressions such as A's in this manner.

The completed flow chart is shown on page 37.

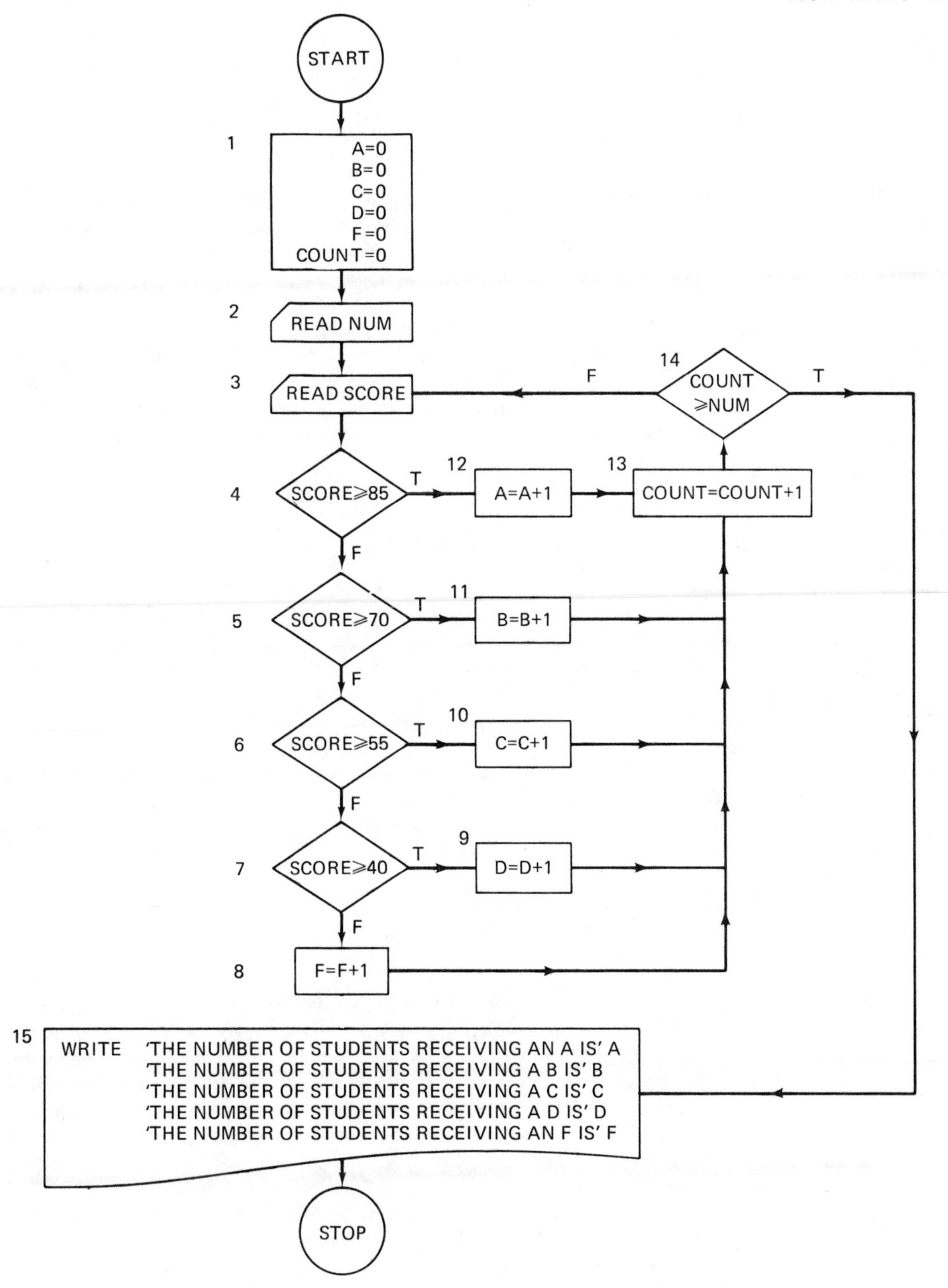
START
1
A=0
B=0
C=0
D=0
F=0
COUNT=0
2
READ NUM
3
READ SCORE
14
COUNT
≥NUM
F
T
4
SCORE≥85
T
12
A=A+1
13
COUNT=COUNT+1
F
5
SCORE≥70
T
11
B=B+1
F
6
SCORE≥55
T
10
C=C+1
F
7
SCORE≥40
T
9
D=D+1
F
8
F=F+1
15
WRITE 'THE NUMBER OF STUDENTS RECEIVING AN A IS' A
'THE NUMBER OF STUDENTS RECEIVING A B IS' B
'THE NUMBER OF STUDENTS RECEIVING A C IS' C
'THE NUMBER OF STUDENTS RECEIVING A D IS' D
'THE NUMBER OF STUDENTS RECEIVING AN F IS' F
STOP

Summary

Let us briefly review the various kinds of statements which can be put into flow charts and the various rules associated with flow charting.

1. The order in which statements are put into the flow chart is crucial. The flow chart begins at the START location, and follows the arrows, performing each statement as it gets to it until it reaches the STOP statement. You must be sure that the order in which the statements are encountered is the order you want. Sometimes very minor changes in the order of the chart make a huge difference in the result which is achieved.
2. Assignment statements: These are always enclosed in rectangles. The contents of the statement look, superficially, like equations. However the "equations" are different from those occurring in algebra in at least two ways. For one thing, the expression to the left of the = sign must always be a single variable or memory location name. It can *never* be something like

   $$A + B = C^2$$

   or

   $$A + 1 = C^2$$

   The effect of the assignment statement is to calculate whatever is called for by the expression to the right of the = sign and place that number in the memory storage location named on the left-hand side of the = sign.
3. Read statements: These are always of the form

   READ __________

   where after READ is a list of one or more variables. The statement is always enclosed in a box shaped as follows.

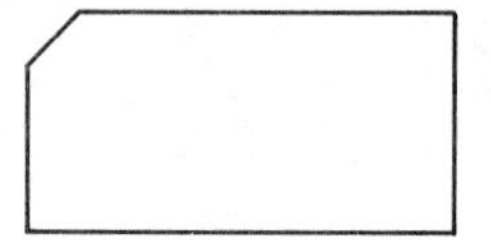

Each time a read statement is encountered the computer looks for a data card. The first time the computer encounters a read statement it reads the first data card. The second time it encounters a read statement it reads the second card, etc. It is therefore crucial that you supply as many data cards as there are encounters with read statements in your flow chart. Moreover the cards have to be in the right order and each card has to contain the same number of numbers as the number of variables in the corresponding read statement. The errors arising from failure to obey these rules vary somewhat from computer to computer but there is always an error. In cases where something other than cards is being used to input the data the basic point about the sequential nature of the reading process still holds. That is, numbers are always read in the order in which they were put on the cards or tape or disk or whatever.

4. Write statements: These allow the computer to write out the content of one or more memory locations. You can also print headings by using the quote device mentioned earlier.
5. Decision statements: These statements are always enclosed in a diamond. The decision is *always* a matter of whether some equality or inequality is true or false. You must indicate where the flow chart is to go if the statement is true, and where it should go if the statement is false.

Exercises Section 4

1 Consider the flow chart:

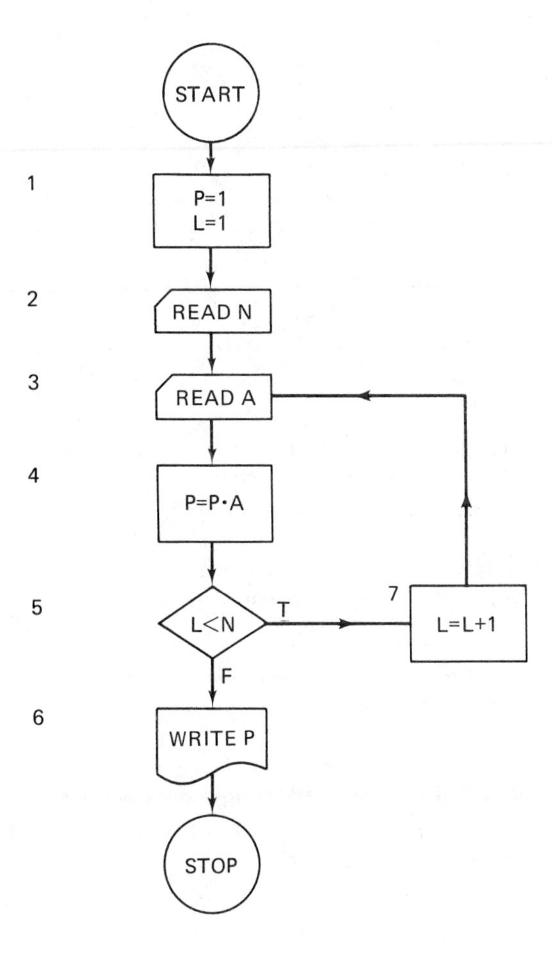

Trace this flow chart for each of the following sets of data cards.

(a)	(b)	(c)
4	2	3
1	4	5
2	3	1
3	1	0
4	2	1

2 Write a flow chart which will read two numbers and print either their sum or their product, depending on which is smaller. If, for example, it reads .5 and 2, the sum, 2.5, is larger than the product, 1, so it should print 1.

3 Consider the following flow chart:

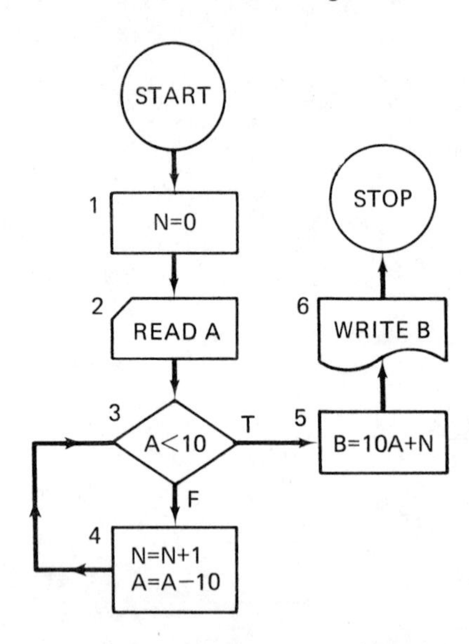

If the number A to be read in is a two-digit number, i.e., one between 10 and 99, what does this flow chart do to A? Try tracing the flow chart for a few values of A. Try reading in A = 17, A = 34, and A = 20, for example.

4 Trace the flow charts shown on page 41 *and* determine what each flow chart does for each of the following sets of data.

(a)	(b)	(c)
3	4	2
2	−1	−1
1	−4	6
4	−2	
	−3	

5 Suppose your computer has only four memory storage locations. Prepare a flow chart which will read in 4 numbers and print out the largest and the smallest number in the set of 4. Trace your flow chart for the following set of numbers:

3, −2, 14, 18

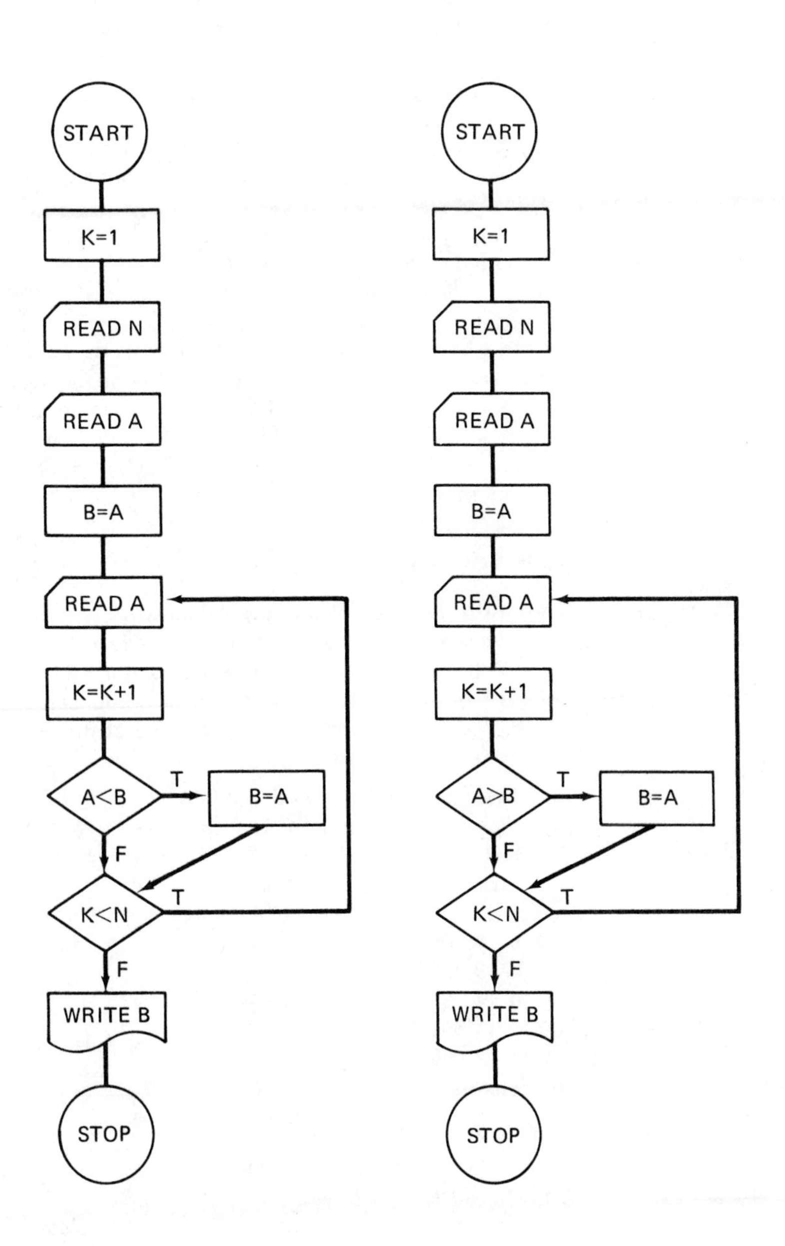
START
K=1
READ N
READ A
B=A
READ A
K=K+1
A<B
T
B=A
F
K<N
T
F
WRITE B
STOP
START
K=1
READ N
READ A
B=A
READ A
K=K+1
A>B
T
B=A
F
K<N
T
F
WRITE B
STOP

6 Prepare a flow chart which will compute and print out both the mean and the standard deviation of a set of numbers. (The most common error in this problem is to forget that the computer reads a data card only once. Try not to make this mistake.) Trace your flow chart for the set of numbers:

1, 2, 3, 4, 5

You will probably have to insert an extra card to tell the machine how many numbers are in the set. Indicate where the card for this should go in the deck of data cards.

7 Prepare a flow chart which will read a length in inches and print out the length accurate to the nearest foot.

8 **(a)** Write a flow chart for printing the squares of the first 10 positive integers.

(b) Write a flow chart for printing the squares of the first 10 odd integers.

9 You have a set of 500 data cards (one number on each card).

(a) Write a flow chart for printing only the number on the tenth card.

(b) Write a flow chart for printing only the numbers on every tenth card (i.e., cards numbered 10, 20, 30, . . . , 500). [*Hint:* Use the flow chart for (a) in a loop.]

Supplementary Exercises for Chapter 1

1 Data from the 1970 Census is given below.
Suppose you wanted to estimate the average value of a house in each county. With a computer or a desk calculator, this would be a routine matter. But by hand the computation is complicated enough to cause problems. You might wish to compare some of the following procedures.

Value of House (in dollars)	Number of Units in Washington County, Ohio	Number of Units in Morgan County, Ohio
less than 5,000	601	259
5,000 to 9,999	1,733	526
10,000 to 14,999	2,257	384
15,000 to 19,999	2,081	216
20,000 to 24,999	1,251	117
25,000 to 34,999	910	62
35,000 to 49,999	336	16
50,000 or more	79	5

(a) Round off each number in the second and third columns of the table to the nearest 100 (i.e., 601 becomes 600, 259 becomes 300, and so on). Now use the class marks for each class in the first column to get estimates of the mean values in each county. (*Note:* You will have to put an upper limit on the last class; perhaps 100,000 would be a reasonable estimate.)

(b) Regroup the data by combining the first two classes into one, the next two classes into one, and so on, ending up with four classes rather than eight. Once again, round off the (revised) second and third column figures to the nearest 100 and compute the estimated means.

(c) Regroup the data into two classes by combining the first four classes into one and the last two classes into one and then proceed as in (a) and (b).

(d) This requires more work but will lead to a more accurate estimate. Round off the data in columns two and three to the nearest 10 (i.e., 601 becomes 600, 259 becomes 260, and so on).

(e) Same as (b) except round off data to nearest 10.

(f) Same as (c) except round off data to nearest 10.

2 Rental costs in two Ohio counties, according to the 1970 Census, appear below.

Monthly rent (dollars)	Number of Units in Washington County, Ohio	Number of Units in Morgan County, Ohio
less than 40	501	193
40 to 59	888	224
60 to 79	967	109
80 to 99	497	29
100 to 119	288	16
120 to 149	383	4
150 to 199	77	1
200 to 299	12	0
300 or more	1	0

(a) Round off the data (columns two and three) to the nearest 100 and use class marks to estimate the average rental cost in each county.

(b) Round off the data to the nearest 10 and use class marks to estimate the average rental cost in each county.

(c) Regroup the data into four classes by combining the first two classes, the next two, and so on (ignoring the last class), round off to the nearest 100, and estimate the average rental costs in each county.

(d) Repeat (c) except round off to the nearest ten.

3 Write out the expanded version of the sums represented by each of the following expressions.

(a) $\sum_{i=4}^{10} X_i$

(b) $\sum_{i=0}^{3} (X_i - 2)$

(c) $\sum_{k=1}^{5} 4X_k$

(e) $\sum_{j=1}^{3} \left(\sum_{i=1}^{2} (i + j) \right)$

(f) $\left(\sum_{i=1}^{3} ai \right)^2$

(g) $\sum_{k=2}^{4} (k^2 + 2)$

(d) $2 \sum_{i=0}^{3} (X + i)$ (h) $\sum_{n=3}^{5} n(n + 1)$

4 Write each of the following expressions using summation notation.

(a) $(X_1 - 2)^2 + (X_2 - 2)^2 + (X_3 - 2)^2$
(b) $3X + 4X + 5X + 6X$
(c) $(X_1 + X_2 + X_3)^2$
(d) $X_1 + 2X_2 + 3X_3 + \cdots + 25X_{25}$
(e) $\frac{1}{10}[(X_1 - \mu)^2 + (X_2 - \mu)^2 + (X_3 - \mu)^2 + \cdots + (X_{10} - \mu)^2]$

5 If $X_1 = 3$, $X_2 = 4$, $X_3 = -2$, $X_4 = 7$, find the numerical value of each of the following expressions.

(a) $\sum_{i=1}^{4} X_i$ (d) $\frac{1}{4} \sum_{i=1}^{4} (X_i - 3)^2$
(b) $\sum_{n=1}^{4} nX_n$ (e) $\sum_{i=2}^{3} (X_i^2 + 5)$
(c) $3 \sum_{k=2}^{4} X_k$

6 Produce a data set of three numbers such that:
(a) the mean is greater than the median.
(b) the mean is less than the median.
(c) the median is twice the mean.

7 If ten items have mean weight 2.5 pounds and five similar items have mean weight 3.0 pounds, what is the mean weight of the fifteen items?

8 A class of 30 students consists of 20 men and 10 women. If the average height of the men is 70 inches and the average height of the women is 65 inches, what is the average height of all students in the class?

9 A common final examination is given to three classes. The first class has 40 students and scored an average grade of 75 on the exam. The second class has 30 students and scored an average grade of 60 on the exam. The third class has 25 students and scored an average grade of 80 on the exam. Find the average grade overall on the exam.

10 Two types of nuts are mixed together to be sold at a price of \$1.50 per pound. If type A is normally sold at \$1.00 per pound and type B is sold at \$1.75 per pound, and if the total weight of the mixed nuts is 100 pounds, then how many pounds of type A nuts were used in the mixture?

11 The following examples illustrate simple but rather extreme situations where one method of computing the standard deviation is far simpler than the other.

(a) Find the standard deviation, using *both* methods, for the data set

1.25, 2.25, 3.25, 4.25, 5.25

(b) Find the standard deviation, using *both* methods, for the data set

1, 2, 2, 4, 5

12 Suppose that a data set

$$X_1, X_2, X_3, X_4$$

has mean 3. Show that the data set

$$X_1 + 5, X_2 + 5, X_3 + 5, X_4 + 5$$

has mean 8 and that both data sets have the same variance.

13 **(a)** State a generalization of the result in Exercise 12.
(b) Verify your generalization.

14 Let the data set

$$X_1, X_2, \ldots, X_{20}$$

have mean 10.

(a) Find the mean of

$$2X_1, 2X_2, \ldots, 2X_{20}$$

(b) Find the mean of

$$-X_1, -X_2, \ldots, -X_{20}$$

(c) Find the mean of

$$3X_1 + 2, 3X_2 + 2, \ldots, 3X_{20} + 2$$

15 If the mean of three numbers $\{X_1, X_2, X_3\}$ is 10, what is the mean of $\{X_1^2, X_2^2, X_3^2\}$? (*Hint:* Pick two data sets having mean 10 and find the mean of the squares of data points.)

16 If the variance of $\{X_1, X_2, X_3\}$ is 5, what is the variance of $\{4X_1, 4X_2, 4X_3\}$?

17 **(a)** Using the data in Exercise 1, estimate the *median* house value in each county.
(b) Using the data in Exercise 2, estimate the *median* rental cost in each county.

18 You have 500 data cards, each of which contains the height (in inches) of a person. Write a flow chart which will do all of the following.
(a) Count the number of people less than 4 feet tall.
(b) Count the number of people at least 4 feet tall but less than 5 feet.
(c) Count the number of people at least 5 feet tall but less than 6 feet.
(d) Count the number of people at least 6 feet tall but less than 7 feet.
(e) Count the number of people at least 7 feet tall.

19 A department store prices its goods in the following way. If the item costs the store less than \$3.00 it adds 25% to the cost to determine the selling price. If the item costs at least \$3.00 but less than \$50.00 it adds 20% to the cost to determine the selling price. If the item costs \$50.00 or more it adds 15%. Once the selling price has been determined by these rules, the state requires a tax of 4 cents on each dollar or fraction thereof which is added to the price. For example a \$4.74 item would sell for \$4.74 + 5(\$.04) = \$4.94. Design a flow chart which will read in the cost to the store of an item and write out the price it charges the customer.

20 A store carries 200 items in stock. Every week a card is punched for each item listing the number of sales for the week and the price for the week. Prepare a flow chart which will read these cards and print out the dollar amount of sales for the week and the average price of an item sold.

CHAPTER

2 Probability and Counting Techniques

0 INTRODUCTION

In Chapter 1 we discussed various methods (histograms, means, standard deviations, etc.) for describing our observations of reality. But understanding is not simply a matter of describing what is; it also involves some sort of analysis or explanation. The phenomena to be described and analyzed in the social and biological sciences seem to be associated with uncertainty in a sense which doesn't arise, on the elementary level, in the physical sciences. Consider the following two problems:

1. You hold a rubber ball in your hand, extend your arm parallel to the floor and release the ball. What happens?
2. John Smith is a 60 year old male. Will he die of a heart attack in the next year?

We have no problem in answering question 1. The ball falls to the floor. But the answer to question 2 isn't so simple. Every ball which is released drops to the floor but some men of 60 die of heart attacks and some don't. Any analysis of Mr. Smith's fate will be an estimate. We could make an estimate by learning what percentage of men his age die from heart attacks within a year. We could get a better estimate by having a heart specialist give him a thorough examination but we will not be able to arrive at certain knowledge until he is dead or has lived through the year.

How can we describe, let alone analyze, phenomena in the midst of this uncertainty? The answer to this question is that we must use probability theory and, in this chapter, we will discuss the rudiments of the subject.

1 SOME DISCRETE EXAMPLES

Let us begin with the experiment of rolling one die. Throughout this and later sections we shall appeal to standard jargon rather than restrict our vocabulary to formal language. A case in point: We could explain that a die (plural dice) is a cube, each of whose six sides or faces contains a certain number of spots or pips, the number ranging from one to six; when the die is rolled on a flat surface, it eventually stops moving in such a way that one face is resting entirely on the surface and the opposite face is visible from above; if the number of pips on the top face is 4, we say that "we rolled a '4'" or "a '4' came up." Our contention is that the preceding description, though perhaps serving as comic relief, is totally unnecessary. The reader will be expected to be equipped with background information concerning such diverse terms as card, bridge deck, club, spade, diamond, jack of hearts; coin, heads, tails; raffle, raffle ticket; survey, Gallup Poll, Nielsen Rating; and many others. Because our examples are apt to be referred to in subsequent discussions, we shall take pains to single them out as follows.

EXAMPLE 1 Rolling One Die

This experiment consists of rolling a die some fixed number of times and recording the number of times each possible outcome (1, 2, 3, 4, 5, 6) occurs.

Let us suppose that we have actually done this experiment by rolling a die 72 times with the results condensed into the following table.

Table 2.1 Results of Rolling a Die 72 Times

Face showing	1	2	3	4	5	6
Absolute Frequency	12	8	15	14	6	17
Relative Frequency	.167	.111	.208	.194	.083	.236

Now the average man in the street would be inclined to draw some conclusions. He might suspect that the die is *loaded*, i.e., not fair or balanced, because if it were fair then we should have roughly as many 5's as 6's. Indeed, if the die were fair, we might *expect* to have

Table 2.2 Theoretical Model for Rolling a Fair Die 72 Times

Face Showing	1	2	3	4	5	6
Absolute Frequency	12	12	12	12	12	12
Relative Frequency	.167	.167	.167	.167	.167	.167

There are already two levels of questions that arise. First of all, where did we get the theoretical model? That's easy; by definition a *fair* die would come up 1 as often as 3, and so forth. Thus if rolled 72 times we would expect

$$\tfrac{72}{6} = 12$$

of each possible face to occur.

Second, while we might propose the above theoretical model, very few people would really expect an experiment of rolling a die 72 times to conform exactly to the model; rather we would expect *approximately* twelve 1's, twelve 2's, and so forth. So our next question is, how far from the theoretical model can we go before we decide our die is *not fair*? In particular, should the presence of only six 5's and then seventeen 6's be sufficient for us to disclaim the theoretical model for this particular die? In time we shall learn how to cope with this kind of question. That is, we shall introduce methods of measuring the discrepancy between the model and reality to see whether it lies within tolerable limits.

It is often helpful to record our experimental results as well as the corresponding theoretical model in the form of a histogram.

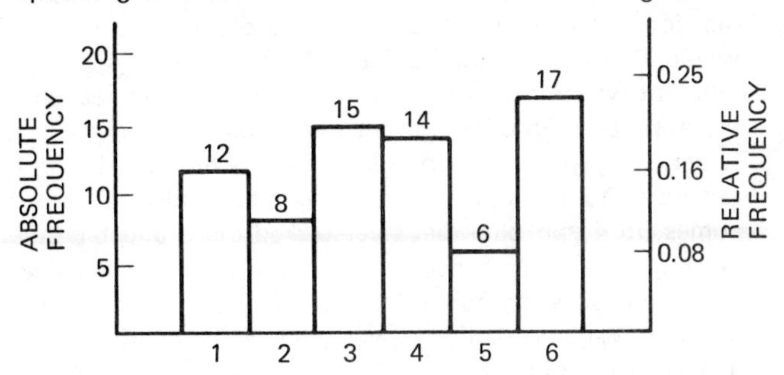

FIGURE 2.1 Results of rolling a die 72 times.

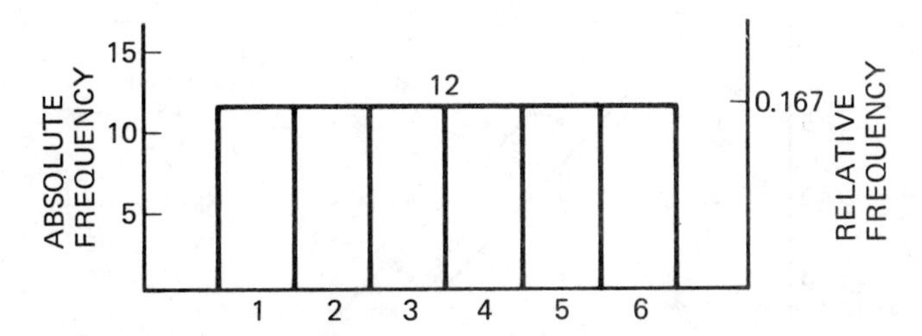

FIGURE 2.2 Model for rolling a *fair* die 72 times.

EXAMPLE 2 Rolling Two Dice

In this experiment we roll a pair of dice some fixed number of times. With each roll we have a pair of faces showing. *Usually* we are interested in the *total* of the two faces (some number from 2 to 12). For purposes of discussion we assume that we have rolled the pair of dice 144 times with the results summarized in the following histogram.

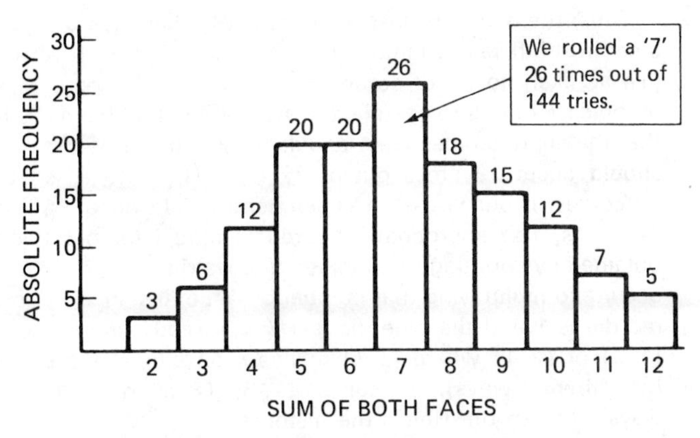

FIGURE 2.3 Results of 144 rolls of a pair of dice.

This time the choice of theoretical model is by no means so obvious. It is pretty clear that we do not want the same kind of model that we used for one die; for instance, we would not expect to roll as many 2's (or 12's) as we would 7's since 2's require that *both dice* show 1's while 7's can occur in many different ways. In fact we can pursue this line of reasoning and come up with the appropriate model without too much difficulty.

Let us suppose that one of our dice is *red* and the other *blue*. This ploy will remind us to distinguish two possible outcomes of our experiment, say red die comes up 4 and blue die 3 versus red die 3 and blue die 4, where otherwise we might casually consider the single outcome: one die 4, the other 3. We list all possible outcomes of our experiment, this time utilizing a graphical device in keeping with the two-dimensional or two-variable (red die, blue die) aspect of the problem:

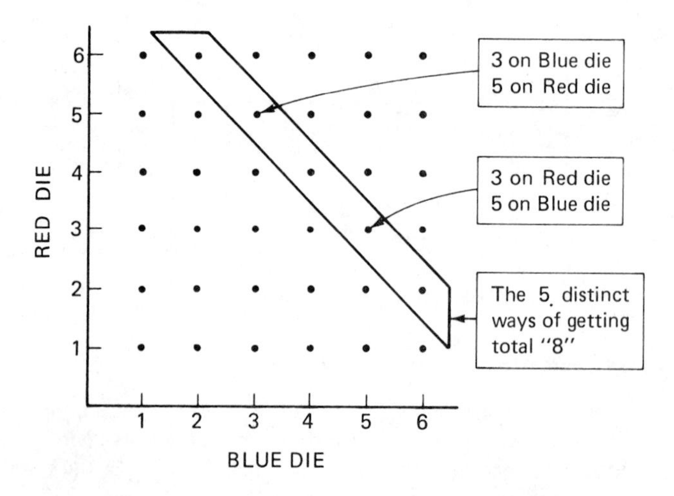

FIGURE 2.4 Possible outcomes of rolling a pair of dice once.

Thus, there are 36 possible outcomes when two dice are rolled. To say the dice are both fair is tantamount to saying that each of these 36 outcomes is just as likely to occur as any other. We say the outcomes are *equally likely*. In the case of rolling one die, the assumption that the die is fair led us to assume that the six outcomes were equally likely; in turn we said each of 1 through 6 should occur 12 times out of 72 rolls. Now we assert that each of the 36 outcomes in our two-dice experiment should occur $\frac{144}{36}$, or four times in the 144 rolls. The appropriate theoretical model for our two-dice experiment is obtained by counting the number of ways (in Fig. 2.4) that a particular total can occur and multiplying that number by 4. For instance "2" occurs only when the red die is 1 and the blue die is 1 (i.e., exactly *one* way); "3" can result from (1, 2) or (2, 1) where (a, b) means a on blue die, b on red (i.e., "3" occurs *two* different ways); "4" can be (1, 3), (2, 2), or (3, 1) (i.e., "4" occurs *three* ways); and so on. Here's the histogram.

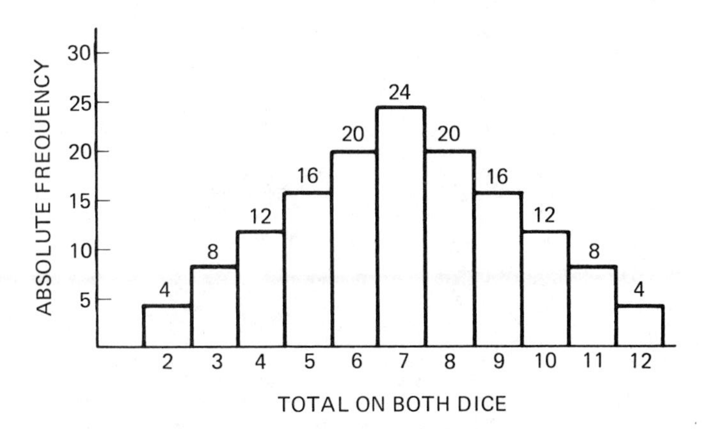

FIGURE 2.5 Theoretical model for rolling a pair of fair dice 144 times.

Once again a question that we could ask, but not yet answer, would be, "Is our experimental data close enough to the theoretical model, or is the discrepancy between theory and practice great enough to arouse our suspicion concerning the fairness of our dice?" When we are able to answer that question, we shall be using more precise language, "What is the probability that there would be as much discrepancy between Fig. 2.3 and Fig. 2.5 if the dice used in the experiment were actually fair?"

Next we examine coin tossing in much the same way that we have been analyzing dice rolling. One objective will be to introduce a type of experiment that can have *infinitely many* different outcomes. We proceed through a sequence of situations each of which extends its predecessor.

EXAMPLE 3 Tossing a Coin

We toss a coin some number of times and record the number of Heads (H) and the number of Tails (T). The theoretical model for this experiment (assuming the coin is fair or balanced so that H and T are equally likely outcomes on each toss) is given in Fig. 2.6.

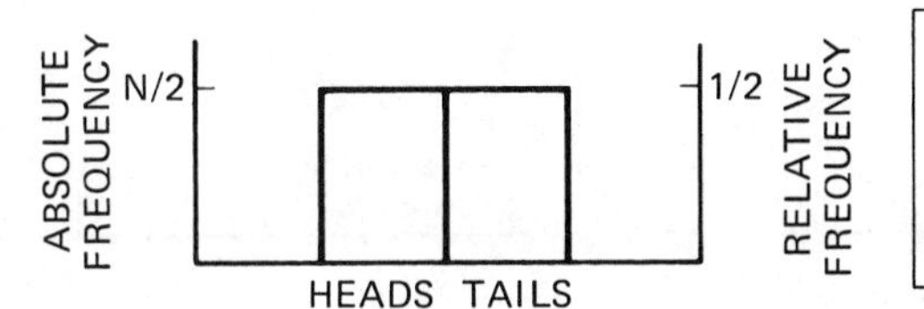

This is a phony histogram since the "classes" are not labeled by numbers. In Example 4 we show how to correct this situation.

FIGURE 2.6 Theoretical model for N tosses of a fair coin.

EXAMPLE 4 Tossing Two Coins

We appeal to the ideas in Example 2. In particular we should think of the coins as being a penny and a dime (or, if you like, one *red* and one *blue* penny).

The possible outcomes of a single toss are

(H, H), (H, T), (T, H), (T, T)

where the first position inside the parentheses refers, say, to the penny and the second to the dime. Again, fair coins imply that these four outcomes are equally likely. While there might be other interpretations of each toss, we, in general, shall be interested in *the number of heads* showing in a toss of two coins. If the pair of coins is tossed repeatedly, we would expect

0 heads about $\frac{1}{4}$ of the time: (T, T)
1 head about $\frac{1}{2}$ of the time: (T, H) or (H, T)
2 heads about $\frac{1}{4}$ of the time: (H, H)

Thus in 36 tosses, say, we would have the following theoretical model.

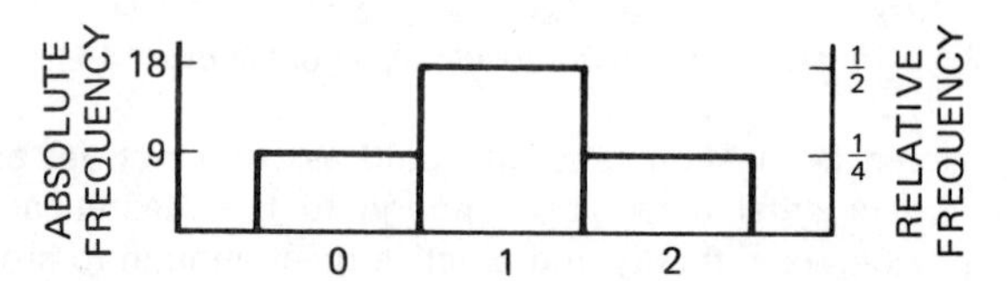

FIGURE 2.7 Theoretical model for tossing 2 fair coins 36 times.

EXAMPLE 5 Tossing Three or More Coins Simultaneously

Our discussion here will be rather sketchy; some aspects of the problem will be dealt with in the exercises, and other aspects will be taken up later in the chapter. If we toss three coins, say a penny, a nickel, and a dime, there are eight possible outcomes:

HHH, HHT, HTH, THH, HTT, THT, TTH, TTT

where HHT means heads on the penny and the nickel, tails on the dime. If the coins are fair, these outcomes are equally likely and we see that the number of heads will be

0 about $\frac{1}{8}$ of the time
1 about $\frac{3}{8}$ of the time
2 about $\frac{3}{8}$ of the time
3 about $\frac{1}{8}$ of the time

Some appropriate histograms will be suggested in the exercises.

Now suppose we have four coins. There are 16 possible outcomes:

HHHH
HHHT, HHTH, HTHH, THHH
HHTT, HTHT, THHT, HTTH, TTHH, THTH
HTTT, THTT, TTHT, TTTH
TTTT

With fair coins it follows that one would expect the number of heads to be 4, 3, 2, 1, 0 in the proportions $\frac{1}{16}$, $\frac{4}{16}$, $\frac{6}{16}$, $\frac{4}{16}$, and $\frac{1}{16}$.

By now the reader may have noticed certain patterns emerging. These patterns will be discussed in detail in the section on the Binomial Distribution in Chapter 3.

Remark: It is worthwhile observing that experiments that are superficially different may lead to the same, theoretical model. For instance, rather than tossing three coins simultaneously, we could toss one coin three times in succession. In either case we would have eight possible outcomes, symbolized

HHH; HHT, HTH, THH; HTT, THT, TTH; TTT

where HTH might refer to either

Heads on the penny, tails on the nickel, heads on the dime

or

Heads on the first toss, tails on the second toss, heads on the third toss

More generally, the same model serves for tossing n coins simultaneously as for tossing one coin n times.

EXAMPLE 6 Tossing a Coin Until it Comes Up Heads

As promised, here is an experiment conveniently viewed as a game that allows any one of *infinitely* many outcomes. A coin is tossed until it comes up heads. The possible outcomes are

H (Heads on the first try)
TH (Tails on the first try, then heads)
TTH
TTTH
TTTTH
⋮ and so on

Rather than set a certain number of rounds for the game to be played, let us record the *relative frequency* of the various possible outcomes, measured by

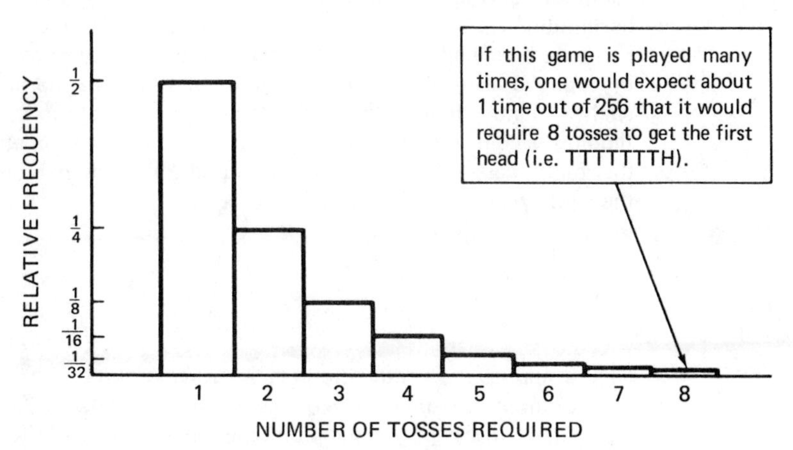

FIGURE 2.8 Tossing a fair coin until it comes up heads.

the number of tosses required to produce the first head (1, 2, 3, . . .). Assuming we have a fair coin, we expect to toss a head on the first attempt one-half of the time. The other half of the time we get tails on the first toss, and then half of those times we would expect heads on the second toss. Continuing in this manner, we get the graph shown in Fig. 2.8, which we still call a *histogram* although the *range* is now infinite.

We use the phrase *infinite but discrete* to describe the above situation. While there are infinitely many possible events or outcomes, they are discrete in the sense that we can enumerate them H, TH, TTH, TTTH, ... In the next section we shall investigate a different type of "experiment" whose outcomes do not lend themselves to this kind of classification.

Summary

We have examined a few simple experiments whose outcomes can be enumerated, that is, experiments for which we can make a *list* of all possible outcomes (even though the "list" in the last example requires the use of three mysterious dots . . .). More importantly, we have constructed a *theoretical model* for each of these experiments. The characteristics of such a model are (1) all possible outcomes are enumerated and (2) to each outcome we assign a number, called its *relative frequency*, that represents the proportion of times the outcome is likely to occur if the experiment is conducted a large number of times. Except for the last example, we can also assign the *absolute frequency* to each outcome. This number represents the number of times the particular outcome is expected to occur when the experiment is performed some fixed number of times. Models of this type (having a list of outcomes, each of which is assigned a positive number for its relative frequency) are called *discrete*. In the next section we will see *continuous* models, characterized by sets of possible outcomes that defy listing.

One thing to bear in mind is that each of our models is an "ideal" representation of the experiment. It arises from making some rather restrictive assumptions about the experiment. For example, we postulate a *fair* die, perfectly balanced so as not to have any propensity to stop rolling with a "6" on top. We also make implicit assumptions such as the ability to roll a die in such a manner that no particular face is more likely to appear, or that when flipping a coin, we are careful not to give the same impetus to different flips that might cause the coin to turn in the air the same number of times and land the same way.

Exercises Section 1

1 Compute the relative frequencies for the outcomes in Example 2.

2 Compute the relative frequencies for the model in Fig. 2.5.

3 **(a)** How many distinct outcomes are there in the experiment consisting of tossing five coins simultaneously? How about n coins?

(b) List the outcomes for five coins in groups having the same number of heads (i.e., all outcomes having five heads, four heads, etc.).

(c) Sketch the histogram for tossing five fair coins 64 times.

4 What follows is a simple device for finding the theoretical model for tossing n fair coins.

(a) (Pascal's triangle.) Fill in the next two rows of the following array.

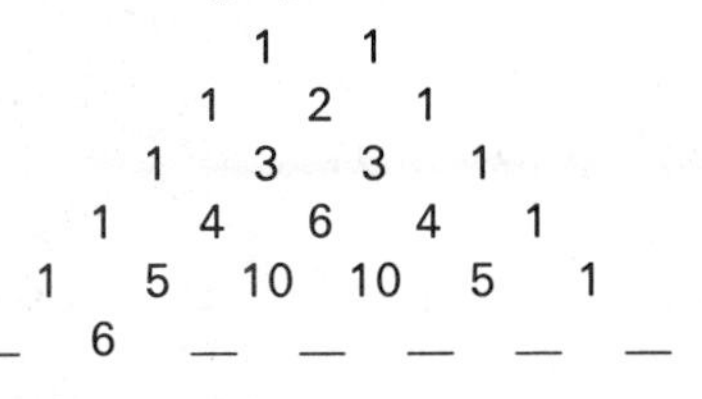

(b) Expand each of the following.

$(T + H)^1$
$(T + H)^2$
$(T + H)^3$
$(T + H)^4$

Note that the coefficients of these expansions (written in decreasing powers of T) are precisely the corresponding rows of Pascal's triangle. Finally, to find the relative frequency of getting three tails and one head in four tosses of a fair coin, you take the coefficient of the term T^3H and divide by 2^4.

(c) Find the relative frequency of getting:

(1) 2 tails and 1 head in three tosses
(2) 1 tail and 2 heads in three tosses
(3) 3 tails and 1 head in four tosses
(4) 2 tails and 2 heads in four tosses

5 Use Pascal's triangle above to draw the theoretical model histograms for tossing:

(a) three fair coins
(b) four fair coins
(c) five fair coins
(d) six fair coins

6 Suppose you wanted to devise the theoretical model for the simultaneous rolling of three dice.

(a) How many distinct outcomes would there be?
(b) How could you represent these different outcomes?
(c) What possible totals (of the three faces) could occur?
(d) Find the relative frequency for a few of these possible totals.

7 **(a)** Find the mean of the data in Example 1.
(b) Find the mean of the set of data that would correspond to the model (Fig. 2.2) of a fair die.

8 **(a)** Find the mean of the data in Example 2.
(b) Find the mean of the data that would correspond to the model (Fig. 2.5) of rolling a pair of fair dice.

9 **(The slot machine.)** A simple slot machine is constructed with three bands, each containing four items: BELL, CHERRY, ORANGE, PLUM. There are five types of payoffs:

Band 1	Band 2	Band 3	Payoff (in dollars)
BELL	BELL	BELL	1.00
CHERRY	CHERRY	CHERRY	.25
CHERRY	CHERRY	NOT CHERRY	.05
ORANGE	ORANGE	ORANGE	.50
PLUM	PLUM	PLUM	.50

(a) How many possible outcomes are there?

(b) Suppose you pay a nickel per play and you play as many times as there are possible outcomes, getting each one once. How much would you win or lose?

10 Repeat the previous problem with each of the following modifications.

(a) A second CHERRY is added to each band, giving five items per band.

(b) A second PLUM is added to each band, giving five items per band.

11 **(a)** Two miniature roulette wheels are spun simultaneously producing a pair of numbers which are then added to yield a number X. If one wheel is numbered from 1 through 4 and the other from 1 through 5, determine the theoretical model for the various values of X. (Assume that for each wheel, the various numbers have equal chances of coming up.)

(b) Repeat the problem, using wheels numbered 1, 3, 5, 7 and 1, 3, 5, 7, 9.

(c) Repeat the problem using wheels numbered 1, 2, 3, 4 and 1, 2, 4, 5, 6.

12 **(a)** Is the sum of the relative frequencies in Example 1 equal to 1?

(b) Rewrite those relative frequencies as fractions and add them.

13 **(a)** Suppose $X_1, X_2, \ldots, X_n$ are distinct data points. Find the relative frequencies associated with each and show that their sum is 1.

(b) Suppose that a data set consists of

X_1 appearing f_1 times
X_2 appearing f_2 times

$\vdots$

X_n appearing f_n times

For example, the set $\{1, 1, 5, 5, 5, 0, 0, 0, -1\}$ has

-1 appearing 1 time
0 appearing 3 times
1 appearing 2 times
5 appearing 3 times

Find an expression for the relative frequency of each X_i and show why their sum is 1.

14 It is interesting to try to show that the "sum" of the relative frequencies in Fig. 2.8 is 1.

(a) Find the sum of the first four relative frequencies in Fig. 2.8.
(b) Find the sum of the first five relative frequencies in Fig. 2.8.
(c) Find the sum of the first six relative frequencies in Fig. 2.8.

15 Consider a data set consisting of

X_1 appearing f_1 times
X_2 appearing f_2 times

$\vdots$

X_n appearing f_n times

(a) Express the mean of this data in terms of the X_i's and the f_i's.
(b) Let P_i be the relative frequency of X_i and express the mean in terms of the X_i's and the P_i's. (*Hint*: Practice with $\{X_1, X_2, X_3\}$.)

16 Compare the concepts of absolute frequency and relative frequency by discussing how Examples 1 and 2 would be affected if the number of tosses were 100 in each case rather than 72 and 144.

2 CONTINUOUS MODELS

The examples presented in the previous section had one common characteristic: In each case the possible outcomes of a given experiment could be enumerated. That is to say, we could make a list of all the outcomes and then assign a relative frequency to each of them. Such phenomena are called *discrete*. In this section we shall discuss the alternative situation, called *continuous*.

We have all seen two kinds of clocks. One kind has no second hand and it "clicks" at the end of each 60 seconds, moving the minute hand ahead one unit. At a given instant, we cannot tell exactly what time it is; the best we can do is to say, for example, that it is between 10:47 and 10:48. In truth, it may be 10:47 *plus* 59 seconds, so that 1 second later the minute hand will move to 48. Such clocks are common in public schools, and variations of them are to be found elsewhere. For instance, the "time and temperature" clocks popular on bank buildings often work on the same principle. The time is usually displayed by a system of lights and it looks like

FIGURE 2.9 A discrete clock.

from 10:47 until 10:48. The temperature is likewise given in units of one degree. A similar situation prevails with the friendly recording available when one dials the number for "time of day." The recording consists of two segments:

The first is a helpful hint about efficient use of the telephone; then a separate recording announces

"Time . . . 10:47"

This portion of the recording changes at the end of each minute.

Clocks of the type just described, that record only minutes, are called *discrete*. Time, in the abstract, however, is a *continuous* phenomenon. That is, time does not "jump" from 10:47 to 10:48 all at once. Rather, time moves from 10:47 to 10:48 in a sort of sweeping fashion. If t is any real number between 0 and 1, there is some time corresponding to precisely t minutes past 10:47. Thus, we could have an actual time of

10:47 plus .03 minutes
10:47 plus .012687532871244 minutes, or
10:47 plus .99999 minutes

At each of these times, the discrete clock would read

10:47

Now, any watch or clock with a second hand can be viewed as a *continuous* phenomenon. At a given instant we might see something like

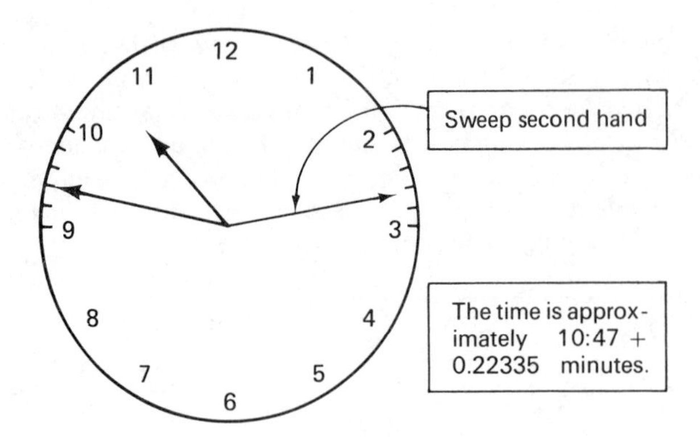

FIGURE 2.10 *A continuous clock.*

Whereas the discrete clock can take on only a finite number of equally spaced values (namely, 720 distinct minutes), the continuous clock can take on *all values in some interval* of, say,

0 to 12 (in hours)

or

0 to 720 (in minutes)

or

0 to 43,200 (in seconds)

and there is no possible way to list all the values of an interval.

Other sources of continuous measurements are the speedometer in an automobile, the mercury thermometer, the bathroom scales, the gauge on a pressure cooker and so forth. In each of these situations the measuring device does not jump from one unit value to the next, but rather it moves steadily throughout its range. When reading any of these instruments, we generally round off the true value to the nearest convenient unit of measurement, thereby introducing an *error*. That is, in the real world we often interpret a continuous measurement as if it were discrete. Again, this is a practical necessity, since no kind of calibration could be made fine enough to exhaust all possible measurements on a continuous scale—they simply cannot be enumerated.

Now, using these two types of clocks, we can devise an "experiment" that has infinitely many possible outcomes. Just as we did in the previous section, we are able to build a theoretical model to allow us to predict the likelihood of certain outcomes. The analogy between discrete and continuous phenomena will be a recurring theme throughout this book.

EXAMPLE 1 Error in a Discrete Clock—A Continous Example

When we record the time from a discrete clock, we introduce an error, which we denote by X. As we have observed, the value of X may range from 0 to 1 (in minutes), and may take on any value in that interval. We list certain possible outcomes and illustrate each of these as a subset of the interval from 0 to 1.

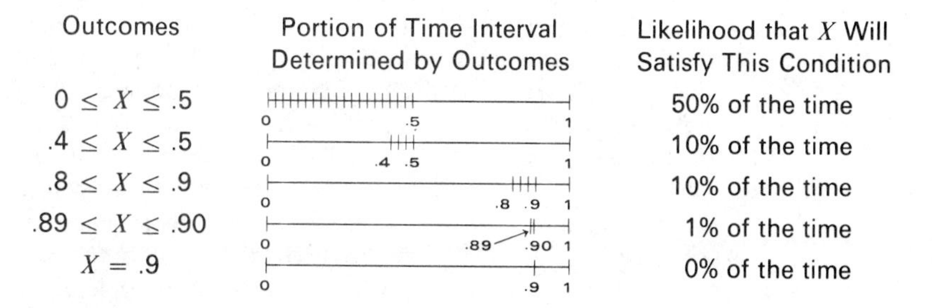

Outcomes	Portion of Time Interval Determined by Outcomes	Likelihood that X Will Satisfy This Condition
$0 \leq X \leq .5$	0 .5 1	50% of the time
$.4 \leq X \leq .5$	0 .4 .5 1	10% of the time
$.8 \leq X \leq .9$	0 .8 .9 1	10% of the time
$.89 \leq X \leq .90$	0 .89 .90 1	1% of the time
$X = .9$	0 .9 1	0% of the time

QUESTION 1 How were these likelihoods computed?

ANSWER We have assumed that time moves steadily through the 1 minute interval, and that the likelihood of X falling into any given subinterval (such as $.4 \leq X \leq .5$) is therefore the ratio of the length of that subinterval to the total length.

QUESTION 2 Why do we assign 0% likelihood to the outcome $X = .9$?
ANSWER This follows from the observation that smaller and smaller sub-intervals represent less and less likely occurrences. We note, for example, that the intervals

$$.8 \le X \le .9, \qquad .89 \le X \le .90, \qquad .899 \le X \le .900,$$

and

$$.8999 \le X \le .9000$$

represent time spans of, respectively,

.1, .01, .001, and .0001

minutes. The value $X = .9$ is common to each of these intervals and therefore its likelihood must be no greater than any of them. The only possibility, as the intervals shrink to the point .9, is that the likelihoods decrease to 0.

QUESTION 3 What precisely is the analogy between this example and one of the discrete examples of Section 1?
ANSWER Let us review the example of rolling two fair dice. We represented the 36 possible outcomes as 36 points. We further agreed that, if the dice were indeed fair, these 36 points represented equally likely outcomes. In the present example, the outcomes are represented by all the points on the interval from 0 to 1. To say that the composite outcome "rolling a 3," represented by the points (1, 2) and (2, 1), is just as likely as "rolling an 11," represented by the points (6, 5) and (5, 6), is analogous to saying that $.4 \le X \le .5$ is just as likely as the outcome $.8 \le X \le .9$.

EXAMPLE 2 Round-Off Error

In many everyday situations we find it necessary to perform a rounding off of a real number. When we report that the length of a table is $48\frac{7}{8}$ inches, what we really mean is that the best our eye can do is to look at the tape measure and see that the table length is closer to $48\frac{7}{8}$ inches than it is to $48\frac{6}{8}$ or 49 inches. That is, we have rounded off the length of the table to the nearest $\frac{1}{8}$ of an inch. Similarly, when a butcher weighs a pot roast and marks a weight of 3 pounds 10 ounces on the package, he is telling us that the true weight is closer to 3 pounds 10 ounces than it is to either 3 pounds 9 ounces or 3 pounds 11 ounces. (In mathematics, one frequently must idealize a situation. Here is a case in point where we have hypothesized an honest butcher.) In any event, this rounding off process provides us with another example of a continuous variable. Moreover, the same theoretical model applies to virtually any situation where we round off to the nearest of two adjacent units, be they tons, ounces, miles, inches, degrees, furlongs, volts, cubits, or whatever. For the sake of simplicity, let us assume that we are doing a forest survey and measuring the circumferences of trees *to the nearest inch*. With each measurement we introduce a round-off error, which we shall denote X. We list a few typical measurements and the corresponding errors as follows:

Actual Circumference (inches)	Recorded Circumference (inches)	Round-off Error (inches)
16.4132 . . .	16	.4132 . . .
21.0123 . . .	21	.0123 . . .
20.9025 . . .	21	−.0974 . . .
10.3162 . . .	10	.3162 . . .
18.5213 . . .	19	−.4786 . . .
24.5600 . . .	25	−.4399 . . .

Thus since our recorded value can be either too large or too small, the round-off error might be either negative or positive. Moreover it should be clear that X will range from $-.5$ to $+.5$ and may take on any value on that interval. Here, then, are some typical outcomes and their likelihoods.

Outcomes	Corresponding Subinterval	Likelihood that X Will Satisfy the Condition
$-.5 \le X \le 0$	−.5 to 0 (on −.5, 0, .5)	50% of the time
$0 \le X \le .1$	0 to .1 (on −.5, 0, .1, .5)	10% of the time
$.09 \le X \le .1$	.09 to .1 (on −.5, .09, .1, .5)	1% of the time
$X = -.3$	−.3 (on −.5, −.3, .5)	0% of the time
X is positive	0 to .5 (on −.5, 0, .5)	50% of the time

Lest the reader be inclined to think that all continuous phenomena have theoretical models like the two just described (the variable X ranges over an interval of length 1 and subintervals of equal length represent classes of equal likelihood), we hasten to point out that we shall encounter a wide variety of situations where not only are the ranges quite different, but also the equally likely aspect no longer holds. To analyze such models, however, requires more precise terminology as well as some rudiments of calculus. The following example, while not explored in detail, should illustrate the potential complexity of continuous phenomena.

EXAMPLE 3 The Roller Coaster

Imagine one hill of a roller coaster track that extends over 200 feet of ground. We denote by X the horizontal distance traveled at any given time as the coaster first ascends then descends on its track (see Fig. 2.11). Now if we randomly select a time while the coaster is on this path, the value of X could be any number between 0 and 200. Yet it should be clear that certain ranges of values of X are much more likely than other ranges of the same length. For instance it would be far more likely for X to fall in the interval from 0 to 100 than it would be for X to be between 100 and 200.

Some typical ranges of values for X and the corresponding intervals are listed below. Although we have no way of finding the likelihood, we have

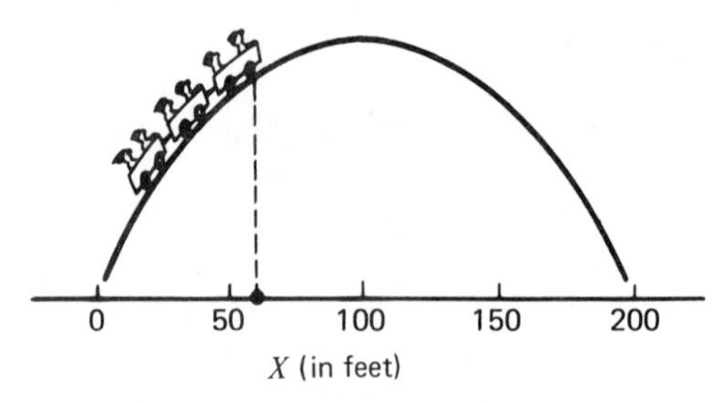

FIGURE 2.11 Roller coaster model.

listed some *guesses* of what their likelihoods might be. It is left to the reader to decide how reasonable these estimates seem.

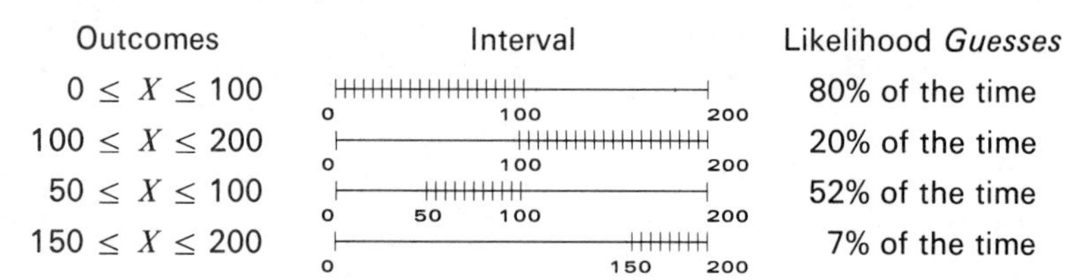

Outcomes	Interval	Likelihood *Guesses*
$0 \leq X \leq 100$	0 100 200	80% of the time
$100 \leq X \leq 200$	0 100 200	20% of the time
$50 \leq X \leq 100$	0 50 100 200	52% of the time
$150 \leq X \leq 200$	0 150 200	7% of the time

Summary

This section provided a contrast to the discrete models of the previous section. Terms such as "experiment" and "outcome," which were appropriate in the discussion of discrete models, were not found here. Nevertheless, we have considered certain phenomena that occur in our everyday life that can be likened to experiments where the results of *observations* (noting the time, noting the position of a roller coaster, finding the round-off error, and so on) play the role of outcomes. And, as before, we constructed theoretical models that help us predict the likelihood of various observations.

The distinguishing feature of the models constructed in this section is that each of them has for its set of possible observations some *interval* of values. As a result, the question of how likely a particular observation might be (the central question in our discussion of discrete models) is replaced by a question of likelihood that an observation falls within a certain subinterval of the interval of possible values. Such models (having an interval of possible "observed" values and supplying information about the likelihood that an observation will fall in a particular subinterval) are called *continuous*. To do justice to them will require some tools from calculus, and while we shall occasionally encounter a continuous model, we must postpone any serious discussion of them for the time being.

Exercises Section 2

1 In each of the following examples, decide whether the variable X is discrete or continuous and specify a reasonable interval over which X would range.

- **(a)** X is the temperature on an outdoor thermometer at the United States Weather Bureau Station in Dallas anytime during August, 1966.
- **(b)** X is the family income of all families on welfare in the United States during 1969.
- **(c)** X is the number of books in the college library for each college in the United States.
- **(d)** X is the exact height of second graders in Chicago.
- **(e)** X is the weight, to the nearest ounce, of babies born in San Francisco during 1968.
- **(f)** X is the average speed of each car during time trials at the 1970 Indianapolis 500.
- **(g)** X is the temperature setting on electric stove burners.
- **(h)** X is the temperature setting on gas stove burners.
- **(i)** X is the setting on television channel selectors.
- **(j)** X is the setting on radio dial selectors.

2 Each of the following situations depicts a continuous variable X. In which cases would subintervals of equal length represent equally likely outcomes for X?

- **(a)** A rock is dropped from a 300-foot cliff. At some randomly selected time during its descent, X is the height of the rock above the ground.
- **(b)** A freely revolving roulette wheel is spun. It has a pointer but no pegs so that it could stop with the pointer pointing anywhere on the wheel. X is the position of this pointer (measured in some convenient unit, say 0 to 360°) when the wheel comes to rest.
- **(c)** A tennis ball is dropped from a height of 5 feet and is allowed to bounce until it comes to rest. At a randomly selected time while it is bouncing, X is the height of the ball above the floor.
- **(d)** X is the number of miles traveled at a given time during flight on a plane from Denver to Chicago.
- **(e)** X is the exact position of an elevator on a typical run in a 10-storey building, measured from the ground floor up in feet.

3 SAMPLE SPACES AND PROBABILITY

In the first two sections of this chapter we examined some "experiments," each involving several possible outcomes, and we compared the frequencies with which these different outcomes occurred. In doing this, we have flirted with the notion of probability and with its companion, sample space. The present section provides the basic definitions and fundamental properties required to pursue a more formal treatment of these ideas.

Suppose that we are interested in analyzing the possible outcomes of some experiment (like rolling one or more dice, flipping coins, drawing lottery tickets, etc.) with an eye towards determining how likely or unlikely certain outcomes might be. Our first task will be to construct an appropriate replica of *all* the possible outcomes of the experiment. We do this by representing the outcomes as a set of points, usually in a plane though sometimes on a line or in three or more dimensions, each point corresponding to a different outcome.

The totality of points is called the *sample space* and the individual points are called *simple events*. Here are some elementary examples.

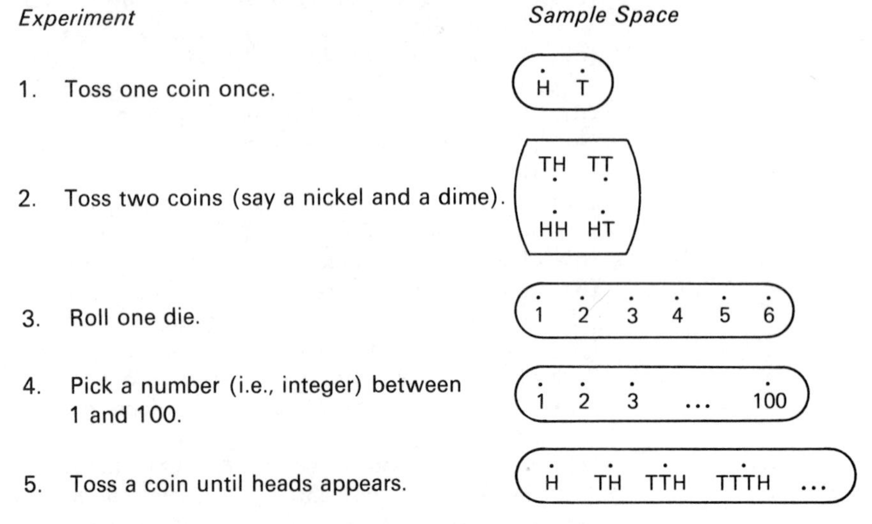

We call these individual points *simple events* to distinguish them from a *compound event* which is any subset of the sample space, i.e., any collection (finite or infinite) of simple events.* For instance we have:

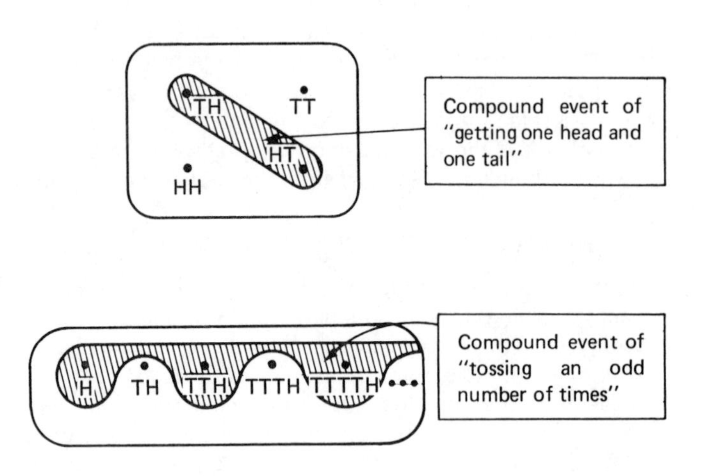

Now the sample space is but half our model for an experiment. The other half is the *probability function*, P, which assigns to each event E (simple and

* So as not to antagonize the purist, we admit here to identifying a simple event, say E, with the corresponding singleton set, $\{E\}$; thus simple and compound events are both *subsets* of the sample space.

compound) some real number $P(E)$ between 0 and 1, i.e.,

$$0 \leq P(E) \leq 1 \qquad \text{for each event } E$$

> $P(E)$ is called the *probability* of the event E.

In terms of our previous experience, $P(E)$ plays the role of the "relative frequency"; it signifies the likelihood or probability that a particular event will occur.

So far we have not given an adequate definition of P. To do so requires a swift excursion through rudimentary set theory. Here are some of the necessary ideas given in formal set-theoretic terms along with their corresponding probability theory interpretations.

A *set* is simply a collection of objects or *elements* specified by some kind of rule or description. For example:

1. The set of all professional basketball players in the National Basketball Association.
2. The set of all National Basketball Association players under 5 feet tall.
3. The set of all even positive integers.
4. The set of all three-letter "words" made from the letters A, E, and R: {AER, ARE, EAR, ERA, RAE, REA}.
5. The sample space associated with the experiment of tossing five coins.

If S is a set (for instance a sample space) and if A is a subset (say some compound event), we write

$$A \subseteq S$$

If s is one element in S, we write

$$s \in S$$

and say s *belongs to* S or s *is a member of* S. If s is not a member of S, we write

$$s \notin S$$

In contrast to the method used above to describe sets (i.e., by words) we frequently utilize some standard symbols. Braces, such as

$$\{\ \}$$

are used to enclose the elements of a set, and thus to distinguish the "list"

$$1, 2, 3, 4, \pi$$

from the set

$$\{1, 2, 3, 4, \pi\}$$

By inserting a vertical bar, |, inside the braces, we have the so-called "set builder notation," which is used as follows. Suppose we use the symbol J to denote the set of positive integers, i.e.,

$$J = \{1, 2, 3, \ldots\}$$

Then various subsets of J can be described by the device of writing braces with a particular property, that defines the subset, written after the vertical bar. For instance we write

$\{n \in J \mid n \text{ is even}\}$ to denote $\{2, 4, 6, \ldots\}$

Sometimes when the context makes clear that we are dealing only with elements from J, we simply write

$$\{n \mid n \text{ is even}\}$$

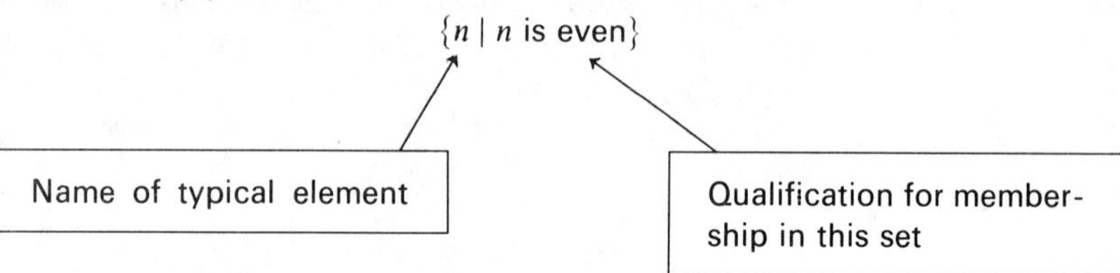

This notational device is particularly useful when we wish to describe subsets formed from other subsets. Let us suppose, for example, that we have a pair of subsets of J:

$$A = \{1, 2, 4, 5, 7, 8\} \quad \text{and} \quad B = \{2, 4, 6, 8, 10\}$$

Then we can define two new sets, C and D, by insisting that C consists of all elements that belong to either A or B (or both); and D consists of all elements that belong to *both* A and B. Thus

$$C = \{1, 2, 4, 5, 6, 7, 8, 10\}$$

$$D = \{2, 4, 8\}$$

We could describe C and D using our set builder notation:

$$C = \{n \in J \mid n \in A \textit{ or } n \in B \text{ (or both)}\}$$

$$D = \{n \in J \mid n \in A \textit{ and } n \in B\}$$

As a matter of fact, these two constructions are fairly common and are referred to by name and symbol. In general, if A and B are subsets of a set S, we define:

Symbol Name Definition

$A \cup B$ = (the *union* of A and B) = $\{x \in S \mid x \in A \textit{ or } x \in B \text{ (or both)}\}$
$A \cap B$ = (the *intersection* of A and B) = $\{x \in S \mid x \in A \textit{ and } x \in B\}$

Two other constructions are likewise worth listing:

$A - B$ = (the *difference* between A and B *or* the *relative complement* of B in A) = $\{x \in S \mid x \in A \text{ and } x \notin B\}$
A^c = (*A complement* or the complement of A in S) = $\{x \in S \mid x \notin A\}$

In many situations, whether related to probability or in another setting, the use of pictures called Venn diagrams can assist our intuition in matters of subsets and the way in which we combine them. The four definitions given above can be realized by shading appropriate regions as follows:

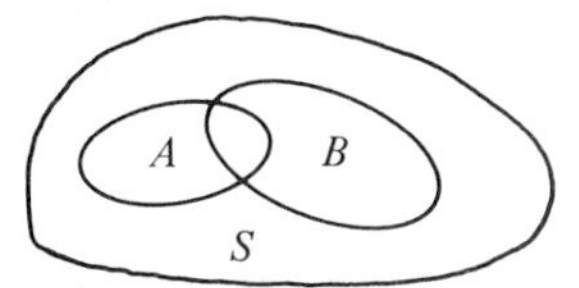

A SET S AND TWO SUBSETS A AND B, OR
A SAMPLE SPACE S AND TWO COMPOUND EVENTS A AND B

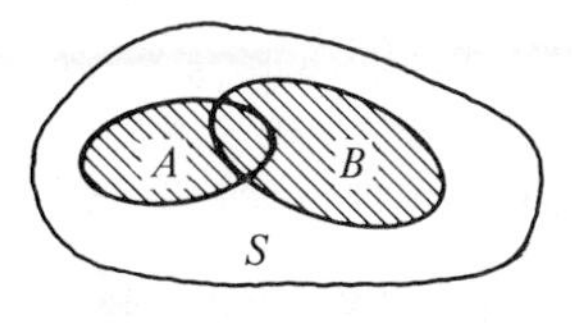

$A \cup B$ or
THE EVENT "EITHER A OR B (OR BOTH)"

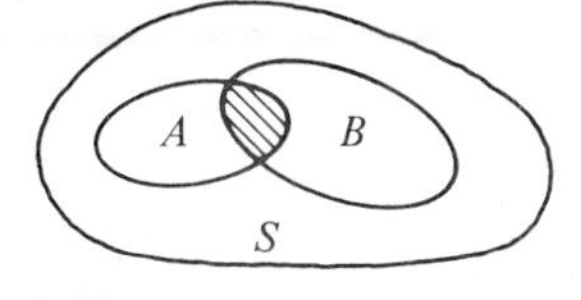

$A \cap B$ or
THE EVENT "A AND B"

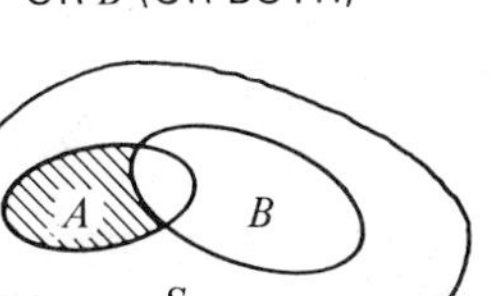

$A - B$ or
THE EVENT "A BUT NOT B"

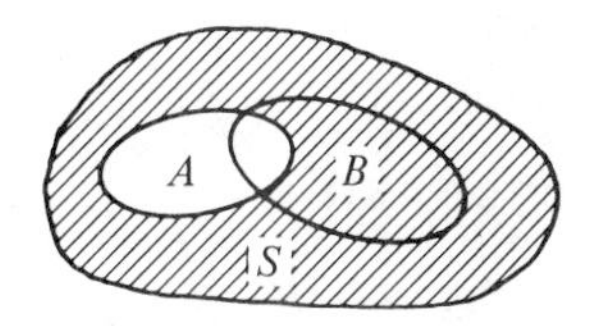

A^c or
THE EVENT "NOT A"

In each case we have indicated how these various new subsets enter into probability discussions.

Finally we adopt the symbol

$$\varnothing$$

to denote *the empty set*, i.e., a set with no elements. Example 2 above is probably the empty set. We often write

$$A \cap B = \varnothing$$

which simply means that A and B are *disjoint;* they share no elements. For instance,

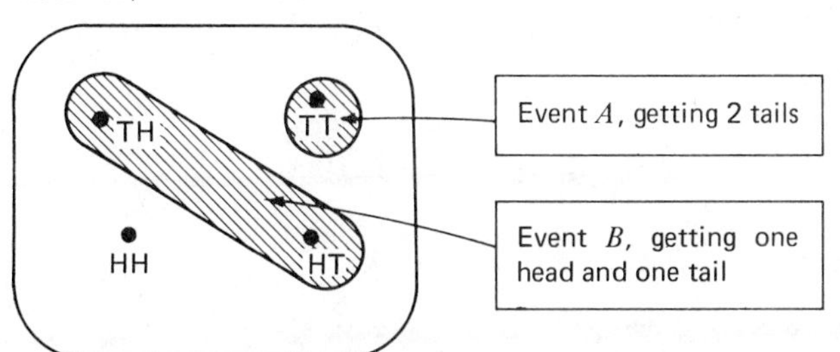

SAMPLE SPACE FOR TOSSING A PAIR OF COINS

More formally, we complete our definition of P.

Definition

A *probability function* on a sample space S is a function P that assigns to each event E of S a real number $P(E)$ such that

1. $0 \leq P(E) \leq 1$, for each $E \subseteq S$
2. $P(S) = 1$
3. $P(A \cup B) = P(A) + P(B)$, if $A \cap B = \varnothing$

The interpretation of these stipulations is fairly clear. We assign to S, the compound event consisting of *all* simple events, the probability 1. In other words, S represents the event, "something occurs," and we give it maximum probability 1 to denote *certainty*. The more unlikely an event, the closer its probability is to 0. The third condition says that if two events cannot both occur (simultaneously), then the probability that *one or the other* will occur is simply the sum of their individual probabilities. We shall return to this via some examples momentarily.

A few immediate consequences of this definition are available. For the most part the proofs would appear easy to the reader who has dabbled with intersections and unions, and somewhat artificial to the reader who hasn't. We therefore omit the formal proofs and encourage the reader to pay particular attention to the examples that follow.

Proposition

Let P be a probability function on a sample space S. Then

1. $P(\varnothing) = 0$. In words, the probability that *nothing* happens is zero.
2. $P(A^c) = 1 - P(A)$. In words, the probability that the event A won't occur equals one minus the probability that A will occur.
3. If $S = \{E_1, E_2, \ldots, E_n\}$ where the E_i's are all simple events of S, then $P(E_1) + P(E_2) + \cdots + P(E_n) = 1$. In words, the sum of the probabilities of all the simple events is one.
4. If $A \subseteq B$, then $P(A) \leq P(B)$. In words, if every simple event of A is also a simple event of B, then the probability of A is at most the probability of B.

EXAMPLE 1 Tossing a Pair of Fair Coins

Tossing a pair of fair coins (say a nickel and a dime) yields the following model.

1. The sample space.

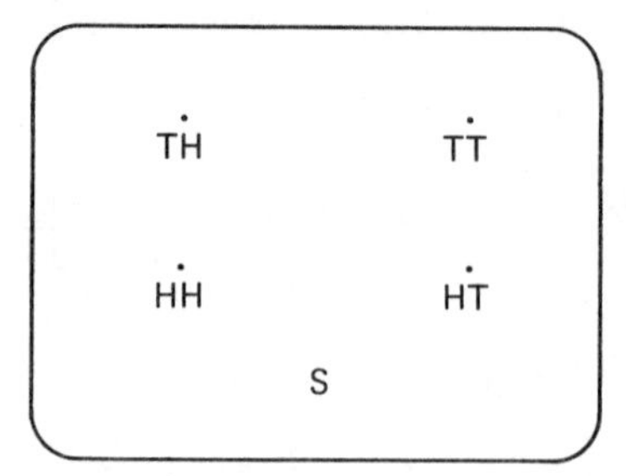

2. The probability function. The several events and their corresponding probabilities are:

$$P(\emptyset) = 0 \quad P(\text{TT}) = \tfrac{1}{4} \quad P(\text{HH, HT}) = \tfrac{1}{2} \quad P(\text{HH, HT, TH}) = \tfrac{3}{4} \quad P(S) = 1$$
$$P(\text{TH}) = \tfrac{1}{4} \quad P(\text{HH, TH}) = \tfrac{1}{2} \quad P(\text{HH, HT, TT}) = \tfrac{3}{4}$$
$$P(\text{HT}) = \tfrac{1}{4} \quad P(\text{HH, TT}) = \tfrac{1}{2} \quad P(\text{HH, TH, TT}) = \tfrac{3}{4}$$
$$P(\text{HH}) = \tfrac{1}{4} \quad P(\text{HT, TH}) = \tfrac{1}{2} \quad P(\text{HT, TH, TT}) = \tfrac{3}{4}$$
$$P(\text{HT, TT}) = \tfrac{1}{2}$$
$$P(\text{TH, TT}) = \tfrac{1}{2}$$

We might wish to give verbal descriptions to some of these events. For instance, the probability of getting at least one head $= P(\text{HH, HT, TH}) = \tfrac{3}{4}$, or the probability of getting more than one head $= P(\text{HH}) = \tfrac{1}{4}$.

It should be observed that *each of the simple events in this sample space has the same probability*. This is the formal counterpart of saying the simple events are *equally likely*. By way of contrast, the sample space for tossing a coin until heads appears has infinitely many simple events and they each have different probabilities:

$$P(\text{H}) = \tfrac{1}{2}, \qquad P(\text{TH}) = \tfrac{1}{4}, \qquad P(\text{TTH}) = \tfrac{1}{8}, \qquad \ldots$$

In the case that the sample space is finite, questions of probability reduce to matters of counting elements in subsets. Though in many respects very simple, such spaces are suggestive under scrutiny so we take time to elaborate.

EXAMPLE 2 Finitely Many Equally Likely Simple Events

1. Our sample space is $S = \{E_1, E_2, \ldots, E_n\}$, which we shall indicate pictorially as

E_1 E_2 E_3 E_4

E_5 ... E_n S

2. $P(E_1) = P(E_2) = \cdots = P(E_n) = 1/n$.

This follows from part 3 of our proposition. Moreover if A is a compound event with k elements,

$$A = \{E_{i_1}, E_{i_2}, \ldots, E_{i_k}\}, \qquad 0 \le k \le n$$

then $P(A) = P(E_{i_1}) + P(E_{i_2}) + \cdots + P(E_{i_k}) = k/n$.

We define $\#(A) =$ the number of elements in A. Then

$$P(A) = \frac{\#(A)}{\#(S)}$$

This is precisely the way in which we obtained *relative frequency* from *absolute frequency* in our informal examples. Let us check that this definition of probability conforms with the three requirements listed in the general definition of a probability function. That is, we must check that

1. $0 \le P(E) \le 1$, for any event E
2. $P(S) = 1$
3. $P(A \cup B) = P(A) + P(B)$ if $A \cap B = \varnothing$

We begin by observing that for any event

$$A = \{E_{i_1}, \ldots, E_{i_k}\}$$

we have

$$P(A) = \frac{\#(A)}{\#(S)} = \frac{k}{n}$$

so that obviously

$$0 \le \frac{k}{n} \le 1 \quad \text{or} \quad 0 \le P(A) \le 1$$

and in particular

$$P(\varnothing) = 0, \qquad P(S) = 1$$

Rather than checking $P(A \cup B)$ only for disjoint sets A and B, let us see what happens in general.

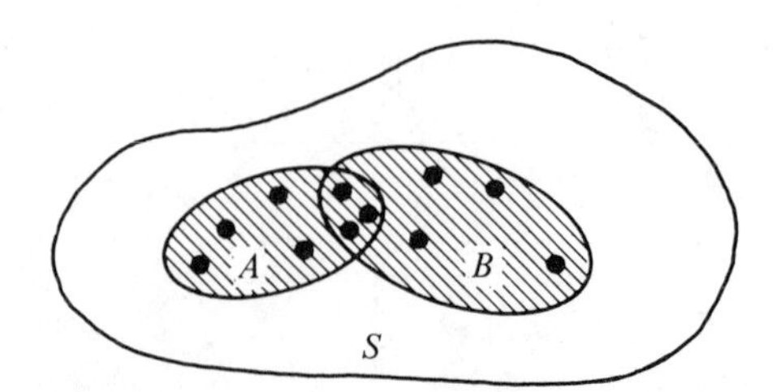

TWO TYPICAL COMPOUND EVENTS, A AND B, OF S

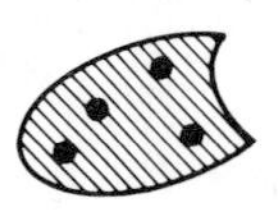

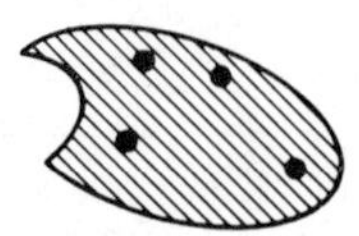

$A - (A \cap B)$ $\qquad A \cap B$ $\qquad B - (A \cap B)$

THREE DISJOINT SUBSETS WHOSE UNION IS $A \cup B$

It should be obvious that

$$\begin{aligned}\#(A \cup B) &= \#(A - (A \cap B)) + \#(A \cap B) + \#(B - (A \cap B)) \\ &= [\#(A) - \#(A \cap B)] + \#(A \cap B) + [\#(B) - \#(A \cap B)] \\ &= \#(A) + \#(B) - \#(A \cap B)\end{aligned}$$

Dividing both sides of the equation by n [$= \#(S)$], we get

$$P(A \cap B) = P(A) + P(B) - P(A \cap B) \quad \text{(for arbitrary } A \text{ and } B\text{)}$$

and in the special case when $A \cap B = \varnothing$

$$P(A \cap B) = P(A) + P(B) \quad \text{(for } A \text{ and } B \text{ disjoint)}$$

showing that 3 holds.

EXAMPLE 3 The Dart Board—A Continous Example

Suppose we have a circular dart board, say 2 feet in diameter. We assume that a dart is thrown in such a way so as to strike the board randomly. (Note that once again we have idealized a situation: The whole point to throwing darts is that as one's skill develops, the randomness of where the dart lands diminishes. Thus we have posited a rank amateur dart thrower rather than a true champion.) At any rate, our model for this experiment is a circular disk of diameter 2 in the plane. The sample space is then all points in that disk (including points in the boundary). The interesting question becomes, "How do we assign a probability function to this sample space?"

Actually, the solution is straightforward. We observe that each event is a region in the plane, and that the event consisting of the entire space must have probability 1. To each region A (i.e., a subset of S) we assign a probability as follows:

$$P(A) = \frac{\text{area of } A}{\text{area of } S} = \frac{\text{area of } A}{\pi}$$

Here are some typical compound events and their respective probabilities.

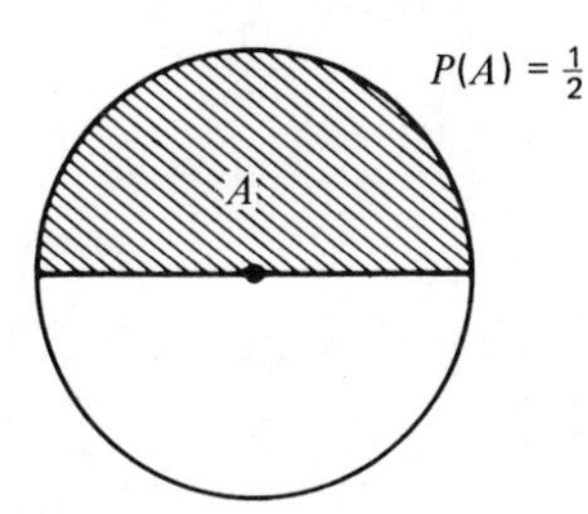

Top half of dart board

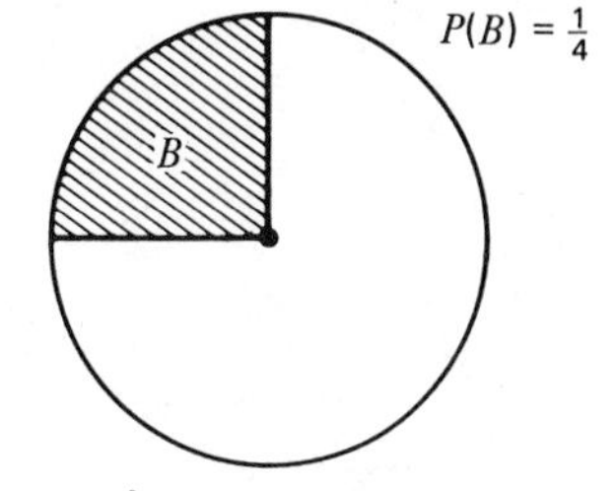

A quarter sector

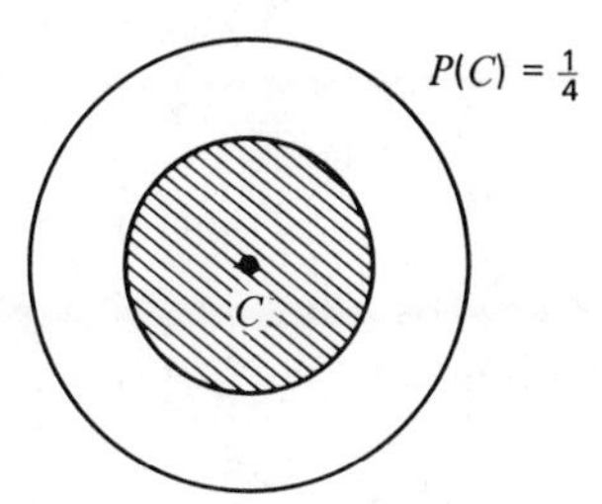

A concentric disk of radius $\frac{1}{2}$

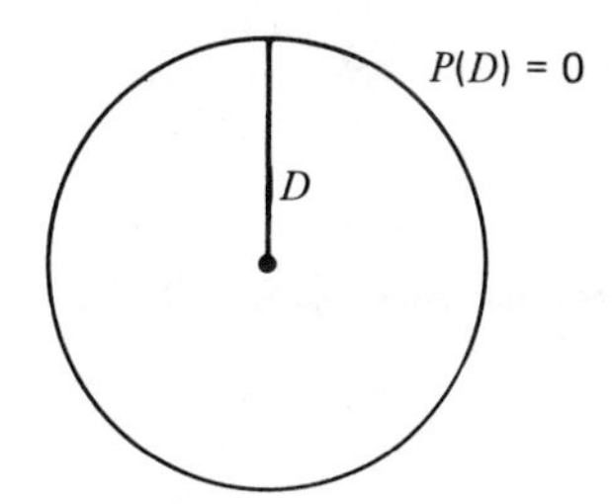

A radius, having area $= 0$

Again it may be instructive to test our probability function against the definition.

1. Since, if A is a region contained in the disk S we have

$$0 \leq \text{area of } A \leq \text{area of } S$$

it follows that

$$0 \leq \frac{\text{area of } A}{\text{area of } S} \leq \frac{\text{area of } S}{\text{area of } S}$$

so that

$$0 \leq P(A) \leq 1$$

2. $$P(S) = \frac{\text{area of } S}{\text{area of } S} = 1$$

3. As before, we examine $P(A \cup B)$ for any two regions (i.e., compound events) A and B. In the following set of figures we show how $A \cup B$ can be decomposed into three disjoint subsets whose aggregate area is the sum of the areas of the three pieces.

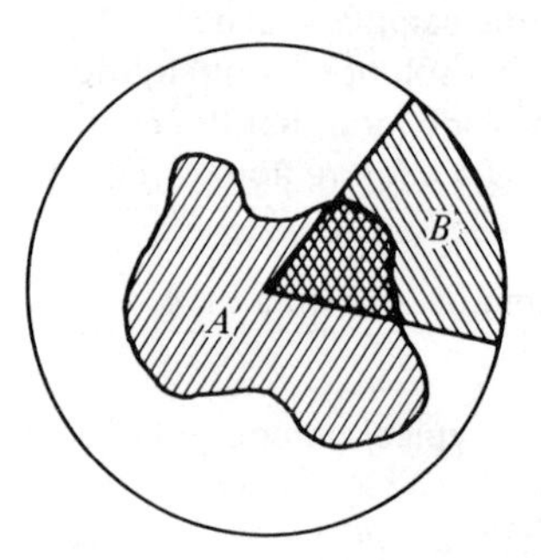

SHADED REGION IS $A \cup B$

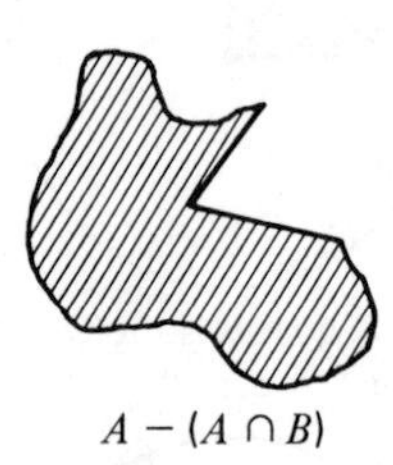

$A - (A \cap B)$

$A \cap B$

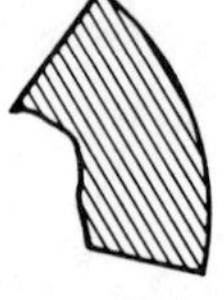

$B - (A \cup B)$

We argue as follows:

$$\begin{aligned}
\text{Area of } (A \cup B) &= \text{area of } [A - (A \cap B)] + \text{area of } (A \cap B) \\
&\qquad + \text{area of } [B - (A \cap B)] \\
&= [\text{area of } A - \text{area of } (A \cap B)] + [\text{area of } (A \cap B)] \\
&\qquad + [\text{area of } B - \text{area of } (A \cap B)] \\
&= \text{area of } A + \text{area of } B - \text{area of } (A \cap B)
\end{aligned}$$

Now dividing both sides of the equation by area of S, we get

$$P(A \cup B) = P(A) + P(B) - P(A \cap B)$$

and in particular,

$$P(A \cup B) = P(A) + P(B), \quad \text{when } A \cap B = \varnothing$$

which completes the verification.

Summary

In this section we have introduced two important definitions that help formalize the mathematical model discussed earlier. The first definition, that of a *sample space*, provides us with the abstract version of the outcomes of an experiment. Here a sample space is simply a set whose elements or, more properly, whose singleton sets are called *simple events* and whose subsets in general are called *events*. The second definition was that of a *probability function*, a function which assigns to each event in a given sample space some number between 0 and 1. Moreover, a probability function, say P on a sample space S, must satisfy three fundamental properties:

1. $0 \leq P(E) \leq 1, \quad$ for all $E \subseteq S$
2. $P(S) = 1$
3. $P(A \cup B) = P(A) + P(B), \quad$ if $A \cap B = \varnothing$

These properties of P guarantee that P is the appropriate abstract version of relative frequency in the sense that

$$\Sigma\, P(E) = 1$$

where the sum is taken over all simple events in a finite sample space.

One class of sample spaces was considered in which all simple events had the same probability. We say the events are *equally likely*, and we found that computation of the probability of any event A reduced to the formula

$$P(A) = \frac{\#(A)}{\#(S)}$$

the ratio of the number of elements in A to the total number of elements (i.e., simple events) in S.

We then considered a *continuous* model where the emphasis shifts from questions of the probability of a simple event to questions of the probability of compound events. The example we chose (a dart board) represents a continuous version of equally likely events in the sense that regions (i.e., events) of the sample space that have equal areas also have equal probabilities.

Throughout the section we used the language of set theory. Notions such as *subsets*, *unions*, *intersections*, and *complements* are indispensable when we speak of sample spaces; thus these concepts as well as the pictorial device of a Venn diagram were reviewed.

Exercises Section 3

1 (a) Write down the sample space for tossing three fair coins (say a penny, a nickel, and a dime).
(b) Next to each simple event write its probability.
(c) Write the set of simple events that make up the compound event A: Heads on the penny.
(d) Write the set of simple events that make up the compound event B: At least two heads.
(e) Find $A \cap B$, $A \cup B$, $A - B$, $B - A$, and A^c.
(f) What is the probability of getting at least one head?
(g) What is the probability of getting at least two tails?

2 (a) Write down the sample space for tossing four fair coins.
(b) Write the simple events that comprise each of the following compound events.
A: at least two heads
B: heads on the first and fourth coin
C: not more than two tails
D: heads on the second and third coin
(c) Find $A \cap B$, $A \cup C$, $C \cap D$, C^c, and $A - B$ and describe each in words.

3 Refer back to Section 1 where you will find, couched in slightly different language, the sample space for rolling a pair of fair dice. Find the probabilities of each of the following compound events.
(a) A: rolling a 7
(b) B: rolling a 12
(c) C: rolling a 2 or a 12
(d) D: rolling a matched pair
(e) E: rolling at least a 9
(f) F: rolling at least an 8
(g) G: not rolling a 7

4 Let a sample space consist of 10 equally likely events. Let A and B be compound events of S containing five and three simple events, respectively. Use Venn diagrams to illustrate how A and B are constructed to satisfy the following.
(a) $A \cup B$ consists of seven simple events
(b) $A \cap B$ consists of two simple events
(c) $A \cap B$ consists of three simple events
(d) $P(A - B) = .5$
(e) $P(A - B) = .2$
(f) $P(B - A) = .2$
(g) $P(S - (A \cup B)) = .3$
(h) $P(S - (A \cup B)) = .5$

5 Let A and B be events of a sample space S with $P(A) = \frac{1}{3}$ and $P(B) = \frac{1}{4}$. Find the minimum and the maximum values of:
(a) $P(A \cap B)$

(b) $P(A \cup B)$
(c) $P(S - (A \cup B))$
(d) $P(A^c)$

6 Let S be the sample space of outcomes from drawing a card at random from a bridge deck. Using notation of the form 2S, JH, AC, 5D to represent the 2 of spades the jack of hearts, the ace of clubs, and the 5 of diamonds, respectively, write down the simple events that comprise each of the following compound events.
(a) A red jack
(b) A black face card
(c) A face card
(d) A red face card or a black 2
(e) A one-eyed jack

7 A box contains 365 balls corresponding to the days in a year. If one ball is drawn at random, find the probability that its date is as follows.
(a) In January
(b) The 31st of some month
(c) The 30th of some month
(d) The 2nd of some month
(e) In either January or December
(f) Prior to July 4th

8 Consider the following diagram.

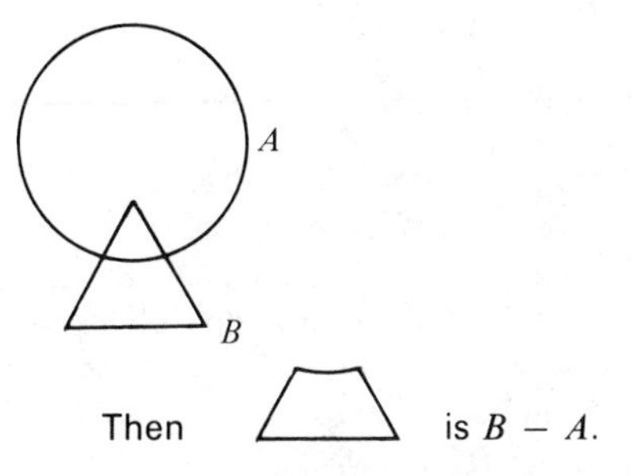

Then [figure] is $B - A$.

Identify each of the following sets in terms of A, B, unions, intersections, complements, and differences.

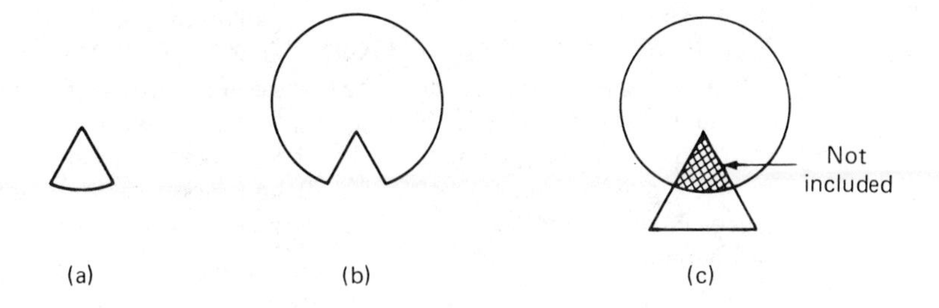

9 The instructions for this exercise are the same as for Exercise 8. The given diagram is:

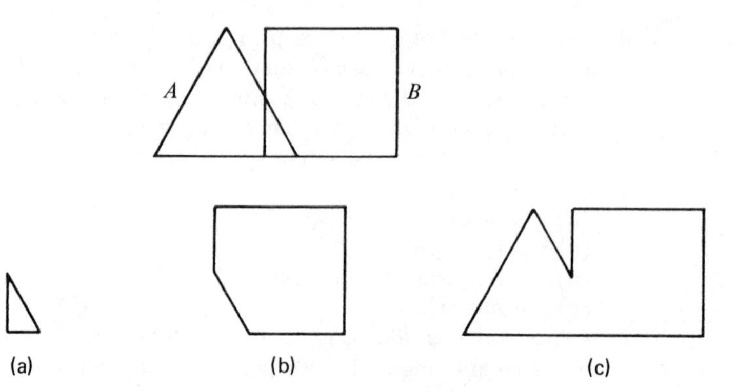

10 Johnny rolled a die three times and noted that the first roll was *larger* than the sum of the second and third. Write out the list of simple events which comprise the compound events S, T, W, where

S = the sample space of all possible ways this can happen
T = the event that the last two rolls were equal
W = the event that the second roll was a 3

Find $P(T)$, $P(W)$, $P(W - T)$, $P(W \cup (T - W))$.

4 CONDITIONAL PROBABILITY

We have, to this point, looked at probability as an absolute. But the determination of probability always depends upon knowledge. For example, the assertion that the probability of dealing a king from a well shuffled deck of cards is $\frac{1}{13}$ depends upon knowing that the deck contains 52 cards, four of which are kings. There are occasions, however, when, in addition to general knowledge, one has specific information which modifies the probability which would be computed on the basis of general knowledge. Let us consider an example.

EXAMPLE 1 A juggler has developed a new trick. He observes that when he is in a good mood he is successful with the trick 9 times out of 10, but when he is not in a good mood he succeeds only 5 times out of 10. If he is in a good mood for 6 performances out of 10, what is the probability that he will successfully perform the trick at a randomly selected performance?

We can analyze this problem by considering 100 performances. The juggler will be in a good mood for 60 and a bad mood for 40 performances. Out of the 60 good mood performances the juggler will be successful $\frac{9}{10}$ of the time, that is, in 54 performances. Out of the 40 bad mood performances will come 20 successful ones. Thus from the 100 performances we should expect 74 successes, and we conclude

$$P(\text{success}) = \tfrac{74}{100} = .74$$

We can modify this problem slightly. Suppose we watch the juggler perform and we see him do the trick successfully. What is the probability that he is in a good mood for that performance? Going back to our calculations with the 100 performances, we see that out of 74 successes, 54 of them occur when he is in a good mood; then the probability that he is in a good mood, knowing that he successfully performed the trick, is $\frac{54}{74}$ or approximately .73.

In this case, we know that, in general, the probability that the juggler is in a good mood is .6. But, when we have the additional information that he successfully performed the trick, we can modify the probability to .73. This is an example of what is known as conditional probability. When we determine the probability of an event A based not only on general knowledge about A as we have been doing, but also knowing that another event B has occurred, we have the conditional probability of A knowing B has occurred, denoted $P(A|B)$. A formal definition of this will be given shortly.

In order to develop a systematic approach to the study of conditional probability problems we consider the following diagram.

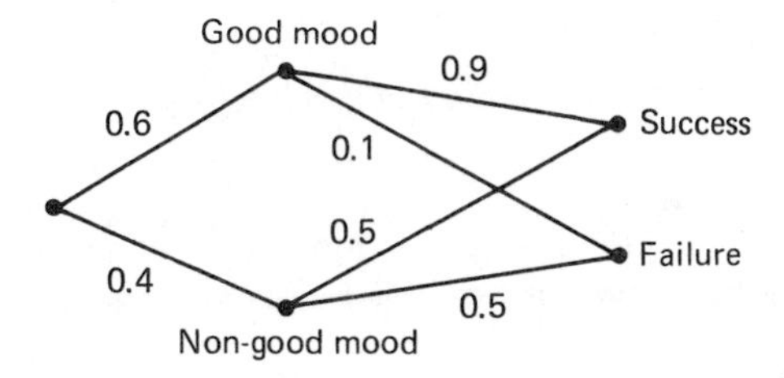

The interpretation of the numbers along each line should be clear. For instance along the line going to "Good Mood" we have the number .6 which is, of course, the probability that, for a given performance, the juggler is in a good mood. The number .5 on the line from "Non-Good Mood" to "Success" is the probability of success knowing that he is not in a good mood. In general the number written along a line in one of these diagrams always represents the probability of going from the left end of the line to the right end. Usually, instead of writing out words to label the points, they would be labeled with letters. One might, for instance, see the diagram as

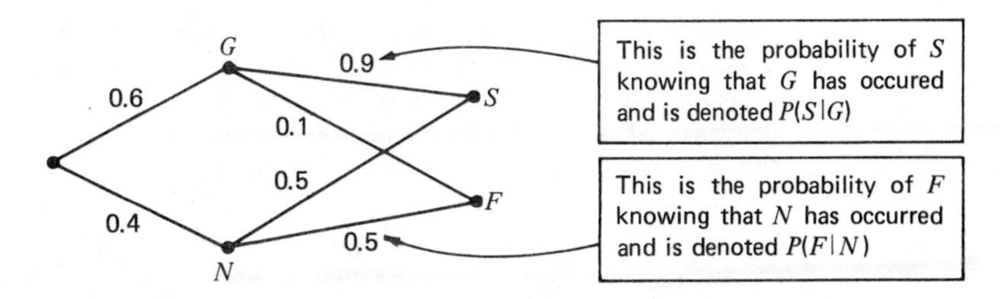

In order to eliminate the confusion caused by the crossing lines, this diagram is often modified to the following.

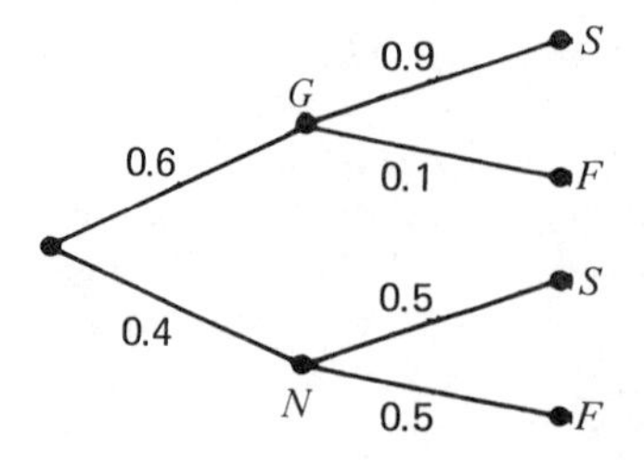

A picture of this sort is called a <u>tree diagram</u>.

It is understood in tree diagrams that all points with the same labels are to be identified. For instance in the above tree diagram the event S consists of *all* points labeled S. Moreover, the way we went about computing $P(S)$ can be described in terms of the diagram as follows:

> To find $P(S)$, we select all paths leading from the starting point to S. On each of these paths we multiply together the probabilities of adjacent branches. The sum of these products is $P(S)$. Thus
>
> $$P(S) = P(G)P(S|G) + P(N)P(S|N)$$

The second half of our argument led to a value for

$$P(G|S)$$

the probability that the juggler was in a good mood knowing he was successful. Our computation could be expressed as

$$P(G|S) = \frac{P(G)P(S|G)}{P(S)}$$

Note that the numerator represents the probability of the particular path (through "Good Mood") and the denominator represents the probability of all paths leading to "Success."

Finally we observe that a particular path along a tree really represents the *intersection* of all events along that path. For instance,

$$\left.\begin{aligned} P(G \cap S) &= P(G)P(S|G) \\ P(N \cap S) &= P(N)P(S|N) \end{aligned}\right\} \quad (\dagger)$$

In words, the first equation tells us that the probability that the juggler was in a good mood *and* that he performed successfully can be found by multiplying the probability that he was in a good mood by the probability that the performance was successful, knowing he was in a good mood.

These remarks should help motivate the following.

Definition

Let A and B be events in some sample space. We define the *conditional probability of B knowing A*, denoted $P(B|A)$, by

$$P(B|A) = \frac{P(B \cap A)}{P(A)} \quad [\text{provided } P(A) \neq 0]$$

Proposition Suppose a sample space S can be written as the union of disjoint events $B_1, B_2, \ldots, B_n$. That is,

$$S = B_1 \cup B_2 \cup \cdots \cup B_n, \qquad B_i \cap B_j = \emptyset, \quad \text{for } i \neq j$$

Then for any event A of S we have

(1) $P(A) = P(A|B_1)P(B_1) + P(A|B_2)P(B_2) + \cdots + P(A|B_n)P(B_n)$

(2) $P(B_i|A) = \dfrac{P(A|B_i)P(B_i)}{P(A)}$ (Bayes' Theorem)

Proof: Rather than give a formal proof of the proposition, we offer a heuristic argument, relying on our experience with the juggler example and our intuition aided by the following diagram.

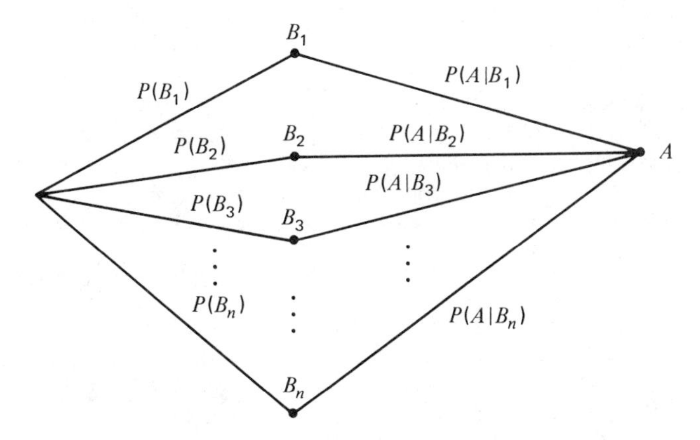

We observe that $P(A)$ should be the sum of the probabilities of *all* paths leading to A. But each path is of the form

$$P(B_i)P(A|B_i) \qquad [\text{or, equivalently, } P(B_i \cap A)]$$

Their sum yields equation (1).

Likewise, $P(B_i|A)$ may be viewed as the probability that we got to A along the path through B_i. As in the juggler problem, this should be

$$P(B_i|A) = \frac{P(A|B_i)P(B_i)}{P(A)}$$

since the numerator on the right represents the probability of taking the path through B_i, and the denominator is the sum of the probabilities of all paths leading to A.

A less precise, but perhaps helpful, way to think of Bayes' Theorem is the following. Imagine that a truck driver periodically delivers produce to the Acme Market (A). However on his way he always (exhaustive) stops at exactly one (mutually exclusive) of three (i.e., n) other markets (B_1, B_2, B_3). Once we know the values of

$P(B_1)$, $P(B_2)$, and $P(B_3)$ (How often he goes to each of the three markets, in terms of relative frequency)

as well as the values of

$P(A|B_1)$, $P(A|B_2)$, and $P(A|B_3)$ [How often he goes from each of the B's to A. (He may go elsewhere!)]

then we compute, first of all, how often he goes to the Acme Market, $P(A)$; moreover, using Bayes' Theorem we can compute, upon his arrival at Acme, the probability that he came there by way of B_1 (or B_2 or B_3). In short, Bayes' Theorem can be viewed as answering the question, "If you end up at a particular spot, what is the probability that you came along a certain route (from among all possible routes)?"

Equation (2), which we have called Bayes' Theorem, while not an earth-shaking result, is used extensively in one form of decision theory, a fairly recently developed branch of mathematics with applications in social and managerial sciences. A brief introduction to decision theory will be given in Chapter 3. For the time being, let us see how Bayes' Theorem can be applied to a few simple problems.

EXAMPLE 2 A book store has observed that, of the people entering the store, 70% are men and 30% are women. However, 60% of all the women make a purchase whereas only 30% of the men make a purchase. What is the probability that a randomly selected person entering the store buys something? If a person makes a purchase, what is the probability that that person is a man?

We can begin by constructing a diagram.

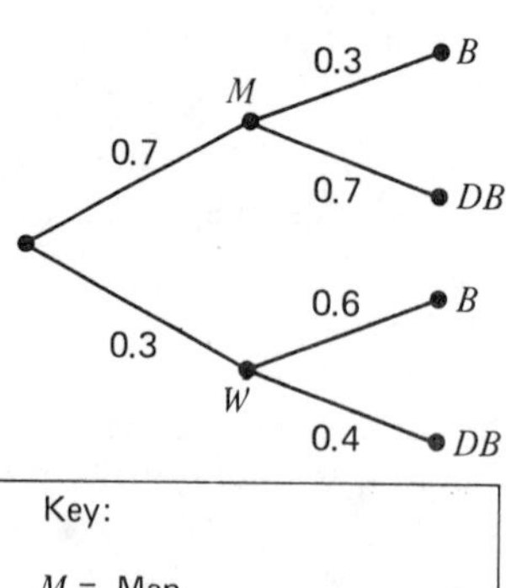

Key:

M = Man
W = Woman
B = Makes a purchase (buys)
DB = Does not buy

We can use the proposition to calculate

$$\begin{aligned}P(B) &= P(B|M)P(M) + P(B|W)P(W)\\ &= (.3)(.7) + (.6)(.3)\\ &= .21 + .18\\ &= .39\end{aligned}$$

and the first question is answered. The answer to the second question is $P(M|B)$. From Bayes' Theorem we have

$$\begin{aligned}P(M|B) &= \frac{P(B|M)P(M)}{P(B|M)P(M) + P(B|W)P(W)}\\ &= \frac{.21}{.39} = .538 \text{ (approximately)}\end{aligned}$$

EXAMPLE 3 The medical research team of Jack Up and Ian Atom has been studying the causes of people being tired in the morning. In addition to a variety of causes best not discussed here, they have discovered a heretofore unnoticed virus which causes this tired feeling and which they have named Winkle.

They have found that about 5% of the population are infected with Winkle. In order to detect its presence they have developed a test. One takes the test by licking a certain type of paper. If it turns blue you are all right but if it turns red, you have Winkle. However, the test is imperfect. The paper turns red only 90% of the time with people who have Winkle and turns red 20% of the time with those who don't. If a person licks the paper and it turns red, what is the probability that he has Winkle?

As usual we begin by constructing a diagram.

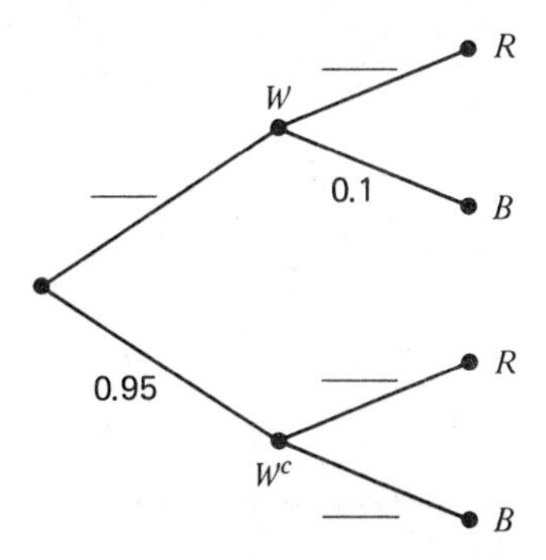

Given that the paper turns red, we want the probability that a person has Winkle, that is, we want to calculate $P(W|R)$. This calls for Bayes' Theorem and so we have

$$P(W|R) = \frac{P(\_|\_)P(\_)}{P(R)}$$

By the proposition,

$$P(R) = P(W)P(\_|\_) + P(\_)P(R|W^c)$$
$$= \_\_\_\_\_\_\_\_$$

Thus $P(W|R) = .19$.

EXAMPLE 4 On a certain par 3 hole of a golf course the green is on top of a hill and cannot be seen from the tee. Suppose that 70% of the golfers are duffers (D), 20% moderately good players (M) and 10% have low handicap (L). Suppose further that on the average these three groups of players can reach the green with their tee shot respectively 25%, 50%, and 90% of the time. If you had just left the green and noted that the player behind you reached the green with his tee shot, would it be more likely that he is a pro or a duffer?

Solution: We have the diagram

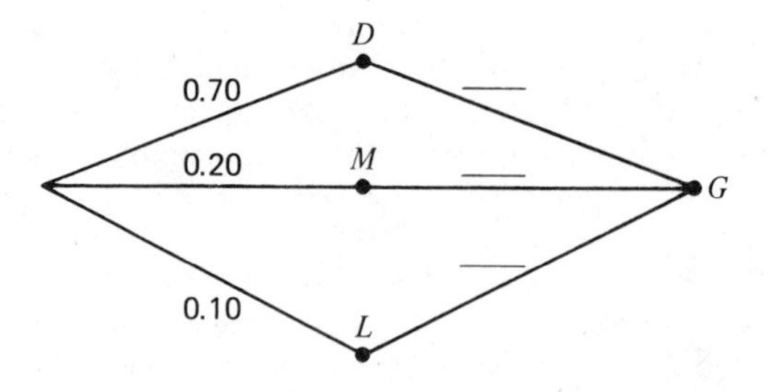

We are interested in

$$P(L|G) \quad \text{and} \quad P(D|G)$$

We first compute

$$P(G) = P(D)P(G|D) + \_\_\_\_\_\_ + \_\_\_\_\_\_$$
$$= (.70)(.25) + (.20)(.50) + \_\_\_\_\_\_$$
$$= .175 + .100 + \_\_\_\_\_\_$$
$$= .365$$

Now

$$P(L|G) = \frac{P(G|L)P(L)}{P(G)} = \frac{\_\_\_}{.365}$$
$$\approx .247$$

whereas

$$P(D|G) = \frac{P(G|D)P(D)}{P(G)} = \frac{\_\_\_}{.365}$$

Summary

Many interesting questions in probability have to do with the relationship between two or more events in a sample space. In particular, one is often concerned with the probability that event A will occur knowing that another event B has occurred. More precisely, we defined the *conditional probability of A knowing B*, denoted $P(A|B)$, by

$$P(A|B) = \frac{P(A \cap B)}{P(B)}$$

One of the most common applications of conditional probability is *Bayes' Theorem*, which gives us a formula for computing $P(B_i|A)$ once we know $P(A|B_i)$ as well as $P(A)$, where the B_j's are *mutually exclusive* ($B_j \cap B_k = \varnothing$ when $j \neq k$) and *exhaustive* (the union of the B_j's is all of S). In such a situation, it follows that

$$P(A) = P(B_1)P(A|B_1) + P(B_2)P(A|B_2) + \cdots + P(B_n)P(A|B_n)$$

and Bayes' Theorem tells us that

$$P(B_i|A) = \frac{P(B_i)P(A|B_i)}{P(A)}$$

Exercises Section 4

1 Suppose that A and B are events in a sample space S and that $P(A|B) = \frac{1}{3}$, $P(B|A) = \frac{1}{4}$, and $P(A) = \frac{1}{2}$. Find $P(B)$ and $P(A \cap B)$.

2 A card is drawn from a deck of cards. Consider the events:

A: the card is a jack
B: the card is red
C: the card is a red jack
D: the card is either red or a jack (or both)

Find:

(a) $P(A)$
(b) $P(B)$
(c) $P(C)$
(d) $P(D)$
(e) $P(A|B)$
(f) $P(B|A)$
(g) $P(A|C)$
(h) $P(B|D)$

3 Let the sample space S consist of the students at a college. Let Y_1 be the set of freshmen; Y_2, sophomores; Y_3, juniors; and Y_4, seniors. Let A be those students whose grade point average is A or better, B those whose grade point average is B, with C, D, and F defined in the obvious manner. Let M be those students who are men and W those who are women.

(a) Express verbally what each of the following probabilities is supposed to be.

(1) $P(Y_2|A)$
(2) $P(C|Y_3)$
(3) $P(M|D)$
(4) $P(M \cap D)$
(5) $P(W \cup Y_3)$
(6) $P(Y_1|W)$

(b) Express symbolically the following.

(1) The probability that a student is a junior knowing that he has a C average.

(2) The probability that a student is a man or a C student.

(3) The probability that a man is a senior.

(4) The probability that a senior is a man.

(5) The probability that a male B student is a junior.

4 Fount of Wisdom University finds that 30% of its students are freshmen, 27% are sophomores, 23% are juniors, and 20% are seniors. It also finds that only 5% of the freshmen make Dean's List, whereas 10% of the sophomores, 15% of the juniors, and 20% of the seniors do. If a randomly selected student is on the Dean's List, what is the probability that he is a junior?

5 A professor observes that 60% of his students are men and 40% are women. If 10% of his class falls asleep during his lectures, but only 5% of the women fall asleep, what is the probability that a man in his class will fall asleep? If you know that a particular student has fallen asleep, what is the probability that he is a man?

6 The Duns Scotus Business Machine Corporation has three factories producing adding machines. The Happy Valley plant produces 60% of the machines, the Purple Meadows plant produces 25% of them, and the remaining 15% are produced at the Aquinas plant. If 5% of the Happy Valley output, 5% of the Purple Meadows output, and 10% of the Aquinas output is defective, find:

(a) the probability that a randomly selected Duns Scotus adding machine is defective.

(b) the probability that a defective Duns Scotus adding machine came from Aquinas.

7 You are given three urns, A, B, and C. Urn A contains one red marble and five blue marbles, urn B contains two red and four blue marbles, while urn C contains three of each color. You choose either A or B randomly, select a marble from the one you have chosen, and drop it into urn C. You then choose a marble from urn C.

(a) If you select a red marble from urn C, what is the probability that you initially chose urn A?

(b) What is the probability that you will select a blue marble from urn C?

8 A group of people, concerned about the possible invasion of the earth by creatures from another planet, have built an alarm to detect the presence of extra-terrestrial life in the earth's atmosphere. Once a minute it scans the atmosphere and if it detects extra-terrestrial life it rings an alarm bell. The machine is so well designed that if extra-terrestrial life is present, the probability is 1 that the alarm will ring. However there is a slight chance (the probability is $10^{-4} = 1/10{,}000$) that the alarm will ring even if no extra-terrestrial life is present. The designers of the machine believe that the probability of this alien presence is 10^{-7}. If the alarm rings, what is the probability that extra-terrestrial life is present?

9 Each of three boxes has two drawers, one has a \$1 bill in each drawer, the second has a \$5 bill in each drawer, and the third has a \$1 bill in one drawer and a \$5 bill in the other. If you choose a box at random and open a randomly selected drawer which contains \$1, what is the probability that the other drawer contains \$5?

10 Suppose that one-tenth of all people are rich and nine-tenths are poor. Suppose further that four-fifths of all rich people own color TV sets and only one-fifth of all poor people do.

(a) Find the probability that a randomly selected person owns a color TV set.

(b) Find the probability that a color TV owner is rich.

11 A group of businessmen consists of 30% Democrats and 70% Republicans. If 20% of the Democrats and 40% of the Republicans smoke cigars, what is the probability that a cigar smoker is a Republican?

12 The statistician for the Swingers Baseball Team has observed over the years that batters can be classified in three categories: superstars, regulars, and glove men. The team generally breaks down into these groups with frequencies .05, .60, and .35, respectively. Of the men currently on the club, 80% of the superstars, 30% of the regulars, and 10% of the glove men managed to bat at least .290 during the first month of their rookie year. If a widely-heralded rookie bats .300 during his first month this season, what are his chances of becoming a superstar?

13 The incidence of diabetes in various age groups shown in the following table was given by the United States Public Health Service:

Age Group	0–24	25–44	45–54	55–64	65 or more
Estimated Number of Diabetics Per 1,000	1.7	10.2	33.1	55.6	68.6

According to the 1970 Census, the United States population in these age groups are as follows:

Age Group	0–24	25–44	45–54	55–64	65 or more
Population in U.S. (in thousands)	95,257	48,393	23,381	18,517	19,799

(a) What percentage of people in the United States have diabetes?

(b) What percentage of diabetics are under 25 years of age?

(c) What percentage of diabetics are under 45 years of age?

5 INDEPENDENT EVENTS

The word "independence" is used in probability theory in a very natural way. On a casual level, if we roll a die and toss a coin, we say that the outcome of the die roll is independent of what happens with the coin. More carefully, if A is the event that the die shows a "4" and B is the event that the coin comes up heads, then when the die is rolled and the coin tossed simultaneously, we should be willing to accept that $P(A) = \frac{1}{6}$ and $P(A|B) = \frac{1}{6}$ (assuming a fair die).

Thus

$$(*) \quad P(A|B) = P(A)$$

Pairs of events, A and B, that satisfy this relationship are called *independent* or we say *A is independent of B*. When $P(A|B) \neq P(A)$, we say that A and B are *dependent*. Let us consider some examples in an effort to develop some feeling for this concept.

EXAMPLE 1 A fair coin is tossed three times. If A represents heads on the first toss and B represents tails on the second toss, then it should be intuitively clear that A is independent of B. Let us check it formally. We have

$$S = \{\text{HHH, HHT, HTH, THH, HTT, THT, TTH, TTT}\}$$
$$A = \{\text{HHH, HHT, HTH, HTT}\}$$
$$B = \{\text{HTH, HTT, TTH, TTT}\}$$
$$A \cap B = \{\text{HTH, HTT}\}$$

By definition of conditional probability,

$$P(A|B) = \frac{P(A \cap B)}{P(B)} = \frac{\frac{2}{8}}{\frac{4}{8}} = \frac{1}{2} = P(A)$$

as we suspected.

EXAMPLE 2 A card is drawn from a bridge deck. If A represents an ace and B represents a black card, then A is independent of B:

$$P(A) = \tfrac{4}{52} = \tfrac{1}{13}$$
$$P(B) = \tfrac{26}{52} = \tfrac{1}{2}$$
$$P(A \cap B) = \tfrac{2}{52} = \tfrac{1}{26}$$

and hence

$$P(A|B) = \frac{P(A \cap B)}{P(B)} = \frac{\frac{1}{26}}{\frac{1}{2}} = \frac{1}{13} = P(A)$$

EXAMPLE 3 Let us modify the previous example by letting A represent drawing an ace, but B represent drawing a 10. Certainly $P(A|B) = 0$ in this case, but $P(A) = \frac{1}{13}$. So we conclude that A and B are *dependent*, as we would suspect.

EXAMPLE 4 Cigarette Smoking and Lung Cancer

A great controversy has raged over the relationship between cigarette smoking and lung cancer. On the one side, the argument is that the incidence of lung cancer among smokers (i.e., the conditional probability that a person has lung cancer knowing that he is a smoker) is "significantly higher" than the incidence of lung cancer among the general public (i.e., the probability that a person has lung cancer). In short, this view claims that the two "events," cancer and smoking, are *dependent*. The opponents of this view argue that even if the events are not independent, one cannot conclude that there is a *causal* relationship.

EXAMPLE 5 Opinion Polls

One of the many characteristics of a properly conducted poll is that the selection of people to be questioned is representative of the population about whom one wishes to generalize. For instance one would hope that being a member of the Young Republicans and being polled regarding a presidential preference would be independent events. That is, the percentage of Young Republicans included in the poll should be roughly equal to the percentage of Young Republicans in the entire population being considered.

Before examining other examples, we offer an alternate and somewhat more comprehensive definition of independence. From the definition of conditional probability, we have

$$P(A|B) = \frac{P(A \cap B)}{P(B)}$$

so that

$$P(A \cap B) = P(A|B)P(B)$$

And in the case where A is independent of B, we can replace $P(A|B)$ by $P(A)$ in the last equation to get

$$(^{**}) \quad P(A \cap B) = P(A)P(B)$$

From now on we shall use (**) to mean that *A is independent of B*. Its advantage over (*) is that when $P(B) = 0$, (**) makes sense whereas (*) does not.

EXAMPLE 6 A Non-Intuitive Example

This example, due to William Feller, illustrates that sometimes our intuition may lead us astray in guessing whether two events are independent. We let S correspond to three-children families:

$$S = \{\text{BBB, BBG, BGB, GBB, BGG, GBG, GGB, GGG}\}$$

and we *assume* that the simple events are *equally likely* (although in reality there seem to be more boys born than girls). Now we consider two events

$$A = \{\text{BBB, BBG, BGB, GBB}\}$$
$$B = \{\text{BBB, GGG}\}^c$$
$$A \cap B = \{\text{BBG, BGB, GBB}\}$$

Hence

$$P(A \cap B) = \tfrac{3}{8}, \qquad P(A) = \tfrac{4}{8}, \qquad \text{and } P(B) = \tfrac{6}{8}$$

So that

$$P(A \cap B) = P(A)P(B)$$

and A *is independent of* B. Lest the reader feel too confident in his ability to perceive independence (in case he guessed correctly), we point out that the corresponding events (A: at most one girl; B: children of each sex) are *not* independent in either two-children families or four-children families, as can easily be checked.

It is possible to define independence for several events $A_1, A_2, \ldots, A_n$, although the formal definition gets cumbersome as n increases. For example, when $n = 3$, we have the following.

Definition The events A_1, A_2, and A_3 are (*mutually*) *independent* if

$$P(A_1 \cap A_2) = P(A_1)P(A_2)$$
$$P(A_1 \cap A_3) = P(A_1)P(A_3)$$
$$P(A_2 \cap A_3) = P(A_2)P(A_3)$$

and

$$P(A_1 \cap A_2 \cap A_3) = P(A_1)P(A_2)P(A_3)$$

For the present, we shall not consider situations where the independence of several events is under question. However, this notion provides the basis for an important context, namely that of "repeated trials" of an experiment. We shall discuss this in detail in Chapter 3. Examples 1 and 6 are both special cases of repeated trials in that a simple experiment (tossing one coin or having one child) is repeated several times under (presumably) identical conditions so that the outcome of any stage (say the second toss or second birth) is not influenced by the outcome of any other stage (say the first toss or the first birth).

EXAMPLE 7 Drawing With or Without Replacement

A great many problems of both practical and theoretical interest can be viewed as drawing a few or several items (called a "sample") from a larger source (called a "population"). A poker hand is simply the result of drawing five cards from a deck of 52. A raffle is drawing one or more tickets from a box. An opinion poll is drawing a few hundred or a few thousand answers to a specific question from a large population, and so on. We can make a basic distinction between two types of drawings in terms of drawing two cards from a deck of 52.

Case 1 Drawing With Replacement

Suppose a first card is drawn, looked at, then replaced in the deck which is then shuffled before a second card is drawn. We refer to this process tersely as "drawing two cards with replacement." The effect of replacing the first card before drawing the second card is that the outcome of the second draw is *independent* of the outcome of the first draw. For instance the events

A: first card drawn is an ace
B; second card drawn is a face card (jack, queen, or king)

are independent; that is,

$$P(A) = \tfrac{4}{52}, \qquad P(B) = \tfrac{12}{52}$$

and

$$P(A \cap B) = \tfrac{4}{52} \cdot \tfrac{12}{52}$$

Case 2 Drawing Without Replacement

This time we simply draw two cards in succession from the deck. If the first card drawn is an ace, then the probability that the second card drawn is a face card becomes

$$\tfrac{12}{51}$$

so that

$$P(A \cap B) = \tfrac{4}{52} \cdot \tfrac{12}{51}$$

In other words, if we are interested in drawing an ace and then drawing a face card, our chances are slightly improved if we draw without replacement. A sharper contrast can be seen by modifying this problem by starting with a "deck" of only two cards: one ace and one face card. Then, drawing with replacement, the probability of getting the ace and then the face card is $\frac{1}{2} \cdot \frac{1}{2} = \frac{1}{4}$, whereas if we draw without replacement, the probability becomes $\frac{1}{2} \cdot 1 = \frac{1}{2}$.

More interesting examples of these two types of drawing will be explored in the next section, where we shall make a further distinction by considering the drawing of two (or more) items simultaneously without regard to order. This type of drawing is necessarily without replacement.

Warning: Although we have tried to point out that in general independence of events is a fairly natural concept, there is a danger in equating "A is independent of B" with "A has nothing to do with B." More precisely, beginning students often confuse independence with *exclusiveness*. That is, when

$$A \cap B = \varnothing$$

we say that A and B are *mutually exclusive* (or disjoint), and whenever this is the case, A and B are most certainly *not independent*, since

$$P(A \cap B) = P(\varnothing) = 0$$

so that

$$P(A \cap B) \neq P(A)P(B)$$

(except of course in a trivial case where one of A or B is empty).

Summary

This section addressed itself to a natural question that arises once the notion of conditional probability has been discussed: Under what conditions will the occurrence of an event B have no effect on the probability of another event A?

In terms of conditional probability, we are interested in the phenomenon

$$P(A|B) = P(A)$$

When this equality holds, we say that A and B are *independent events*. From the definition of conditional probability,

$$P(A|B) = \frac{P(A \cap B)}{P(B)}$$

we derived a consequence of the independence of A and B, namely that

$$P(A \cap B) = P(A)P(B) \qquad \text{(whenever } A \text{ and } B \text{ are independent)}$$

Indeed, we took this last equality, called the *multiplication rule*, as the definition that A and B are independent. A generalization of the idea to more than two events led us to postulate that if three or more events (say A, B, C) are to be independent, then

$$P(A \cap B \cap C \cdots) = P(A)P(B)P(C) \ldots$$

It is important to note, however, that we did not assert that this more general multiplication rule by itself characterizes the independence of several events—it is but part of a more complicated definition.

To return to the original question, we examined some common situations that gave rise to independent events. One context in particular, *repeated trials*, not only provided us with a simple example (successive births) but heralded a topic deserving a full-blown discussion that we postpone until Chapter 3. A special case of repeated trials occurs in *drawing with replacement*. For instance we may draw a card from a deck, a number from a hat, a marble from an urn, and so on. Once the denomination (or number or color) is recorded, the object is replaced and a second drawing is performed, and so on. The result of the second drawing is *independent* of the result of the first drawing. By way of contrast, we can draw *without replacement*, say a card from a deck. Then when a second card is drawn, the probabilities of various events are different from those of the first drawing. For instance if we first draw a heart (with probability $\frac{13}{52}$), then the probability or drawing a second heart would be $\frac{12}{51}$.

In many instances, our intuition can properly determine whether two events are independent. However one example illustrated how we can, on occasion, be misled unless we rely on the definition of independence.

Exercises Section 5

1 In each of the following situations decide whether you think events A and B should be independent.

(a) S: drawing one card from a deck
A: drawing an ace
B: drawing a black card

(b) S: selecting a male student at UCLA
A: he is over 6 feet tall
B: he plays varsity basketball

(c) S: votes on a referendum to allow 18 year olds to drink
A: vote is yes
B: voter is member of WCTU

(d) S: selecting a female student at Ohio State University
A: she is less than 5 feet tall
B: she is left-handed

2 An urn contains three red marbles and five blue marbles.

(a) Two marbles are drawn in succession. Find the probability that both are red if they are drawn with replacement.

(b) Same as (a) except drawing is without replacement.

(c) Same as (a) except four marbles are drawn.

(d) Same as (c) except drawing is without replacement.

3 A red die and a blue die are tossed. For $i = 1, 2, \ldots, 6$, let R_i be the probability that the red die comes up i, B_i be the probability that the blue die comes up i, and, for $i = 2, 3, \ldots, 12$, let S_i be the probability that the sum is i. Determine whether each of the following pairs of events are independent or dependent.

(a) B_2, S_6

(b) R_3, S_7

(c) $R_2 \cup B_2, S_7$

(d) $R_1 \cup B_1, S_2$

(e) $R_1 \cap B_1, S_2$

4 A red die and a blue die are rolled. Let A be the event that the red die comes up 6, B the event that the blue die comes up 6, and C the event that the sum of the two dice is odd. Show that any two of these three events are independent but the three events together are not independent. (This exercise should tell you something about the difficulties connected with the independence of more than two events.)

5 Tom, Dick, and Harry are about to take an examination. The probability that Tom gets an A is .3, the probability that Dick gets an A is .5, but the probability that Harry gets an A is .8. Assuming that the three marks are independent events (that's why exams are proctored) calculate the probabilities that none of them get an A, exactly one of them does, exactly two of them do, and that all three of them get an A.

6 A student has the probability .99 of answering any given question correctly. How many questions must there be on a test before the probability that he gets a perfect score is less than .9? You may find the computer useful for this problem.

7 Look at the figure. Each of the dots represents an event in the sample

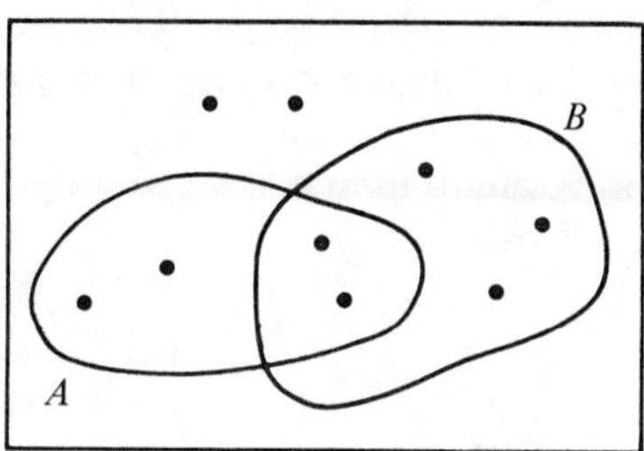

space and all of these dots are equally likely events. Events A and B are compound events composed of the dots inside the curves. As it stands, A and B are not independent events. Add dots to the picture (continuing to assume that all dots are equally likely) in such a way that A and B are independent events.

Going back to the given figure, make A and B independent events by moving dots around.

Going back to the given figure, make A and B independent by the deletion of points.

8 Let A, B, and C be the three dart board events depicted below. Which of the pairs (A, B) and (A, C) are independent?

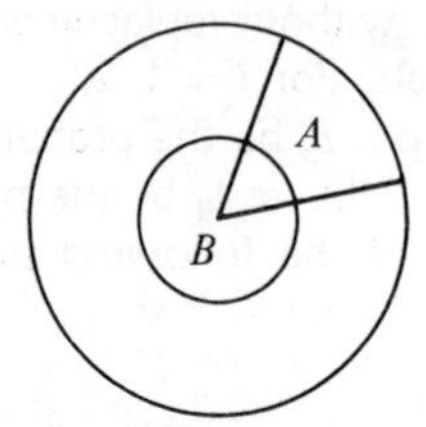

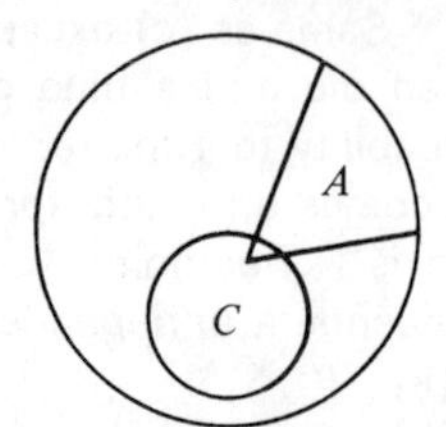

9 Show that if $P(A|B) = P(A)$, then $P(B|A) = P(B)$.

10 Let S correspond to three-children families and assume equally likely simple events.

Let A: first child is a boy
B: at most one boy
$C = (A \cap B) \cup \{\text{GBB, BBG, BGB}\}$

Show that $P(A \cap B \cap C) = P(A)P(B)P(C)$, *but none* of the other three conditions for independence of A, B, and C are satisfied. (See Exercise 4 for another pathological example.)

6 COUNTING TECHNIQUES

The concept of probability, when applied to discrete sample spaces, involves counting. In order to expand the domain of problems we can analyze, it is necessary to develop some standard techniques in counting that help alleviate the burden of writing down an exhaustive list of simple events. Our first three examples are essentially prototypes of common situations that demand these techniques.

EXAMPLE 1 **The Theater Queue** (Permutations of n Objects)

Suppose that several people are about to form a line to purchase theater tickets. The question we raise is, "How many different lines (or queues) can be formed?" Obviously, the answer depends upon the number of people available; so we begin with

Case 1 Two people, A and B, are to form a line.

Solution The only possible lines are

AB and BA

two possible lines with two people

(where we think of the left-most letter representing the first person in the line and the right-most letter representing the last person).

Case 2 Three people, A, B, and C are to form a line.

Solution: We observe that any one of the three people could be the first in line. Then the remaining two people could line up behind him in two different ways (by Case 1). Thus we have

(A 1st)	(B 1st)	(C 1st)
ABC	BAC	CAB
ACB	BCA	CBA

Six possible lines with three people

Before continuing, we note that the six lines are the result of

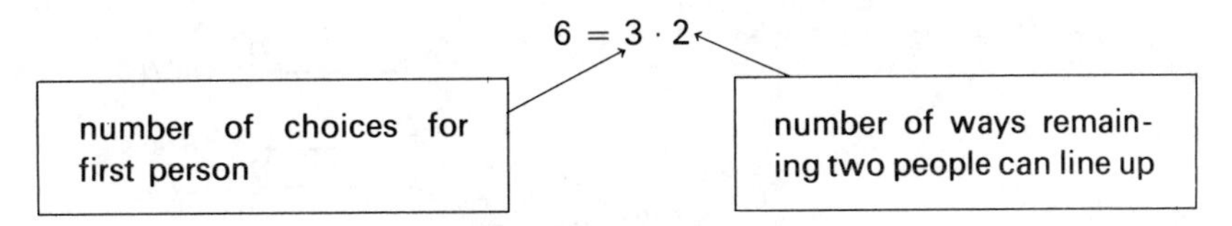

Case 3 Four people A, B, C, and D are to form a line.

Solution: Again we can place any one of four people at the front of the line; then the remaining three can line up behind him in $3 \cdot 2 = 6$ ways (Case 2). Thus there should be

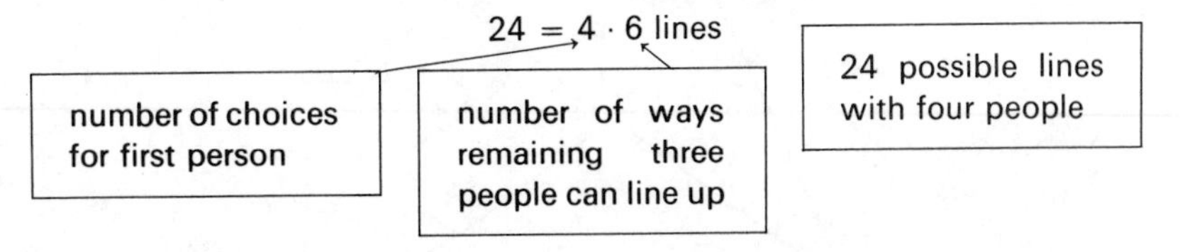

The reader should be able to supply the missing letters to the 24 lines:

(A 1st)	(B 1st)	(C 1st)	(D 1st)
ABCD	BACD	CABD	DABC
ABDC	B____	C____	D____
ACBD	B____	C____	D____
ACDB	B____	C____	D____
ADBC	B____	C____	D____
ADCB	B____	C____	D____

Definition The symbol

$n!$ (read "n factorial")

is defined by the equations

$$1! = 1$$
$$n! = n \cdot (n - 1)! = n \cdot (n - 1) \cdot (n - 2) \cdots 2 \cdot 1$$

Thus, for example, we have

$2! = 2 \cdot 1 = 2$
$3! = 3 \cdot 2 \cdot 1 = 6$
$4! = 4 \cdot 3 \cdot 2 \cdot 1 = 24$
$8! = 8 \cdot 7 \cdot 6 \cdot 5 \cdot 4 \cdot 3 \cdot 2 \cdot 1 = 40{,}320$

General Solution to Theater Queue Problem:

> There are precisely $n!$ ways of ordering n people in a line.

An alternate method of arriving at the 24 queues in Case 3 is to use the following tree diagram.

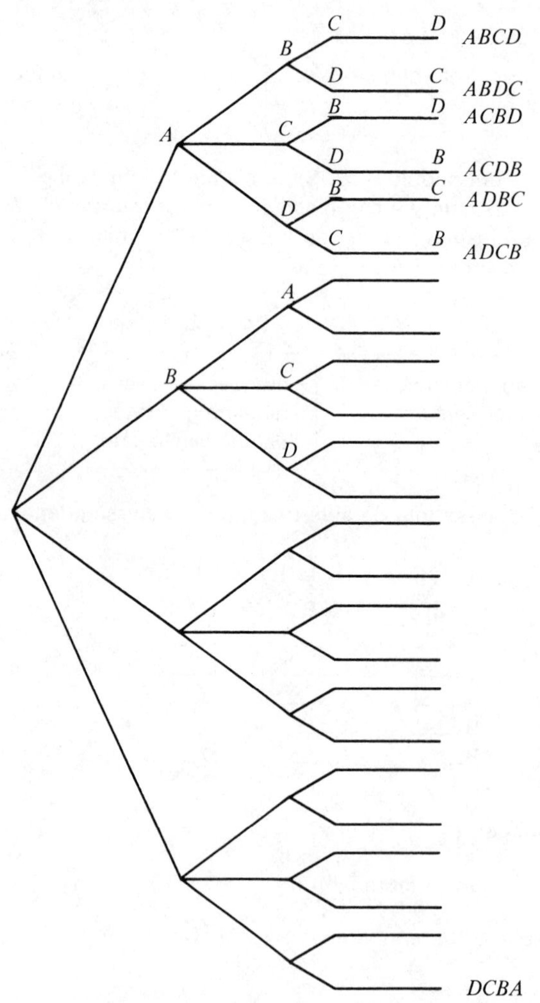

Remark: Here are some other problems that have the same solution as the theater queue problem.

1. How many ways can five textbooks be arranged on a shelf?
2. How many three-letter "words" can be formed with the letters A, B, C?
3. How many ways can first prize, second prize, and third prize be awarded to three entrants in a decathlon?

As the word permutation would suggest, the example of the theater queue was presented to emphasize a problem where we count the *number of different arrangements* of n objects, be they people, books, letters, or whatever. We now present a generalization of this notion.

EXAMPLE 2 The Beauty Contest (Permutations of n Objects Taken x at a Time)

Suppose in a beauty contest there are n contestants, x of which will be awarded prizes: 1st prize, 2nd prize, . . . , xth prize (i.e., x may be 1 or 2 or any number up to n). How many distinct outcomes can there be to this contest?

Case 1 Suppose there are five contestants vying for only two prizes, first prize and second prize. Let us call the girls A, B, C, D, E (say for Abigail, Beatrice, etc.).

Solution: We represent a final judgment in the form

AB

meaning A wins first prize and B second. We might think of a preliminary step in the judgment of ordering all five girls then selecting only the first two. Thus the outcome AB might correspond to any of

ABCDE
ABCED
ABDCE
ABDEC
ABECD
ABEDC

These are the 3! arrangements of five letters with AB in the first two positions.

In general, then, to each outcome there corresponds exactly 3! arrangements of all five girls in this manner. Thus we argue

$$\begin{pmatrix}\text{number of arrange-}\\ \text{ments of all five girls}\end{pmatrix} = \begin{pmatrix}\text{number of contest}\\ \text{outcomes}\end{pmatrix} \cdot \begin{pmatrix}\text{number of arrange-}\\ \text{ments of remaining}\\ \text{three girls}\end{pmatrix}$$

or

$$5! = \begin{pmatrix}\text{number of contest}\\ \text{outcomes}\end{pmatrix} \cdot 3!$$

So that

$$\text{Number of contest outcomes} = \frac{5!}{3!}$$

More generally, if there are n contestants of which x are to be arranged in order for prizes, then there are

$$\frac{n!}{(n-x)!}$$

such arrangements.

Definition The symbol

$P(n, x)$ (read "the number of permutations of n objects taken x at a time")

is defined by

$$P(n, x) = \frac{n!}{(n-x)!}$$

You should keep in mind this interpretation:

$P(n, x)$ is the number of arrangements of x objects chosen out of a set of n objects where two arrangements are considered different if they are differently ordered.

The emphasis we have placed on *order* in the above examples if to contrast them with problems such as the following where subsets are to be chosen from a larger set without regard to any order of selection.

EXAMPLE 3 The Committee (Combinations of n Objects Taken x at a Time)

Suppose we wish to select a committee of x people from a group of n candidates (i.e., x can be 1 or 2 or any number up to n). How many such committees can be formed?

Case 1 Suppose there are five people in the group and the committee is to consist of three people.

Solution: We denote by A, B, C, D, E the five group members. We can use the symbol

ABC

to denote one possible committee of three, *provided* we observe that the various arrangements,

ACB, BCA, BAC, CAB, and CBA

all represent *the same committee*, ABC, since we are not assigning any order to the members. In other words, to each committee of three chosen there are 3! arrangements of that committee. Thus

$$\begin{pmatrix}\text{Number of 3-member}\\ \text{committees}\end{pmatrix} \cdot (3!) = \begin{pmatrix}\text{Number of arrangements of 5}\\ \text{things taken 3 at a time}\end{pmatrix}$$

or

$$\begin{pmatrix}\text{Number of 3-member}\\ \text{committees}\end{pmatrix} = \frac{P(5, 3)}{3!} = \frac{5!}{2!3!}$$

General Case The number of x-member committees that can be chosen from a group of n members is

$$\frac{n!}{(n-x)!x!}$$

Definition

The symbol

$C(n, x)$ (read "the number of *combinations* of n objects taken x at a time")

is defined by

$$C(n,x) = \frac{n!}{(n-x)!x!}$$

$C(n, x)$ is the number of ways x objects can be selected from a set of n objects without regard to the order of selection.

It is frequently useful to be able to write

$C(n, 0), C(n, n)$, and $P(n, n)$

all of which, according to their definitions, involve the as yet undefined symbol

$0!$

We therefore extend our factorial definition by putting

$$\boxed{0! = 1}$$

Now for example we have

$$\begin{pmatrix}\text{The number of permutations of} \\ n \text{ objects taken } n \text{ at a time}\end{pmatrix} = P(n,n) = \frac{n!}{0!} = n!$$

$$\begin{pmatrix}\text{The number of combinations of} \\ n \text{ objects taken } n \text{ at a time}\end{pmatrix} = C(n,n) = \frac{n!}{0!n!} = \frac{n!}{n!} = 1$$

$$\begin{pmatrix}\text{The number of combinations of} \\ n \text{ objects taken 0 at a time}\end{pmatrix} = C(n,0) = \frac{n!}{n!0!} = \frac{n!}{n!} = 1$$

Of course each of the above facts should seem obvious. If we are to choose n objects from n objects, there is only one way to do it: You take them all. Similarly if we are to choose 0 objects from n objects, there is only one way to do it: You don't take any of them; and so on. Before working some illustrative examples, let us notice how certain expressions involving factorials can be simplified.

PROBLEM 1 Simplify

$$\frac{7!}{5!}$$

SOLUTION

$$\frac{7!}{5!} = \frac{7 \cdot 6 \cdot 5!}{5!} = 7 \cdot 6 = 42$$

PROBLEM 2 Simplify

$$\frac{C(7, 2)}{C(6, 3)}$$

SOLUTION

$$\frac{C(7, 2)}{C(6, 3)} = \frac{\frac{7!}{5!2!}}{\frac{6!}{3!3!}} = \frac{7!3!3!}{5!6!2!}$$

$$= \frac{7 \cdot 6! \cdot 3 \cdot 2!3!}{6! \cdot 5 \cdot 4 \cdot 3!2!}$$

$$= \frac{7 \cdot 3}{5 \cdot 4} = \frac{21}{20}$$

PROBLEM 3 Simplify $P(10, 2)$.
SOLUTION

$$P(10, 2) = \frac{10!}{8!} = 10 \cdot 9 = 90$$

EXAMPLE 4 A committee of three is to be chosen from the membership of a group of 10. Tom and Dick, who are both in the group of 10 are very interested in the work the committee will be doing. What is the probability that at least one of them will be on the committee if the committee is chosen randomly?

What we have to do is clear. We want to count the number of committees that include Tom or Dick or both and then divide this number by the total number of three-person committees which can be formed. The total number of committees which can be formed is $C(10, 3)$. (We could work this out and get a number, but, as you will shortly see, it's usually a good idea to hold off on the arithmetic for a while.) We can break the committees containing at least one of Tom or Dick into three disjoint sets.

A—those committees containing Dick but not Tom
B—those committees containing Tom but not Dick
C—those committees containing both Tom and Dick

How many elements are there in the set A? If the committee is to contain Dick but not Tom, it must consist of Dick and two people chosen from among the other eight group members. In how many ways can these two be chosen? The answer is $C(8, 2)$. Similarly the number of elements in B is $C(8, 2)$. How many elements are there in C? The committee would consist of Tom, Dick, and one person chosen from the remaining eight, so we have $C(8, 1)$ elements in C.

Therefore the probability that the chosen committee contains at least one of Tom or Dick is

$$\frac{C(8,2) + C(8,2) + C(8,1)}{C(10,3)}$$

which is

$$\frac{\frac{8!}{6!2!} + \frac{8!}{6!2!} + \frac{8!}{7!1!}}{\frac{10!}{7!3!}}$$

Note that

$$\frac{7!}{6!} = 7 \quad \text{and} \quad \frac{3!}{2!} = 3 \qquad \text{(Look at the definition of } n! \text{ if this confuses you.)}$$

Therefore, if we multiply through this fraction by $7!3!$ in both numerator and denominator, we have

$$\frac{7 \cdot 3 \cdot (8!) + 7 \cdot 3 \cdot (8!) + 3!8!}{10!}$$

Noting that $10!$ divided by $8!$ is $10 \cdot 9$, and dividing both numerator and denominator by $8!$, we have

$$\frac{7 \cdot 3 + 7 \cdot 3 + 3 \cdot 2}{10 \cdot 9} = \frac{48}{90} \approx .533$$

There is an alternative, and somewhat easier, way to do the problem. If we let D be the event, "Either Tom or Dick (or both) is on the committee," then we note that

$$P(D) = 1 - P(D^c)$$

But D^c is the event "neither Tom nor Dick is on the committee." What is the probability that neither of them is on the committee? We calculate the number of committees which can be formed from the other eight members of the group, $C(8, 3)$ and divide by the total number of committees. The answer to our problem is, therefore,

$$1 - \frac{C(8,3)}{C(10,3)} = 1 - \frac{\frac{8!}{5!3!}}{\frac{10!}{7!3!}}$$

$$= 1 - \frac{7!3!8!}{5!3!10!}$$

$$= 1 - \frac{7 \cdot 6}{10 \cdot 9}$$

$$= \frac{90 - 42}{90} = \frac{48}{90} \approx .533$$

EXAMPLE 5 A bridge hand consists (hopefully) of a random selection of 13 cards out of the deck of 52. What is the probability that a bridge hand contains all black cards?

There are 26 black cards (13 spades and 13 clubs) so there are ___ hands containing all black cards. There are a total of ___ bridge hands all together. Therefore the probability of a bridge hand containing only black cards is

$$\frac{C(26, 13)}{C(52, 13)}$$

After simplification (which you should do yourself) this gives approximately .000017.

This example should convince you that even though we have a good way to write down the formula for these probabilities, the actual calculations can get burdensome as soon as n becomes at all large.

EXAMPLE 6 If six coins are tossed, what is the probability of getting four heads?

We could, presumably, write out the sample space for this experiment and solve the problem that way. Fortunately we now have easier ways to do the problem. There are many distinct ways in which four heads can be obtained, for instance, HHTHTH. The probability can easily be calculated.

$$\begin{aligned} P(\text{HHTHTH}) &= \text{P(H)P(H)P(T)P(H)P(T)P(H)} \\ &= (\tfrac{1}{2})(\tfrac{1}{2})(\tfrac{1}{2})(\tfrac{1}{2})(\tfrac{1}{2})(\tfrac{1}{2}) \\ &= \tfrac{1}{64} \end{aligned}$$

Similarly, any point in the sample space with exactly four heads will have probability $\frac{1}{64}$. The question that must be answered is, "How many such points are there?" We can view each point in the sample space as a six-letter "word" made up of H's and T's. We can form such a word by filling up six slots

— — — — — —

Our problem then becomes to count the number of words with four H's and two T's, and we simply note that such a word is determined by choosing four of the six slots in which to place H's. But this can be done in

$$C(6, 4) = 15$$

possible ways. Hence the answer to our original question is

$$C(6, 4) \cdot \tfrac{1}{64} = \tfrac{15}{64} \approx .234$$

There is an alternate approach to the solution of this problem which is often useful so we will work the problem a second time. We can think of calculating the probability of getting four heads on the six coins by dividing the number of ways four heads can show by the total number of ways the coins can land. We have already found the number of ways we can get four heads. It is $C(6, 4)$. Now, how many ways can the six coins land? The answer is 2^6. Let's see if we can find some justification for this assertion. One way to look at it is in terms of a tree diagram. As the first coin falls, we have two possibilities:

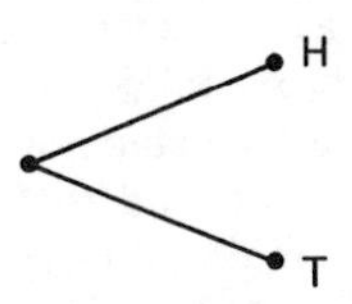

Now the second coin falls, lands either heads or tails, and adds two possibilities to each previous possibility. So we have

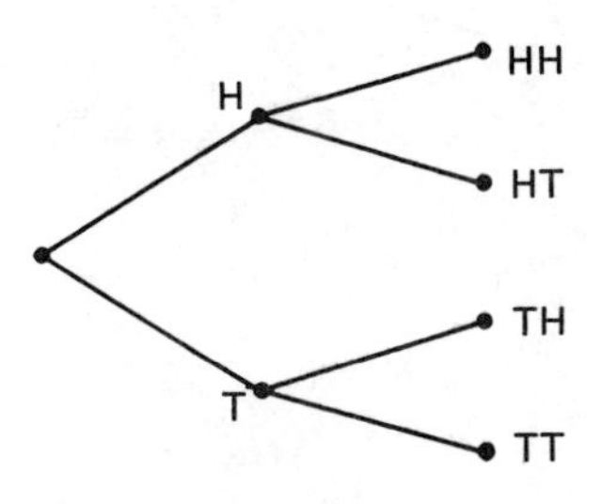

Clearly the number of possibilities doubles as each coin falls. For six coins then, we have $2 \cdot 2 \cdot 2 \cdot 2 \cdot 2 \cdot 2 = 2^6$ possibilities. Our answer is, once again, $\frac{15}{64}$.

EXAMPLE 7 Three cards are drawn from a standard deck of 52. What is the probability that all three cards are face cards (jacks, queens, or kings)? Consider the problem in the case of drawing either with replacement or without replacement.

Case 1 Drawing with Replacement

If we are drawing from a full deck of cards, 12 of the 52 cards are face cards so the probability of drawing a face card is $\frac{12}{52}$ or $\frac{3}{13}$. Therefore the required probability is

$$\left(\tfrac{3}{13}\right)\left(\tfrac{3}{13}\right)\left(\tfrac{3}{13}\right) = \tfrac{27}{2197} \approx .0123$$

Case 2 Drawing without Replacement

There are $C(52, 3)$ ways to draw three of the 52 cards. There are $C(12, 3)$ ways to draw three face cards out of the 12 and so the probability is

$$\frac{C(12,3)}{C(52,3)} = \frac{\dfrac{12!}{9!3!}}{\dfrac{52!}{49!3!}} = \frac{49!3!12!}{52!9!3!}$$

$$= \frac{12 \cdot 11 \cdot 10}{52 \cdot 51 \cdot 50}$$

$$= \frac{33}{3315} \approx .0099$$

To better understand the relationship between drawing with and without replacement, let us consider the problem of drawing without replacement from a slightly different point of view. When we draw the first card, the probability of getting a face card is $\frac{12}{52}$. But when we draw the second card, knowing that a face card was drawn on the first draw, the probability of drawing a face card is $\frac{11}{51}$. On the third draw, knowing that we have face cards on the first two draws, the probability of drawing a face card has declined to $\frac{10}{50}$. We can utilize a portion of the tree diagram to illustrate these conditional probabilities:

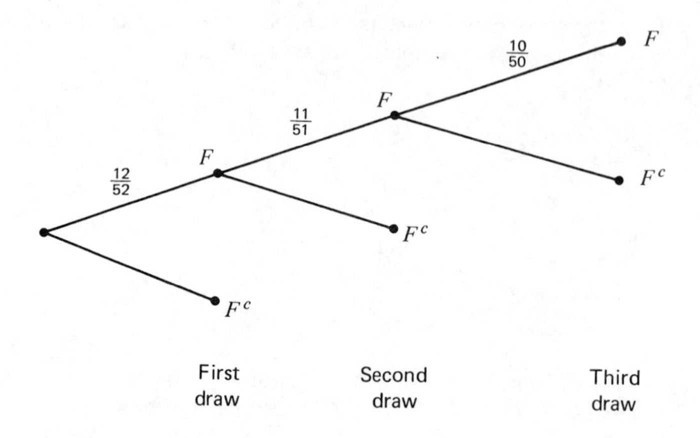

where F refers to "face card" (and therefore F^c refers to "not a face card"). Thus the probability of drawing three face cards in succession can be found by multiplying the probabilities of the three branches on the appropriate path,

$$\frac{12}{52} \cdot \frac{11}{51} \cdot \frac{10}{50}$$

which is the same answer we got before.

Throughout this section we have used implicitly a fact that perhaps ought to be stated:

The Multiplicative Principle
If A can be performed in m ways and B can be performed in n ways, then together A and B can be performed in $m \cdot n$ ways.

It is important to exercise judgment when this principle is invoked. It has purposely been stated in a simple fashion and as such it can be misinterpreted. One usually thinks of A and B as being acts that (1) are to be performed in succession, and (2) have no effect upon one another. This is how we have used it in our illustrative examples. For instance, if a first card can be drawn in 52 ways and a second card can be drawn in 51 ways, then we conclude that both cards can be drawn in $52 \cdot 51$ ways.

Another feature of this principle is that it extends from two acts to several in the obvious fashion:

Generalized Multiplicative Principle

If A_1 can be performed in n_1 ways, A_2 in n_2 ways, . . . , A_k in n_k ways, then together (i.e., in succession) $A_1, A_2, \ldots, A_k$ can be performed in $n_1 \cdot n_2 \cdot \cdots \cdot n_k$ ways.

To illustrate the power of this rule, we consider the popular but deceptively difficult problem of determining the probability of being dealt a poker hand with "one pair," that is five cards in all, exactly two of which are of the same denomination. We list four factors that completely determine such a hand:

1. The denomination of the matching pair must be specified.
   This can be done in $\binom{13}{1}$ ways.
2. The two suits of the matching pair must be specified.
   This can be done in $\binom{4}{2}$ ways.
3. The denominations of the three odd cards must be specified.
   This can be done in $\binom{12}{3}$ ways.
4. The suits of the three odd cards must be specified.
   This can be done in 4^3 ways.

Thus the number of possible poker hands having *exactly* one pair is

1,098,240

Summary

This section has been devoted to the development of three useful counting techniques. Given a collection of n objects, we can ask three questions.

1. In how many ways can these n objects be arranged in a row?
2. In how many ways can k of these n objects be arranged in a row? (Here $0 \leq k \leq n$.)
3. In how many ways can k objects be selected (without regard to order of selection and without arranging them in any way) from these n objects? (Again, $0 \leq k \leq n$.)

The answers to these questions in symbolic as well as verbal terms are, respectively:

1. The number of *permutations* of n objects $= n! = n(n-1)(n-2) \ldots 2 \cdot 1$
2. The number of *permutations of n objects taken k at a time* $= P(n,k) = \dfrac{n!}{(n-k)!}$
3. The number of *combinations of n objects taken k at a time* $= C(n,k) = \dfrac{\cdot\, n!}{(n-k)!k!}$

The symbol $n!$ is read "*n factorial.*" Consistent with the above definitions is the stipulation

$$0! = 1$$

Exercises Section 6

1 Evaluate each of the following.

(a) $P(7, 3)$
(b) $P(7, 4)$
(c) $C(7, 3)$
(d) $C(7, 4)$
(e) $C(5, 2) + C(5, 3)$
(f) $C(6, 3)$
(g) $C(7, 5) + C(7, 6)$
(h) $C(8, 6)$

2 A group of four men and three women select a committee of three. Assuming that the selection is random, what is the probability that the majority of the committee will be women? What is the probability that the majority of the committee will be men?

3 An urn contains 10 marbles, seven of which are red and three of which are blue. Three marbles are drawn at random.

(a) If the drawing is with replacement, calculate the probability that the majority of the marbles are blue.

(b) If the drawing is without replacement, calculate the probability that the majority of the marbles are blue.

4 Assuming that the last four digits of a telephone number are a random selection from among the 10 digits 0, 1, 2, 3, 4, 5, 6, 7, 8, 9, what is the probability that a telephone number has at least two digits the same?

5 What is the probability that a bridge hand contains no 2's? What is the probability that a bridge hand contains no 3's? Are these two events independent?

6 A group of eight men and six women choose a committee of three. Assuming that the selection is random, find the probability that the majority of the committee members are women. Find the probability that the majority of the committee members are men. Compare these results with your answer to Exercise 2.

7 You are stranded on a desert island with five other people. It turns out that five of the six can get back to civilization. If this group is selected randomly, what is the probability that you will make the trip?

8 A box contains 50 light bulbs, four of which are defective. If a random sample of three light bulbs is chosen, what is the probability that it contains at least one defective bulb? What is the smallest sample size such that the probability that the sample contains at least one defective is at least .5?

9 Write (and run) a program which will cause the computer to print out values of $n!$ for $n = 1$ to 60. (*Remark:* This problem brings out some of the limitations of computers. On every computer there is some maximum number that is handled and 60! may well be too large for your machine. You may have to print out something such as $n!/10^n$ to get around this limitation.)

10 Show that:

(a) $C(n, x) = C(n, n - x)$

(b) $C(n, x) = C(n - 1, x - 1) + C(n - 1, x)$

11 Find the probabilities of being dealt the following poker hands.

(a) One pair (already discussed in the text)

(b) Two pairs: a pair of one denomination, a pair of a different denomination, and one odd card of a third denomination

(c) Three-of-a-kind: three cards of one denomination and two odd cards of other denominations
(d) A straight: five cards in sequential order (ace, 2, 3, 4, 5 or 2, 3, 4, 5, 6 or . . . or 10, jack, queen, king, ace) with no regard to suits *except* they may not *all* be of the same suit (for then the hand would be called a straight flush)
(e) A flush: any five cards of the same suit
(f) A full house: a pair of one denomination and a triple of another denomination
(g) Four-of-a-kind: all four cards of some denomination and an odd card
(h) A straight flush: the intersection of the set of straights with the set of flushes. The special case where the lowest card in the sequence is "10" is called a *royal straight flush*.
(i) None of the above

12 (a) In how many ways can eight boys be divided into two teams of four?
(b) What if Pat is the captain of one team and Mike is the captain of the other?

13 A child has a box of red blocks and a box of blue blocks. How many color patterns can he make by forming a row of seven blocks if:
(a) he places a red block at each end?
(b) he uses only two red blocks?
(c) he alternates colors?
(d) he places the red blocks next to each other?

14 How many batting orders are possible on a baseball team (nine players) if it is known that:
(a) the pitcher will bat last?
(b) the three left-handed batters will bat fourth, fifth, and sixth and the pitcher will bat last?
(c) the two speedsters will bat first and second, the three left-handed batters will bat fourth, fifth, and sixth, and the pitcher will bat last?

Supplementary Exercises for Chapter 2

1 A die is loaded in such a way that $P(1) = \frac{4}{21}$, $P(2) = \frac{1}{21}$, $P(3) = \frac{5}{21}$, $P(4) = \frac{2}{21}$, $P(5) = \frac{6}{21}$, $P(6) = \frac{3}{21}$. Find the probability of rolling each of the following.
(a) An odd number
(b) A number greater than 2
(c) A number divisible by 3

2 A pair of dice are loaded like the one in Exercise 1. If both dice are rolled, find the probability of getting a total
(a) of 7.
(b) exceeding 10.
(c) of 4.

3 (a) A sample space S consists of two mutually exclusive events A and B. When we say that *the odds in favor of A are 3 to 1*, what probability do we assign to B? to A?

(b) More generally, when we say that the odds in favor of A are x to y, what odds do we assign to A and to A^c?

4 The instructions on a certain examination direct the students to select any four of seven given problems to solve. In a class of 30 students, could each student select a different set of problems to work on?

5 While browsing in the library a student removes two books from a shelf containing 10 other books and retires to a secluded carrell. Later he tries to replace the books only to find that in his absence a librarian had straightened the shelf so that there are no gaps where his two books should fit. Not understanding the cataloguing system, he guesses where to place the books. Find the probability that both books are returned to their original position assuming that he remembers each of the following sets of conditions.

(a) His books were adjacent to each other and neither was at the end of the row.

(b) Same as (a) but in addition he knows which of his books came first (i.e., to the left of the other).

(c) His books were originally separated by one other book, neither was at an end of the row, and he remembers which came first.

(d) Same as (c) except that two other books separated his.

(e) Same as (c) except that five other books separated his.

(f) One of his books was at an end of the row.

(g) One of his books was at the left end of the row.

(h) One of his books was at the left end of the row and he knows which one.

(i) He remembers only which of his books came first.

(j) He remembers nothing.

6 A six-member college committee has two students, two faculty members, and two administrators. Find the probability that a randomly selected subcommittee of three will have

(a) exactly one student, one faculty member, and one administrator.

(b) no students.

(c) no administrators.

7 **(a)** Write out the list of subsets of three elements chosen from 1, 2, 3, 4, 5, 6.

(b) Find the average sum of each of the subsets in (a).

8 A committee of four teachers has the responsibility of selecting a text for adoption. Preliminary screening has resulted in three books to be considered. Each teacher writes his first choice on a slip of paper with the understanding that if a majority agrees on one text, then it will be adopted. Otherwise further discussion will be necessary.

(a) How many distinct outcomes to the vote can there be?

(b) How many outcomes would lead to an adoption of book A?

(c) What is the probability (assuming that the voting is independent) that a decision to adopt some text will be reached on the first vote?

9 Modify Exercise 8 by assuming that three teachers must choose from four texts.

10 A cafeteria advertises "over 1000 different meals to choose from." How can this be?

11 In how many ways can four people (A, B, C, and D) be seated at a bridge table if

(a) A and B are to be partners?

(b) A and B are to be opponents?

12 Three marbles are drawn (without replacement) from a bag containing four red marbles and two blue marbles. Write out the corresponding sample space and the probabilities of each event.

13 Modify Exercise 12 by drawing with replacement.

14 Modify Exercise 12 by assuming that there are eight red marbles and four blue marbles in the bag.

15 If E and F are independent events, and if F and G are independent events, then does it follow that E and G are independent events?

16 If E and F are independent events, are E and F^c also independent? How about E^c and F^c?

17 If A is an event in a sample space S, can A and S be independent?

18 If A and B are distinct events with $A \subset B$, can A and B be independent?

19 Prior to a national election a political analyst narrows down the list of candidates to three men for president and four for vice-president in the Republican party and five men for president and three men for vice-president in the Democratic party. (*Note:* the incumbent President is not seeking re-election.)

(a) How many ballots can emerge?

(b) How many ballots can emerge if two of the candidates for the Democratic presidential nomination are also candidates for the vice-presidency?

20 Six men and two women are seated in a row. Find the probability that

(a) the women are seated next to each other.

(b) one woman is at an end of a row.

(c) the women are separated by two men.

21 Modify Exercise 20 by assuming that there are 12 men and 2 women.

22 A pair of fair dice are rolled. Find the probability that the total is 7 if you are told that

(a) the total is odd.

(b) the dice do not match.

(c) the total is even.

(d) the total is less than 8.

(e) one die came up 4.

(f) one die came up 1.

23 A quiz contains nine multiple choice questions, each having three possible answers (A, B, and C). Rumor has it that the professor always arranges to have the correct answers equally divided among A's, B's, and C's. If a student guesses the answer to each question, being careful to have an equal number of A's, B's, and C's, what are his chances of getting a perfect paper?

24 Modify Exercise 23 so that there are only six questions on the test, but there is no information about the number of A's, B's, and C's that occur as correct answers.

25 A professor announces that each student in his class will be given an oral examination. The ground rules are as follows: Prior to the exam, the class is given a list of 10 problems. When a student arrives for his exam, the professor randomly selects four problems from the list and the student may then choose any two of these four on which to discourse. How many problems must a student be prepared to solve to guarantee that he can handle at least two of the four given him.?

26 Referring to Exercise 25, suppose that a student is able to solve only five problems on the list of 10.

(a) What are his chances of solving at least two of the four given him?

(b) What are his chances of solving at least one of the four given him?

27 Five red marbles and five blue marbles are to be distributed between two urns so that each urn contains at least one marble. Then an urn is selected at random and a marble is selected at random from that urn. How should the marbles be distributed so as to maximize the probability that a red marble is chosen?

28 An instructor notices that a text book has been left in his classroom. Since the text has no name in it, he decides to make an educated guess as to whether its owner is a man or a woman. He knows that $\frac{3}{5}$ of his class are men and he conjectures that only $\frac{1}{3}$ of the men but $\frac{2}{3}$ of the women normally bring their books to class. At the next class meeting he announces that a book was left behind and that it *probably* belongs to a woman. Was his presumption justified?

29 Al and Dick spend their lunch hours playing tennis or handball. Al wins at handball about 80% of the time, but Dick wins at tennis 70% of the time. They play handball twice as often as they play tennis. One day Al and Dick appear at the cafeteria after their competition and Dick is wearing a victor's smile. What is the probability that they played tennis?

30 A man has a sensitive knee that aches, on the average, 50% of the time. When it rains, his knee hurts 70% of the time and when it doesn't rain, the knee hurts only 20% of the time. How often does it rain?

31 A probability function P on a sample space S must assign a number to each subset of S. Using the properties of a probability function, we can unambiguously define such a function by specifying its value for each of the simple events in a given finite sample space. This is fortunate, indeed, for the number of *subsets* of a finite set is quite large compared to the number of *elements* in the set. For instance, the set

$$S = \{a, b, c\}$$

has three elements but eight subsets:

$$\varnothing, \{a\}, \{b\}, \{c\}, \{a, b\}, \{a, c\}, \{b, c\}, \{a, b, c\}$$

(a) Make a list of all the subsets of the set $\{1, 2, 3, 5\}$.

(b) Suppose a sample space $S = \{A, B, C\}$ has a probability function P such that

$$P(A) = \tfrac{1}{4} \quad \text{and} \quad P(C) = \tfrac{1}{3}$$

Make a list of all the events in S and the corresponding value of P for each one.

32 Specify a sample space corresponding to each of the following experiments. In some cases you may not wish to list all the simple events, but you should.

(a) Two partnerships are formed from four bridge players.
(b) Two partnerships and one kibitzer are formed from five bridge players.
(c) A boy selects a pair of shoes from a mixture of three identical pairs.
(d) Six pennies are simultaneously tossed.
(e) A die is rolled until a "6" appears.
(f) A boy selects a pair of socks from a mixture of three identical pairs.
(g) Two coins are tossed repeatedly until they both come up heads.
(h) Two coins are tossed repeatedly until either they both come up heads or they have been tossed three times.

33 Use tree diagrams to help answer each of the following.

(a) How many ways can the letters A, B, C, D, E be arranged so that A occurs somewhere before B and C occurs immediately before E?
(b) A football team scored 10 points in the first quarter. In each of the remaining quarters it scored either 0, 3, 6, 7, or 9 points, but it scored successively less in each period. List all the possible outcomes quarter by quarter.
(c) Two teams play a "best of seven" series. That is, the first team to get four victories wins the series. List all the possible victory patterns.
(d) A die is rolled until either two even numbers in succession are rolled or the last roll fails to exceed the preceding one. How many ways can this happen?
(e) A coin is flipped until either three heads occur or two tails in succession occur. How many ways can this happen?

34 Refer to Exercise 33(c) to find the probability that evenly matched teams would have a series lasting at least six games.

CHAPTER

3 Discrete Statistics

0 INTRODUCTION

Thus far we have learned to analyze and organize data using means, standard deviations, histograms, etc. We have also learned the rudiments of probability theory. It is now time to begin the process of putting these together in such a way that we can tackle the kinds of problems which arise in the real world.

The application of mathematics to practical problems always involves a model. You may not be familiar with the term model, but you are certainly familiar with certain models. For example, in high school you developed a model for motion at a constant speed in a straight line. You let d stand for distance, r for speed, and t for the time, and you expressed this model by the formula

$$d = r \cdot t$$

In the preceding chapter we developed a model for what happens when two fair dice are thrown. The model was a histogram indicating that the probability of throwing a 2 is $\frac{1}{36}$, the probability of throwing a 3 is $\frac{2}{36}$, etc.

The point in having a model is that it allows us to predict certain things, but the nature of the two models mentioned is quite different in one respect. Our model for straight-line motion allows us to predict that if we go in a straight line at a speed of 50 miles per hour for 4 hours we will cover a distance of 200 miles. We can make this prediction without making the trip. But what kind of prediction do we get from our model for rolling two fair dice? One reasonable prediction is that if you roll a pair of (fair) dice 72 times there should be four occurrences of throwing a 3. But this prediction is quite different from the first one.

We expect the answer 200 miles to be exact (or at least as exact as the measurement of time and speed). You would not expect to find that you had really traveled 203 miles. Indeed were you to find that you have travelled 203 miles you would assume that either the speed or the time had been measured inaccurately.

On the other hand, if the dice are rolled 72 times and on five occasions you get a 3, you are not at all surprised. You don't assume that you have made an error in counting either the number of times the dice were rolled or the number of times a 3 occurred. You don't assume that the dice are not fair. This is typical with models based upon probability. There is a bit of fluctuation involved in the results. The fact that you expect to get a 3 on the average 2 times out of 36 doesn't mean that on every 36 rolls, it will happen exactly twice.

The distance formula is *deterministic* in that it determines an exact answer to the problem, whereas the model for rolling the pair of fair dice predicts an answer which, in any finite number of trials, will be approximately correct.

This creates a certain difficulty in deciding whether or not the model is any good for practical purposes. It is fairly easy to decide whether or not the distance formula is a good model because it is presented with the claim that it always works (whenever it applies). If it ever failed to work, if for example you really did travel 203 miles in 4 hours at a speed of 50 miles per hour, and if all those measurements were correct, then the distance formula would have to be discarded.

But where the model is only approximate, where we expect there might be a little difference between what the model predicts and what actually happens, how do we decide whether or not the model is a good one? We shall see that there is a decision procedure. We can never have absolute certainty that our model is right, but we can make the probability that it is right very close to 1. Therefore probability comes into this chapter in two ways. It is used in designing the models, or distributions, that we study and is, at the same time, used to assess the validity of the model as a representation of the real world.

1 RANDOM VARIABLES AND FREQUENCY FUNCTIONS

This section is concerned with two kinds of functions associated with a sample space. The first of these functions is called a *random variable*; the other, which is intimately related to the random variable, is the *frequency function*. Much of what we have to say is simply a formalization of ideas we have already encountered.

EXAMPLE 1 Rolling a Pair of Fair Dice

Let us review the sample space S consisting of all ordered pairs of the integers from 1 to 6, i.e.,

$$S = \{(a, b) \mid a, b \text{ are integers and } 1 \leq a, b \leq 6\}$$

This set, together with the assignment of probabilities

$$P((a, b)) = \tfrac{1}{36}, \qquad \text{for all } (a, b) \in S$$

serves as a model for the rolling of a pair of fair dice. When we roll a pair of dice, we are interested typically in the sum of the two faces showing. Thus if we roll (5, 2), we associate the number 7; if we roll (4, 5), we associate the number 9, and so on. More formally we can define a function X that associates to each simple event (a, b) in the sample space the real number $a + b$:

$$X((a, b)) = a + b$$

This function X is called a *random variable*. We can write

$$X((3,2)) = 5$$
$$X((3,5)) = 8$$
$$X((5,3)) = 8$$
$$X((4,4)) = 8$$

and so forth for each of the 36 simple events in S. Notice that, as often happens, several different simple events may be assigned the same value by the random variable.

Definition Let S be any sample space. A *random variable* on S is any function X that assigns a real number to each simple event s of S. If all the values of X can be enumerated or listed, we call X *discrete*; if X takes on all values in some interval, we call X *continuous*.

The dichotomy of continuous versus discrete has been mentioned earlier. The present definition will suffice for our purposes, but it should be mentioned that there exist sets of numbers which cannot be enumerated but which also do not contain an interval. A classic example of such a set, called the *Cantor Set*, is a popular source of examples in the branches of mathematics known as topology and analysis. An interested reader should have no trouble finding a description of this set in the literature.

EXAMPLE 2 Tossing a Coin n Times

We shall find it convenient to call a string of letters of the form

HTH

a *three-letter* word made up of H's and T's. As we have already seen, the outcomes of coin tossing experiments can be expressed symbolically as such words. We let S be the sample space of all n-letter words made up of H's and T's. That is,

$$S = \{\text{HHH} \cdots \text{H}, \text{HTH} \cdots \text{H}, \text{THT} \cdots \text{T}, \ldots (n \text{ letters each})\}$$

We define a random variable X on S by

$$X(\text{HTH} \cdots \text{H}) = \text{the number of H's in the word}$$

Thus S represents the possible outcomes of tossing a coin n times and X merely counts the number of heads that occur. (If the coin is fair, we would assign a probability function to make the simple events equally likely.) In this case, as in Example 1, X is *discrete* since its possible values are

$$0, 1, 2, \ldots, n$$

EXAMPLE 3 The Continuous Clock vs. The Discrete Clock

Let S be the sample space consisting of all possible times from noon to midnight. Suppose we have a continuous clock synchronized with a discrete clock.

At any given time t we let X be the difference between the time on the continuous clock and the time on the discrete clock, say in minutes. Thus at time

$$t = 2{:}14 + .872165\cdots \quad \text{minutes}$$

we have

$$X(t) = .872165\ldots \quad \text{minutes}$$

Here X can take on all values between 0 and 1, and therefore X is a *continuous* random variable.

In general, we are interested in the likelihood that X have certain values. For the present we will *ignore continuous variables* and concentrate on the following.

For a *discrete random variable* X on a sample space S we define the *frequency function* f for X by the formula

$$f(x) = P(\{s \in S | X(s) = x\})$$

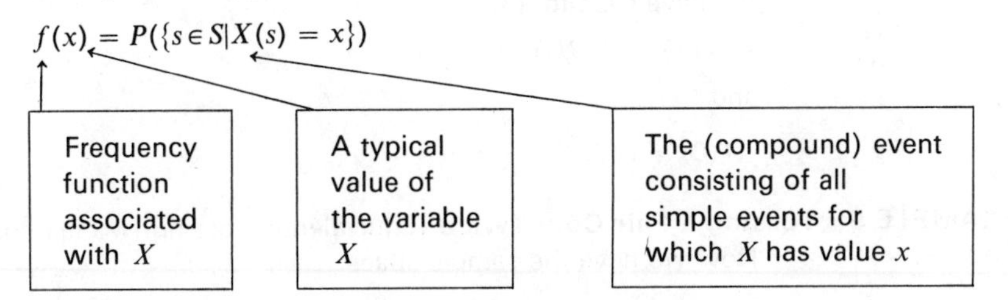

This perhaps awesome looking formula is not so bad as it might appear at first glance. In fact we shall always abbreviate it with the symbols

$$f(x) = P(X = x)$$

The important features to be noted are:

1. The *domain* of the frequency function (i.e., all values x to which f assigns values) consists of all possible values of the random variable X.
2. Since X can assign a given value x to several different simple events, we collect together all such events into one compound event $E = \{s \in S | X(s) = x\}$. This compound event has some probability $P(E)$ associated with it.
3. $f(x) = P(E)$ using the notation in (1) and (2).

Thus we always have

$$0 \leq f(x) \leq 1$$

since probabilities always lie between 0 and 1.

Some other important characteristics of f can best be seen by treating familiar sample spaces in some detail.

EXAMPLE 4 Tossing a Fair Coin Once

The sample space in this case is

$$S = \{H, T\}$$

One random variable we could define is

$$X(H) = 1$$
$$X(T) = 0$$

Note that all we must insure is that each simple event gets assigned some real number. It so happens that X can be described simply in words:

$X(s) =$ the number of heads occurring (either 0 or 1)

The associated frequency function is f, given by

$$f(0) = \tfrac{1}{2}$$
$$f(1) = \tfrac{1}{2}$$

Note here that f must be defined on all possible values of X, and since

$$\{s \in S | X(s) = 1\} = \{H\} \quad \text{and} \quad \{s \in S | X(s) = 0\} = \{T\},$$

we have seen to it that

$$f(1) = P(\{H\})$$

and

$$f(0) = P(\{T\})$$

EXAMPLE 5 Tossing a Fair Coin Twice (Equivalently, tossing two fair coins once)

Now we have the sample space

$$S = \{\text{HH, HT, TH, TT}\}$$

We can define the random variable X by

$$X(\text{HH}) = 2$$
$$X(\text{HT}) = 1$$
$$X(\text{TH}) = 1$$
$$X(\text{TT}) = 0$$

If we think of one coin tossed twice, HT would represent getting heads on the first toss and tails on the second. If we consider two coins, say a penny and a nickel, being tossed once, then HT would represent heads on the penny and tails on the nickel.

Once again we have arranged that

$X(s) =$ the number of heads that come up

This time we have three possible values of X: 0, 1, and 2, so we must define f on each of these. We have

$$f(0) = P(X = 0) = P(\{\text{TT}\}) = \tfrac{1}{4}$$
$$f(1) = P(X = 1) = P(\{\text{HT, TH}\}) = \tfrac{1}{2}$$
$$f(2) = P(X = 2) = P(\{\text{HH}\}) = \tfrac{1}{4}$$

since we assume that the four simple events are equally likely.

EXAMPLE 6 Rolling a Pair of Fair Dice Once

The sample space has 36 simple events. We list them in a two-dimensional array. The diagonal lines separate compound events determined by the random variable X, where $X(a, b) = a + b$.

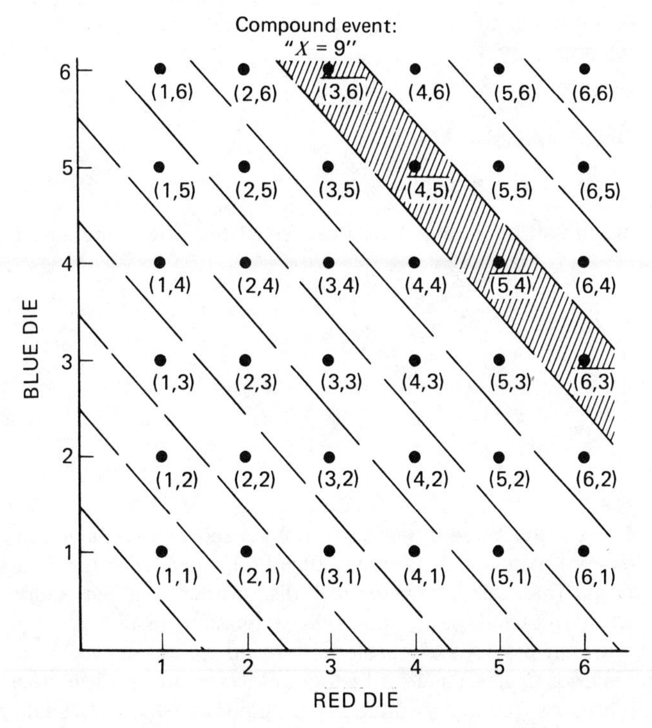

FIGURE 3.1 Thirty-six equally likely outcomes. The sample space for rolling a pair of fair dice once.

Each of the compound events determined by adjacent pairs of diagonal lines is made up of simple events that share a common value of the random variable X. The shaded region is

$$\{s \in S | X(s) = 9\}$$

Since the simple events are equally likely, we can prescribe the frequency function f associated with X. The values of X are 2, 3, 4, 5, 6, 7, 8, 9, 10, 11, and 12. Thus

$$f(2) = P(X = 2) = P(\{(1, 1)\}) = \tfrac{1}{36}$$
$$f(3) = P(X = 3) = P(\{(1, 2), (2, 1)\}) = \tfrac{2}{36}$$
$$f(4) = P(X = 4) = P(\{(1, 3), (2, 2), (3, 1)\}) = \tfrac{3}{36}$$

and so on. The other values are easily found to be

$$f(5) = \tfrac{4}{36}, \quad f(6) = \tfrac{5}{36}, \quad f(7) = \tfrac{6}{36}, \quad f(8) = \tfrac{5}{36}$$
$$f(9) = \tfrac{4}{36}, \quad f(10) = \tfrac{3}{36}, \quad f(11) = \tfrac{2}{36}, \quad f(12) = \tfrac{1}{36}$$

EXAMPLE 7 Using the same sample space of 36 ordered pairs, we define a different random variable Y by

$$Y((a, b)) = a - b$$

Now Y can take on values

$$-5, -4, -3, -2, -1, 0, 1, 2, 3, 4, 5$$

and it can be shown (see Exercise 8) that the associated frequency function is

$$f(-5) = f(5) = \tfrac{1}{36}$$
$$f(-4) = f(4) = \tfrac{2}{36}$$
$$f(-3) = f(3) = \tfrac{3}{36}$$
$$f(-2) = f(2) = \tfrac{4}{36}$$
$$f(-1) = f(1) = \tfrac{5}{36}$$
$$f(0) = \tfrac{6}{36}$$

By now the student has no doubt realized that we have not introduced any new examples or computations, but rather we have put matters on a more formal basis. The "relative frequency" discussions in earlier sections covered virtually all the material we are currently working with.

In our previous discussion of absolute and relative frequency we employed a pictorial device called a histogram. We shall find it increasingly convenient to represent frequency functions by pictures. Not only shall we continue to use histograms, but we shall also want to incorporate an alternative representation, the *graph* of a frequency function. Like histograms, graphs are drawn with the aid of a horizontal scale (or axis or number line) on which the various values of a random variable X are marked off, and a vertical scale on which we mark off representative values of $f(x)$. The graph consists of a collection of dots, one above each possible value x of X at a height equal to $f(x)$. A more careful discussion of functions and graphs will be presented later in this chapter. For the time

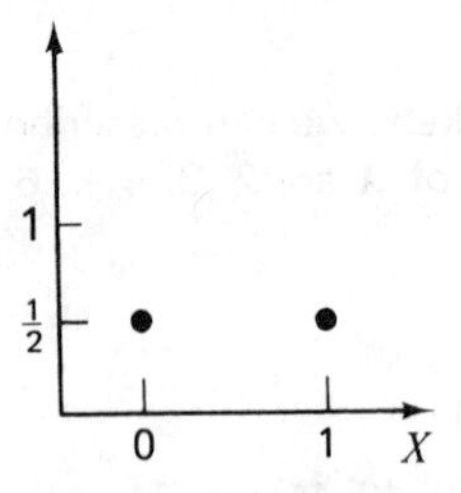

FIGURE 3.2 Frequency function for X, the number of heads on 1 toss of a fair coin.

being we illustrate the idea in Figs. 3.2, 3.3, and 3.4, where graphs of the frequency functions from Examples 4, 5, and 6, respectively, are sketched. *Note:* A fact which is easily checked in each of our examples is that for any frequency function f,

$$\Sigma f(x) = 1$$

where this summation is taken over all possible values x of X.

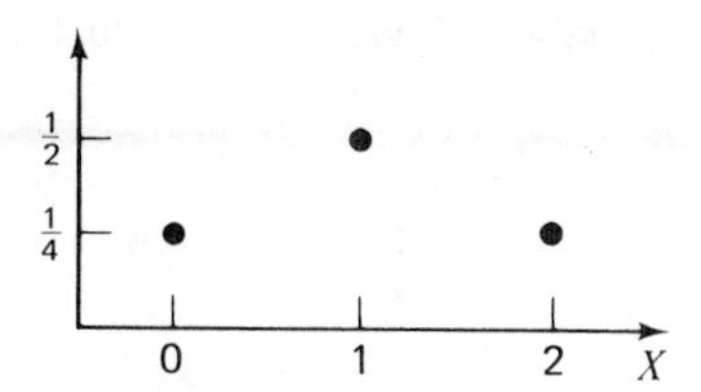

FIGURE 3.3 Frequency function for X, the number of heads on 2 tosses of a fair coin.

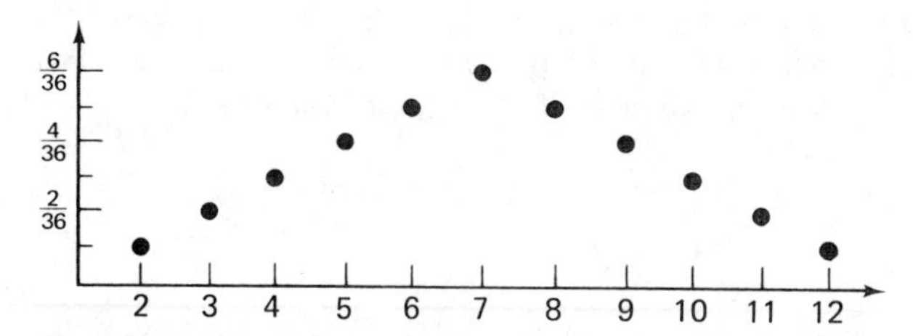

FIGURE 3.4 Frequency function for *discrete* random variable X, the total of two faces on 1 toss of a pair of fair dice.

Summary

The language of probability, introduced in Chapter 2, helps us describe the experiment of rolling a pair of dice. Abstractly, the 36 possible outcomes can be represented as 36 points that comprise a sample space, on the subsets of which a probability function is defined. Yet, a moment's thought should convince us that this abstract representation of the dice throwing experiment is somewhat incomplete. Generally when we roll a pair of dice, we are less interested in the particular ordered pair of numbers (such as a 3 on the red die and a 5 on the blue die) than we are in the *sum* of these numbers. The purpose of this section has been to broaden the concept of an abstract model to include the notion of a *random variable* and its partner, a *frequency function*.

A random variable is simply a function defined on a sample space. That is, a random variable assigns a number to each simple event in a sample space. In the aforementioned case of the pair of dice, the function X, defined by

$$X(a, b) = a + b$$

assigns to each possible outcome the sum of the numbers showing on the dice. Another easily defined random variable is the one that counts the number of heads that occur in, say, three tosses of a fair coin. Yet a third example of a familiar random variable would be the function that assigns to each person his height (here the sample space might be the set of all freshmen at a particular university).

Random variables are called *discrete* if their full set of values can be enumerated; otherwise, they are called *continuous*. While there are certain similarities, this section concentrated on discrete random variables, postponing a discussion of the continuous case till later.

To each discrete random variable X, we associate a frequency function f, defined by

$$f(x) = P(\{E|X(E) = x\})$$

or more tersely

$$f(x) = P(X = x)$$

All this says is that for a given value x of X, we construct a set A consisting of all the simple events E in the sample space for which $X(E) = x$. This set A, now a compound event, has a probability $P(A)$, which is the number we call $f(x)$. That is, $f(x)$ is the probability that a randomly selected simple event will give rise to the value x of the random variable X.

Exercises Section 1

1 Pick, at random, a white page in a telephone book. Let X be the first digit and Y be the last digit of the street addresses of the first 40 listings on the page. Now write down the frequency functions for X and Y.

2 An urn contains three red and three blue marbles. Suppose you draw four marbles (*without replacement*) and let X be the number of red ones you get. Write the frequency function for X and sketch its graph.

3 An urn contains 36 marbles, six each with the numbers 1, 2, 3, 4, 5, and 6 on them. If you draw two *with replacement* and let X be the sum of their numbers, what kind of frequency function would X have? Sketch its graph.

4 Refer back to Example 7. Draw the 36 points representing the sample space and indicate the grouping into compound events effected by the random variable Y.

5 Consider the experiment of tossing a fair coin three times in succession. Let X be the discrete random variable that counts the total number of heads that come up.

(a) Write down the sample space S for this experiment.

(b) List the values of X on each simple event.

(c) Draw the graph of the frequency function f for X.

6 Repeat Exercise 5 where the coin this time is unbalanced so that the probability that heads occurs on a single toss is $\frac{1}{3}$.

7 Draw the graph of the frequency function g for the discrete random variable Z defined for the pair of dice experiment as follows

$$Z((a,b)) = a \cdot b$$

8 A man has three keys (say A, B, and C) exactly one of which will open a certain lock. He tries one key at random; if it is not the right one, he discards it and tries a second; if it is not correct, he tries the third.

(a) Write the sample space for this experiment.

(b) Compute the values of the random variable X representing the number of keys the man tries.

(c) Draw the graph of the frequency function corresponding to X.

9 Do Exercise 8 with four keys, one of which works.

10 Suppose in Exercise 8 the man has only two keys, but after trying one key and finding it doesn't fit the lock, he shuffles up the two keys and tries again, and continues until he gets the right key. Now find the sample space, the random variable, and its frequency function. Is this random variable discrete?

11 Do Exercise 10 with three keys.

12 Verify the following property of discrete random variables for those given in Examples 4, 5, and 6.

$$\Sigma f(x) = 1$$

where the sum is taken over *all* possible values of the random variable X.

13 **(a)** Draw the graph for the frequency function whose histogram is depicted below.

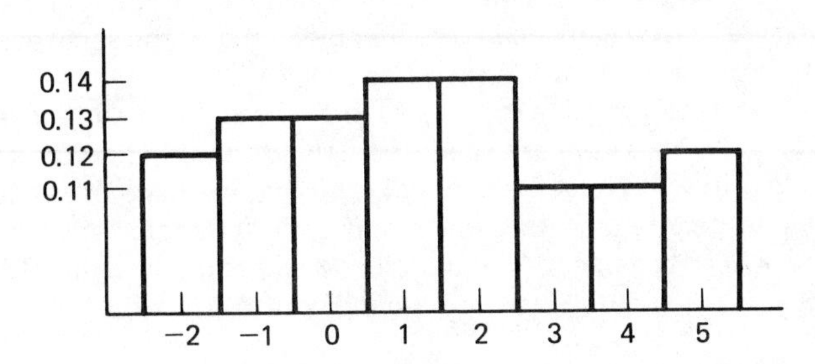

(b) Draw the histogram for the frequency function whose graph appears below.

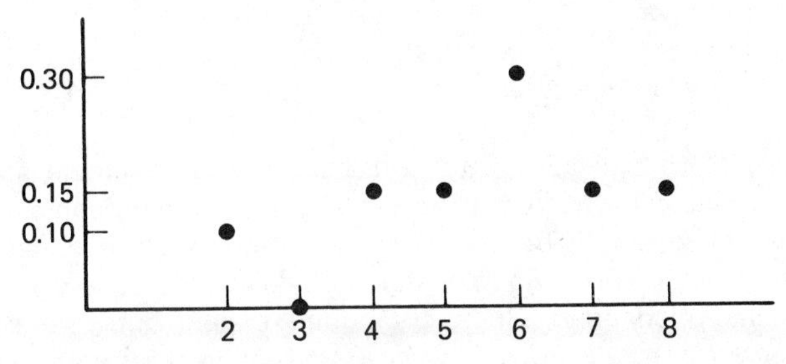

14 Five cards are drawn without replacement from a standard deck.

(a) Let X be the number of aces drawn. Find the frequency function f corresponding to X.

(b) Let Y be the number of red cards drawn. Find the frequency function corresponding to Y.

(c) Let Z be the number of face cards (jacks, queens, and kings) drawn. Find the frequency function corresponding to Z.

15 Two evenly matched cribbage players are finalists in a contest. If they play until one person wins two games and if X is the number of games they play, find its frequency function. (Note: Ties cannot occur in cribbage.)

16 Modify the above exercise so that a person must win three games to be declared champion.

17 Three coins are selected (without replacement) at random from a box containing two pennies, three nickels, one dime, and one quarter. Let X be the aggregate value of the coins drawn. Find its frequency function.

2 THE BINOMIAL DISTRIBUTION

In a significant number of experiments, the outcome may be viewed as falling into one of two mutually exclusive categories. For instance, when we flip a coin, the outcome is either heads or tails (i.e., not heads); when we take a poll, we expect to get an answer of either "yes" or "no" (although some polls are more elaborate); when we test for evidence of a particular disease, the outcome will be either positive or negative. This sort of dichotomy can also be imposed on many experiments: when we roll a die we can get either a "6" or "not a 6"; when we draw a card it can be either red or black; the same card could be viewed as being either a heart or not a heart.

When dealing with examples such as these it is often convenient to label one of the outcomes *success* (denoted S) and the other *failure* (denoted F).

Warning: The reader should note that we have used the symbol S in two ways: to denote a sample space and now to denote "success."

It is further convenient (and traditional) to assign the symbol p to the probability of S and the symbol q to the probability of F. Thus

$$P(\mathrm{S}) = p, \qquad P(\mathrm{F}) = q$$

and since F and S are mutually exclusive events whose union is the whole sample space

$$1 = p + q \quad \text{or} \quad q = 1 - p$$

What we really want to consider now is a sample space representing n successive performances of an experiment where outcomes satisfy this success-failure dichotomy. That is, we want to repeat such an experiment n times under identical conditions so that the probability of success on each performance remains unchanged. Such compound experiments are called *Bernoulli* (or *repeated* or *independent*) *trials.* By assuming that the outcome of a particular trial is in no way influenced by the results of other trials, we assert that the probability of a particular sequence of events is the *product* of the component, single trial outcomes. Again, it is convenient to use the device of n-letter words.

We can write our sample space for n repeated trials as

$$\{\text{SSS} \ldots \text{S, SFS} \ldots \text{S, FSF} \ldots \text{F, etc. } (n \text{ letters in each word})\}$$

and our independence hypothesis gives us*

$$P(\text{SSFS} \ldots) = P(\text{S})P(\text{S})P(\text{F})P(\text{S}) \ldots$$
$$= p \cdot p \cdot q \cdot p \ldots$$

We define the discrete random variable X as follows:

$$X(\text{SSFS} \ldots) = \text{the number of S's in the event SSFS} \ldots$$

Our primary goal in this section will be to describe the frequency function associated with this random variable. For reasons which will soon become evident, we denote this frequency function, or more accurately its value at some number x in the range of X, by

$$B(x; n, p)$$

and we call the totality of such frequency functions, where n and p may vary from one situation to another, the *binomial distribution*. The variables n and p are called *parameters*.

In other words any random variable X that measures the number of "successes" in a given number of repeated trials will be said to be *binomially distributed*. Once the parameters n and p have been fixed, the job is to describe

* Although the equations that follow are in a form that is common in the literature, they do involve both an abuse of notation and a slight obfuscation of what we mean by *independent trials*. If we consider a special case where $n = 3$, we can offer a more formal argument in behalf of the assertion

$$P(\text{SSF}) = P(\text{S})P(\text{S})P(\text{F})$$

Note first of all that on the left-hand side we have the probability of a particular sequence of occurrences, SSF, that represents a simple event in the sample space corresponding to three repeated trials. By contrast, each factor in the right-hand side appears to be a probability of an event in a *different* sample space, namely the one corresponding to a *single* trial. Closer examination, however, reveals that we may translate symbolically

$$\text{SSF} = (\text{S}\_\_) \cap (\_\text{S}\_) \cap (\_\_\text{F})$$

where S_ _ is the *compound* event, "success on the first trial," in the sample space corresponding to three repeated trials. To put it more explicitly,

$$\text{S}\_\_ = \{\text{SSS, SSF, SFS, SFF}\}$$

Similarly, _S_ and _ _F are compound events. Now we can call upon our definition of independent events that tells us

$$P(\text{SSS}) = P(\text{S}\_\_) \cdot P(\_\text{S}\_) \cdot P(\_\_\text{F})$$

Finally we observe that the product on the right, by our assumption that on the various trials the probability of success (p) [and hence of failure (q)] remains constant, becomes

$$p \cdot p \cdot q$$

Needless to say, we shall continue to use the customary simplification

$$P(\text{SSF}) = P(\text{S})P(\text{S})P(\text{F})$$

trusting that no confusion will arise in practice.

the *frequency function* $B(\_\_; n, p)$ for the variable X. Thus we shall be exploring a whole class of frequency functions that come under the heading binomial distribution.

For fixed values of n and p, the frequency function can be adequately described by brute force: We simply list *all* possible n-letter words containing S and F; separate into groups (i.e., compound events) those words with a given x number of S's; and compute the probabilities of these compound events. For instance if $n = 3$ and $p = .4$, we have the following simple events with their corresponding probabilities. Remember that $q = 1 - p = .6$ in this case.

Sample Space for Three Bernoulli Trials with $P(\mathrm{S}) = .4$

X	Simple Event	Probability
0	FFF	$(.6)(.6)(.6) = .216$
	FFS	$(.6)(.6)(.4) = .144$
1	FSF	$(.6)(.4)(.6) = .144$
	SFF	$(.4)(.6)(.6) = .144$
	FSS	$(.6)(.4)(.4) = .096$
2	SFS	$(.4)(.6)(.4) = .096$
	SSF	$(.4)(.4)(.6) = .096$
3	SSS	$(.4)(.4)(.4) = .064$

For convenience we have grouped the events by common values of the random variable X, the number of successes. Using our new notation $B(x; n, p)$ we have

$$B(0; 3, .4) = 1 \cdot (.216) = .216$$

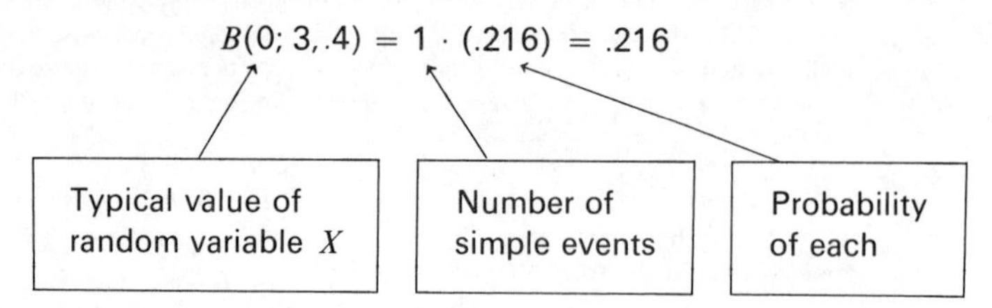

$$B(1; 3, .4) = 3 \cdot (.144) = .432$$

$$B(2; 3, .4) = 3 \cdot (.096) = .288$$

$$B(3; 3, .4) = 1 \cdot (.064) = .064$$

As a check, we add all the values of the frequency function to make sure we get 1 for a total:

$$.216 + .432 + .288 + .068 = 1.000$$

We give two graphical interpretations: the graph (consisting of four isolated points in this case) of $B(x; n, p)$ and the corresponding histogram.

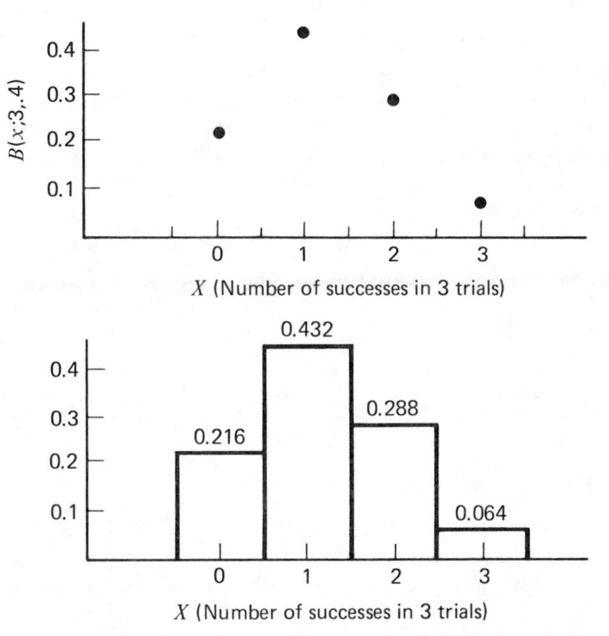

FIGURE 3.5 Frequency function graph and histogram of $B(x;\, 3,\, .4)$.

EXAMPLE 1 Tossing an Unfair Coin

Suppose that a coin is malformed so that it comes up heads on the average 4 times in 10 tosses. We then assign probability .4 to getting heads (success) on a single toss and our model with $B(x;\, 3,\, .4)$ serves to predict the likelihood of various outcomes if we toss the coin three times in succession. For instance, we would not deem it very likely to toss three consecutive heads; the probability of doing this is .064. Thus if 100 people each tossed this coin three times in succession, we might expect 6 or 7 of them to get all three heads. Other simple inferences could be drawn: The most likely event would be to get two tails (failures) and one head in three tries, the probability of this being .432. Also, getting no heads (probability $=$ .215) is nearly as likely as getting two heads (probability $=$ .288).

A second prototype of experiments that yield a random variable with a binomial distribution is *drawing with replacement*. We discussed earlier situations where an urn contains a certain number of balls of different colors. The probability of drawing a ball of a certain color, say red, is given by

$$p = P(\text{red ball}) = \frac{\text{number of red balls in the urn}}{\text{total number of balls in the urn}}$$

Moreover if after the ball is drawn and its color recorded we replace the ball and mix thoroughly the contents of the urn, then the probability of drawing a red ball on the second try is still p. Thus we encounter the notion of repeated (Bernoulli) trials in counting the number of red balls drawn.

EXAMPLE 2 Drawing with Replacement

Suppose an urn contains four red balls, five white balls, and one blue ball. An experiment consists of drawing a ball, recording its color, replacing the ball, mixing the balls, drawing again, recording color, replacing, mixing, then drawing a third time and recording the color. Since there are four red balls and 10 balls all together, the probability of drawing a red ball (this will denote a success) on any given turn is .4. We lump together as failures on a given turn drawing either a white or a blue ball. Thus $B(x; 3, .4)$ is the appropriate frequency function for the discrete random variable X equalling the number of red balls drawn in three tries. Comments concerning likelihood of various outcomes would be similar to those suggested in Example 1.

Returning now to the basic problem, we suppose that p (some real number, $0 \le p \le 1$) and n (some positive integer) are fixed. Our job is to find all the values

$$B(0; n, p), B(1; n, p), \ldots, B(n; n, p)$$

of the frequency function $B(\_; n, p)$ for the random variable X that counts the number of successes in n Bernoulli trials, where success on a single trial has probability p. Recall that by definition of frequency function

$$B(x; n, p) = P(X = x)$$

The expression on the right denotes the probability of the compound event consisting of all words (i.e., simple events)

SSFS . . .

containing exactly x S's [and hence $(n - x)$ F's].

Observation 1 The number of n-letter words containing exactly x S's and $(n - x)$ F's is

$$C(n, x)$$

since we can think of filling any x of the n possible spaces with S's. For example, when $n = 3$ we had

SSF, SFS, and FSS

which amounts to $3 = C(3, 2)$ words with 2 S's and

SSS

which amounts to $1 = C(3, 3)$ words with 3 S's.

Observation 2 The probability of any given word with x S's is

$$p^x q^{n-x}$$

That is, if SSFS . . . has x S's and $(n - x)$ F's, then

$$\begin{aligned} P(\text{SSFS}\ldots) &= P(\text{S})P(\text{S})P(\text{F})P(\text{S})\ldots \\ &= ppqp\ldots \qquad \text{[A product with } x\ p\text{'s and } (n-x)\ q\text{'s]} \\ &= p^x q^{n-x} \end{aligned}$$

The order in which the letters occur does not affect this probability since the real numbers p and q can be rearranged in the final product. Thus, for example, we saw that in the case $n = 3$,

$$P(\text{SSF}) = P(\text{SFS}) = P(\text{FSS}) = p^2q^1$$

From these two observations we conclude that

$$B(x; n, p) = \begin{pmatrix} \text{number of } n\text{-letter words} \\ \text{containing exactly } x \text{ S's} \\ \text{and } (n - x) \text{ F's} \end{pmatrix} \cdot \begin{pmatrix} \text{probability of an} \\ \text{event containing exactly} \\ x \text{ S's and } (n - x) \text{ F's} \end{pmatrix}$$
$$= C(n, x) \cdot p^x q^{n-x}$$

An important corollary can be recorded at this point. Recall that for any n and x,

$$C(n, x) = C(n, n - x)$$

[For example, $C(6, 2) = C(6, 4)$ and $C(10, 3) = C(10, 7)$.] Thus we have

$$B(x; n, p) = C(n, x)p^xq^{n-x} = C(n, n - x)q^{n-x}p^x = B(n - x; n, q)$$

Thus whenever we calculate the values

$$B(0; n, p), B(1; n, p), \ldots, B(n; n, p)$$

we automatically get, reading backwards,

$$B(0; n, q), B(1; n, q), \ldots, B(n; n, q)$$

In essence we get two binomial frequency functions for the price of one. This property will be explored in the exercises.

Just as we hastened to check that the sum of the values of the frequency function was 1 in the case $n = 3$, $p = .4$, we should confirm our formula for $B(x; n, p)$ in general by showing that

$$\sum_{x=0}^{n} B(x; n, p) = 1$$

This equation holds for any choice of parameters n and p. The standard method for establishing this fact was actually hinted at in Exercise 4 of Section 1, Chapter 2, where the algebraic expressions

$$(\text{H} + \text{T})^1, (\text{H} + \text{T})^2, (\text{H} + \text{T})^3, (\text{H} + \text{T})^4$$

were to be expanded.

Binomial Expansion Theorem

For any non-negative integer n,

$$(x + y)^n = x^n + C(n, 1)x^{n-1}y^1 + C(n, 2)x^{n-2}y^2 + \cdots + C(n, n - 1)x^1y^{n-1} + y^n$$

or, more succinctly,

$$\boxed{(x + y)^n = \sum_{k=0}^{n} C(n, k)x^{n-k}y^k}$$

Proof: Rather than give a rigorous proof (which incidentally is fairly straightforward if the reader is familiar with mathematical induction), we offer a heuristic argument. Let us examine the product

$$(x + y)^n = (x + y)(x + y)(x + y) \cdots (x + y) \qquad [n \text{ factors of } (x + y)]$$

When such an expression is expanded, the result is a sum of terms, each of which is a product of x's and y's. Then those alike terms are collected together as follows:

$$\begin{aligned}(x + y)^2 &= (x + y)(x + y) = x \cdot x + x \cdot y + y \cdot x + y \cdot y \\ &= x^2 + xy + xy + y^2 \\ &= x^2 + 2xy + y^2\end{aligned}$$

Similarly,

$$\begin{aligned}(x + y)^3 &= (x + y)(x + y)(x + y) \\ &= x(x + y)(x + y) + y(x + y)(x + y) \\ &= x[x^2 + xy + yx + y^2] + y[x^2 + xy + yx + y^2] \\ &= x^3 + x^2y + xyx + xy^2 + yx^2 + yxy + y^2x + y^3\end{aligned}$$

(At this stage, there are eight terms, each is a product of x's and y's, and each has three factors.)

$$= x^3 + 3x^2y + 3xy^2 + y^3$$

(Now the terms have been grouped together: x^2y, xyx, and yx^2, all being equal to x^2y, were lumped into $3x^2y$.)

The important thing to note is that each of the eight terms (before the final grouping) can be viewed as being comprised of one term from each of the factors $(x + y)$. For instance, if we underline as follows

$$(\underline{x} + y)(x + \underline{y})(\underline{x} + y)$$

we arrive at the term

$$xyx$$

Similarly,

$$(\underline{x} + y)(\underline{x} + y)(\underline{x} + y) \quad \text{yields} \quad x^3$$

$$(x + \underline{y})(\underline{x} + y)(\underline{x} + y) \quad \text{yields} \quad yx^2$$

and so on. In other words, each selection of one term from every factor of $(x + y)$ produces a term in the expansion. Since there are $2 \cdot 2 \cdot 2$ such choices, there are 2^3 or eight terms in the expansion. Thus the description is analogous to our analysis of how n-letter words are formed from S's and F's. For an arbitrary positive integer n it follows that

$(x + y)^n$ will have $C(n, k)$ terms of the form $x^k y^{n-k}$

After grouping we get

$$(x+y)^n = x^n + C(n, n-1)x^{n-1}y^1 + C(n, n-2)x^{n-2}y^2 + \cdots + C(n, 1)x^1y^{n-1} + y^n$$

which is precisely what we wanted to show, once we replace each $C(n, n-k)$ with $C(n, k)$, since they are equal. Note also that the coefficient of x^n, namely 1, can be viewed as $C(n, n)$ and the coefficient of y^n can be viewed as $C(n, 0)$.

One of the important corollaries of the Binomial Expansion Theorem is seen by letting $x = p$ and $y = q = 1 - p$. Then we have

$$1 = (p+q)^n = \sum_{k=0}^{n} C(n, k)p^k q^{n-k}$$

But since $B(k; n, p) = C(n, k)p^k q^{n-k}$, we have

$$1 = \sum_{k=0}^{n} B(k; n, p)$$

> The sum of the binomial frequencies is 1.

EXAMPLE 3 Comparing the Frequency Functions $B(x; 6, .4)$, $B(x; 10, .4)$ and $B(x; 20, .4)$

In time we shall be concerned with experiments involving large values of n. As one might imagine, the computations of $B(x; n, p)$ soon become unwieldy as n grows large. For instance, we have

$$\begin{aligned} B(0; 10, .4) &= C(10, 0) \cdot (.4)^0(.6)^{10} \\ &= 1 \cdot 1 \cdot (.0060466176) \\ &\approx .0060 \end{aligned}$$

To do this by hand would look something like:

$$(.6)^2 = .36$$
$$(.6)^4 = (.6)^2(.6)^2 = (.36)(.36) = .1296$$
$$(.6)^{10} = (.6)^4(.6)^4(.6)^2 = (.1296)(.1296)(.36) = .0060466176.$$

Clearly one must seek better ways to find the remaining values of $B(x; 10, .4)$. Fortunately it is not difficult to find tables for $B(x; n, p)$ for reasonably small values of n (say 1 to 20) and certain values of p (say $p = .1, .2, .25, .3, .4, .5, .6, .7, .75, .8, .9$). It is also possible to appeal to the computer for these and other cases. Moreover, as soon as n gets much larger than 20, it becomes advisable to use *approximate* models where suitable tables are readily accessible. One of these models, called the Poisson distribution, will be discussed later in this chapter. Another, a continuous model called the normal distribution, will be developed once we get acquainted with some elementary calculus.

Here we present in tabular form the three frequency functions $B(x; 6, .4)$, $B(x; 10, .4)$, and $B(x; 20, .4)$, complements of a handy table. In addition we include the companion tables which we get free by recalling that $B(x; n, p) = B(n-x; n, q)$. Values of $B(x; n, p)$, for selected n and p, are given in the Appendix in Table B.

Some Binomial Frequency Functions for $p = .4$ and $p = .6$

x	$B(x; 6, .4)$	x	$B(x; 6, .6)$
0	.0467	0	.0041
1	.1866	1	.0369
2	.3110	2	.1382
3	.2765	3	.2765
4	.1382	4	.3110
5	.0369	5	.1866
6	.0041	6	.0467

x	$B(x; 10, .4)$	x	$B(x; 10, .6)$
0	.0060	0	.0001
1	.0403	1	.0016
2	.1209	2	.0106
3	.2150	3	.0425
4	.2508	4	.1115
5	.2007	5	.2007
6	.1115	6	.2508
7	.0425	7	.2150
8	.0106	8	.1209
9	.0016	9	.0403
10	.0001	10	.0060

x	$B(x; 20, .4)$	x	$B(x; 20, .6)$
0	.0000	0	.0000
1	.0005	1	.0000
2	.0031	2	.0000
3	.0123	3	.0000
4	.0350	4	.0003
5	.0746	5	.0013
6	.1244	6	.0049
7	.1659	7	.0146
8	.1797	8	.0355
9	.1597	9	.0710
10	.1171	10	.1171
11	.0710	11	.1597
12	.0355	12	.1797
13	.0146	13	.1659
14	.0049	14	.1244
15	.0013	15	.0746
16	.0003	16	.0350
17	.0000	17	.0123
18	.0000	18	.0031
19	.0000	19	.0005
20	.0000	20	.0000

Summary

We have examined a special kind of sample space, each of whose elements represents the result of n performances (or trials) of an experiment that has exactly two outcomes, one of which is called success (S) and the other failure (F). As such, we can represent a simple event by an n-letter "word" made up of S's and F's.

We are particularly interested in the random variable X that assigns to each such word the number of S's in the word. Thus X can take on each of the integer values from 0 through n. We say that X is a *binomially* distributed random variable or that X has a *binomial* distribution. Moreover we were able to show that the frequency function for X satisfies a formula involving combinations of n objects taken x at a time. If we denote by p the probability of success on a single trial, and by $q(= 1 - p)$ the corresponding probability of failure, then the frequency function for X is denoted $B(\_; n, p)$ and it satisfies

$$B(x; n, p) = C(n, x)p^x q^{1-x}, \qquad \text{for } x = 0, 1, \ldots, n$$

Note that to single out a particular binomial frequency function $B(\_; n, p)$, we must specify values for n and p. We call n and p the *parameters* of the binomial distribution.

A prototype of the repeated trials phenomenon is the experiment of *drawing with replacement*. If an urn is filled with a mixture of red and black marbles in such a way that the ratio of red marbles to *all* marbles is p (this must be a number satisfying $0 \leq p \leq 1$), then if we draw n marbles with replacement and denote by S the selection of a red marble and by F the selection of a black marble, then the probability of drawing exactly x red marbles ($0 \leq x \leq n$) is given by

$$B(x; n, p)$$

Exercises Section 2

1 Write out the values of $B(x; n, p)$ for each of the following choices of parameters n and p. Then sketch the corresponding graphs. (The results of this exercise will be needed for other exercises.)
(a) $n = 3, p = .5$
(b) $n = 4, p = .5$
(c) $n = 5, p = .5$
(d) $n = 4, p = .2$
(e) $n = 4, p = .1$

2 Use parts (d) and (e) of Exercise 1 to sketch the graphs for $B(x; n, p)$ when:
(a) $n = 4, p = .8$
(b) $n = 4, p = .9$ [*Hint:* Recall that $B(x; n, p) = B(n - x; n, 1 - p)$.]

3 Use parts (a), (b), and (c) of Exercise 1 to find the probability:
(a) that at least one head will occur in three tosses of a fair coin.
(b) that at least one head will occur in four tosses of a fair coin.
(c) that at least one head will occur in five tosses of a fair coin.
(d) that at least two heads will occur in five tosses of a fair coin.

4 Compare the probabilities of getting:
(a) exactly one head in two tosses of a fair coin.
(b) exactly two heads in four tosses of a fair coin.
(c) exactly three heads in six tosses of a fair coin.
(d) exactly four heads in eight tosses of a fair coin.

5 How many times must a fair die be rolled in order that the probability of getting at least one "4" is greater than .5? [*Hint:* Use P(at least one "4") = 1 − P(no "4").]

6 A student has to take a true-false test with 10 questions. Suppose he guesses on every question (so that the probability of answering any given question correctly is .5). Find the probability that he gets at least five correct answers. [*Hint:* Compute $B(5; 10, .5)$ directly, then use the fact that $B(6; 10, .5) = B(4; 10, .5)$, $B(7; 10, .5) = B(3; 10, .5)$, and so on, so that

$$\sum_{x=6}^{10} B(x; 10, .5) = \sum_{x=0}^{4} B(x; 10, .5)$$

and finally

$$1 = B(5; 10, .5) + \sum_{x=6}^{10} B(x; 10, .5) + \sum_{x=0}^{4} B(x; 10, .5)]$$

7 **(a)** Reduce to lowest terms:

$$\frac{B(4; 6, .3)}{B(3; 6, .3)}$$

(b) Simplify:

$$\frac{B(x; n, p)}{B(x - 1; n, p)}$$

(c) When is the expression in (b) less than 1? equal to 1? greater than 1?

(d) Using (b) and (c), how can you tell for which value of x the quantity $B(x; n, p)$ will be maximum?

8 An important committee is being established by the students at Mudlark College. Conscious of the possibility of periodic conflicts of meeting times with other campus activities, the organizers wish to minimize the probability of not having a quorum (i.e., majority) present for a meeting. Which of the following committee sizes would be optimal? $n = 3, 5, 7$. [*Hint:* (1) Assume that the probability of conflict for each student is fixed at say $p = \frac{1}{4}$ and that whether any particular student has a conflict is independent of other conflicts; (2) see if you can do the problem with a fixed but unspecified p.]

9 An urn contains four red marbles, two blue marbles, and four white marbles. If six marbles are drawn with replacement, find the probability that:

(a) three marbles are red.

(b) at least four marbles are red.

(c) at least three marbles are either red or blue.

(d) at most two marbles are white.

10 For each of the tables for $B(x; n, p)$ given in the text, construct a new column headed

$$\sum_{k=0}^{x} B(k; n, p)$$

These values represent *cumulative* binomial probabilities.

3 EXPECTATION—THE DISCRETE CASE

As we encounter more and more examples of random variables and their associated frequency functions, it behooves us to devise methods of describing them so as to distinguish one random variable from another. In this section we borrow two descriptive tools from our earlier work with numerical data, the mean and the standard deviation, and apply these measurements to random variables. Our first example shows how the mean of a random variable arises rather naturally in the framework of a game of chance.

EXAMPLE 1 Heads I Win, Tails You Lose

Suppose you are invited to play the following game. You toss two fair coins. If two heads come up, you win 50 cents; if two tails come up, you win 25 cents;

and if one head and one tail come up, you pay your opponent 50 cents. How can you determine whether to play? More precisely, if you do play the game, how much do you expect to win or lose on the average?

We can answer this question by letting S be the sample space {HH, HT, TH, TT} of the four equally likely events corresponding to tossing two coins. We define a random variable X on S to represent the payoff to you for each event:

$$X(\text{HH}) = 50$$

$$\left.\begin{aligned} X(\text{HT}) &= -50 \\ X(\text{TH}) &= -50 \end{aligned}\right\}$$ These negative payoffs to you mean that you *pay* your opponent.

$$X(\text{TT}) = 25$$

Now imagine that on four successive tosses, each of the four equally likely events occurred. You would "win"

$$50 + (-50) + (-50) + 25 = -25 \text{ cents}$$

and your *average* winnings per play would be

$$(-25)/4 \approx -6 \text{ cents}$$

In other words, you could anticipate losing a little more than 6 cents per play. Of course, it could be argued that, in spite of the fact that the four events are equally likely, it might be unusual for each of these events to occur once in four successive plays. On the other hand, in the long run, we should expect about as many HH's as HT's as TH's as TT's, and our analysis should therefore seem reasonable.

Let us examine this average payoff more closely. We have

$$\begin{aligned} \text{Average payoff} &= \frac{(50 + (-50) + (-50) + 25)}{4} \\ &= (50)\tfrac{1}{4} + (-50)\tfrac{1}{4} + (-50)\tfrac{1}{4} + (25)\tfrac{1}{4} \\ &= (50)\tfrac{1}{4} + (-50)\tfrac{1}{2} + (25)\tfrac{1}{4} \end{aligned}$$

and each term in the sum is a product of some value of the random variable X with its frequency. Thus we may write

$$\text{Average payoff} = \sum_{i=1}^{3} x_i f(x_i)$$

where x_1, x_2, and x_3 are the values of X, and f is the associated frequency function. In general, we have the following.

Definition Let X be a discrete random variable on some sample space and let f be the associated frequency function. The *mean* of X (also called the *expected value* or the *average value* or the *expectation* of X) is denoted by μ_X [also by $E(X)$] and is defined by

$$\mu_X = \sum x_i f(x_i)$$

where the sum runs over *all* values of x_i of X.

EXAMPLE 2 Tossing a Fair Coin Three Times—The Mean

Let us compute the expected number of heads in three tosses of a fair coin. If X is the number of heads, then we know that X is binomially distributed with $n = 3$ and $p = .5$. Thus

$$\begin{aligned}\mu_X &= \sum_{x=0}^{3} xB(x; 3, .5)\\ &= 0 \cdot B(0; 3, .5) + 1 \cdot B(1; 3, .5) + 2 \cdot B(2; 3, .5) + 3 \cdot B(3; 3, .5)\\ &= 0 + 1 \cdot C(3, 1)\,(.5)^1(.5)^2 + 2 \cdot C(3, 2)\,(.5)^2(.5)^1\\ &\quad + 3 \cdot C(3, 3)\,(.5)^3(.5)^0\\ &= 0 + 1 \cdot \frac{3}{1} \cdot (.5)\,(.25) + 2 \cdot \frac{3 \cdot 2}{1 \cdot 2} \cdot (.25)\,(.5) + 3 \cdot 1 \cdot (.125)\,(1)\\ &= 0 + .375 + .750 + .375\\ &= 1.5\end{aligned}$$

Conclusion: If we toss a fair coin three times, the expected number of heads occurring is 1.5. *Note:* It is not possible to get 1.5 heads. That is, 1.5 is *not* one of the possible values of X; yet it is our expected value.

EXAMPLE 3 Rolling a Fair Die

Let X be the value showing on a single roll. To say the die is fair says that X is a *uniformly distributed discrete* random variable; that is, the values of X are equally likely:

$$f(1) = f(2) = f(3) = f(4) = f(5) = f(6) = \tfrac{1}{6}$$

So

$$\begin{aligned}E(X) = \sum_{x=1}^{6} xf(x) &= 1 \cdot \tfrac{1}{6} + 2 \cdot \tfrac{1}{6} + 3 \cdot \tfrac{1}{6} + 4 \cdot \tfrac{1}{6} + 5 \cdot \tfrac{1}{6} + 6 \cdot \tfrac{1}{6}\\ &= (1 + 2 + 3 + 4 + 5 + 6)\,(\tfrac{1}{6})\\ &= 21\,(\tfrac{1}{6})\\ &= 3.5\end{aligned}$$

Once again, 3.5 is not a possible value for X, yet it represents the *expected value* of X.

It is easy to justify calling $E(X)$ the *mean* value of X. We simply represent f by a histogram and pretend we are dealing with a set of numerical data; for instance if X is the number showing on a fair die, we have

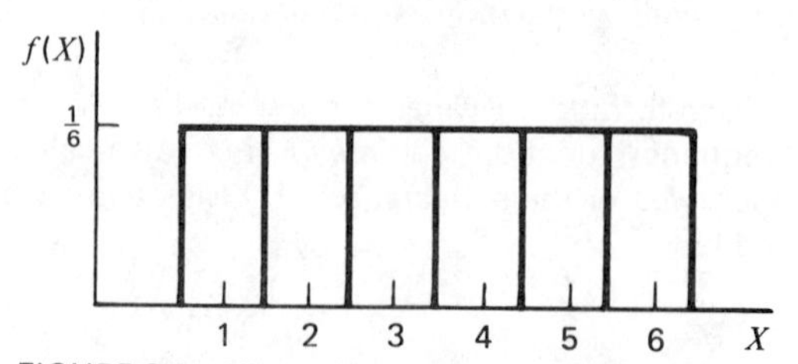

FIGURE 3.6 Histogram representation for rolling a fair die.

The mean value of the "data" associated with this histogram would be

$$\bar{x} = m_x = (\tfrac{1}{6})(1 + 2 + 3 + 4 + 5 + 6) = 3.5$$

Indeed, this theoretical mean μ_X is one characteristic of random variables that helps describe them. Another is the *variance* of a random variable. Based upon the analogy with numerical data sets, we shall define variance to correspond with the earlier formula for variance, s_x^2, of a data set

$$\{x_1, x_2, \ldots, x_n\}$$

which was given in two forms

$$s_x^2 = \frac{1}{n}\sum_{i=1}^{n}(x_i - \bar{x})^2 = \frac{1}{n}\sum_{i=1}^{n}x_i^2 - (\bar{x})^2$$

Now the factor $1/n$ can be thought of as representing "relative frequency" of the x_i's, so we could rewrite

$$s_x^2 = \sum_{i=1}^{n}\frac{1}{n}(x_i - \bar{x})^2$$

It is this form we copy:

Definition Let X be a *discrete* random variable and let f be its frequency function. The *variance* of X, denoted σ_X^2 [or sometimes var (X)] is given by

$$\sigma_X^2 = \Sigma\,(x - \mu_X)^2 f(x)$$

where the sum is taken over *all* values x of X. The positive square root σ_X of the variance is called the *standard deviation* of X.

Once again we have an alternate formula for variance that can be derived from the one given here; namely,

$$\sigma_X^2 = \Sigma\, x^2 f(x) - (\Sigma\, x f(x))^2$$

or simply

$$\sigma_X^2 = \Sigma\, x^2 f(x) - (E(X))^2$$

EXAMPLE 4 Tossing a Fair Coin Three Times—The Variance

We compute the variance for the random variable X that counts the number of heads in three tosses of a fair coin. From Example 3 we have

$$E(X) = \mu_X = 1.5$$

Thus

$$\begin{aligned}\sigma_X^2 &= \Sigma\,(x - \mu_X)^2 f(x)\\ &= (0 - 1.5)^2 B(0; 3, .5) + (1 - 1.5)^2 B(1; 3, .5)\\ &\quad + (2 - 1.5)^2 B(2; 3, .5) + (3 - 1.5)^2 B(3; 3, .5)\\ &= (2.25)\cdot(.125) + (.25)\cdot(.375) + (.25)(\_\_\_) + (\_\_\_)(\_\_\_)\\ &= .75\end{aligned}$$

The standard deviation σ_X is therefore $\approx .866$.

Note that we could also compute σ_X^2 by the alternate formula, using $\Sigma\, x^2 f(x)$. In this case we have

$$\begin{aligned}\Sigma\, x^2 f(x) &= 0^2 \cdot (.125) + 1^2(.375) + 2^2(.375) + 3^2(.125)\\ &= 0 + .375 + 1.50 + 1.125\\ &= 3.00\end{aligned}$$

so that

$$\begin{aligned}\sigma_X^2 &= \Sigma\, x^2 f(x) - (\mu_X)^2\\ &= 3.00 - 2.25\\ &= .75 \qquad \text{(as we computed earlier)}\end{aligned}$$

Now we tackle the problem of finding both the mean and the variance for *any* binomially distributed random variable. To that end, let n and p be given, and let X be a random variable whose frequency function is

$$f(x) = B(x; n, p)$$

We wish to find formulas for μ_X and σ_X^2.

The first step in our argument is to guess at μ_X. Using the histogram analogy and Example 1, it shouldn't be difficult to convince yourself that

$$\mu_X \quad \text{should be } n \cdot p$$

Why? Well, p is the probability of success on a single trial; therefore it should represent the percentage of successes in the long run. In n trials we should expect to get n times that percentage of total successes.

Now let us precede the general algebraic argument with the special case: $n = 3$, p arbitrary. By definition,

$$\begin{aligned}\mu_X &= \sum_{x=0}^{3} x \cdot B(x; 3, p)\\ &= \Sigma\, x \cdot C(3, x) p^x q^{3-x}\\ &= 0 \cdot C(3,0)p^0q^3 + 1 \cdot C(3,1)p^1q^2 + 2 \cdot C(3,2)p^2q^1 + 3 \cdot C(3,3)p^3q^0\\ &\overset{(1)}{=} 0 + 1 \cdot \frac{3 \cdot 2 \cdot 1}{1 \cdot 2 \cdot 1}p^1q^2 + 2 \cdot \frac{3 \cdot 2 \cdot 1}{2 \cdot 1 \cdot 1}p^2q^1 + 3 \cdot \frac{3 \cdot 2 \cdot 1}{3 \cdot 2 \cdot 1 \cdot 1}p^3q^0\\ &\overset{(2)}{=} (3p)\left(1 \cdot \frac{2 \cdot 1}{1 \cdot 2 \cdot 1}p^0q^2 + \frac{2 \cdot 1}{1 \cdot 1}p^1q^1 + \frac{2 \cdot 1}{2 \cdot 1 \cdot 1}p^2q^0\right)\\ &= (3p)\left(\frac{2!}{0!2!}p^0q^2 + \frac{2!}{1!1!}p^1q^1 + \frac{2!}{2!0!}p^2q^0\right)\\ &\overset{(3)}{=} (3p) \sum_{x=0}^{2} C(2, x)p^xq^{2-x}\\ &= (3p) \cdot 1 = 3p\end{aligned}$$

The important steps to note are:

1. The first summand is 0 since 0 is a factor of it.
2. np [in this case $(3p)$] is factored out of each remaining term.
3. Once $(3p)$ is factored out, what's left is simply the sum of the binomial frequencies for $n-1$ and p, i.e.,

$$\sum_{x=0}^{n-1} B(x; n-1, p)$$

and this equals 1.

When n and p are arbitrary, we utilize the following factorization of the term $xB(x; n, p)$ when $x \neq 0$:

$$\begin{aligned} xB(x; n, p) &= x\frac{n!}{(n-x)!x!}p^x q^{n-x} \\ &= (np)\frac{(n-1)!}{(n-x)!(x-1)!}p^{x-1}q^{n-x} \\ &= \underbrace{(np)\frac{(n-1)!}{(n-x)!(x-1)!}p^{x-1}q^{n-1-(x-1)}}_{B(x-1;\, n-1,\, p)} \end{aligned}$$

Thus for $x = 1, 2, 3, \ldots, n$ we get from

$$1 \cdot B(1; n, p) + 2 \cdot B(2; n, p) + \cdots + n \cdot B(n; n, p)$$

the n new terms

$$(np)B(0; n-1, p) + (np)B(1; n-1, p) + \cdots + (np)B(n-1; n-1, p)$$

and factoring out the (np) we have left

$\sum_{x=0}^{n-1} B(x; n-1, p)$ which equals 1. Thus

$$\mu_X = \sum_{x=0}^{n} xB(x; n, p) = np.$$

The mean of any binomially distributed variable is np.

The formula for σ_X^2 can be deduced by repeated use of this factorization technique. From the definition we have

$$\sigma_X^2 = \Sigma\, x^2 B(x; n, p) - (\Sigma\, xB(x; n, p))^2$$

Since we know the last term is $(np)^2$, we concentrate on the sum $\Sigma\, x^2B(x; n, p)$. In particular, for all but the case $x = 0$, we have

$$\begin{aligned} x^2B(x; n, p) &= x \cdot xB(x; n, p) \\ &= x(np)B(x-1; n-1, p) \\ &= (np)xB(x-1; n-1, p) \end{aligned}$$

Now we use a gimmick! We write the factor x as $(x - 1) + 1$. This produces a term $(x-1)B(x-1; n-1, p)$ which can itself be factored (pretending $x - 1$ is x and $n - 1$ is n in our earlier argument):

$$\begin{aligned} (np)xB(x-1; n-1, p) &= (np)[(x-1)+1]B(x-1; n-1, p) \\ &= (np)(x-1)B(x-1; n-1, p) + (np)B(x-1; n-1, p) \\ &= (np)((n-1)p)B(x-2; n-2, p) + (np)B(x-1; n-1, p) \end{aligned}$$

Finally we note that when $x = 1$, the term involving $(x - 1)$ is 0. It follows that the sum

$$E(X^2) = \Sigma\, x^2B(x; n, p) = 0 + 1^2B(1; n, p) + 2^2B(2; n, p) + \cdots + n^2B(n; n, p)$$

can be rewritten:

$$\begin{aligned} &0 + [0 + (np)B(0; n-1, p)] + [(np)(n-1)pB(0; n-2, p) + (np)B(1; n-1, p)] + \cdots \\ &\qquad + [(np)(n-1)pB(n-2; n-2, p) + (np)B(n-1; n-1, p)] \end{aligned}$$

Collecting together terms with common factors, we get

$$\begin{aligned} E(X^2) &= (np)(n-1)p \sum_{x=0}^{n-2} B(x; n-2, p) + (np) \sum_{x=0}^{n-1} B(x; n-1, p) \\ &= np(n-1)p + np \\ &= n^2p^2 - np^2 + np \end{aligned}$$

Finally

$$\begin{aligned} \sigma_X^2 &= E(X^2) - (E(X))^2 \\ &= (n^2p^2 - np^2 + np) - n^2p^2 \\ &= np - np^2 = np(1-p) = npq \end{aligned}$$

The variance of any binomially distributed variable is npq.

The variance of a random variable X is a measure of dispersion of that variable. It tells us something about the spread of values of X away from its mean. In fact, the standard deviation of X serves as a unit measure; one speaks of values of X that lie two standard deviations to the right (i.e., above) the mean. Later on we shall study a continuous distribution called the normal distribution for which the standard deviation is a very convenient indicator: If a variable is normally distributed, then an arbitrarily chosen value of X falls within one standard deviation of the mean about 68% of the time, within two standard deviations about 95% of the time, and within three standard deviations about 99% of the time. Unfortunately, such sweeping statements do not apply to *all* random variables. Instead one must develop a feel for the standard deviation and interpret it in terms of the particular distribution under consideration.

The reader has been asked to accept the analogy between data sets and values of a discrete random variable as sufficient motivation for our definition of mean and variance of random variables. Moreover, the two alternative definitions of variance have been introduced as before, and in Example 4 both forms were used to get the same result. In drawing this analogy, however, we have managed to neglect an important concept concerning random variables, which we shall now explore.

One formula for variance involves the expression

$$\Sigma x^2 f(x)$$

where the sum runs over all possible values x of some random variable X. While it may not be apparent, this expression is in its own right a *mean value* of *some* random variable, not X but a related random variable called X^2, which is defined by

$$X^2(A) = [X(A)]^2, \qquad \text{for any event } A$$

In other words, *given any random variable* X, we can define a new random variable X^2 by the above equality.

More generally we can construct innumerable new random variables from a given X. For instance if X is the variable defined by

$$X(A) = \text{number of heads in } A$$

where A is any event in the sample space

$$S = \{\text{HH, HT, TH, TT}\}$$

then we could define new random variables

$$X^2, X^3, 3X, (-2)X, X + 3, 2X + 5, (X - 1)^2$$

in a natural way as indicated in the chart that follows:

Event (A)	X	X^2	X^3	$3X$	$(-2)X$	$X+3$	$2X+5$	$(X-1)^2$
HH	2	4	8	6	−4	5	9	1
HT	1	1	1	3	−2	4	7	0
TH	1	1	1	3	−2	4	7	0
TT	0	0	0	0	0	3	5	1

In general then, we could define a new random variable

$$Y = g(X)$$

by the rule

When X has value x, Y has value $g(x)$.

An important feature of this definition is that:

The probability that $g(X)$ takes on the value $g(x)$ is precisely the probability that X takes on the value x, namely $f(x)$.

Hence to find the *expected value* of the new random variable $g(X)$, we have

$$E(g(X)) = \Sigma\, g(x)\, f(x)$$

The value of $g(X)$ when $X = x$	The probability that $X = x$

where the sum runs over all possible values x of X.

In particular, the expression

$$\Sigma\, x^2 f(x)$$

is simply the *expectation* of the random variable X^2. Returning to our two definitions of variance, we can write

$$\sigma_X^2 = E((X - \mu)^2) = \Sigma\,(x - \mu)^2 f(x)$$

or

$$\sigma_X^2 = E(X^2) - E(X)^2 = \Sigma\, x^2 f(x) - (\Sigma\, x f(x))^2$$

EXAMPLE 5 Find the mean and the variance for the random variable X with frequency function f given by

X	−1	0	1	2	3	4
$f(x)$	.1	.1	.2	.3	.2	.1

We compute the mean or expected value of X by

$$\begin{aligned} E(X) &= (-1)(.1) + (0)(.1) + (1)(.2) + (2)(.3) + (3)(.2) + (4)(.1) \\ &= 1.7 \end{aligned}$$

And we compute the variance of X by

$$\begin{aligned} \sigma_X^2 &= E(X^2) - E(X)^2 \\ &= \Sigma\, x^2 f(x) - \mu^2 \\ &= (-1)^2(.1) + (0)^2(.1) + (\_)^2(\_) + (\_)^2(\_) \\ &\quad + (\_)^2(\_) + (\_)^2(\_) - (1.7)^2 \\ &= 4.90 - 2.89 \\ &= 2.01 \end{aligned}$$

Note: In the formula just used for variance, there is a subtle notational distinction between

$E(X^2)$ and $E(X)^2$

the former being $\Sigma x^2 f(x)$ and the latter being $(\Sigma xf(x))^2$.

This symbol E can be viewed as a function that assigns to each random variable X some real number. As such, E has two important properties which together say that *E is linear*:

1. $E(cX) = c \cdot E(X)$
2. $E(X + Y) = E(X) + E(Y)$

These equations hold true for any constant c and for any two random variables X and Y.

Summary

In Chapter 1, we introduced the notions of mean and standard deviation to aid us in comparing different sets of data. The mean of a set of numbers gave us a *measure of central tendency*, a sort of typical or representative number for the set, while the standard deviation gave us a *measure of dispersion*, i.e., it reflected the extent to which the data was spread out from its mean.

In the present section we have drawn an analogy between data sets and the set of values of a discrete random variable X with frequency function f. We defined the *mean* (or *expected* or *average*) *value* of X, denoted $E(X)$, by

$$E(X) = \Sigma xf(x)$$

where the sum runs over all possible values x of X. We think of $E(X)$ as being, once again, a measure of central tendency or a representative value of X [although it must be borne in mind that $E(X)$ may not be one of the possible values of X]. Likewise we defined the *variance* of X and the *standard deviation* of X by

$$\text{Variance of } X = \text{var}(X) = E(X^2) - (E(X))^2$$

$$\text{Standard deviation of } X = \sigma_X = \sqrt{\text{var}(X)}$$

In the formula for variance, we see the expression $E(X^2)$, which is defined by

$$E(X^2) = \Sigma x^2 f(x)$$

More generally, we extended the original definition of $E(X)$ to include $E(g(X))$, where g is a function of X:

$$E(g(X)) = \Sigma g(x)f(x)$$

We applied our definitions of mean and standard deviation to the special case where X is a binomially distributed random variable, and we found: If X has frequency function $B(\_\_; n, p)$, then

$$\mu_X = np$$

and

$$\sigma_X = \sqrt{npq}$$

We also observed that the expectation function [obtained when we think of E as a "function" defined on random variables and producing "values" of the form $E(X)$] satisfies the following important properties:

$$E(X + Y) = E(X) + E(Y)$$

and

$$E(aX) = aE(X)$$

Exercises Section 3

1 A random variable X takes on the values 1, 2, 3, 4, 5 with respective probabilities .18, .25, .25, .20, .12. Find the mean and the variance of X.

2 For the variable X in Exercise 1, find:
(a) $E(2X)$
(b) $E(X + 3)$
(c) $E(2X + 1)$
(d) $E(1)$
(e) $E(X^2)$

3 Compare the values of μ and σ for $B(x; n, p)$ when
(a) $p = \frac{1}{2}$; $n =$ 4, 8, 16, 32, 64
(b) $p = \frac{1}{3}$; $n =$ 9, 18, 27, 36, 54, 72

4 A man has two fair coins and two trick coins (each with both sides heads).
(a) Suppose he flips one fair and one trick coin. Find the expected number of heads.
(b) Suppose he picks one coin randomly and flips it twice. Find the expected number of heads.
(c) Suppose he picks two coins randomly and flips them. Find the expected number of heads.
(d) Suppose he picks a coin randomly, flips it, and it lands heads. What is the probability it was one of the trick coins?
(e) If he flips two coins and the first lands tails, what is the probability the second will land heads?

5 An urn contains four balls labeled "1" and six balls labeled "3."
(a) One ball is drawn. Let X_1 be the number on the ball and compute $E(X_1)$ and $\sigma_{X_1}^2$.
(b) Two balls are drawn with replacement. Let X_2 be the average of the two numbers drawn, and compute $E(X_2)$ and $\sigma_{X_2}^2$.
(c) Three balls are drawn with replacement. Let X_3 be the average of the three numbers drawn and compute $E(X_3)$ and $\sigma_{X_3}^2$.
(d) Four balls are drawn with replacement. Let X_4 be the average of the four numbers drawn and compute $E(X_4)$ and $\sigma_{X_4}^2$.

6 Suppose 60% of the voters plan to vote "yes" on a particular referendum. A voter is selected at random and a random variable X is given value 0 if the voter says he will vote "no" and 1 if he will vote "yes." Find μ_X and σ_X^2.

7 In the referendum of the previous problem, suppose 10 voters are selected at random and X is the number who say they will vote "yes."

(a) Find μ_X and σ_X^2.
(b) Find $P(X \leq 5)$, $P(X \leq 4)$, and $P(X \leq 3)$.
(c) Find $P(|X - \mu_X| > \sigma_X)$.
(d) Find $P(|X - \mu_X| > 2\sigma_X)$.
(e) Find $P(|X - \mu_X| > 3\sigma_X)$.

8 If X is a random variable with mean 10 and standard deviation 5, find

(a) $E(X^2)$
(b) $E(3X)$
(c) $E(X + 4)$
(d) $E\left(\dfrac{X - 10}{5}\right)$

9 Suppose X is a random variable with mean μ and standard deviation σ. Define a new random variable Y by

$$Y = \frac{X - \mu}{\sigma}$$

Find the mean and the standard deviation of Y.

10 (Roulette.) A standard roulette wheel contains 38 numbers (0, 00, 1, 2, 3, . . . , 36). For each dollar bet, the player wins \$35.00 if his number comes up and loses \$1.00 if another number comes up.

(a) Find the expected value of the random variable X representing the winnings (or losses) corresponding to betting on one number, say 19.
(b) If a player bets on 1,000 consecutive spins, how much money can he expect to win or lose?

11 (Chuck-a-Luck.) Three dice are rolled simultaneously. Prior to the roll, a player places a bet of one dime (or one dollar, etc.) on one of the numbers 1, 2, 3, 4, 5, or 6. If none of the three dice shows his number, he loses his dime; if exactly one die shows his number, he wins one dime; if exactly two dice show his number, he wins two dimes; and if all three dice show his number, he wins three dimes. Let X be the number of dice that show his number.

(a) Find the frequency function for X.
(b) Find the expected value of X.
(c) Determine the average winnings (or losses) per play the player is faced with in this game.

12 By testing various values of p between 0 and 1, guess where $\sigma = npq$ will be largest for a fixed value of n. Can you prove your conjecture?

13 (A glimpse of an infinite series.) Consider the experiment of tossing a fair coin until it comes up heads the first time. Recall the model involved the random variable X with values

$$X = 1, 2, 3, 4, 5, \ldots$$

where $f(1) = \frac{1}{2}, f(2) = \frac{1}{4}, f(3) = \frac{1}{8}$, and in general $f(n) = (\frac{1}{2})^n$. Find $E(X)$. (*Hint:* $E(X) = \Sigma\, xf(x) = 1 \cdot \frac{1}{2} + 2 \cdot \frac{1}{4} + 3 \cdot \frac{1}{8} + 4 \cdot \frac{1}{16} + 5 \cdot \frac{1}{32} + \cdots$).

Let

$$S_1 = 1 \cdot \tfrac{1}{2}$$
$$S_2 = 1 \cdot \tfrac{1}{2} + 2 \cdot \tfrac{1}{4}$$
$$S_3 = 1 \cdot \tfrac{1}{2} + 2 \cdot \tfrac{1}{4} + 3 \cdot \tfrac{1}{8}$$

and so on. First guess at the answer by computing the first five or six terms. Then write a computer program that prints out S_n for $n = 1, 2, \ldots, 20$ and guess again.

4 CHEBYCHEV'S THEOREM AND THE LAW OF LARGE NUMBERS

We have mentioned both in the context of analyzing data and in the context of discrete random variables that the variance and the standard deviation are measures of dispersion. It has also been stated that in order to interpret these measures, the precise data set or the specific random variable must be known. In this section we consider a theorem that tells us how to use the standard deviation in a very general situation. Because of this generality, we must bear in mind that the result, when applied to a particular situation, may provide us with very crude information that can be improved upon by taking advantage of further properties of the random variable under consideration. We shall also use the theorem to establish an important result about binomially distributed random variables, a result that helps justify an assumption implicit in our discussion of repeated trials.

To set the stage for our theorem, we consider any discrete random variable X with mean μ and standard deviation σ. One way to measure the distance between μ and x (a typical value of X) is in terms of σ. More precisely, to say that

$$|x - \mu| \geq 3\sigma$$

simply means that x is at least 3σ units away from μ [either to the left of μ or to the right of μ (see Fig. 3.7)].

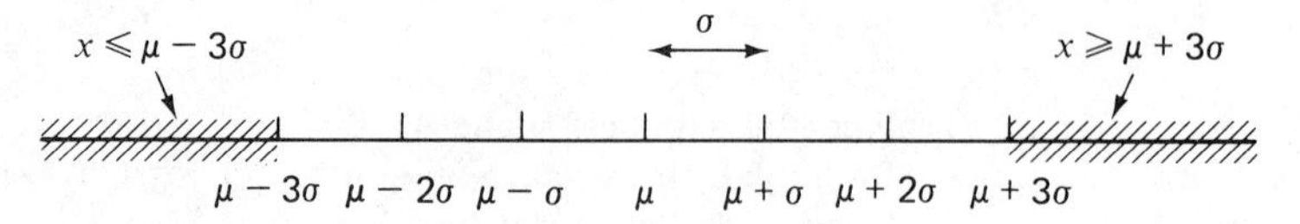

FIGURE 3.7 Measuring distance from μ in terms of σ.

Chebychev's Theorem

In the above context, if k is any positive real number, the probability that X takes on a value at least $k\sigma$ units away from μ is less than or equal to $1/k^2$.

More succinctly, Chebychev's Theorem says that

$$P(|X - \mu| \geq k\sigma) \leq \frac{1}{k^2}, \qquad \text{for any } k > 0$$

In particular, this result tells us that for any discrete random variable X with mean μ and standard deviation σ,

$$P(|X - \mu| \geq 2\sigma) \leq \tfrac{1}{4}$$
$$P(|X - \mu| \geq 3\sigma) \leq \tfrac{1}{9}$$
$$P(|X - \mu| \geq \tfrac{5}{2}\sigma) \leq \tfrac{4}{25}$$

and so on.

EXAMPLE 1 A fair coin is tossed 100 times. How likely is it that 80 heads occur?

This question as put is poorly phrased. The simple answer is $B(80; 100, .5) =$ ____. Although this number is small, it should be borne in mind that the probability of getting any particular number of heads in 100 tosses is relatively small. Indeed, the chance of getting exactly 50 heads is $B(50; 100, .5) =$ ____. However, questions of this sort are commonly asked and part of what this example illustrates is how to phrase the question properly. Rather than ask for the probability of *exactly* 80 heads in 100 tosses of a fair coin, we might ask one of the following:

1. What is the probability of getting *at least* 80 heads in 100 tosses of a fair coin?
2. What is the probability that the number of heads in 100 tosses differs from the mean (50) by as much as 30?

We answer question 2 first, by letting X be the random variable that denotes the number of heads in 100 tosses of a fair coin. Then X has frequency function $B(\_; 100, .5)$ and hence has mean and standard deviation given by

$$\mu = n \cdot p = 100 \cdot (.5) = 50$$
$$\sigma = \sqrt{npq} = \sqrt{(100)(.5)(.5)} = \sqrt{25} = 5$$

From Chebychev's Theorem we can compute

$$\begin{aligned} &P(X \text{ differs from 50 by as much as 30}) \\ &= P(|X - 50| \geq 30) \\ &= P(|X - 50| \geq 6\sigma) \leq (\tfrac{1}{6})^2 \approx .028 \end{aligned}$$

To answer question 1, we take advantage of the *symmetry* of the frequency function $B(\_; 100, .5)$, i.e.,

$$B(x; 100, .5) = B(100 - x; 100, .5)$$

so that we can split up the event

"X differs from 50 by at least 30"

into 2 *disjoint* events

A: X is less than or equal to 20
B: X is greater than or equal to 80

The symmetry tells us that $P(A) = P(B)$. Since

$$P(A \cup B) = P(A) + P(B)$$

and

$$P(A \cup B) \leq .028$$

we have

$$P(A) \leq .014 \quad \text{and} \quad P(B) \leq .014$$

To return to the original problem, we can assert that the probability of getting as many as 80 heads in 100 tosses is *at most .014*. (*Note:* The reader was warned that in practice Chebychev's Theorem can give very crude results. Here is a case in point. By checking the table for $B(x; 100, .5)$, we can find

$$P(\text{at least 80 heads in 100 tosses of a fair coin})$$
$$= \sum_{x=80}^{100} B(x; 100, .5)$$
$$= \underline{\qquad\qquad}$$

a value much smaller than the *estimate* of .014 found by the theorem. We repeat: In many situations where the frequency function of X is known, direct calculation of probability estimates of the distance away from the mean will usually give sharper results than those obtained from Chebychev's Theorem.) One of the nice features of this theorem is that its proof is "elementary." That is, in spite of its generality, we can prove the result without appealing to any high-powered mathematics. The tools used in the argument that follows are few: the definition of mean and standard deviation; the distributive law for finite summations ($\Sigma\, af(x) = a\, \Sigma\, f(x)$); and the trick of breaking up a sum into two pieces depending on the values of the summands (for example,

$$\sum_{k=1}^{n} k = \sum_{k\,\text{odd}} k + \sum_{k\,\text{even}} k).$$

Proof of Chebychev's Theorem:

We have

$$\sigma^2 = \Sigma\, (x - \mu)^2 f(x), \qquad \text{by definition}$$

and we split up the sum into the sum over all terms where $|x - \mu| < k\sigma$ plus the sum over all terms where $|x - \mu| \geq k\sigma$. Thus

$$\sigma^2 = \sum_{|x-\mu|<k\sigma} (x - \mu)^2 f(x) + \sum_{|x-\mu|\geq k\sigma} (x - \mu)^2 f(x)$$

Since all the summands are positive, we have

$$\sigma^2 \geq \sum_{|x-\mu|\geq k\sigma} (x - \mu)^2 f(x)$$

But replacing each $(x - \mu)^2 f(x)$ by $(k\sigma)^2 f(x)$, we get

$$\sigma^2 \geq \Sigma\, (x - \mu)^2 f(x) \geq \Sigma\, (k\sigma)^2 f(x) = (k\sigma)^2\, \Sigma\, f(x)$$

Finally dividing both sides of this inequality

$$\sigma^2 \geq k^2\sigma^2 \Sigma f(x)$$

by $k^2\sigma^2$ yields

$$\sum_{|x-\mu| \geq k\sigma} f(x) \leq \frac{1}{k^2}$$

But this sum is taken over all x for which $|x - \mu| \geq k\sigma$. Thus

$$P(|X - \mu| \geq k\sigma) = \sum_{|x-\mu| \geq k\sigma} f(x) \leq \frac{1}{k^2}$$

and the proof is complete.

Before continuing to a major application, we consider one example that shows how a random variable *can* be such that Chebychev's Theorem gives the best possible estimate for the probability of values away from the mean.

EXAMPLE 2 A Distribution that Shows Chebychev's Theorem is the Best Possible One

We consider the random variable X with frequency function f given by

X takes on values $-2, 0, 2$

$f(-2) = f(2) = \frac{1}{8}$

$f(0) = \frac{3}{4}$

We have $\mu = 0$ (by symmetry or by direct calculation) and

$$\sigma^2 = \tfrac{1}{8} \cdot 2^2 + \tfrac{1}{8}(-2)^2 = 1 \quad \text{so that} \quad \sigma = 1$$

Choosing $k = 2$ in the theorem, we have

$$P(|X - \mu| \geq 2 \cdot 1) \leq (\tfrac{1}{2})^2 = \tfrac{1}{4}$$

But

$$\begin{aligned} P(|X - \mu| \geq 2) &= P((X = 2) \quad \text{or} \quad (X = -2)) \\ &= \tfrac{1}{8} + \tfrac{1}{8} \\ &= \tfrac{1}{4} \end{aligned}$$

Thus the value produced by Chebychev's Theorem is the exact value.

To see the power of Chebychev's Theorem as a theoretical tool, let us consider a very basic question, namely, what do we really mean when we say that the probability of getting heads on one toss of a coin is $\frac{1}{3}$? Certainly if we toss that coin once, we get either heads or tails, and to postulate that $P(H) = \frac{1}{3}$ would seem to say that tails is *more likely* to occur. To really try to answer the question could lead us, as it has many others, into a complex philosophical discussion. Yet on one level most of us would agree that in a large number of tosses, roughly $\frac{1}{3}$ of the trials would result in heads and roughly $\frac{2}{3}$ of the tosses should result in tails. Indeed, this position is taken as the so-called *relative frequency definition* of probability. With this interpretation in mind, let us ask a related question. Suppose we build a probability model of repeated trials for flipping this coin several times. In what sense will this model bear out our suspicion that the ratio of successes (i.e., heads) to trials should be roughly $\frac{1}{3}$? Here is a question we can answer in a precise form called the *Law of Large Numbers*. To do so requires that we first set the stage with some careful definitions and then compute some expected values.

Let X be a random variable with frequency function $B(\_\_; n, p)$, and let Y be the random variable defined by

$$Y = \frac{1}{n} X$$

That is, when X takes on value x with frequency $f(x)$, Y takes on value $(1/n)x$ with frequency $f(x)$. Hence Y is simply the ratio of successes to trials. As we observed in our discussion of expectation, we have

$$E(X) = np$$
$$E(X^2) = npq + n^2p^2$$

Hence

$$E(Y) = E\left(\frac{1}{n} X\right) = \frac{1}{n} E(X) = \frac{1}{n}(np) = p$$

and

$$E(Y^2) = E\left(\frac{1}{n^2} X^2\right) = \frac{1}{n^2} E(X^2) = \frac{1}{n^2}[npq + n^2p^2]$$
$$= \frac{pq}{n} + p^2$$

Thus we can compute the variance of Y:

$$\sigma_Y^2 = E(Y^2) - E(Y)^2$$
$$= \frac{pq}{n} + p^2 - p^2$$
$$= \frac{pq}{n}$$

Now let us examine the probability that Y, the ratio of successes to trials, differs from p, its mean, by as much as some small number c; that is,

$$P(|Y - p| \geq c)$$

Notice that to say

$$|Y - p| \geq c$$

That is, when X takes on value x with frequency $f(x)$, Y takes on value $(1/n)x$ is equivalent to saying that

$$|Y - p| \geq \frac{c}{\sigma_Y}\sigma_Y$$

But by Chebychev's Theorem

$$P\left(|Y - p| \geq \frac{c}{\sigma_Y}\sigma_Y\right) \leq \frac{1}{(c/\sigma_Y)^2} = \frac{\sigma_Y^2}{c^2}$$

and

$$\frac{\sigma_Y^2}{c^2} = \frac{pq}{nc^2}$$

Thus

$$(*) \quad P(|Y - p| \geq c) \leq \frac{pq}{nc^2}$$

and the expression on the right can be made as small as desirable, for fixed values of p, q, and c, simply by letting n get large. In other words, we have proved the following.

Theorem (Law of Large Numbers)

Let p be the probability of success on a single trial. Let c be any positive number (no matter how small). By choosing n sufficiently large, the probability that the ratio of successes to trials differs from p by as much as c,

$$P\left(\left|\frac{X}{n} - p\right| \geq c\right)$$

can be made as small as desirable.

EXAMPLE 3 To get some feeling for how large n must be in a particular situation, let us suppose that $p = q = \frac{1}{2}$, and that we wish to make small the probability

$$P\left(\left|\frac{X}{n} - p\right| \geq .01\right)$$

that the ratio of successes to trials differs from p by as much as .01. From (*) we have

$$\begin{aligned} P\left(\left|\frac{X}{n} - \frac{1}{2}\right| \geq .01\right) &\leq \frac{(\frac{1}{2})(\frac{1}{2})}{n(.01)^2} \\ &= \frac{1}{4 \cdot n \cdot (.0001)} \\ &= \frac{10{,}000}{4n} \\ &= \frac{2{,}500}{n} \end{aligned}$$

Thus, if we take

$$n = 25{,}000$$

we make the probability less than or equal to $\frac{1}{10}$ and if we take

$$n = 250{,}000$$

we guarantee that the probability in question does not exceed $\frac{1}{100}$. To put it another way, the Law of Large Numbers tells us that in 250,000 flips of a fair coin, we can expect the ratio of heads to tosses to be within the interval .49 to .51 at least 99 times out of 100, and that ratio should be in the same interval at least 9 times out of 10 if we only flip the coin 25,000 times.

Warning: The aforementioned weakness of Chebychev's Theorem, namely that it generally gives a very crude estimate, carries over to our formulation of the Law of Large Numbers. That is, the inequality (*) can be expected to provide only a very conservative estimate of the true probability that the ratio of successes to trials differs from p by the prescribed amount c or more. Exercise 6 addresses itself to this issue.

Finally we point out that the probability estimate in Chebychev's Theorem can be restated in a form that will be useful to us in Chapter 5. We note that

$$|x - \mu| \geq k\sigma$$

means the same as

$$\left|\frac{x - \mu}{\sigma}\right| \geq k$$

Thus we have

$$P\left(\left|\frac{X - \mu}{\sigma}\right| \geq k\right) \leq \frac{1}{k^2}$$

In this latter form, we can think of a new random variable:

$$Z = \frac{X - \mu}{\sigma}$$

which has two nice features, namely

$$\mu_Z = 0 \quad \text{and} \quad \sigma_Z^2 = 1$$

The conversion X to Z is referred to as *converting to standard units* (the values z of Z), and this process will be explored in detail when X is a continuous variable that satisfies the normal distribution.

Summary

For a discrete random variable X with mean μ and standard deviation σ, *Chebychev's Theorem* asserts that the probability is at most $1/k^2$ that X will take on a value as great as $\mu + k\sigma$ or as small as $\mu - k\sigma$ (where k is any positive number). While this result holds in all generality and while it is relatively simple to prove, as a practical matter the estimates it yields are usually crude compared to the true probability values.

On the other hand, Chebychev's Theorem is a useful theoretical tool as can be seen by its corollary, the *Law of Large Numbers*. This law asserts that the relative frequency of successes in repeated trials affords a good approximation of the probability p of success on a single trial. More precisely, we have

$$P\left(\left|\frac{X}{n} - p\right| \geq c\right) \leq \frac{p(1-p)}{nc^2}$$

which tells us that by increasing n we can make as small as we like the probability that the relative frequency differs from p by some fixed number c.

Exercises Section 4

1 A student guesses at each of 100 questions on a true-false exam. Compare the estimate from Chebychev's Theorem to the actual probability that he will get:

(a) at least 60 or at most 40 correct answers.
(b) at least 70 correct answers.
(c) at least 80 correct answers.

2 In each of the distributions $B(x; 6, .4)$, $B(x; 10, .4)$, and $B(x; 20, .4)$:

(a) find μ and σ.
(b) find $P(|X - \mu| \geq 2\sigma)$ and $P(|X - \mu| \geq 3\sigma)$ and comment on the applicability of Chebychev's Theorem.

3 Explain the following inequality.

$$P(|X - \mu| < k\sigma) \geq 1 - \frac{1}{k^2}$$

4 Let X be the number showing in one roll of a fair die.

(a) Compute μ_X and σ_X.
(b) Find the estimate from Chebychev's Theorem for the probability that X differs from μ by as much as 2σ.
(c) Find the exact probability that X differs from μ by as much as 2σ.

5 If the probability that a certain thumb tack lands point up (success) is .3, how many times must it be dropped to guarantee that the probability that the ratio of successes to trials differs from .3 by at least .05 is no more than .01?

6 **(a)** Use equation (*) to find an estimate for

$$P\left(\left|\frac{X}{100} - .5\right| \geq .1\right)$$

where X represents the number of heads in 100 tosses of a fair coin.

(b) Compute the exact value of the above probability.

7 Use the properties of expectation to show that the random variable defined by

$$Z = \frac{X - \mu}{\sigma}, \qquad \text{where } \mu = \mu_X \text{ and } \sigma = \sigma_X$$

has mean 0 and standard deviation 1.

8 (Buffon needle problem) Several parallel lines, 10 inches apart, are drawn on a floor and a 5-inch needle is dropped. "Success" means that the needle comes to rest crossing a line. It can be shown that $p = P(S) = (1/\pi)$ (where $\pi \approx 3.1416$) if the needle is dropped "randomly." State carefully how you could use equation (*) to get an estimate for $1/\pi$ that is likely to be within .01 of the true value, provided you performed the experiment a sufficient number of times.

5 HYPOTHESIS TESTING WITH THE BINOMIAL DISTRIBUTION

We are now prepared to look at one of the basic techniques in applied statistics. We have seen how to use the binomial distribution to answer the question, "If a coin is fair, what is the probability of getting 12 heads in 20 tosses?" What we want to do now is learn how to apply statistical techniques to see if the coin is fair.

Let us begin by looking intuitively at the problem of deciding whether or not a coin is fair. What is needed is some sort of test. Suppose the following test were proposed to you. We will toss the coin 100 times and if we get 100 heads or 100 tails we will decide that the coin is not fair.

Your natural reaction to this test is that, indeed, if the coin comes up heads every time or tails every time in 100 tosses, one is justified in thinking that the coin is unfair. Of course, such a result is possible with a fair coin. It would come up heads 100 consecutive times only once out of every 2^{100} trials. But this result is so unlikely that you would feel quite justified in asserting that the coin is unfair.

However you also have to realize that if you use such a test, unfair coins are frequently going to be accepted as fair. For if the probability of heads on any one toss is .75, one would only expect 75 heads out of 100 tosses and 100 consecutive heads would still be an unlikely outcome.

Observing this you might be inclined to think about modifying the test. We might decide to reject the fairness of the coin unless we get exactly 50 heads in 100 tosses. This test has brought us to the opposite extreme from the first test. Now we are quite likely to reject the fairness of a fair coin.

But the basic ideas involved in these tests have some merit. First of all, the reasonable thing to do, to decide whether or not the coin is fair, is to toss it several times and see what happens. It certainly should, if it is fair, come up heads approximately half of the time. The serious question is, how far away from half the time must heads come up before you reject the fairness of the coin? We shall, in this section, give a precise answer to this and similar questions.

A related, and somewhat more practical application of these ideas, is connected with opinion polling. Suppose you are conducting a poll to determine the winner of an upcoming election between candidates named, imaginatively enough, Smith and Jones. If you poll 100 people, what result would lead you to predict a victory for Jones? Your answer might well be that it depends upon the size of the voting population. If there are only 100 voters, then your sample is really the same as the election (except perhaps for timing) so if a majority of the people pick Jones, he will win. But the interesting cases are where the group sampled is small relative to the total voting population. Let's assume that our sample of 100 is drawn from among 1,000,000 voters.

Now the answer is a little more complicated. Suppose it happens to be the case that 40% of the people are for Jones and 60% for Smith. Then in an election Jones would get 400,000 votes and Smith 600,000 so Smith would win easily. But your sample of 100 might happen to be composed entirely of people who intended to vote for Jones. So your poll would come out 100 for Jones, 0 for Smith and you would probably believe on this basis that Jones was going to win. This might make any forecasting seem dubious. However, with a few precautions, this is really a very unlikely situation. Indeed, assuming you make your sample reasonably random (that is, try not to do all your sampling in a Citizens for Jones meeting), we can deal with this problem in much the same way we deal with the tossing of a coin. The assumed breakdown of voters (400,000 for Jones and 600,000 for Smith) makes the probability .4 that a randomly selected voter will be for Jones. Therefore the probability that 100 out of 100 sampled would be for Jones is given by

$$B(100; 100, .4)$$

a very small number indeed.

So we would be fairly safe in predicting Jones' victory if all 100 people said they were going to vote for him. Even 99 would be fairly convincing evidence. But what about 51? That would and probably does strike you as being rather flimsy evidence that he will win. The question is, how many more than half must say they will vote for Jones before you can reasonably predict his victory?

The major difference between this problem and the coin tossing problem is that while you would decide that the coin is not fair, either because too many or too few heads come up, you will predict Jones' victory only if "too many" people say they will vote for him. You certainly won't predict his victory because too few people are going to vote for him.

In order to tackle these and other problems we need to introduce some common statistical terms.

Definition A *statistical hypothesis* is an assumption that a given random variable has a particular distribution (e.g., the binomial) and that the unknown parameters (e.g., n and p) have particular values.

Remark: At the moment we have only one distribution to use, the binomial distribution, so it will be used for all the work in this section. Later as we come to know other distributions we shall have a real choice. The binomial distribution has two parameters n and p. The value of n is usually determined by the situation (the size of the sample or the number of times the coin is to be tossed, etc.), so for our purposes, at present the part of the hypothesis which we shall be testing is an assumption about the value of p.

Definition A *statistical test* is a procedure for deciding whether to accept or reject a statistical hypothesis.

Remark: The definition is, with some reason, left vague. If your hypothesis is that a particular coin is fair, that is, the probability p of getting heads on a given toss is $\frac{1}{2}$, then the definition would allow you to test this hypothesis by drawing a card from a standard deck of cards and saying that you accept the hypothesis if the card is red and reject it if the card is black. Such tests are, however, considered frivolous by serious practitioners of statistics since it is difficult to assess their reliability. The test would normally consist of deciding to toss the coin a certain number of times and accept or reject the hypothesis that the coin is fair on the basis of the outcome of this experiment.

Definition Associated with a test are a null hypothesis and an alternative hypothesis. These hypotheses are complements in the sense that either one or the other but not both are true. The *null hypothesis* is the hypothesis that the statistical test actually tests. If the test leads you to reject the null hypothesis, then the *alternative hypothesis* is accepted.

Let us take a look at these definitions in the context of the coin tossing and polling problems which we just considered. For the coin tossing problem, our null hypothesis would be that the coin is fair, that is, $p = .5$, where p is the probability of getting heads on any individual toss. The null hypothesis is usually designated by H_0 and one would write

$$H_0 : p = .5$$

The alternative hypothesis has to be what is true if the null hypothesis isn't. In this case if $p = .5$ is not true, then $p \neq .5$ must be true. So the alternative hypothesis, usually denoted by H_1, would be written

$$H_1 : p \neq .5$$

The test is to toss the coin 100 times. We could have chosen any number of times to toss the coin. There is nothing magical about the number 100; but we have been using it and since some number has to be chosen, we shall stick with it. Note that the decision concerning the test determined the parameter n in the binomial distribution. This is typical of hypothesis testing with the binomial.

Let's consider the polling problem. We want to know if we can predict that Jones will win. The null hypothesis is that he won't win, that is, he will lose or there will be a tie. We let p be the probability that a randomly selected voter (out of the entire "population" of 1,000,000 voters) will vote for Jones. In other words, p is the proportion of all voters who will vote for Jones. Then the appropriate null hypothesis is

$H_0 : p \leq .5$ (This corresponds to Jones' losing or tying.)

And the alternative hypothesis is

$H_1 : p > .5$ (This corresponds to Jones' winning.)

Incidentally, the term "null" hypothesis historically meant a "no difference between" hypothesis because it was used in problems where the means of two random variables were being compared and the conjecture to be tested was $H_0 : \mu_1 = \mu_2$. Nowadays, the adjective "null" is used, as we have indicated, in connection with any conjecture that is to be tested.

To resume the definitions, we have a statistical hypothesis and a test. Presumably there are certain outcomes of this test which will lead us to reject the null hypothesis.

Definition The *critical region* of a statistical test consists of those outcomes of the test which will cause us to reject the null hypothesis (and hence accept the alternative hypothesis).

In our discussion of the coin, we agreed that if the number of heads attained in the 100 tosses was 0 or 100 we would reject the hypothesis that the coin is fair. The point at which we left the discussion was, in essence, what numbers (of heads) belong in the critical region besides 0 and 100. We suggested that it wasn't a very good idea to put all numbers other than 50 in the critical region.

One important thing to have, for any statistical test, is some notion of how valid it is. As we have pointed out, these tests are not foolproof. A fair coin *can* come up heads 100 times in a row. There is some chance of being wrong. We want to know how likely it is that we will be wrong. The following definitions allow us to discuss this important issue.

Definition Associated with any test of a statistical hypothesis there are two kinds of error. *Type I error* is rejecting a true hypothesis. If, for example, the coin were really fair but we rejected this because it came up heads 100 times in a row, we would be committing a type I error. *Type II error* is accepting a false hypothesis. As an example, suppose that we made the critical region for our test of the coin the outcomes 0 and 100. If it is the case that the coin is really not fair but we only got 90 heads, we would, since 90 isn't in the critical region, accept the null hypothesis that the coin is fair. We would then have committed a type II error by accepting this false hypothesis.

These definitions are summarized in the following table:

Decision	*Reality*: H_0 is true	*Reality*: H_0 is false
Reject H_0	Type I Error	Correct Decision
Accept H_0	Correct Decision	Type II Error

There is a relation between the sizes of these two types of error. In the example we have been discussing, where we toss a coin 100 times and make the critical region consist of 0 and 100 (heads), the probability of type I error is very small indeed but the probability of type II error is very large. We could diminish the probability of the type II error by making the critical region larger. If, for example, we made the critical region consist of 0–25 and 75–100, then many fewer unfair coins would slip through the test. On the other hand, increasing this critical region to make the probability of type II error smaller does increase the probability of the type I error. This is generally the case. For a fixed sample size, anything you do to decrease the probability of one type of error will increase the other type.

The standard response to this is to control the probability of type I error by the way in which the critical region is chosen and to deal with type II error by adjusting the size of the sample. We shall learn how to adjust the sample size in Chapter 5 in the section on the Central Limit Theorem. For the moment, we shall content ourselves with controlling the probability of the type I error.

Definition The *significance level* of a statistical test is a number α where $0 \leq \alpha \leq 1$, such that the probability of committing a type I error is less than or equal to α.

The choice of a significance level is to a large degree arbitrary. Typical choices of α are .10, .05, and .01.

The basic idea is that once the test is performed, if the number X (of heads on coin tosses, or of people who say they will vote for Jones) falls in the critical region, we can conclude one of two things. Either H_0 is false or an *unlikely* event has occurred—an event with probability less than or equal to α. In practice the value of α may well depend on the penalty to be paid for making a type I error.

Now that we have discussed the terminology of hypothesis testing, we should see how to put it into practice. Here is a procedure to follow:

Hypothesis Testing Algorithm

Step 1: Choose a significance level α.
Step 2: Choose a sample size n.
Step 3: Determine the null hypothesis.

For hypothesis testing with the binomial distribution the null hypothesis always consists of the choice of a number p_0 such that $0 \leq p_0 \leq 1$ and the assertion

(a) $H_0{:}p = p_0$
(b) $H_0{:}p < p_0$
(c) $H_0{:}p \leq p_0$
(d) $H_0{:}p > p_0$
(e) $H_0{:}p \geq p_0$

Step 4: Determine the critical region R.

For hypothesis testing with the binomial distribution the critical regions associated with each of the above null hypotheses are

for (a): Determine the largest integer k such that $P(\{0, 1, \ldots, k\}) \leq \alpha/2$ and the smallest integer m such that $P(\{m, m + 1, \ldots, n\}) \leq \alpha/2$. Then

$$R = \{0, 1, \ldots, k, m, m + 1, \ldots, n\}$$

for (b) and (c): Determine the smallest integer m such that $P(\{m, m + 1, \ldots, n\}) \leq \alpha$. Then

$$R = \{m, m + 1, \ldots, n\}$$

for (d) and (e): Determine the largest integer k such that $P(\{0, 1, \ldots, k\}) \leq \alpha$. Then

$$R = \{0, 1, \ldots, k\}$$

Now let's look at a few problems. We shall begin with the coin problem but, because of the amount of arithmetic involved, we will not toss it 100 times.

EXAMPLE 1 A coin is to be tossed 20 times. Can you conclude, at a .05 significance level, that it is not a fair coin if it comes up heads 18 times?

Solution: Steps 1 and 2 are already done for us by the statement of the problem. We are told to let $\alpha = .05$ and $n = 20$. What about the null hypothesis? As we said earlier, it should be that the coin is fair, i.e.,

$$H_0{:}p = \tfrac{1}{2}$$

To do Step 4 we refer to the histogram for $B(x; 20, \frac{1}{2})$ on page 159. We want the largest value of k such that

$$P(\{0, 1, \ldots, k\} = \sum_{x=0}^{k} B(x; 20, .5) \leq \frac{.05}{2} = .025$$

The histogram tells us that

$$P(\{0, 1, 2, 3\}) = .000 + .001 = .001$$

$$P(\{0, 1, 2, 3, 4\}) = .000 + .001 + .005 = .006$$

$$P(\{0, 1, 2, 3, 4, 5\}) = .006 + .015 = .021 < .025$$

and

$$P(\{0, 1, 2, 3, 4, 5, 6\}) = .021 + .036 = .057 > .025$$

and thus $k = 5$.

Similar calculations yield that $m = 15$ and the critical region R is given by

$$R = \{0, 1, 2, 3, 4, 5, 15, 16, 17, 18, 19, 20\}$$

The number 18 is in the critical region so we can reject, at the .05 significance level, the null hypothesis

$$H_0{:}p = .5$$

and, therefore, accept the alternative hypothesis

$$H_1 : p \neq .5$$

that is, the coin is not fair.

This example helps to explain why we shaped the critical region corresponding to a null hypothesis of type (a) as we did. The idea is that we assign as much of the critical region to the possibility that $p < p_0$ as we do to the possibility $p > p_0$. In the problem we just did, where $p_0 = .5$, this amounted to assigning as many outcomes to the left-hand tail of the critical region as to the right-hand tail. If we had assumed p equal to something other than .5, the two tails would not necessarily have the same number of outcomes but would be approximately equal in probability. Critical regions of type (a) are known as *two-tailed regions* and a test of this type as a *two-tailed test*.

EXAMPLE 2 A thumbtack can land either point up or point down. Find the critical region for a test, at the .10 significance level, that the probability of landing up is $\frac{1}{3}$, where the test is to consist of 20 tosses of the tack.

Once again the statement of the problem gives us the results of Steps 1 and 2. We are to let $\alpha = .10$ and $n = 20$. The null hypothesis is

$$H_0 : p = \tfrac{1}{3}$$

Referring to the histogram for $B(x; 20, \frac{1}{3})$ on page 160 we see that

$$P(\{0, 1, 2\}) = .000 + .003 + .014 = .017 < \frac{.10}{2} = .05$$

and

$$P(\{0, 1, 2, 3\}) = .017 + .043 = .060 > .05$$

Therefore $k = 2$. Working on the right-hand tail, we see that $m = 11$ and the critical region is

$$R = \{0, 1, 2, 11, 12, 13, 14, 15, 16, 17, 18, 19, 20\}$$

Any of these outcomes would lead us to reject, at a significance level of .10, the null hypothesis $H_0 : p = \frac{1}{3}$ and accept the alternative hypothesis $H_1 : p \neq \frac{1}{3}$.

EXAMPLE 3 An election is to be held between Smith and Jones. If you take a poll of 20 randomly selected voters, how many must say that they will vote for Jones before you can predict his victory with a significance level of .05?

Solution: Once again the statement of the problem gives us $\alpha = .05$ and $n = 20$. Next we must select the null hypothesis. Just as we have been deciding that a coin is unfair by rejecting the hypothesis that it is fair, we predict Jones' victory by rejecting the hypothesis that he won't win. If we let p be the probability that a randomly selected voter will vote for Jones, then $p < .5$ means less than half the voters will vote for him, so he will lose. If $p = .5$, then exactly half the voters will vote for him and there will be a tie, so he will not win.

Therefore the null hypothesis is

$$H_0 : p \leq .5$$

Now we have a null hypothesis of the type which was labeled (c) so we want the smallest integer m such that

$$P(\{m, m + 1, \ldots, 20\}) \leq .05$$

This time it may not be so clear how to compute the probabilities, since we do not have a single value for p but rather we have a range of values. Both $p = \frac{1}{3}$ and $p = \frac{1}{2}$ are in the range of values. Using $p = .5$, we have

$$P(\{14, 15, 16, 17, 18, 19, 20\}) = .058 > .05$$

and

$$P(\{15, 16, 17, 18, 19, 20\}) = .021 < .05$$

We might note that as p gets larger, the critical region corresponding to it is getting smaller. In order to have a critical region R such that we are sure that $P(R) \leq .05$ no matter what $p \leq .5$ we choose, we must choose the smallest possible critical region. This leads us to the critical region

$$R = \{15, 16, 17, 18, 19, 20\}$$

corresponding to $p = .5$. Therefore, if 15 or more people in our sample say that they will vote for Jones, we can reject the null hypothesis $H_0 : p \leq .5$ and accept the alternative hypothesis $H_1 : p > .5$, i.e., Jones will win.

Note that the way in which we defined the critical region for a null hypothesis of type (b) or (c) corresponds to our earlier observation that it would make no sense to predict that a man was going to win an election because too few people say that they are going to vote for him. A test of this type is called a *one-tailed test*. It might also be noted that in dealing with the one-tailed tests arising from null hypothesis of the form $p < p_0$, $p \leq p_0$, $p > p_0$, $p \geq p_0$, one always uses the probability $p = p_0$ in calculations for the critical region. You should check through the logic of this for yourself in the manner in which we did it for Example 3.

Now that we have seen some examples it may be useful to make a general remark about the choice of a null hypothesis in actual practice. In the first place, the null hypothesis usually represents an assumption we would like to discredit. Moreover rejecting a null hypothesis is tantamount to making some sort of decision. The natures of the possible decisions and their relative importance are factors that can influence the choice of null hypothesis.

Let us imagine a merchant who is faced with a decision of whether to buy a shipment of oranges, some of which are rotten. Imagine further that he knows he can afford to make the purchase if at least 80% of the oranges are good. That is, if 80% or more are good, he is likely to make a profit, whereas, if less than 80% are good, he is likely to incur a loss. He wishes to select a sample of

20 oranges and let X be the number of good oranges in the sample. What should his null hypothesis be?

As it turns out, a case could be made for either of the following choices:

Choice 1: $H_0: p < .8$ (where p = proportion of good oranges in the entire shipment)

Choice 2: $H_0: p \geq .8$

Using choice 1, the merchant would determine the appropriate right-tailed critical region R and if the value of X were big enough to fall into R, he would reject H_0 and buy the oranges. This decision procedure would help protect him from buying a poor shipment.

On the other hand, choice 2 would lead to a left-tailed critical region R' and if X were small enough to fall into R', the buyer would reject H_0 and decide not to buy the shipment. This decision procedure would help protect him from passing up a good shipment.

We could paraphrase choices 1 and 2 as follows. Choice 1 says to buy only if the shipment is very likely to have at least 80% good oranges. Choice 2 says to pass up the chance to buy only if the shipment is very likely to have less than 80% good oranges. The ultimate choice between 1 and 2 would depend on the attitude of the merchant and how costly it would be to him to get stuck with a bad shipment versus passing up the opportunity to buy a good shipment.

Before concluding this section, we should mention a minor distinction between our treatment of hypothesis testing and that found in other books. Many authors allow for an alternate hypothesis that does not represent simply the negation or the complement of H_0. For example some texts would have

$$\begin{Bmatrix} H_0: & p = \frac{1}{2} \\ H_1: & p < \frac{1}{2} \end{Bmatrix} \quad \text{or} \quad \begin{Bmatrix} H_0: & p = \frac{1}{2} \\ H_1: & p = \frac{1}{3} \end{Bmatrix}$$

In our discussions, both now and later, we shall always make the simplification that H_1 is "everything other than H_0." That is, we would have either

$$\begin{Bmatrix} H_0: & p \geq \frac{1}{2} \\ H_1: & p < \frac{1}{2} \end{Bmatrix} \quad \text{or} \quad \begin{Bmatrix} H_0: & p = \frac{1}{2} \\ H_1: & p \neq \frac{1}{2} \end{Bmatrix}$$

to cover the above two situations.

We conclude now with four histograms of the binomial distribution, some of which have been referred to in the examples and all of which will be needed in the exercises.

Summary

The basic premise of this section is that we are given a random variable X, we know that it is binomially distributed (this temporary restriction is imposed because of our limited experience with specific types of random variables), and the parameter n is known. Our job is to gain information about the other parameter p. A *statistical hypothesis* is a statement about the value of p, such as "$p = \frac{1}{2}$" or "$p \leq \frac{1}{3}$" or the like. Indeed, these statistical hypotheses always come

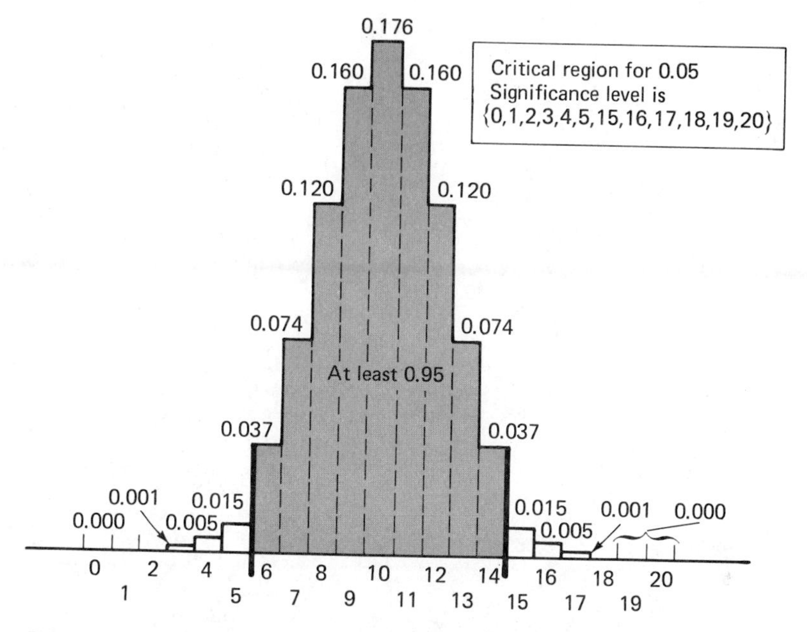

FIGURE 3.8 Histogram for $B(x; 20, \frac{1}{2})$.

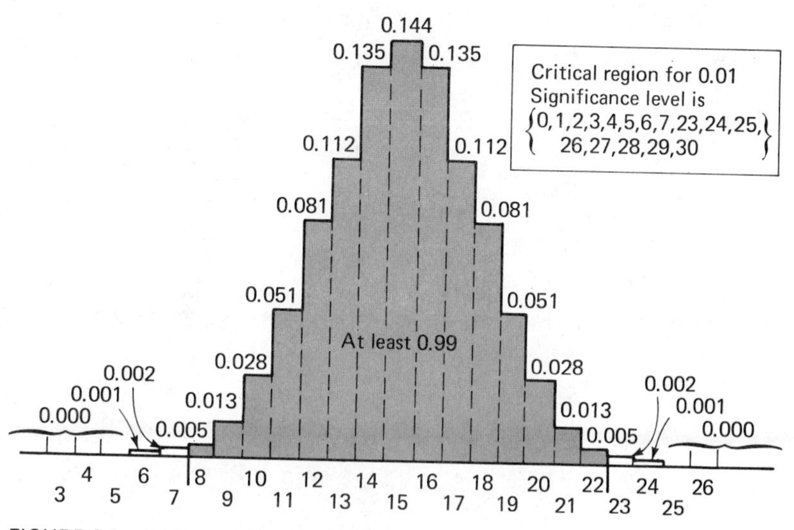

FIGURE 3.9 Histogram for $B(x; 30, \frac{1}{2})$.

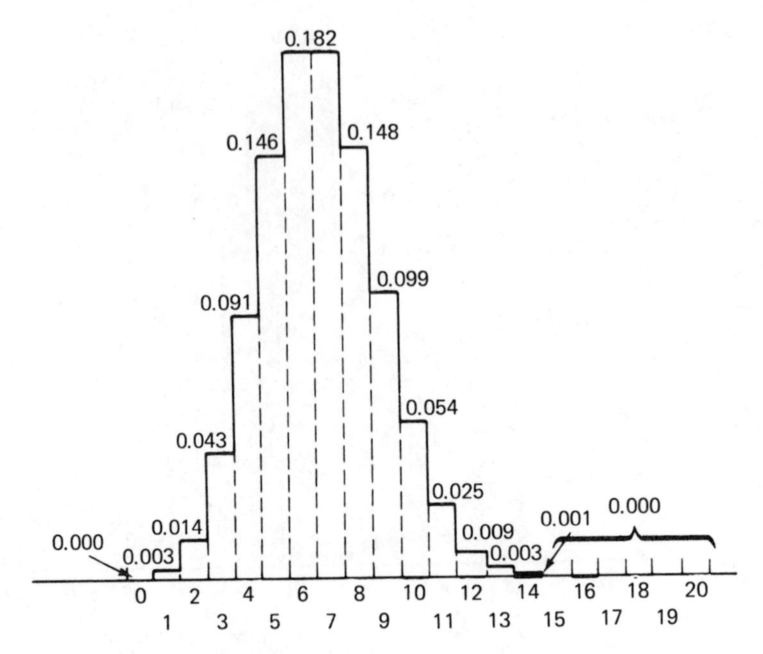

FIGURE 3.10 Histogram for $B(x; 20, \frac{1}{3})$.

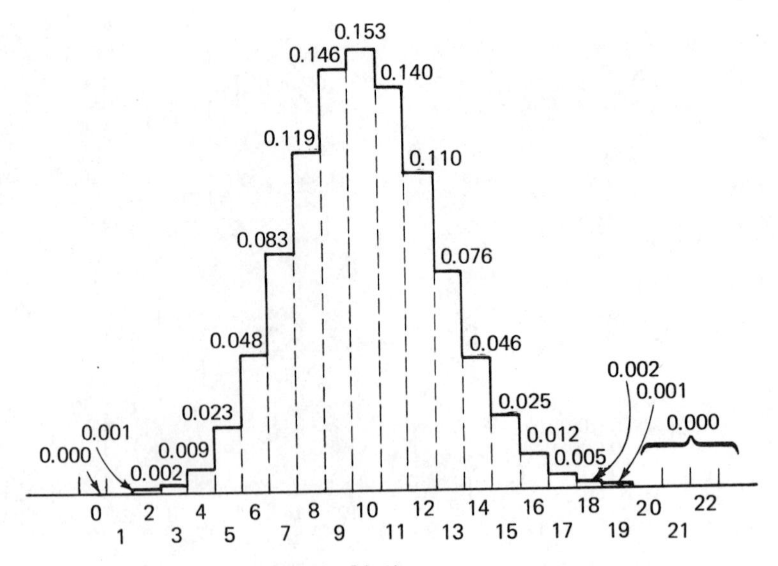

FIGURE 3.11 Histogram for $B(x; 30, \frac{1}{3})$.

in pairs: The *null hypothesis* denoted H_0, is accompanied by its denial, the *alternative hypothesis*, denoted H_1. Once these hypotheses are specified, we proceed to conduct a *statistical test*, which amounts to taking a random sample of n values of X and checking whether the outcome is consistent with H_0 (in which case, we *accept* H_0) or inconsistent with H_0 (in which case, we *reject* H_0 in favor of H_1). The meaning of "consistent" is made more precise as follows. Prior to the sampling, we decide upon a *level of significance* α, which is a number between 0 and 1—typically either .10, .05, or .01. We then determine a *critical region*, which is simply a subset of $\{0, 1, \ldots, n\}$ of the following forms:

$R_1 = \{0, 1, \ldots, k\}$ (left sided, one-tail region)

$R_2 = \{m, m + 1, \ldots, n\}$ (right sided, one-tail region)

$R_3 = \{0, 1, \ldots, k, m, m + 1, \ldots, n\}$ (two-tail region)

In any case, we insist that the critical region R is as large as possible subject to $P(R) \leq \alpha$. The exact type of critical region is dictated by the nature of H_0. In particular, a null hypothesis of the form

$$H_0: p = \tfrac{1}{4}$$

would dictate a two-tail region; whereas

$$H_0: p < \tfrac{1}{3} \qquad (\text{or } p \leq \tfrac{1}{3})$$

and

$$H_0: p > \tfrac{4}{5} \qquad (\text{or } p \geq \tfrac{4}{5})$$

would dictate critical regions of the form R_2 and R_1, respectively.

Once H_0 and α are specified and the critical region is determined, the test is conducted. If the number x of successes (in the sample of size n) falls into the critical region, then we *reject* H_0; otherwise, we accept H_0. The basis for our reasoning is that the probability of x being in the critical region, *if* H_0 *is true*, is relatively small (namely α).

The following table summarizes two kinds of errors we can make in hypothesis testing. In practice, one must exercise some judgment in determining the significance level α, since the lower the value of α, the smaller the probability of making a type I error, *but* the larger the probabilities of making a type II error.

		Reality	
		H_0 is true	H_0 is false
Decision	Reject H_0	Type I Error	Correct decision
	Accept H_0	Correct decision	Type II Error

Exercises Section 5

1 For each situation, decide what distribution we need, what H_0 is, and whether a one- or two-tailed test is called for.
   (a) We want to prove a coin is not fair by tossing it 100 times.
   (b) We want to prove a coin lands heads more than $\frac{3}{4}$ of time (250 tosses).
   (c) We want to prove at least 90% of the people are right-handed (test 500 people).
   (d) We want to prove at least 10% of people are left-handed (test 500 people).

2 The XYZ presidential poll has just completed a survey of 20 people asking whether they will vote Republican or Democratic in 1976. How many "Republican" responses would be necessary before they could predict a Republican victory at a .10 significance level? At what significance level could they predict victory with 11 "Republican" responses? (Assume only two political parties.)

3 Find the one- and two-tailed critical regions corresponding to .10, .05, and .01 significance levels in each of the following histograms.
   (a) $B(x; 12, .5)$
   (b) $B(x; 20, .5)$
   (c) $B(x; 30, .5)$
   (d) $B(x; 20, \frac{1}{3})$
   (e) $B(x; 30, \frac{1}{3})$

4 A student takes a language placement examination consisting of 20 true-false questions. His score is 25%. Should the grader become suspicious? Explain by carefully formulating a statistical hypothesis and testing at various levels. You should probably use a one-tail test in this case. Why?

5 Suppose the examination in Exercise 4 consisted of 30 multiple choice questions each with three possible answers. This time the student gets only five problems correct. Was he deliberately giving wrong answers?

6 A new drug is claimed to substantially reduce the number of deaths resulting from a certain disease. In the past, one-third of all people having the disease died. In a sample of 30 patients, only six people having the disease, but treated with the drug, died. Does this experiment convince you that the drug is effective? Explain in terms of accepting or rejecting a suitable null hypothesis.

7 Referring to Exercise 6, suppose another manufacturer develops a drug and runs an experiment on 20 diseased people, only four of whom die. Would these results be more or less convincing than the six out of 30 deaths considered earlier?

8 Under normal conditions $\frac{2}{3}$ of the students taking Math 789 pass each year. Last year's class, taught by the same instructor, using the same text and exams, produced only five failures out of 30 students. To what extent could that group of students be considered superior to those of other years?

6 THE POISSON DISTRIBUTION

In Section 2 we introduced the binomial distribution. We shall now introduce a distribution which, in addition to being important in its own right, is a useful tool for approximating the binomial distribution in certain situations where the binomial probabilities are hard to calculate or to find in tables.

Let us begin by reformulating the binomial formula. We have

$$\begin{aligned}B(x; n, p) &= C(n, x)p^x q^{n-x}\\ &= \frac{n!}{x!(n-x)!}p^x\frac{q^n}{q^x}\\ &= \frac{1}{x!}\frac{n^x}{n^x}p^x\frac{n!}{(n-x)!}\frac{(1-p)^n}{(1-p)^x}\\ &= \frac{(np)^x}{x!}(1-p)^n\frac{n!}{n^x(n-x)!(1-p)^x}\end{aligned}$$

Let $np = \lambda$ (so $p = \lambda/n$). We have

$$B(x; n, p) = \frac{\lambda^x}{x!}\left(1 - \frac{\lambda}{n}\right)^n \frac{n!}{(n-x)!n^x(1-p)^x}$$

Note that

$$\frac{n!}{(n-x)!n^x(1-p)^x} = \frac{n(n-1)(n-2)\dots(n-(x-1))}{n^x(1-p)^x}$$

Note that this numerator has x factors.

$$= \frac{n}{n(1-p)}\cdot\frac{n-1}{n(1-p)}\cdot\frac{n-2}{n(1-p)}\cdots\frac{n-(x-1)}{n(1-p)}$$

$$(\dagger) \qquad = \frac{1}{1-p}\cdot\frac{1-(1/n)}{1-p}\cdot\frac{1-(2/n)}{1-p}\cdots\frac{1-((x-1)/n)}{1-p}$$

If n is very large and p is close to 0 and if x isn't large relative to n, then each of the factors in (†) is close to 1. Therefore, for large n and small p

$$B(x; n, p) \approx \frac{\lambda^x}{x!}\left(1 - \frac{\lambda}{n}\right)^n$$

Your initial reaction to this approximation is apt to be, "So what?" But $[1 - (\lambda/n)]^n$ is a familiar quantity to mathematicians. What happens to this expression as n becomes large? It might, at first glance seem that it becomes 1, since the quantity

$$1 - \frac{\lambda}{n}$$

inside the parentheses is close to 1 when n is large. However it turns out that this is not the case. Below we have included a computer printout which gives us some insight as to what happens.

The program calls for the computer to print out values of

$$\left(1 - \frac{\lambda}{n}\right)^n$$

for $\lambda = -1, -2, -3, 1, 2, 3$, and for selected values of n from 1 to 5,000. This printout makes it fairly clear that the expression does not get close to 1 for large n.

Look, for example, at the first column which gives values of $[1 + (1/n)]^n$. When n changes from 1 to 2, the value of the expression changes from 2.000 to 2.250, a difference of .25. When n changes from 2 to 3, the expression changes in value by .12. But when n changes from 4,500 to 5,000, the change in value of the expression is less than .0001. The values are getting closer and closer to a fixed number, customarily denoted by the letter e, and approximately equal to 2.7183.

```
PAGE    1     ERW

//  JOB

LOG DRIVE      CART SPEC      CART AVAIL      PHY DRIVE
  0000           0B0E           0B0E            0000

V2  M09     ACTUAL     8K     CONFIG     8K

// FOR
*LIST SOURCE PROGRAM
*IOCS(CARD,1132PRINTER)
*EXTENDED PRECISION
       WRITE(3,10)
    10 FORMAT('1        N          (1+1/N)**N          (1+2/N)**N          (1+3/N)**N
      1(1-1/N)**N          (1-2/N)**N          (1-3/N)**N')
       DO 30 K=1,15
       N=K
       X=N
       A=(1+1/X)**N
       B=(1+2/X)**N
       C=(1+3/X)**N
       D=(1-1/X)**N
       E=(1-2/X)**N
       F=(1-3/X)**N
       WRITE (3,20)N,A,B,C,D,E,F
    30 CONTINUE
       DO 500 K=3500,5001,100
       N=K
       X=N
       A=(1+1/X)**N
       B=(1+2/X)**N
       C=(1+3/X)**N
       D=(1-1/X)**N
       E=(1-2/X)**N
       F=(1-3/X)**N
       WRITE (3,20)N,A,B,C,D,E,F
```

```
    500 CONTINUE
     20 FORMAT(1X,15,2X,6(F13,10,3X))
        CALL EXIT
        END

FEATURES SUPPORTED
  EXTENDED PRECISION
  IOCS

CORE REQUIREMENTS FOR
  COMMON       0  VARIABLES      30  PROGRAM      382

END OF COMPILATION

// XEQ
```

N	(1+1/N)**N	(1+2/N)**N	(1+3/N)**N	(1−1/N)**N	(1−2/N)**N	(1−3/N)**N
1	2.0000000018	3.0000000018	4.0000000037	0.0000000000	−1.0000000009	−2.0000000018
2	2.2500000018	4.0000000037	6.2500000037	0.2500000002	0.0000000000	0.2500000002
3	2.3703703712	4.6296296399	8.0000000074	0.2962962971	0.0370370369	0.0000000000
4	2.4414062518	5.0625000037	9.3789062574	0.3164062502	0.0625000000	0.0039062500
5	2.4883199920	5.3782400023	10.4857599921	0.3276800016	0.0777600004	0.0102400000
6	2.5216263663	5.6186556816	11.3906250074	0.3348979785	0.0877914955	0.0156250000
7	2.5464996984	5.8077950961	12.1426567807	0.3399166775	0.0948645064	0.0198945289
8	2.5657845167	5.9604644812	12.7767849676	0.3436089161	0.1001129150	0.0232830643
9	2.5811747936	6.0862751435	13.3182949386	0.3464394171	0.1041597134	0.0260122950
10	2.5937424516	6.1917363759	13.7858491800	0.3486784416	0.1073741834	0.0282475249
11	2.6041990239	6.2814128547	14.1934788450	0.3504939003	0.1099886999	0.0301071967
12	2.6130352756	6.3585995305	14.5519152395	0.3519956300	0.1121566557	0.0316763520
13	2.6206008708	6.4257346224	14.8694939985	0.3532585033	0.1139833279	0.0330169100
14	2.6271515339	6.4846606813	15.1527828238	0.3543353129	0.1155433478	0.0341747304
15	2.6328787133	6.5367960222	15.4070213809	0.3552643682	0.1168910886	0.0351843725
3500	2.7178901927	7.3848156072	20.0597019940	0.3678272044	0.1352582278	0.0497231029
3600	2.7178947152	7.3849556427	20.0604289248	0.3678292466	0.1352602325	0.0497249606
3700	2.7179087363	7.3850439600	20.0611107796	0.3678303475	0.1352625039	0.0497266262
3800	2.7179172988	7.3851661197	20.0617677047	0.3678320350	0.1352641920	0.0497282662
3900	2.7179248705	7.3852383848	20.0624103769	0.3678328226	0.1352663232	0.0497296827
4000	2.7179395826	7.3853323515	20.0629488751	0.3678340631	0.1352680671	0.0497311302
4100	2.7179478248	7.3854431342	20.0634347423	0.3678350850	0.1352695275	0.0497325921
4200	2.7179452655	7.3855280913	20.0639902800	0.3678368835	0.1352711788	0.0497338192
4300	2.7179600317	7.3855964820	20.0644903033	0.3678371098	0.1352728023	0.0497349909
4400	2.7179738190	7.3856888394	20.0649782866	0.3678381510	0.1352740458	0.0497362872
4500	2.7179796826	7.3857465870	20.0654125884	0.3678390339	0.1352754276	0.0497374379
4600	2.7179821860	7.3858202900	20.0658466741	0.3678394991	0.1352766047	0.0497384335
4700	2.7179797571	7.3858869411	20.0663081333	0.3678413446	0.1352779774	0.0497394421
4800	2.7179900249	7.3859587591	20.0666839256	0.3678414033	0.1352791975	0.0497405436
4900	2.7179852072	7.3859983421	20.0669865906	0.3678434245	0.1352806476	0.0497415733
5000	2.7180001689	7.3860854860	20.0674714893	0.3678434657	0.1352816943	0.0497422917

The situation which has occurred here arises frequently in mathematics. For each positive integer n we have a number, call it s_n (in the case we just looked at $s_n = [1 + (1/n)]^n$). We find that there is a number s (in the case we just looked at, the part of s was played by e) with the property that as n gets large, s_n gets close to s. The phrase "s_n gets close to s" can be (and, if we were to do a careful theoretical job on this concept, would have to be) replaced by the statement that, given any amount of error, no matter how small, we can find a point in the sequence s_n such that for every number in the sequence beyond that point, the difference between s_n and s is less than the prescribed amount of error. When this situation occurs, we write

$$\lim_{n \to \infty} s_n = s$$

(read the limit of s_n as n becomes infinite is s). We shall make a careful study of this limit concept in Chapter 4.

We hope that the various columns in the computer printout will convince you that for each number x there is a number which is

$$\lim_{n \to \infty} \left(1 + \frac{x}{n}\right)^n$$

By taking this limit, for each value of x we get a function of x. This function is of considerable importance in mathematics both pure and applied. We write the function as exp (x). You should keep in mind that, for each x

$$\exp(x) = \lim_{n \to \infty} \left(1 + \frac{x}{n}\right)^n$$

Your first reaction to this function may be that, while it is possibly of some theoretical importance, you will never be able to work with it because you can't calculate with it. It's certainly true that it is more complicated than working with a function such as $f(x) = x^2 + 1$. With this kind of function, if you are asked for $f(2)$, you simply write $f(2) = 2^2 + 1 = 5$. If you are asked for exp (2) the situation isn't quite so easy. However most of the situations where we need explicit values of the function exp will arise in work we do on the computer and the function is important enough to be on most computers. If, in a computer program you write EXP(X) the computer will calculate the value of the function exp for whatever is the value of x. [At least it is called EXP(X) in FORTRAN, BASIC, AND ALGOL. If you are using a computer with a different language, you should check to see if it also uses EXP(X).] Also, Table A in the Appendix has the value of exp (x) for selected values of x.

We have mentioned that

$$\left(1 + \frac{1}{n}\right)^n$$

gets close to

$$e = 2.71828\ldots$$

Thus we have

$$e = e^1 = \exp(1)$$

The table might suggest that

$$\exp(2)$$

which is the value approached by

$$\left(1 + \frac{2}{n}\right)^n$$

is really equal to

$$e^2$$

[compute $(2.718)^2$ and compare it to the values of $[1 + (2/n)]^n$ for large n]. Indeed, this is true and, in general, for any positive integer n, we have

$$e^n = \exp(n)$$

But let us return to the original problem, the approximation of the binomial distribution. We have argued that for large n and small p

$$B(x; n, p) \approx \frac{\lambda^x}{x!} \exp(-\lambda)$$

where $\lambda = np$. This yields

$$B(0; n, p) \approx \exp(-\lambda)$$

$$B(1; n, p) \approx \frac{\lambda}{1} \exp(-\lambda)$$

$$B(2; n, p) \approx \frac{\lambda^2}{2 \cdot 1} \exp(-\lambda)$$

$$B(3; n, p) \approx \frac{\lambda^3}{3 \cdot 2 \cdot 1} \exp(-\lambda)$$

etc.

We define the function

$$\mathbf{P}(x; \lambda) = \frac{\lambda^x}{x!} \exp(-\lambda)$$

where λ is any positive number, and x is any non-negative integer. While it may not be obvious, $\mathbf{P}(x; \lambda)$ is a frequency function for each choice of λ. The totality of such functions is called the *Poisson distribution*. It exists as a distribution in its own right and can be used for a variety of problems which have nothing to do with the binomial distribution. We shall shortly look at some of these, though much of the explanation of its significance must await the development of calculus in the next chapter. It is also useful, in certain situations, as an approximation to the binomial distribution.

It must be emphasized that the Poisson distribution can only be used as an approximation to the binomial distribution when n is large, when p is small, and when np is of moderate size. For the purposes of this course we shall assume that the phrase "moderate size" means $np \leq 5$ and $n \geq 50$. In this case we make the approximation

$$B(x; n, p) \approx \mathbf{P}(x; np)$$

Let us illustrate this by an example.

EXAMPLE 1 Among the population of the United States, it is known that only 1 out of 10,000 people prefer fried ants to steak. If 20,000 people are selected at random in the United States, what is the probability that three of them prefer fried ants to steak?

Since a person either prefers fried ants to steak or doesn't, we have a problem that calls for the binomial distribution. We have $n = 20{,}000$ and $p = 1/10{,}000 = .0001$ and the answer to the question is

$$B(3; 20{,}000, .0001).$$

Since $np = 2$ in this case, we can use the Poisson distribution as an approximation and have

$$B(3; 20{,}000, .0001) \approx \mathbf{P}(3; 2)$$

Consulting Table C in the Appendix we find

$$\mathbf{P}(3; 2) = .1804$$

and this gives us the answer.

Having seen something of how to use the Poisson distribution, let us investigate some of its properties. First of all we should ask whether or not it really is a distribution. We must show that

$$\sum_{x=0}^{\infty} \mathbf{P}(x; \lambda) = 1$$

In order to see this we must go back to the way in which the function exp is defined. We said that for any number z,

$$\exp(z) = \lim_{n \to \infty} \left(1 + \frac{z}{n}\right)^n$$

We expand a typical product.

$$\begin{aligned}
\left(1 + \frac{z}{n}\right)^n &= \sum_{k=0}^{n} C(n, k) \left(\frac{z}{n}\right)^k \\
&= \sum_{k=0}^{n} \frac{n!}{k!(n-k)!} \cdot \frac{z^k}{n^k} \\
&= \sum_{k=0}^{n} \frac{n(n-1)(n-2)\ldots(n-k+1)}{n \cdot n \cdot n \cdots n} \cdot \frac{z^k}{k!} \\
&= \sum_{k=0}^{n} (1)\left(1 - \frac{1}{n}\right)\left(1 - \frac{2}{n}\right)\cdots\left(1 - \frac{k-1}{n}\right)\frac{z^k}{k!}
\end{aligned}$$

As before, when n is very large the factors $1 - (1/n), 1 - (2/n), \ldots, 1 - [(k-1)/n]$ are all essentially 1. Therefore

$$\exp(z) = \lim_{n\to\infty}\left(1 + \frac{z}{n}\right)^n = \sum_{k=0}^{\infty}\frac{z^k}{k!}$$

The summation from $k = 0$ to ∞ has not been defined. What is meant is that we take the sum from $k = 0$ to $k = n$ and then take the limit as n becomes larger. We shall discuss this more completely in Chapter 4.

With this equation in hand we obtain

$$\begin{aligned}\sum_{x=0}^{\infty}\mathbf{P}(x;\lambda) &= \sum_{x=0}^{\infty}\frac{\lambda^x}{x!}\exp(-\lambda)\\ &= \exp(-\lambda)\sum_{x=0}^{\infty}\frac{\lambda^x}{x!}\\ &= \exp(-\lambda)\exp(\lambda)\end{aligned}$$

We now need to point out one more property of the function exp. The name, exp, was chosen to suggest the word exponent. You recall from high school that

$$a^u a^v = a^{u+v}$$

that is, multiplication of exponential quantities is done by adding exponents. Also $a^0 = 1$. The same properties hold for the function exp. That is,

$$\exp(u)\cdot\exp(v) = \exp(u+v)$$

and

$$\exp(0) = 1$$

When we deal with calculus, we will in fact show that

$$\exp(x) = e^x$$

From this we see that

$$\begin{aligned}\exp(-\lambda)\exp(\lambda) &= \exp(-\lambda+\lambda)\\ &= \exp(0)\\ &= 1\end{aligned}$$

and the Poisson distribution is really a frequency function.

Now we can ask for the theoretical mean of the Poisson distribution, that is, for $E(X)$.

$$\begin{aligned}E(X) &= \sum_{x=0}^{\infty}x\mathbf{P}(x;\lambda)\\ &= \sum_{x=1}^{\infty}x\frac{\lambda^x}{x!}\exp(-\lambda)\\ &= \sum_{x=1}^{\infty}\frac{\lambda^x}{(x-1)!}\exp(-\lambda)\end{aligned}$$

Letting $y = x - 1$

$$= \lambda \sum_{y=0}^{\infty} \frac{\lambda^y}{y!} \exp(-\lambda) = \lambda \cdot 1 = \lambda$$

A similar calculation will show that

$$\sigma^2 = E(X^2) - [E(X)]^2 = \lambda$$

Now let us look at an application of the Poisson distribution which is not, directly, related to the binomial distribution. A careful explanation of why this problem should lead us to the Poisson distribution must await our discussion of calculus.

EXAMPLE 2 A business office is about to install a new telephone switchboard. It has been observed that the times when people call in or out are random and that the length of each conversation is also random. However it has been noted that at any given time there are, on the average five lines in use. If they install eight lines into the office, what is the probability that when someone wants to call in or out, all the lines will be in use?

The first thing you should note about this problem is that there is no apparent way to apply the binomial distribution to it. However, for this and a number of similar problems we can (and will later) show that the Poisson distribution applies. The mean of the Poisson function $\mathbf{P}(x; \lambda)$ is λ. The mean number of lines in use is five, so we should use the Poisson distribution with $\lambda = 5$. Then $\mathbf{P}(x; 5)$ will represent the probability that x people are using the phones (or would if they could when $x > 8$). The solution to our problem is then

$$\sum_{x=9}^{\infty} \mathbf{P}(x; 5) = .0363 + .0181 + .0082 + .0034 + .0013 + .0005$$
$$+ .0002 + 0, \qquad \mathbf{P}(x; 5) = 0 \quad \text{for} \quad x \geq 16$$

and with a bit of addition we obtain .0680 as the answer.

It is somewhat difficult, at this point, to explain the circumstances under which the Poisson distribution applies. Roughly speaking, the Poisson distribution applies to what are known as traffic problems, that is, you are told that the average number of phone lines in use is some number, or the average number of cars passing a certain point per unit of time is something, or the average number of people coming into a store in a given time interval is something. Then, using that average as the value of λ, you can calculate the probability of some particular number, using the Poisson distribution.

While a real justification of the use of the Poisson distribution in traffic problems must await calculus, one can, for some problems, devise a relation to the binomial distribution which makes the application plausible.

Suppose that we have been told that the average number of cars passing a certain point per half hour is three and we wish to calculate the probability that no cars pass it in a particular half hour. To keep things simple, we will assume that we are dealing with only one lane of traffic.

Divide a 30-minute time interval up into 36,000 intervals of length $\frac{1}{20}$ of a second. A car traveling at 60 miles per hour goes 88 feet per second or 4.4 feet during one of our little time intervals and therefore we can assume that at most one car can pass its front wheels over a line in the road during such an interval. Thus, regarding each $\frac{1}{20}$-second interval as an experiment we either have a success (a car's front wheels pass over the line) or a failure.

What is p, the probability of success? If the average number of cars passing is three per half hour, we would seem to have three successes per 36,000 trials and a probability

$$p = \frac{3}{36{,}000} = \frac{1}{12{,}000}$$

The probability of no cars in a 30-minute time interval would be

$$B\left(0; 36{,}000, \frac{1}{12{,}000}\right)$$

which is approximately

$$\mathbf{P}\left(0; 36{,}000 \cdot \frac{1}{12{,}000}\right) = \mathbf{P}(0; 3)$$

It is hoped that this discussion will give you more confidence in the application of the Poisson distribution to traffic problems. However, if you try to interpret other traffic problems via the binomial distribution (consider the difficulties with a multiple lane highway or with the telephone problem of Example 2), you will appreciate the need for an alternate approach such as the one we adopt in Chapter 4.

In many applications of the Poisson distribution, the value of λ will turn out to be such that our table (or indeed most other tables) cannot be used without rounding off λ to a suitable value. An alternative technique is to go to a table for values of e^{-x} (such as Table A in the Appendix listing e^{-x} for $x = 0$ to 5 in steps of .01) and computing $\mathbf{P}(x; \lambda)$ from the formula

$$\mathbf{P}(x; \lambda) = \frac{\lambda^x e^{-\lambda}}{x!}$$

For instance, we can evaluate

$$\begin{aligned}\mathbf{P}(2; 2.5) &= \frac{(2.5)^2 e^{-2.5}}{2!} \\ &= \left(\frac{6.25}{2}\right) e^{-2.5} \\ &= .2565625 \\ &\approx .2566\end{aligned}$$

Notice the potential danger of crude round-off, say of λ to 2 or to 3 where

$$\mathbf{P}(2; 2) = .2707 \quad \text{and} \quad \mathbf{P}(2; 3) = .2240$$

As a final example of the Poisson distribution in action, we shall use it for hypothesis testing.

EXAMPLE 3 Professors Dickens and Dostoievsky of the Physics Department are debating the level of knowledge of the average student. Dickens asserts that not more than one student in 100 knows Newton's Third Law. Dostoievsky, while no great optimist, feels that this estimate is somewhat too low. They agree to test it by questioning a randomly selected group of 300 students. What is the critical region for the test.

If we let p be the probability that a randomly selected student will know Newton's Third Law, then Dickens' hypothesis is that $p \leq .01$. Since they are going to question 300 people, this means that $np \leq 3$, and so we can use the Poisson distribution to approximate the binomial distribution. Our null hypothesis H_0 is that $\lambda \leq 3$. Let us assume that we want to test at a .05 significance level.

It is apparent that we are not going to reject the null hypothesis because too few people know the law. We shall only reject it if too many know it. We have therefore a situation which calls for a one-tailed test.

If you look at Table C for the Poisson distribution, you will see that for a fixed x, $\mathbf{P}(x; \lambda)$ increases as λ increases. This tells us that the larger λ is, the smaller will be the corresponding critical region for a test at the .05 significance level. Therefore, if we take the largest value of λ in the range of our hypothesis ($\lambda = 3$ for our problem), we will obtain a certain critical region. The critical region for any smaller λ will be even larger and it will contain the one which we have found. So if we use the critical region for the largest value of λ we will have a critical region which may be valid for all $\lambda \leq 3$. We calculate

$$\sum_{x=7}^{\infty} \mathbf{P}(x; 3) = .0216 + .0081 + .0027 + .0008 + .0002 + .0001 + 0$$

$$= .0335$$

This is less than .05 and if we put in $x = 6$ we get too large a critical region. Therefore the critical region is

$$\{7, 8, 9, 10, \ldots\}$$

Thus if, out of 300 students, at least seven of them know Newton's Third Law, we conclude that Dickens is wrong and Dostoievsky is right: The average number of students knowing the law is greater than 1 out of 100.

Summary

The object of this section has been three-fold. First, we sought a suitable approximation to the values $B(x; n, p)$ of a binomial frequency function. Second, in our search we encountered the important exponential function exp which is defined in terms of limits. Third, we saw how the approximation to $B(x; n, p)$ is a frequency function in its own right. We shall briefly review the key definitions and formulas here in a logical order.

The function exp has for its domain all real numbers. It is called the *exponential function* and is defined by

$$\exp(x) = \lim_{n\to\infty}\left(1+\frac{k}{n}\right)^n$$

This definition makes more precise the following notation

$$e^x = \exp(x), \qquad \text{where } e = 2.71828\ldots$$

Thinking of $\exp(x)$ in this alternate form, we can establish the usual rules governing exponents

$$e^{a+b} = e^a e^b \quad \text{or} \quad \exp(a+b) = \exp(a)\exp(b)$$

$$e^0 = 1 \quad \text{or} \quad \exp(0) = 1$$

$$e^{-a} = 1/e^a \quad \text{or} \quad \exp(-a) = \frac{1}{\exp(a)}$$

With this definition at hand, we can show the following approximation. When n is large and p is small and np is moderate,

$$B(x; n, p) \approx \mathbf{P}(x; np)$$

where

$$\mathbf{P}(x; \lambda) = \frac{\lambda^x}{x!}\exp(-\lambda)$$

In addition to its role in approximating $B(x; n, p)$, the function $\mathbf{P}(\_; \lambda)$ is a frequency function for an *infinite but discrete* random variable X that takes on values 0, 1, 2, ... Such a random variable is said to have a *Poisson* distribution. Moreover the single parameter λ plays a multiple role in that

$$\mu_X = \lambda,$$

and

$$\sigma_X^2 = \lambda, \quad \text{when } X \text{ has frequency function } \mathbf{P}(\_; \lambda)$$

Exercises Section 6

1. If your phone rings an average of three times per day, how many days per year would you expect to receive no calls at all?
2. A particular stretch of road has an average of one accident per day. What is the probability of no more than one accident during an entire week?
3. In Section 5 we gave an algorithm for hypothesis testing with the binomial distribution. Write out an analogous algorithm for hypothesis testing with the Poisson distribution.
4. You are told that no more than one person in fifty can name both his state's United States senators. If you take a random sample of 100, how many people must know both in order to lead you to reject the assertion at the .05 significance level?

5 If only one person in a hundred is more than 74 inches tall, find the probability that in a randomly selected sample of 400 no one is over 74 inches tall.

6 100 raisins are mixed in with some cookie dough to make 50 raisin cookies. Assuming that the probability that a cookie contains a given number of raisins is given by the Poisson distribution, what is the probability that if you eat a cookie it will have four or more raisins?

7 If a 300-page book contains 900 misprints, how many pages do you expect to have no misprints, one misprint, two misprints, etc.? You may assume that the probability that a page has a certain number of misprints is distributed according to the Poisson distribution.

8 The Prussian Cavalry used to keep track of the number of fatalities due to being kicked by horses. The data for 200 corps-years was as follows.

Number of deaths per corps-year	Frequency
0	109
1	65
2	22
3	3
4	1

If we assume that the random variable representing the number of deaths per corps-year (i.e., the number of deaths in one cavalry corps in one year) is distributed according to the Poisson distribution, what value should we give λ? What is the probability of five deaths in a corps-year?

9 Show that the variance of $\mathbf{P}(x; \lambda)$ is λ.

10 It was asserted in the text that

$$\exp(x) = \sum_{k=0}^{\infty} \frac{x^k}{k!}$$

It can be shown that if, instead of taking the sum from 0 to ∞ we take the sum from 0 to n, the error is no greater than the last term included, that is, the error in the approximation is less than $x^n/n!$. What is the smallest value of n which will allow us to calculate $e = \exp(1)$ accurate to six decimal places? Get an approximation of e accurate to six decimals from this.

11 Using the table for e^{-x}, compute the following probabilities:

(a) $\mathbf{P}(0; 2.5)$

(b) $\mathbf{P}(1; 2.2)$

(c) $\mathbf{P}(2; .75)$

(d) $\sum_{k=0}^{2} \mathbf{P}(k; 2.1)$

7 A GLIMPSE AT DECISION THEORY

In Chapter 2 it was mentioned that Bayes' Theorem is used in a recently developed branch of mathematics called *decision theory*. While we shall not attempt to present a full-blown discussion of the subject, we can illustrate one aspect of decision-making with a simple problem. A number of texts are available to the reader who wishes to make a more detailed study.

First of all, decision theory embodies a wide variety of techniques, of which the so-called *Bayesian* method is only one. Moreover, we shall confine ourselves to a rudimentary aspect of the Bayesian approach, namely how to revise certain probability estimates, thereby neglecting a significant portion of the overall "solution" of a given problem.

The following (admittedly contrived) problem provides us with a vehicle for illustrating the above terminology as well as for examining how the Bayesian process can be used.

Suppose that a professor named Joe is an inveterate thumbtack dropper who has noticed over the years that there are basically three kinds of tacks, types A_1, A_2, A_3, characterized by their respective probabilities of landing point up: $\frac{1}{3}$, $\frac{1}{2}$, and $\frac{2}{3}$. Suppose further that experience has led him to believe that of all tacks manufactured, 60% are type A_1, 30% are type A_2, and only 10% are type A_3. Finally let us imagine that Joe wishes to purchase a box of tacks for demonstration in his probability class. While shopping, he finds a sale on thumbtacks, but he is curious about what type they are, for in point of fact, he really wants the rare type A_3 for his demonstration. Surreptitiously, Joe opens a box and drops 20 tacks on the floor, and observes that 13 tacks land point up. The question becomes, should he conclude that he has stumbled upon a box of the rare type A_3 tacks or should he seek further information?

Before we proceed with the solution to Joe's problem, a few comments are in order. First, we should note that there are two distinct kinds of information involved and the Bayesian method takes advantage of both. Not only do we have the experimental information that 13 out of 20 tacks landed point up, but we also have the benefit of Joe's experience that tells us how likely we are to find each of the three types of tacks. By way of contrast, suppose that one of Joe's students is asked to purchase the tacks and knows only that there are three types. If he performs the experiment of dropping 20 tacks and counts 13 with points up, he would be more apt to conclude that he had found type A_3 tacks than Joe would. Alternately, if the student were told the respective likelihoods of finding each type of tack and if he did not perform the experiment of dropping some tacks, he would think it likely that type A_1 tacks, being most plentiful, would be the kind he buys. Finally, we should note that in addition to the fact that determining the type of thumbtack we purchase is a matter of limited practical significance, this problem also suffers from a presumption seldom applicable in the real world. Rather than knowing that only three types of tacks exist, it would be far more reasonable to assume that the probability of landing point up can range throughout an infinite set of values. (For example, it might be more realistic to suppose that p can take on *any* value from .28 through .74.) In other words, we have once again sacrificed realism for simplicity in our mathematical model. Moreover, decision theory is far more powerful than we are able to discern at this stage, and quite complex and realistic assumptions can often be incorporated.

Now let us return to Joe's problem. In a nutshell, he wishes to estimate what type of thumbtack he has found. In the terminology of Bayesian decision techniques, Joe wishes to estimate the probability of three possible *states of nature*, which we shall symbolize by

A_1: the tacks are of type A_1

A_2: the tacks are of type A_2

A_3: the tacks are of type A_3

From his past experience, he can assign *prior probabilities*

$$P(A_1) = .60$$
$$P(A_2) = .30$$
$$P(A_3) = .10$$

Moreover, he can use his knowledge of probability theory to find that, if X represents the number of tacks that land point up out of 20,

$$P(X = 13|A_1) = .003$$
$$P(X = 13|A_2) = .074$$
$$P(X = 13|A_3) = .182$$

These values are, respectively,

$$B(13; 20, \tfrac{1}{3}),\ B(13; 20, \tfrac{1}{2}),\ \text{and}\ B(13; 20, \tfrac{2}{3})$$

since to say for example that A_1 is the existing state of nature implies that dropping 20 tacks would correspond to 20 repeated trials with a probability of success (point up) of $\frac{1}{3}$ on a single trial.

We can represent the information gathered thus far schematically as follows:

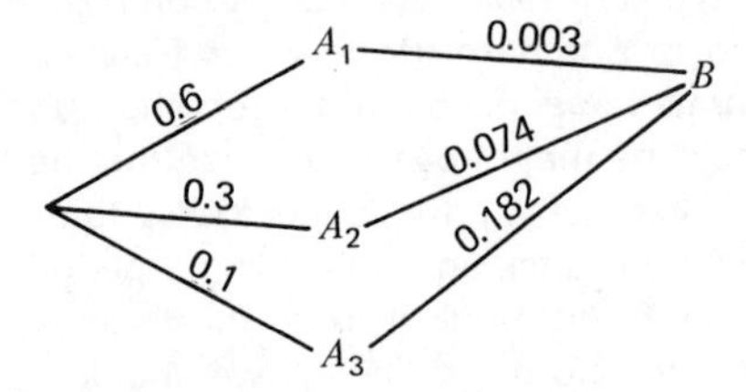

where B signifies the event "13 tacks landed point up." At this point, we invoke Bayes' Theorem to compute

$$P(A_1|B) \approx .043$$
$$P(A_2|B) \approx .526$$
$$P(A_3|B) \approx .431$$

These are the *posterior probabilities.* They represent the revised estimates of likelihoods A_1, A_2, and A_3 are the prevailing states of nature. At this point, the elementary Bayesian decision method is complete. Joe would probably conclude that the tacks in question are very likely either type A_2 or type A_3, but almost definitely not type A_1.

It is at this juncture that more sophisticated aspects of decision theory may enter the picture. It may be possible for Joe to bring to bear many other factors in making his decision. For instance, he may have plenty of time to spare and know of several other sources of tacks or he may be shopping at closing hour and not have the opportunity to shop around. Having type A_3 tacks for his demonstration might be of little value or of considerable value to him. Even the price of the tacks or the opportunities to return them may enter his decision process.

Once again we hasten to point out that all these auxiliary factors may seem contrived and hardly worth incorporating. Yet if we translate our problem to one in which a surgeon is attempting to determine the likelihood of certain states of nature (for example, the type of tumor that might be present in a patient) and combining his experience with some available experimental evidence (symptoms, results of tests and X rays), then the importance of his decision of whether and when to operate or what to expect is such that any additional information would seem worthwhile.

Summary

The usual context for decision theory is the following. A sample space S is partitioned into a finite number of mutually exclusive events:

$$S = A_1 \cup A_2 \cup \ldots \cup A_n, \qquad A_i \cap A_j = \varnothing \quad \text{for} \quad i \neq j$$

These A_i's are called *states of nature*. Our job is to find estimates for the probabilities of each possible state of nature. Generally the process of estimation is two-fold. First we make initial estimates, called *prior probabilities*, based on most anything from experience with similar situations to out-and-out guesswork. Second, we gather additional information about the sample space, usually by a sampling device, and use this information to make revised estimates, called *posterior probabilities*.

Exercises Section 7

1 FBN Airlines decides to inaugurate a special student flight from Columbus to New York to coincide with the beginning of Thanksgiving vacation. One of the executives at FBN is concerned about the possible large number of last minute cancellations. He makes a study of other leading airlines' experiences with student flights. He finds that the cancellation rates generally fall into four categories: 2%, 5%, 10%, and 20% with respective relative frequencies of .05, .15, .60, and .20. In an effort to predict their future cancellation pattern, FBN decides to start the first year with one small flight, accommodating only 40 passengers. Moreover, they decide to accept exactly 40 reservations rather than be faced with the possibility of not having room for some

customers with reservations. Using Bayesian methods, help the officials at FBN determine the probabilities of the four possible cancellation rates for their future flights, if on the day of their trial flight the number of cancellations is
(a) 2
(b) 4
(c) 6
(d) 8

2 Modify the thumbtack problem by assuming that of the 20 tacks dropped in the store,
(a) only 10 landed point up.
(b) only 7 landed point up.

3 A history professor always gives a final examination consisting of 100 true-false questions. Scrutiny of previous exams in a fraternity file indicates that particular patterns are used. Fifty percent of the previous exams were designed so that $\frac{3}{4}$ of the correct answers were "true"; in 40% of the exams, half the correct answers were "true"; and in 10% of the exams, only $\frac{1}{4}$ of the correct answers were "true." A math major realizing that it would be advantageous to know the percent of "true" answers on this exam, decides that he could easily locate 25 questions where he would know the correct answer. Explain how he could then make a good guess as to the distribution of "trues" and "falses" on the exam. In particular, what should he conclude in case his sample of 25 questions had
(a) 19 "true" answers?
(b) 12 "true" answers?
(c) 10 "true" answers?
(d) 7 "true" answers?

8 A SAMPLING OF DISCRETE DISTRIBUTIONS

We have now discussed in some detail the binomial and Poisson distributions. In this section we shall introduce, in considerably less detail, several other commonly used discrete distributions. We begin by compiling a list of these distributions, along with the binomial and Poisson, and indicating in each case the kind of problem it is designed to be applied to.

1 *Binomial distribution:* The frequency function for this distribution is

$$B(x; n, p) = C(n, x)p^x(1 - p)^{n-x}$$

The mean and variance are given by

$$\mu = np, \qquad \sigma^2 = np(1 - p)$$

$B(x; n, p)$ is the probability that in n trials of an experiment you have x successes, where the probability of success on any single trial is p and where the outcomes of the trials are independent events.

2 *Poisson distribution:* The frequency function for this distribution is

$$\mathbf{P}(x; \lambda) = \frac{e^{-\lambda}\lambda^x}{x!}$$

The mean and variance of this distribution are respectively

$$\mu = \lambda \quad \text{and} \quad \sigma^2 = \lambda$$

This distribution is applied in two types of situations. The first is in problems where the random variable can, on a given trial of an experiment, take on any integer 0, 1, 2, . . . as a value, and where λ is the average value. A typical example of this would be the probability that a business firm receives x calls in a certain time interval when they receive an average of λ calls during that period.

The second application of the Poisson distribution is as an approximation of the binomial distribution under certain circumstances. It serves as an approximation when n is large and np is of moderate size, say $np < 5$. In this case, $\lambda = np$.

3 *Hypergeometric distribution:* The frequency function for this distribution is

$$H(x; N, n, a) = \frac{C(a, x)C(N - a, n - x)}{C(N, n)}$$

The mean and variance of the hypergeometric distribution are given by

$$\mu = n \cdot \frac{a}{N} \quad \text{and} \quad \sigma^2 = n \cdot \frac{a}{N} \cdot \frac{N - a}{N} \cdot \frac{N - n}{N - 1}$$

You should note that the hypergeometric distribution is defined only when $x \geq 0$ and $x \leq a$, $x \leq n$.

This distribution is for use in sampling without replacement. (The binomial distribution copes with sampling with replacement.) In this distribution N represents the size of the population, n represents the size of the sample, a represents the number of elements in the population with a certain property, and x is, of course, the number of elements in the sample with that property. In cases where N is large relative to n, say $N > 10n$, it is common to use the binomial distribution for sampling without replacement since the error will be small.

EXAMPLE

A box of 10 flashbulbs contains three defectives. If you test a sample of size two, what is the probability that you will find at least one defective?

We have, as population size, $N = 10$. Our sample size is $n = 2$. The property that we are interested in is that of being defective and the number of defectives is $a = 3$. The answer to the problem is

$$H(1; 10, 2, 3) + H(2; 10, 2, 3) = \frac{C(3, 1)C(7, 1)}{C(10, 2)} + \frac{C(3, 2)C(7, 0)}{C(10, 2)}$$

$$= \frac{3 \cdot 7}{45} + \frac{3 \cdot 1}{45} = \frac{24}{45} \approx .533$$

4 *Multinomial distribution:* This distribution is analogous in many ways to the binomial distribution, but differs from it in that it deals with situations where you are interested in more than two outcomes. To be more precise, the binomial distribution is used for repeated independent trials in the situation where on any given trial one of two things happens. The multinomial distribution is used where more than two outcomes are possible on any given trial. It would be used, for example, if you were going to draw, with replacement, cards from a standard deck of 52 and see whether they were diamonds, hearts, spades, or clubs.

In the case where there are k outcomes to be considered, the frequency function is

$$M(x_1, x_2, \ldots, x_k; n, p_1, p_2, \ldots, p_k) = \frac{n!}{(x_1!) \ldots (x_k!)} p_1^{x_1} \ldots p_k^{x_k}$$

where $x_1 + \cdots + x_k = n$ and $p_1 + \cdots + p_k = 1$. This is interpreted as follows. If the experiment has k mutually exclusive but exhaustive outcomes, if the first occurs with probability p_1 and, in general, the ith occurs with probability p_i, if successive trials of the experiment are independent events, and if the experiment is repeated n times, then the probability that the first of the k outcomes occurs x_1 times and in general the ith outcome occurs x_i times is given by the above formula.

EXAMPLE

Four cards are drawn, with replacement, from a standard deck of 52. What is the probability that two of them are hearts, one is a club, and one is a diamond?

Let the four outcomes be, first drawing a heart, second drawing a diamond, third drawing a spade, and fourth drawing a club. We have $n = 4$, $x_1 = 2$, $x_2 = 1$, $x_3 = 0$, $x_4 = 1$, and $p_1 = p_2 = p_3 = p_4 = \frac{1}{4}$. The answer to the problem is, therefore,

$$M\left(2, 1, 0, 1; 4, \frac{1}{4}, \frac{1}{4}, \frac{1}{4}, \frac{1}{4}\right) = \frac{4!}{2!1!0!1!}\left(\frac{1}{4}\right)^2\left(\frac{1}{4}\right)^1\left(\frac{1}{4}\right)^0\left(\frac{1}{4}\right)^1$$

$$= \frac{24}{2}\left(\frac{1}{4}\right)^4 = \frac{3}{4^3} = \left(\frac{3}{64}\right) \approx .0469$$

5 *Generalized hypergeometric:* The generalized hypergeometric distribution bears the same relation to the multinomial distribution that the hypergeometric distribution bears to the binomial distribution. To be more precise it relates to a situation where you have drawings without replacement and wish to consider more than two mutually exclusive, exhaustive outcomes. The frequency function is

$$GH(x_1, \ldots, x_k; N, n, a_1, \ldots, a_k) = \frac{C(a_1, x_1)C(a_2, x_2) \ldots C(a_k, x_k)}{C(N, n)}$$

where:

(a) The total population has N elements.

(b) A sample of size n is drawn (without replacement).

(c) Of the N elements of the population, a_i have the ith property, where i goes from 1 to k, and $N = a_1 + \cdots + a_k$.

(d) Of the n elements of the sample, x_i have the ith property, where i goes from 1 to k, and $x_1 + \cdots + x_k = n$.

EXAMPLE

Suppose the drawing in the example on the multinomial distribution was done without replacement. Now $N = 52$, $n = 4$, $x_1 = 2$, $x_2 = 1$, $x_3 = 0$, $x_4 = 1$, and $a_1 = a_2 = a_3 = a_4 = 13$. The required probability is

$$GH(2, 1, 0, 1; 52, 4, 13, 13, 13, 13) = \frac{C(13, 2)C(13, 1)C(13, 0)C(13, 1)}{C(52, 4)}$$

$$= \frac{78 \cdot 13 \cdot 1 \cdot 13}{\frac{52 \cdot 51 \cdot 50 \cdot 49}{4 \cdot 3 \cdot 2}}$$

$$= \frac{78 \cdot 13}{17 \cdot 25 \cdot 49} \approx .0497$$

It should be noted that this is only about .003 away from the answer to the earlier problem. The multinomial can be used as an approximation to the generalized hypergeometric when the sample is small relative to the population, just as we used the binomial as an approximation to the hypergeometric.

6 *Geometric distribution:* This distribution is just a generalization of the problem we have considered in the past of tossing a coin until the first time we get tails. The geometric distribution is used whenever we have a binomial experiment and want to calculate the probability that the first success is on the xth trial when the probability of success on a single trial is p. We have the frequency function

$$G(x; p) = p(1 - p)^{x-1}$$

The mean and variance of this distribution are respectively

$$\mu = \frac{1}{p} \quad \text{and} \quad \sigma^2 = \frac{1 - p}{p^2}$$

7 *Negative binomial distribution:* Suppose once again we have a binomial experiment to be repeated several times. But also suppose that we don't want the kind of information given to us directly by the binomial distribution (that is, the probability of x successes in n trials). Instead suppose that we want the probability that the kth success is on the $(x + k)$th trial. Assuming that the probability of success on any given trial is p, we have

$$NB(k, x; p) = C(k + x - 1, x)p^k(1 - p)^x$$

EXAMPLE

If you roll a die repeatedly, what is the probability that you will roll your second 6 on the fourth roll?

In this problem $k = 2$, $x + k = 4$, and so $x = 2$, $p = \frac{1}{6}$ and $1 - p = \frac{5}{6}$. The answer is therefore

$$NB(2, 2; \tfrac{1}{6}) = C(3, 2)\,(\tfrac{1}{6})^2(\tfrac{5}{6})^2$$

$$= \frac{(3 \cdot 25)}{6^4} \approx .0579$$

To complete the discussion we point out that the mean and variance of the negative binomial distribution are given by

$$\mu = \frac{k(1 - p)}{p} \quad \text{and} \quad \sigma^2 = \frac{k(1 - p)}{p^2}$$

There are a number of other distributions which could be discussed. The point to be made here is that there are a wide variety of distributions which have been studied. Working statisticians have several at their fingertips and, given a problem, they can usually find a model which will serve.

Exercises Section 8

Most of the problems in this set do not refer to a specific distribution. In each case a part of the problem is to select, from among the distributions discussed in the text, the most suitable one for the problem.

1 Write out algorithms for hypothesis testing for each of the five new distributions introduced in this section. Writing out the general algorithm for the multinomial and hypergeometric distributions would be overly tedious, so assume that at most two variables among the k are involved in the hypothesis.

2 A factory has three buildings labeled A, B, and C. Of its employees 20 work in A, 30 in B, and 50 in C. A grievance committee of three is chosen from among the workers. Find the probability that there will be one member of the committee from each building.

3 An urn contains three white balls and a single red ball. Balls are drawn from the urn without replacement, until the red ball is drawn. Find the probability that the red ball is drawn on the first draw, second draw, third draw, and fourth draw.

4 Modify Exercise 3 by having the drawing with replacement.

5 A photographer buys flash bulbs in boxes of 40. As a quality control device he tries a sample of a certain size and rejects the box if any of the sample prove defective. He wants this test to be good enough so that if there are 10 defectives in the box he has a probability of .95 of rejecting the box. What is the smallest sample size he can use?

6 A college has 2,000 male students and 1,000 female students. A sociology major, preparing a term paper, is interviewing randomly selected students to discover their preferences in baroque music. What is the probability that it is not until his fifth interview that he interviews a female?

7 In Exercise 6, what is the probability that the second female occurs at the sixth interview?

8 A basketball player makes 70% of his free throws. Suppose he makes a wager that he can make at least four free throws in a row.

(a) What is the probability of accomplishing this feat?

(b) What is the probability that he will make at least three in a row?

9 Let X be a variable whose frequency function is

$$f(k) = q^{k-1}p, \qquad \text{for } k = 0, 1, 2, \ldots$$

Show that for any two positive integers m and n

$$P(X > m + n | X > m) = P(X > n)$$

and interpret this result.

10 Suppose that a cereal manufacturer claims to put one valuable coupon in every five boxes of cereal. What is the probability that you won't get a coupon if you buy four boxes of cereal?

9 GRAPHS OF FUNCTIONS AND EQUATIONS

As we begin to refine our ideas about probability and statistics, it becomes important to be able to draw from a larger source of examples. Whereas the counting techniques discussed earlier help us cope with discrete phenomena, the bulk of the material covered in the remainder of this chapter and virtually all of the next chapter involves functions of one or more continuous variables. Thus the object of this section is to develop some familiarity with some of the more elementary functions and their graphs.

The idea of the graph of a function plays an important role in mathematics. Through it, geometry and algebra are merged and each gains from the other. The basic idea, commonly ascribed to Descartes, though others had used the concepts before him, is to make each equation in x and y correspond to a curve in the plane and each curve in the plane correspond to an equation. Thus an algebra problem involving one or more equations can be represented as a picture and the nature of the picture often gives an insight into the solution of the algebraic problem. Conversely a geometry problem involving one or several curves can be translated to an algebra problem involving one or several equations and the solution of the algebraic problem will answer the geometry problem.

We begin by reviewing the notion of a *coordinate plane*. Two number lines are drawn, one horizontal (called the *x-axis*) and one vertical (called the *y-axis*) such that they meet at their respective points labeled "0" (see Fig. 3.12). These two lines determine a plane (called a *coordinate* or *xy-plane*), every point P of which is assigned an ordered pair of real numbers (x, y), where x (called the *x-coordinate* or *first coordinate*) is the label of the point on the x-axis directly above or below P and y (called the *y-coordinate* or *second coordinate*) is the point on the y-axis directly to the left or right of P. Once we have a coordinate plane, we can sketch *graphs*. Usually we want to sketch either *graphs of functions* or *graphs of equations*. Each graph is simply a set of points, perhaps

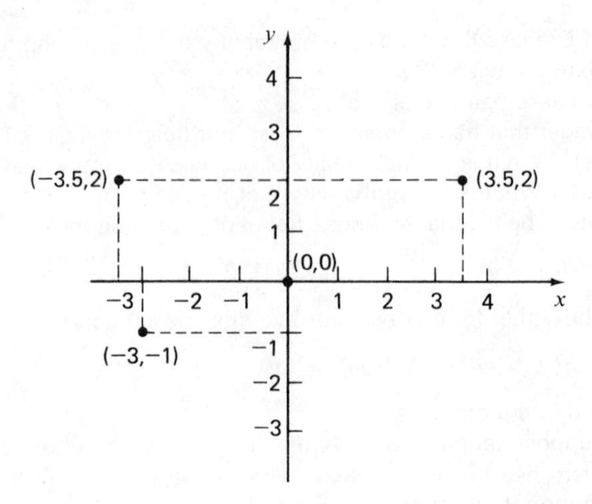

FIGURE 3.12 A coordinate plane with a few representative points.

a line or a circle or a parabola or some other curve. We have already sketched graphs of frequency functions that consisted of a finite set of points. Before we sketch more function graphs, let us review the definition of a function, for the time being restricting our attention to a particular kind of function (a real-valued function of one real variable).

Definition Let D be a subset of the real numbers (e.g., D might be a finite set, or some interval, or all the positive real numbers, or all the real numbers). A *function f with domain D* is a rule that assigns to each x in D a unique real number, denoted $f(x)$.

EXAMPLE 1 Let $D = \{0, 1\}$. We define a function f by

$$f(0) = \tfrac{1}{2}$$
$$f(1) = \tfrac{1}{2}$$

This function f has been studied previously. It represents the frequency function for the (discrete) random variable that counts the number of heads in one toss of a fair coin.

EXAMPLE 2 Let $D = \{1, 2, 3, 4, 5, 6\}$. We define f by

$$\left.\begin{aligned} f(1) &= \tfrac{1}{6} \\ f(2) &= \tfrac{1}{6} \\ f(3) &= \tfrac{1}{6} \\ f(4) &= \tfrac{1}{6} \\ f(5) &= \tfrac{1}{6} \\ f(6) &= \tfrac{1}{6} \end{aligned}\right\}$$

This function is also familiar. It is the frequency function for the random variable that records the outcome of one roll of a fair die.

So long as the domain D of the function is finite, it is possible to write out all the assignments. Of course if D contained one million points, we might hope for some more efficient method of prescribing the rule f. Moreover when D is infinite, it becomes necessary to discover a pattern that f assumes and to specify the assignments in some abbreviated fashion.

EXAMPLE 3 Let $D = \{x|x \geq 0\}$, the non-negative real numbers. We define a function f with domain D by

$$f(x) = \sqrt{x}, \qquad \text{for all } x \in D$$

Thus, for example,

$$f(4) = 2, \qquad f(0) = 0, \qquad f(16) = 4, \qquad f(26) = \sqrt{26}, \qquad \text{etc.}$$

EXAMPLE 4 Let $D =$ the set of all real numbers, i.e., the interval $(-\infty, \infty)$. We define f by

$$f(x) = 2x + 1, \qquad \text{for } x \in (-\infty, \infty)$$

Thus for example

$$f(0) = 1, \qquad f(-3) = -5, \qquad f(100) = 201, \qquad \text{etc.}$$

Definition Let f be a function with domain D. The *graph of* f is the set of all points in a coordinate plane of the form $(x, f(x))$ where $x \in D$.

The graphs of the four functions in the above examples are sketched below.

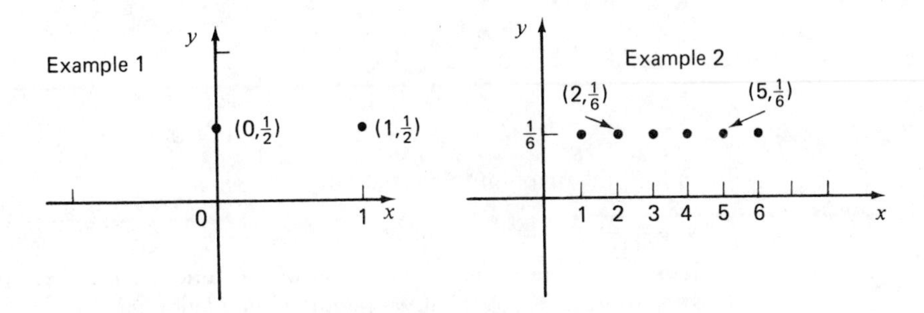

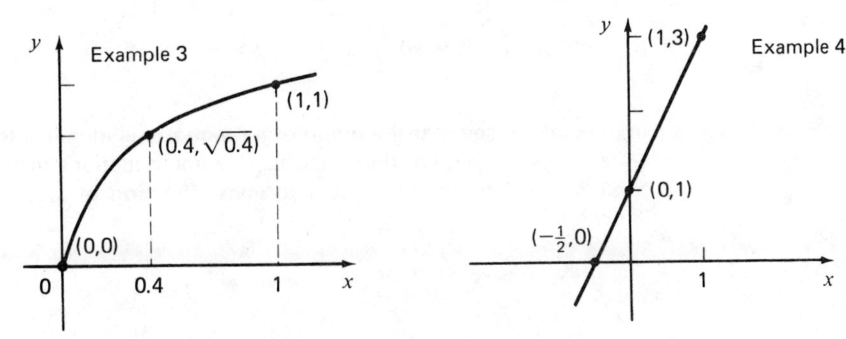

FIGURE 3.13 Graphs of the functions in Examples 1–4.

Another way of viewing these graphs, especially the last two, is in terms of an equation. Since the graph of the function f defined by

$$f(x) = \sqrt{x}, \qquad \text{for } x \geq 0$$

is the collection of all ordered pairs of the form

$$(x, \sqrt{x}) \qquad (\text{where } x \geq 0)$$

We can think of these ordered pairs as satisfying the equation

$$y\text{-coordinate} = \sqrt{x\text{-coordinate}}$$

or more tersely, $y = \sqrt{x}$. Equivalently, we can write

$$y - \sqrt{x} = 0$$

a form which is a special case of the following.

Definition Let $G(x, y) = 0$ be an equation with two variables x and y. The *graph* of this equation is the set of all points (x, y) such that $G(x, y) = 0$.

EXAMPLE 5 The graph of the equation

$$y - 2x - 1 = 0$$

consists of all points whose coordinates satisfy the equation. Some representative points are

$(0, 1)$ since $1 - 2 \cdot 0 - 1 = 0$

$(1, 3)$ since $3 - 2 \cdot 1 - 1 = 0$

$(-4, -7)$ since $-7 - 2 \cdot (-4) - 1 = 0$

In fact, the graph of this equation,

$$y - 2x - 1 = 0$$

is precisely the same as the graph of the function f of Example 4, namely a straight line. Of course, if we rewrite the equation as

$$y = 2x + 1$$

then we see that the equation simply says

$$y = f(x)$$

In general we see that the graph of a function f is identical to the graph of the equation $y = f(x)$. On the other hand, some equations in x and y are not the result of merely setting y equal to some function of x, as the next example illustrates.

EXAMPLE 6 The graph of the equation

$$x^2 + y^2 = 5$$

is a circle of radius $\sqrt{5}$ whose center is at $(0, 0)$

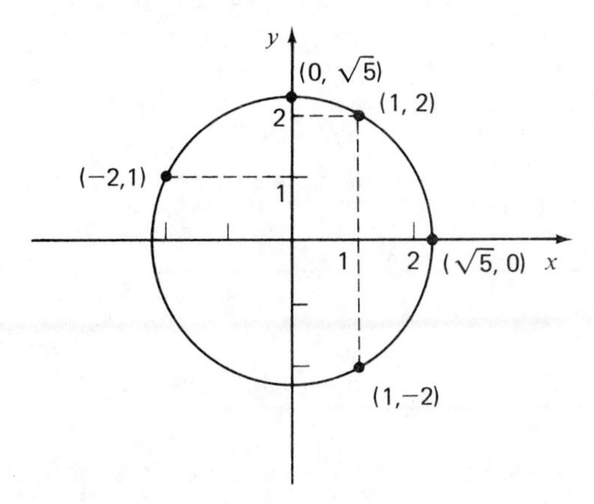

FIGURE 3.14 The graph of the equation $x^2 + y^2 = 5$.

This circle could not be the graph of a function of the form $y = f(x)$ because it contains, for example, the points $(1, 2)$ and $(1, -2)$, and were it a function graph we would conclude

$$2 = f(1), \qquad \text{since } (1, 2) \text{ is on the graph}$$

and

$$-2 = f(1), \qquad \text{since } (1, -2) \text{ is on the graph}$$

Such ambiguity [$f(1)$ having two values] is not tolerated with functions.

We have seen one function

$$f(x) = 2x + 1$$

whose graph is a line. More generally we have:

For any choice of real numbers a and b, the graph of the function

$$f(x) = ax + b$$

is a line.

Several examples of graphs of these functions are shown in Fig. 3.15. Notice how we can generate various "families" of lines by leaving the coefficient a fixed and changing b, or vice versa. These numbers a and b are called *parameters*. It should be clear from looking at these pictures of lines that the value of a in $ax + b$ regulates the degree of steepness of the line, whereas the value of b determines where the graph cuts across the y-axis. (That is, when $x = 0$, we have $y = a \cdot 0 + b = b$, so that the point $(0, b)$ lies on the graph of $y = ax + b$.) We call a the *slope* and b the *y-intercept* of the line $y = ax + b$.

In high school geometry, we learned that a line is determined by specifying two points on it. Let us see how to find the equation of a line determined by two points, say (x_1, y_1) and (x_2, y_2). First we observe that if the equation of

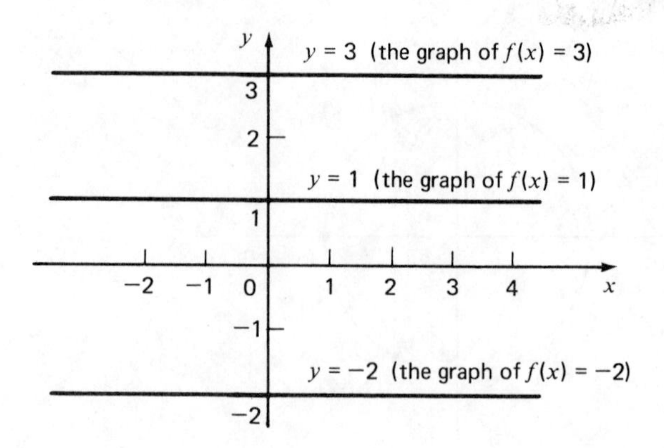

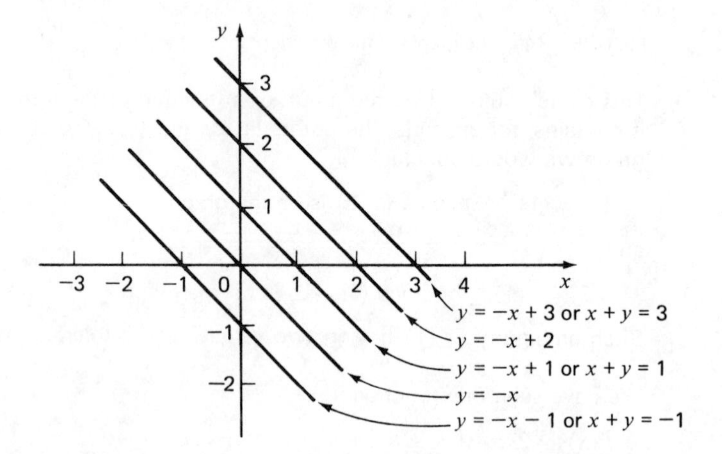

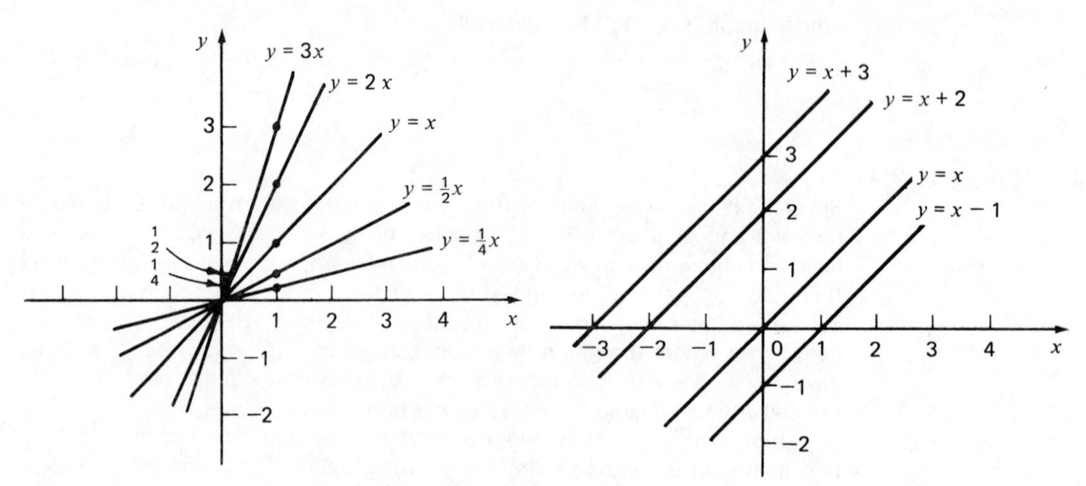

FIGURE 3.15 Graphs of functions of the form $f(x) = ax + b$.

the line is to be of the form

$$y = ax + b$$

then we must have

$$y_1 = ax_1 + b$$
$$y_2 = ax_2 + b$$

Subtracting, we get

$$\begin{aligned} y_1 - y_2 &= ax_1 - ax_2 \\ &= a(x_1 - x_2) \end{aligned}$$

Now (assuming for the moment that $x_1 \neq x_2$) we divide by $(x_1 - x_2)$

$$\boxed{\frac{y_1 - y_2}{x_1 - x_2} = a}$$

The slope of a line can be found by taking the ratio of the differences of the respective of any two points on the line.

Moreover, if (x, y) is any other point on the line, then

$$\frac{y - y_1}{x - x_1} = a \qquad \text{(Again we assume } x \neq x_1.\text{)}$$

or

$$\boxed{y - y_1 = a(x - x_1)}$$

This is called the *point-slope formula* for a line.

PROBLEM Find the equation for the line passing through the points (1, 3) and (−1, −1).

SOLUTION We first compute the slope of the desired line.

$$\frac{y_1 - y_2}{x_1 - x_2} = \frac{3 - (-1)}{1 - (-1)} = \frac{4}{2} = 2$$

Next we use the point-slope formula, substituting in the coordinates of the point (1, 3);

$$y - y_1 = a(x - x_1)$$

gives

$$y - 3 = 2(x - 1)$$

or

$$y - 3 = 2x - 2$$

which can be rewritten in any of several ways:

$$y = 2x + 1, y - 2x = 1, y - 2x - 1 = 0$$

Note that we could have used *either* of the given points when substituting in the point-slope formula. Had we used the point $(-1, -1)$ instead of $(1, 3)$, the computation would have been

$$y - (-1) = 2(x - (-1))$$

or

$$y + 1 = 2x + 2$$

or

$$y = 2x + 1$$

as we found.

PROBLEM Find the equation of a line passing through the point $(2, 1)$ and having slope (-2).

SOLUTION Since the slope is already given, we use the point-slope formula:

$$y - y_1 = a(x - x_1)$$

to get

$$y - 1 = (-2)(x - 2)$$

or

$$y = -2x + 5$$

Special case of lines: We have omitted the case where two points on a line have the same first coordinate. When this happens, the line must be *vertical*, and *all* its points have the same first coordinate. The equation of such a line is of the form

$$x = c$$

Equation of a vertical line

and some examples are sketched in Fig. 3.16.

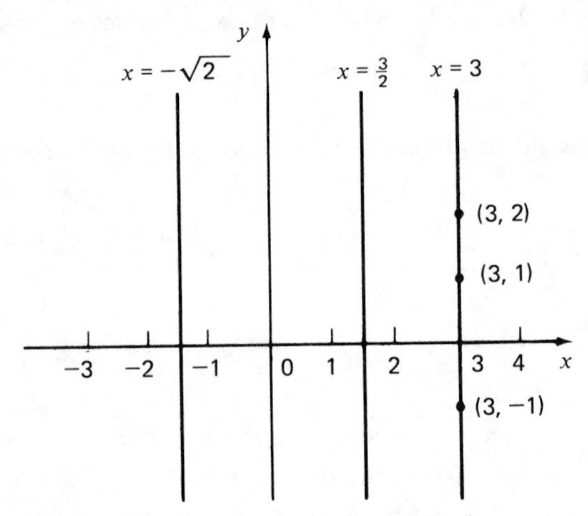

FIGURE 3.16 Graphs of equations of the form $x = c$.

Now we turn our attention to another family of graphs called parabolas. (Actually we won't study all possible parabolas in the plane—only those that are graphs of functions.) In Fig. 3.17 we illustrate various graphs of functions of the form $f(x) = a(x - b)^2 + c$ or, alternately, of equations of the form

$$y = a(x - b)^2 + c$$

From these pictures we can see how the choice of the three parameters a, b, and c in

$$y = a(x - b)^2 + c$$

affect the location and shape of the curve.

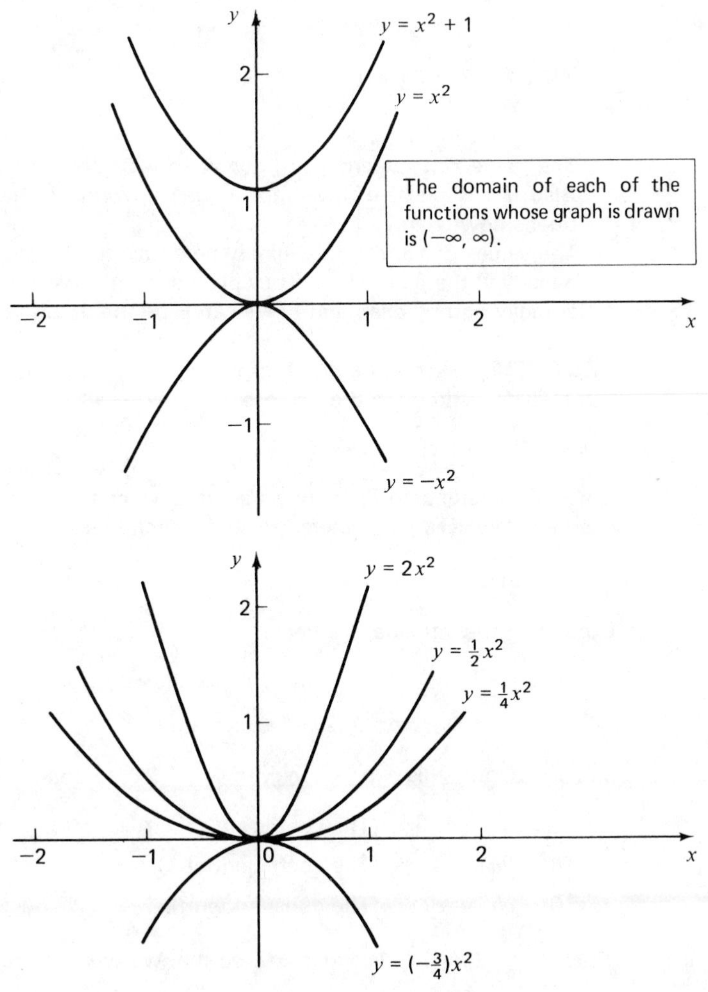

FIGURE 3.17 Graphs of equations of the form $y = a(x - b)^2 + c$. *(continued overleaf)*

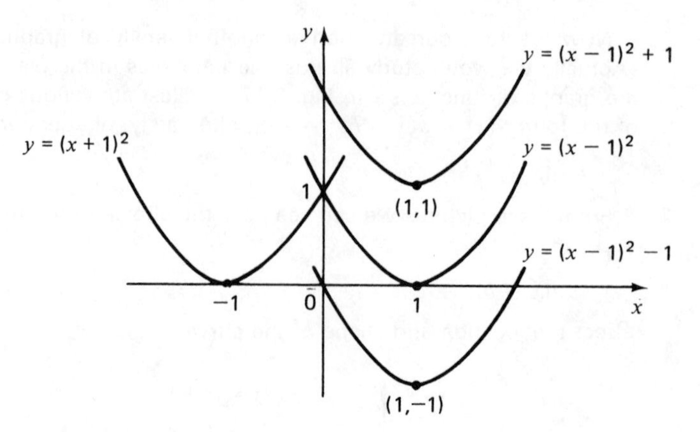

FIGURE 3.17 *(continued)*

1 The value of a determines how narrow or how spread out the graph is. Also if a is positive, the graph opens upward; if a is negative, the graph opens downward.
2 The values of b and c determine where the "turning point" of the graph lies, namely at the point (b, c). Thus by changing b, we can slide the graph horizontally and by changing c, we can slide the graph vertically.

PROBLEM Sketch the graph of $y = 4x - 2x^2$.
SOLUTION Although the expression

$$4x - 2x^2$$

is not in the standard form for parabolas, we can, by the following technique, rewrite it. We seek parameters a, b, and c such that

$$a(x - b)^2 + c = 4x - 2x^2$$

Expanding the left side, we get

$$\begin{aligned} a(x - b)^2 + c &= a(x^2 - 2bx + b^2) + c \\ &= ax^2 - 2abx + (ab^2 + c) \end{aligned}$$

and we identify this expression with $4x - 2x^2$ to conclude

$$\begin{aligned} -2 &= a && \text{(The coefficient of } x^2\text{)} \\ 4 &= 2ab && \text{(The coefficient of } x\text{)} \\ 0 &= ab^2 + c && \text{(The constant term)} \end{aligned}$$

Thus $a = -2$, $b = -1$, and $c = 2$, so that we want to graph the equation

$$y = -2(x - 1)^2 + 2$$

From the comments above about the role of the parameters, we see that the turning point of the graph will be at (1, 2), the graph will open downward, and it will be twice as steep as (and hence narrower than) the parabola $y = x^2$. Now the graph can be sketched by selecting a few points that satisfy the equation:

When $x = 0$, $\quad y = -2(0 - 1)^2 + 2 = 0$

When $x = 1$, $\quad y = -2(1 - 1)^2 + 2 = 2$

When $x = 2$, $\quad y = -2(2 - 1)^2 + 2 = 0$

When $x = -\frac{1}{2}$, $\quad y = -2(-\frac{1}{2} - 1)^2 + 2 = -\frac{5}{2}$

When $x = \frac{5}{2}$, $\quad y = -2(\frac{5}{2} - 1)^2 + 2 = -\frac{5}{2}$

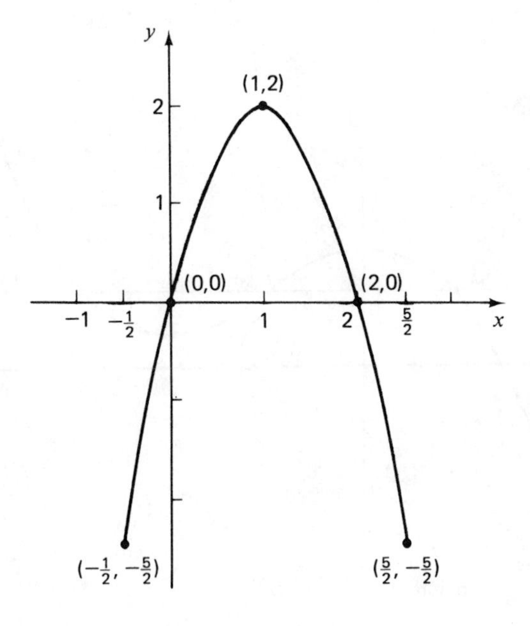

FIGURE 3.18 The graph of $y = 4x - 2x^2$.

EXAMPLE 7 A Cubic Curve

Our discussion about parabolas (that is, graphs of quadratic expressions in x) can be generalized to graphs of higher degree polynomials although the roles of the parameters are more difficult to characterize. We present one example of a third degree polynomial and we plot some representative points and "connect the dots" to get a sketch of its graph.

Consider the equation

$$y = x^3 - 3x^2 - x + 3$$

We list a table of values

x	y
0	3
−1	0
1	0
−2	−15
−3	−48
2	−3
3	0
4	15
5	48

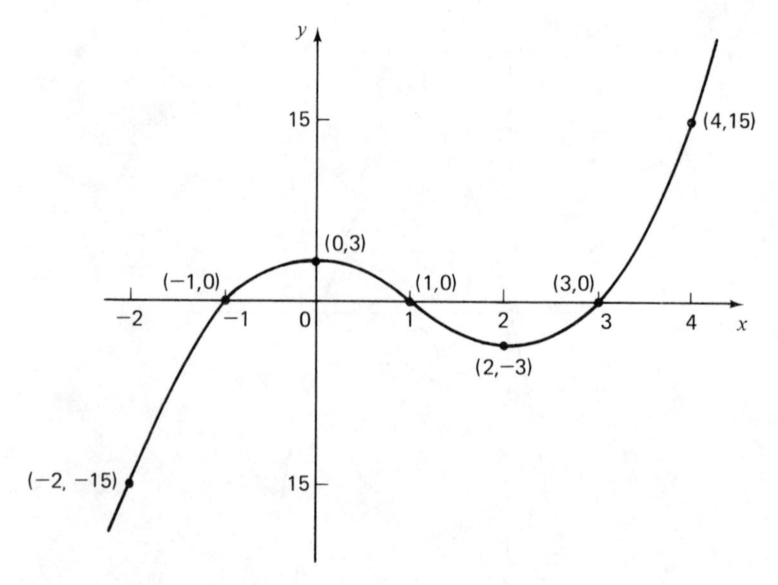

FIGURE 3.19 The graph of $y = x^3 - 3x^2 - x + 3$.

Throughout our discussion of lines and parabolas, we have ignored the *domain* of the functions

$$f(x) = ax + b \quad \text{and} \quad f(x) = a(x - b)^2 + c$$

The reason for this is that both functions have for their domain the entire set of real numbers. In our next set of examples, we may still have all real numbers in the domain, but the rules for the functions demand that we pay attention to the location of the point x.

EXAMPLE 8 The Postage Stamp Function

For each value $x > 0$, there corresponds a cost to mail x ounces of mail by first class. If we denote this cost by $C(x)$, then we have a function where the domain is the set of positive real numbers and whose rule looks like

$$C(x) = \begin{cases} 8, & \text{if } 0 < x \leq 1 \\ 16, & \text{if } 1 < x \leq 2 \\ 24, & \text{if } 2 < x \leq 3 \\ 32, & \text{if } 3 < x \leq 4 \\ \text{and so on} \end{cases}$$

A function of this type is called a *step function* because of the shape of its graph.

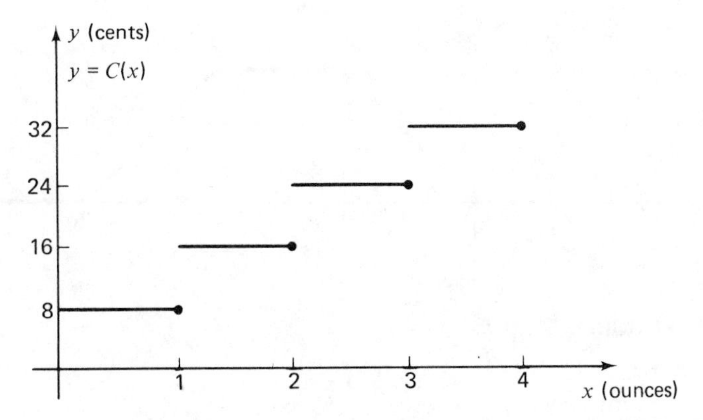

FIGURE 3.20 The graph of the postage function.

Other examples of step functions are taken up in the exercises. One important feature of a function whose rule changes for different parts of the domain is that the different portions of the domain do not overlap. For example it would not be proper to define

$$F(x) = \begin{cases} 3, & \text{if } 1 \leq x \leq 2 \\ 4, & \text{if } 2 \leq x \leq 3 \\ 0, & \text{otherwise} \end{cases}$$

since we would introduce an ambiguity about the value of

$$F(2)$$

That is, one part of the rule tells us $F(2) = 3$, whereas another part tells us

$$F(2) = 4$$

Another useful source of examples comes from piecing together various graphs we know.

EXAMPLE 9 Let us find the graph of

$$h(x) = \begin{cases} 0, & \text{if } x < 0 \\ x, & \text{if } 0 \le x \le 1 \\ 1, & \text{if } 1 < x \le 2 \\ 3 - x, & \text{if } 2 < x \le 3 \\ 0, & \text{if } 3 < x \end{cases}$$

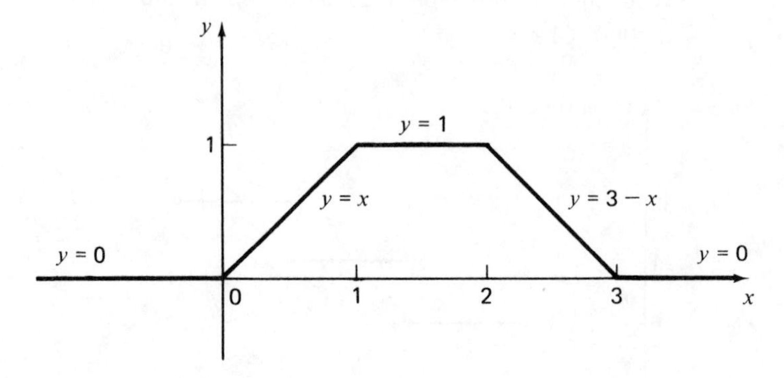

FIGURE 3.21 Graph of the function h.

Summary

The object of this section has been to explain with some precision how to graph elementary functions of one real variable and to provide some examples of elementary functions and their graphs. All of our graphs were collections of points in a *coordinate plane*, i.e., a collection of points with labels of the form (x, y), where x represents the position on the horizontal axis directly above or below the point and y the position on the vertical axis directly to the left or the right of the point.

We defined a *function* f with domain D to be a rule which assigns a number denoted $f(x)$ to each x in D. The *graph of f* consists of all points of the form $(x, f(x))$ where $x \in D$. This set of points is also the *graph of the equation* $y = f(x)$. We examined the family of lines, or the collection of all graphs of equations of the form

$$y = ax + b \qquad \text{(Non-vertical line)}$$
$$x = c \qquad \text{(Vertical line)}$$

where the numbers a, b, and c are called *parameters*. In particular, a is called the *slope* of the line $y = ax + b$ and b is called its *y-intercept*; the slope is a measure of the rate at which the line rises (if $a > 0$) or falls (if $a < 0$) as x increases in value, and the y-intercept tells us where the line meets the vertical axis. When two points (x_1, y_1) and (x_2, y_2) are given, the line determined by them (i.e., the line passing through both of them) has the equation

$$y - y_1 = m(x - x_1), \qquad \text{where the slope } m = \frac{y_2 - y_1}{x_2 - x_1}$$

except in the special case where $x_1 = x_2$ when the line joining (x_1, y_1) to (x_2, y_2) is the vertical line $x = x_1$.

We also considered a *family of parabolas* consisting of graphs of equations of the form

$$y = a(x - b)^2 + c$$

where the parameters a, b, and c gave us information about the shape and position of the curves.

Finally we sketched the graph of a function (the postage stamp function) whose rule is given in several parts and whose graph reflects this property by being a collection of line segments, called a *step function*. In point of fact, many of the functions we shall encounter will have rules given in this form and will have graphs that look like several segments of graphs have been pasted together.

Exercises Section 9

1 Find the equation of and graph the line:

(a) passing through the points (1, 2) and (3, 4).
(b) passing through the points (0, 1) and (1, 0).
(c) passing through (1, 3) and having slope -2.
(d) with y-intercept 3 and slope -1.
(e) parallel to the y-axis and passing through (2, 5).

2 Sketch the graph of each of the following equations.

(a) $y = 1 - x^2$
(b) $y = 2 - x^2$
(c) $y = x^2 - 2$
(d) $y = 4x^2$
(e) $y = (x - 1)^2 + 2$
(f) $y = (x - 2)^2 + 1$
(g) $y = -(x - 2)^2 + 1$

3 Plot a few points to sketch the graph of each of the following functions.

(a) $F(x) = x - x^2$
(b) $F(x) = x^2 - x$
(c) $F(x) = 3 - x + x^2$
(d) $F(x) = x^3 - x^2 - 4x + 4$
(e) $G(x) = \begin{cases} -x, & \text{if } x < 0 \\ x, & \text{if } x \geq 0 \end{cases}$
(f) $G(x) = x^3$
(g) $G(x) = x^3 - 1$
(h) $G(x) = (x - 1)^3$

4 Find the slope of each of the following lines.

(a) $2x - 3y = 7$
(b) $7 + 3x = 5y$
(c) $-x - y - 2 = 0$
(d) $6x = 7 - 2y$

5 A parking lot advertises:
50 cents for the first half-hour
25 cents for each additional hour or any part thereof
Sketch the graph of the function C defined by
$C(x)$ = cost of parking for x hours

6 A taxi cab advertises a fare of
35 cents for the first mile
5 cents for each additional $\frac{1}{10}$ mile
Sketch the graph of the function F defined by
$F(x)$ = fare for traveling x miles

7 Sketch the graphs of each of the following functions.

(a)
$$F(x) = \begin{cases} 1, & \text{if } x < 0 \\ 1 - x, & \text{if } 0 \leq x \leq 1 \\ (x-1)^2, & \text{if } 1 < x \end{cases}$$

(b)
$$G(x) = \begin{cases} -1, & \text{if } x < 0 \\ 1, & \text{if } x \geq 0 \end{cases}$$

8 A copying machine at a small college in the midwest has the following rate schedule for making x copies of a single page:

$$C(x) = \begin{cases} 5.5x \text{ (cents)}, & \text{if } x = 1, 2, 3 \\ 3.5x & \text{if } x = 4,5,6,7,8,9 \\ 2.5x & \text{if } x \geq 10 \end{cases}$$

(a) Sketch the graph of this cost function.
(b) Find all situations where it would be more economical to make more copies than you want.

10 DISTRIBUTION FUNCTIONS

Now that we have explored the elementary properties and encountered several typical examples of discrete random variables and their associated frequency functions, we turn our attention to continuous random variables. The first observation we should make is that the continuous case demands a slightly different treatment. In particular, the standard properties of a frequency function f for a discrete variable X are

$$f(x) = P(X = x), \qquad \text{for all } x \text{ in the range of } X$$

and

$$\Sigma f(x) = 1, \qquad \text{where the sum ranges over all possible values } x \text{ of } X$$

To assert these properties for a continuous variable X would be silly. For one thing, the only continuous variables X we have studied so far (see our earlier discussion of round-off error or continuous clocks) have had the property that

$$P(X = x) = 0, \qquad \text{for every } x$$

so that the frequency function would be identically zero. Second, continuous variables, by definition, take on all values in some interval, so that there would be no way to give meaning to the expression

$$\Sigma f(x)$$

Happily, there is an appropriate modification of the frequency function to the continuous case. The analogy can perhaps best be developed by first introducing the notion of a distribution function for a discrete random variable, which we do by way of a now familiar example.

EXAMPLE 1 The Distribution Function for $B(x; 3, \frac{1}{2})$

Recall the frequency function $B(x; 3, \frac{1}{2})$ for a random variable X. This function is defined for x values of 0, 1, 2, 3 and its graph is given in Fig. 3.22. In Fig. 3.23 we have the graph of a companion function F, called the *distribution function* or the *cumulative frequency function* for X, and defined by

$$F(x) = P(X \leq x)$$

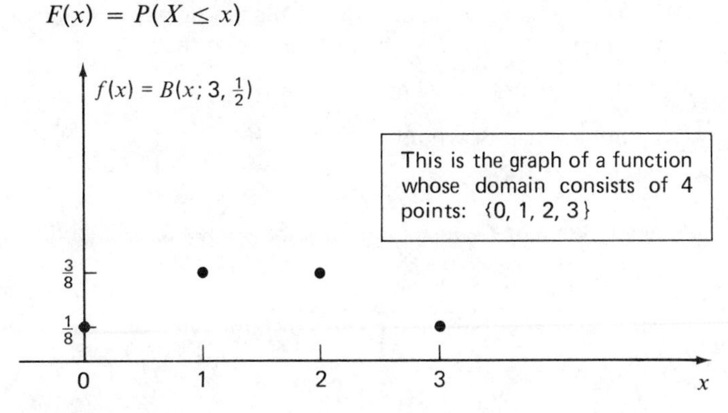

FIGURE 3.22 Graph of a frequency function.

The following table illustrates some typical values of these two functions. Note that, while $B(x; 3, \frac{1}{2})$ is defined only for $x = 0, 1, 2,$ and 3, $F(x)$ is defined for all real numbers x.

	Frequency function for X	Distribution function for X
x	$B(x; 3, \frac{1}{2})$	$F(x)$
-7	not defined	0
0	$\frac{1}{8}$	$\frac{1}{8}$
.5	not defined	$\frac{1}{8}$
.9	not defined	$\frac{1}{8}$
1	$\frac{3}{8}$	$\frac{4}{8}\ (= \frac{1}{8} + \frac{3}{8})$
1.4	not defined	$\frac{4}{8}$
2	$\frac{3}{8}$	$\frac{7}{8}\ (= \frac{1}{8} + \frac{3}{8} + \frac{3}{8})$
3	$\frac{1}{8}$	$1\ (= \frac{1}{8} + \frac{3}{8} + \frac{3}{8} + \frac{1}{8})$
5	not defined	1
127.263	not defined	1

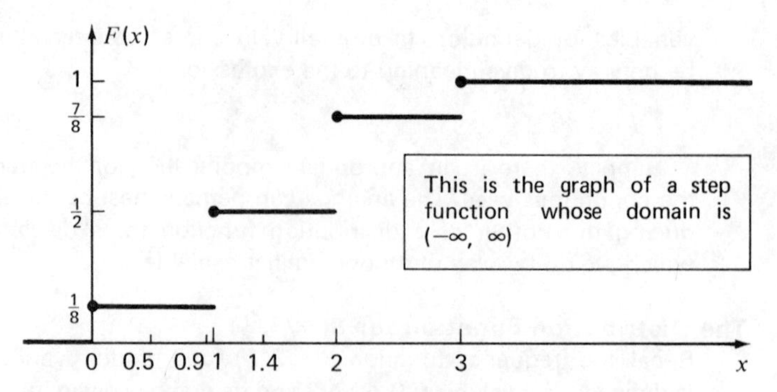

FIGURE 3.23 Graph of a distribution function.

Thus, from a frequency function we can construct its distribution function. In like manner, we can recover the frequency function from the distribution function as follows. We first observe that the event

$$\{s|X(s) \le 2\} \quad \text{or simply} \quad (X \le 2)$$

can be broken down into

$$\{s|X(s) < 2\} \cup \{s|X(s) = 2\}$$

Moreover since X takes on only the values 0, 1, 2, and 3,

$$\{s|X(s) < 2\} = \{s|X(s) \le 1\}$$

Thus we have the union of two disjoint events

$$\{s|X(s) \le 2\} = \{s|X(s) = 2\} \cup \{s|X(s) \le 1\}$$

and the probability of this union is

$$P(X \le 2) = P(X = 2) + P(X \le 1)$$

Rearranging this equation,

$$P(X = 2) = P(X \le 2) - P(X \le 1)$$

or in terms of the functions f and F,

$$f(2) = F(2) - F(1)$$

In like manner, we can show that

$$f(3) = F(3) - F(2)$$
$$f(1) = F(1) - F(0)$$
$$f(0) = F(0) - F(-1)$$

EXAMPLE 2 Rolling One Die

We offer another example of a distribution function corresponding to a familiar frequency function. Here X is the uniformly distributed discrete random variable having values

1, 2, 3, 4, 5, 6

representing the faces of a fair die.

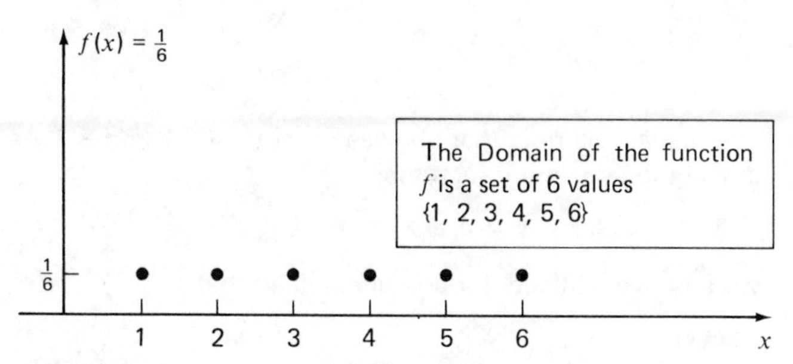

FIGURE 3.24 The graph of a frequency function.

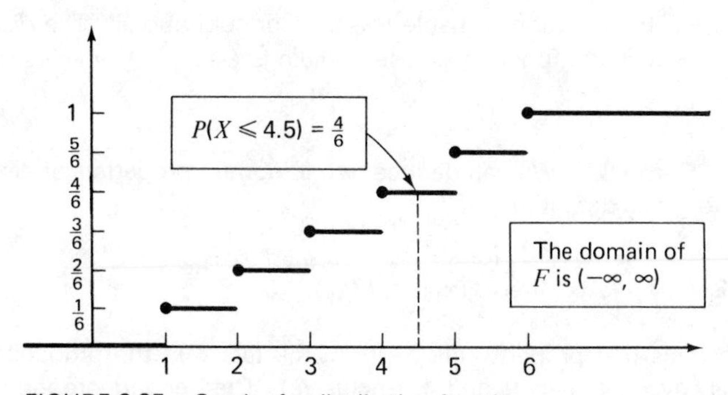

FIGURE 3.25 Graph of a distribution function.

We can write the formula for F:

$$F(x) = \begin{cases} 0, & \text{if } x < 1 \\ \frac{1}{6}, & \text{if } 1 \leq x < 2 \\ \frac{2}{6}, & \text{if } 2 \leq x < 3 \\ \frac{3}{6}, & \text{if } 3 \leq x < 4 \\ \frac{4}{6}, & \text{if } 4 \leq x < 5 \\ \frac{5}{6}, & \text{if } 5 \leq x < 6 \\ 1, & \text{if } 6 \leq x \end{cases}$$

And we could work backwards from this formula to discover the frequency function f of X. Indeed, we could even deduce that X is *discrete*, given only the function F. For instance, it is clear that

$$P(X = x) = 0, \qquad \text{unless } x \text{ is } 1, 2, 3, 4, 5, \text{ or } 6$$

We see this for $x = 3.2$ as follows:

$$\begin{aligned} P(X = 3.2) &\le P(3.1 < X \le 3.3) \\ &= P(X \le 3.3) - P(X \le 3.1) \\ &= F(3.3) - F(3.1) \\ &= \tfrac{3}{6} - \tfrac{3}{6} \\ &= 0 \end{aligned}$$

The reader will note that choices other than 3.1 and 3.3 were available. Indeed any numbers x and y such that

$$3 < x < 3.2 < y < 4$$

would have sufficed, for this guaranteed that

$$F(x) = F(y)$$

The following definition sums up our intuitive discussion:

Definition Let X be a random variable (discrete or continuous). The *distribution function* F *for* X is the function whose domain is $(-\infty, \infty)$ which satisfies

$$F(x) = P(X \le x)$$

Remark: We can deduce two important properties of distribution functions from this definition.

1 $0 \le F(x) \le 1$, for all x.
2 If $x_1 < x_2$, then $F(x_1) \le F(x_2)$.

This first property follows from the fact that the probability of any event is always between 0 and 1 (inclusive). The second property follows from the observation that if $x_1 \le x_2$, then

$$\{x|x \le x_1\} \subseteq \{x|x \le x_2\}$$

so that the probabilities of these events, more succinctly written

$$X \le x_1 \quad \text{and} \quad X \le x_2$$

satisfy

$$P(X \le x_1) \le P(X \le x_2)$$

Geometrically, (1) tells us that the graph of F always lies in the region bounded by the x-axis (the line $y = 0$) and the horizontal line $y = 1$; then (2) tells us that the graph of F is always either horizontal or rising from left to right.

EXAMPLE 3 Round-Off Error—Its Distribution Function

Let X be the random variable representing round-off error for some measurement. We know

1 $P(X < -.5) = 0$

2 $P(X > .5) = 0$
3 For any subinterval $[a, b]$ of $[-.5, .5]$

$$P(a \leq X \leq b) = b - a$$

Now we want to define a function F such that

$$F(x) = P(X \leq x)$$

In particular, then

$$F(x) = 0, \qquad \text{if } x < -.5$$

$$F(x) = 1, \qquad \text{if } x > .5$$

Moreover we know that for any x in $[-.5, .5]$

$$\begin{aligned} F(x) = P(X \leq x) &= P(X < -.5) + P(-.5 \leq X \leq x) \\ &= 0 + x - (-.5) \\ &= x + .5 \end{aligned}$$

Thus we have shown

$$F(x) = \begin{cases} 0, & \text{if } \quad x < -.5 \\ x + .5, & \text{if } -.5 \leq x \leq .5 \\ 1 & \text{if } \quad .5 < x \end{cases}$$

The graph of F is shown in Fig. 3.26.

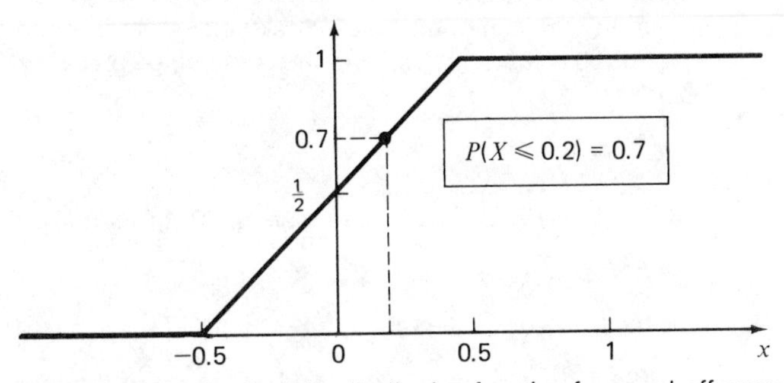

FIGURE 3.26 Graph of the distribution function for round-off error.

EXAMPLE 4 Another Continous Distribution

This time we begin with a distribution function (see Fig. 3.27) for some random variable X and we see what we can deduce about the behavior of X. From the formula for F we can deduce, for example,

$$\begin{aligned} P(.5 < X \leq .6) &= P(X \leq .6) - P(X \leq .5) \\ &= F(.6) - F(.5) \\ &= .6 - .5 \\ &= .1 \end{aligned}$$

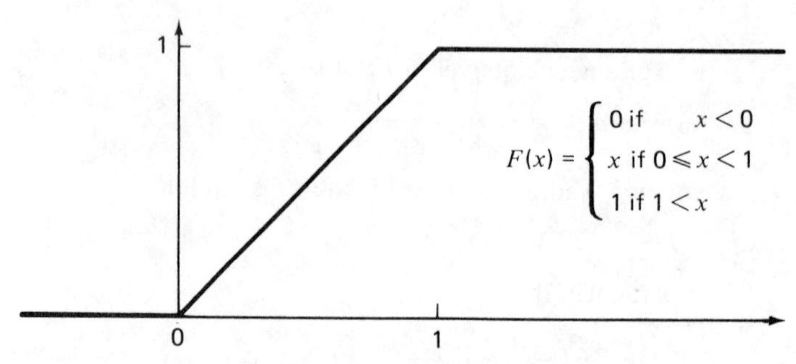

FIGURE 3.27 The graph of a distribution function.

Similarly, we could show that

$$P(0 < X \leq .1) = .1$$
$$P(.1 < X \leq .2) = .1$$
$$P(.2 < X \leq .3) = .1$$

and so on. In fact any subinterval of $[0, 1]$, regardless of how small, will yield a *positive* probability that X falls into it. Moreover, any two subintervals of $[0, 1]$ of equal length (just as we have observed for some subintervals of length .1) will yield equal probabilities.

It is also not hard to see that any interval $[a, b]$ outside the interval $[0, 1]$, whether lying to the left of 0 or to the right of 1, will yield probability 0.

These facts should remind us of a random variable studied previously; in connection with the continuous clock versus the discrete clock, we introduced a continuous random variable X satisfying precisely the aforementioned properties:

1 $P(X < 0) = 0$
2 $P(X > 1) = 0$
3 If $[a, b]$ and $[c, d]$ are subintervals of $[0, 1]$ with equal length, then

$$P(a \leq X \leq b) = P(c \leq X \leq d)$$

In fact, for such intervals

$$P(a \leq X \leq b) = b - a \qquad \text{(The length of the interval)}$$

Summary

Whereas the frequency function f for a random variable X allows us to compute the probability that X takes on a particular value x, we are often interested in the probability that X takes on a value *less than or equal to* x. To this end, we defined the *distribution function* F corresponding to X by

$$F(x) = P(X \leq x)$$

Note: the word "distribution" has been used previously to refer to a class of random variables and their associated frequency functions. We spoke of the "binomial distribution" and the "Poisson distribution." Our present use of the word as an adjective in *distribution function* has a different meaning.

One of the features of a distribution function is that it applies to *both continuous and discrete* random variables. Indeed, by studying distribution functions, we shall be able to discover the appropriate analogy of the frequency function for the case where X is continuous.

Another important feature of F is that its domain is always the entire real line, $(-\infty, \infty)$, regardless of the range of values of X (discrete or continuous). Two further properties of distribution functions are

$$0 \leq F(x) \leq 1$$

and

$$\text{If } x_1 < x_2, \text{ then } F(x_1) \leq F(x_2)$$

Exercises Section 10

1. Sketch the graphs of the distribution functions for the random variable X having the following frequency functions.
   (a) $f(x) = B(x; 2, \frac{1}{2})$
   (b) $f(x) = B(x; 2, \frac{1}{3})$
   (c) $f(x) = B(x; 4, \frac{1}{2})$
   (d) $f(x) = B(x; 4, \frac{1}{3})$
2. Write out the formulas for each of the distributions found in Exercise 1.
3. Each of the following graphs represents a frequency function for some discrete random variable X. Draw the associated distribution function.

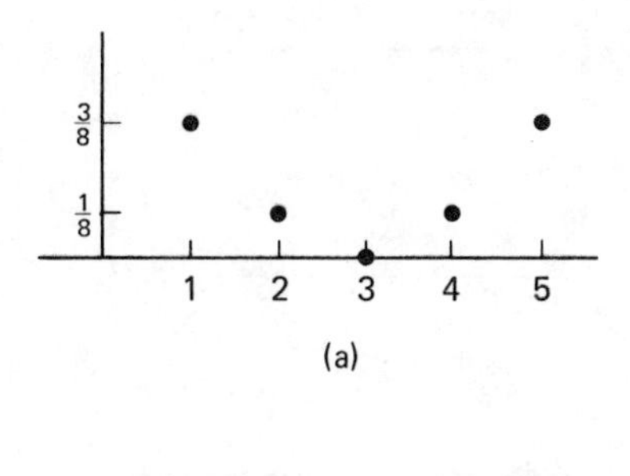

(a)

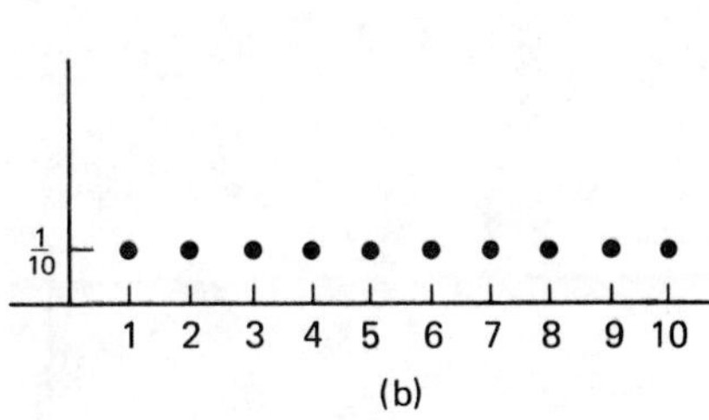

(b)

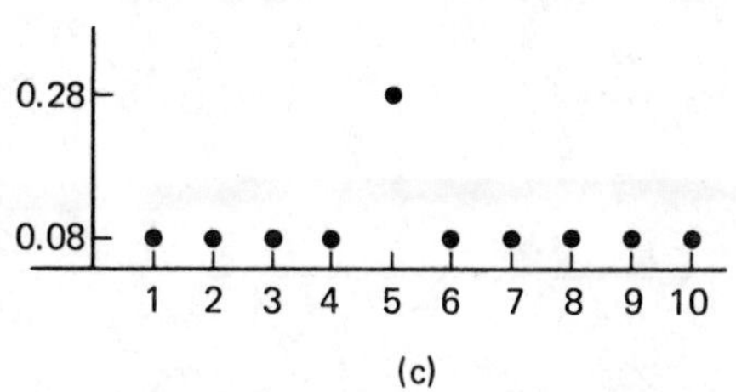

(c)

4 Each of the following functions represents a distribution function for some random variable X. Find the associated frequency functions.

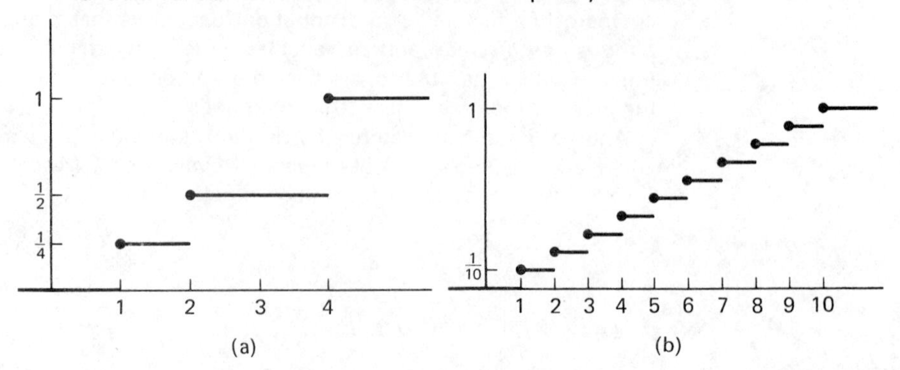

5 The following three pictures represent distribution functions for continuous random variables. In each case find the probabilities requested.

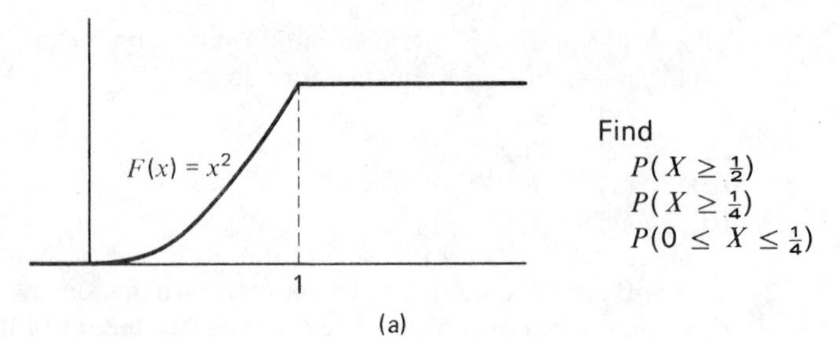

(a)

Find
$P(X \geq \frac{1}{2})$
$P(X \geq \frac{1}{4})$
$P(0 \leq X \leq \frac{1}{4})$

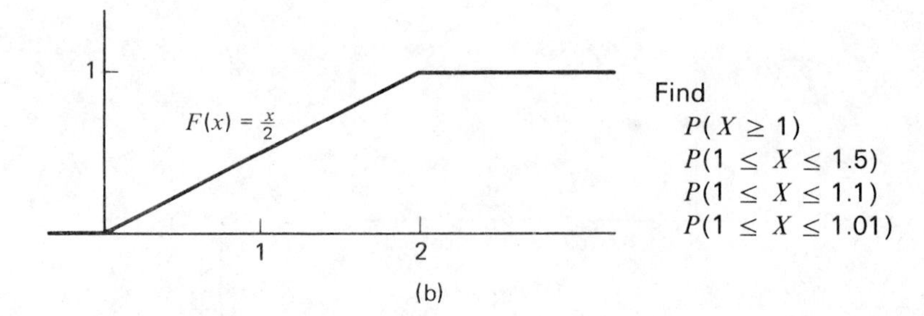

(b)

Find
$P(X \geq 1)$
$P(1 \leq X \leq 1.5)$
$P(1 \leq X \leq 1.1)$
$P(1 \leq X \leq 1.01)$

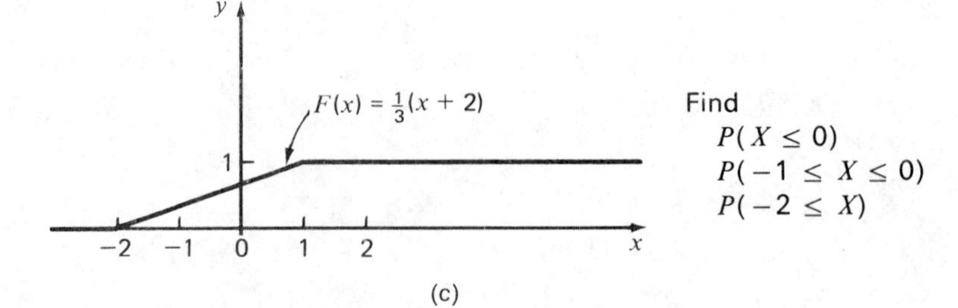

(c)

Find
$P(X \leq 0)$
$P(-1 \leq X \leq 0)$
$P(-2 \leq X)$

6 Sketch the graphs of the distribution functions for the Poisson random variable with the following.
(a) $\lambda = 2$
(b) $\lambda = 5$

11 PROBABILITY AS AREA—DENSITY FUNCTIONS

We still need to fill in a gap in the analogy between discrete and continuous random variables; namely, what, if anything, is the counterpart in the continuous case to the frequency function for discrete variables? We recap our progress to date in tabular form.

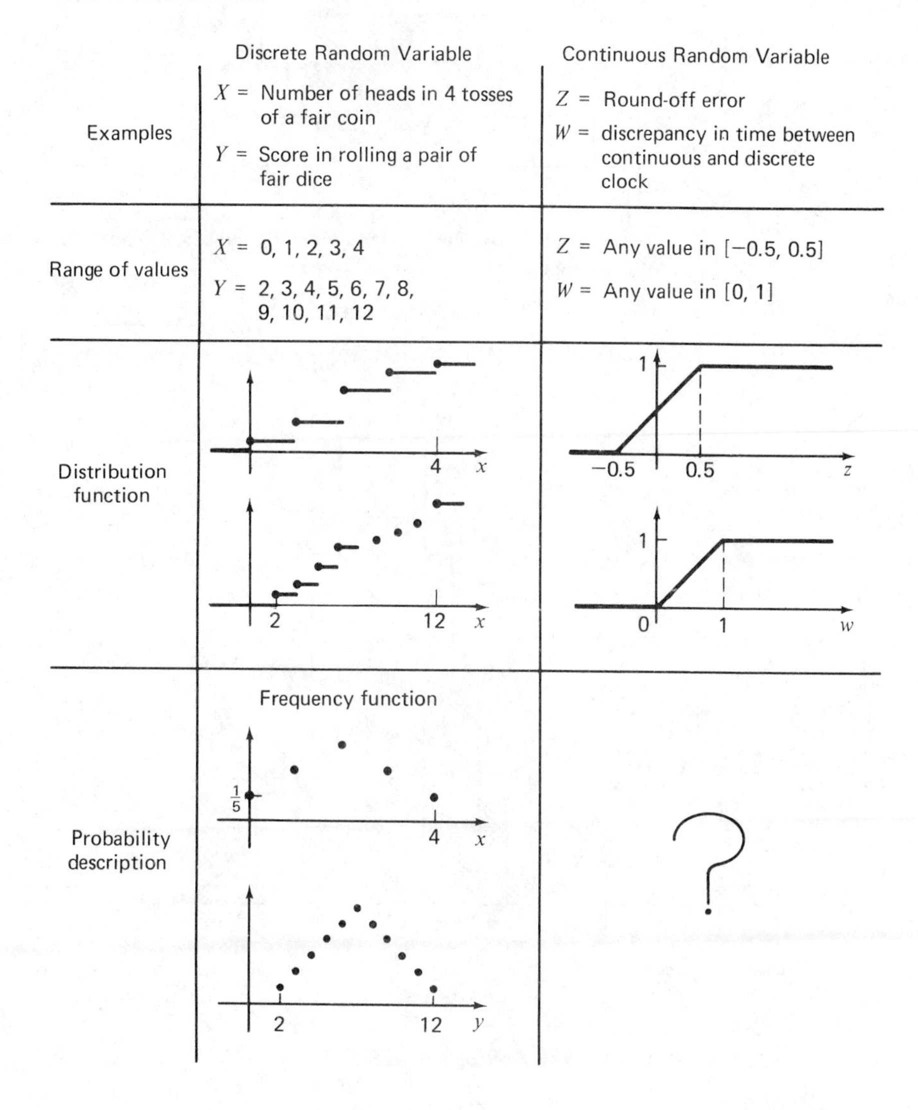

Now what precisely is the role of the frequency function f for a discrete random variable X? We can list a few of its properties.

1 $f(x)$ is defined for all values x in the range of X.
2 $f(x) = P(X = x)$ and thereby allows us to compare the likelihood of certain values of X.
3 $\Sigma f(x) = 1$; the sum of all the probabilities is 1.

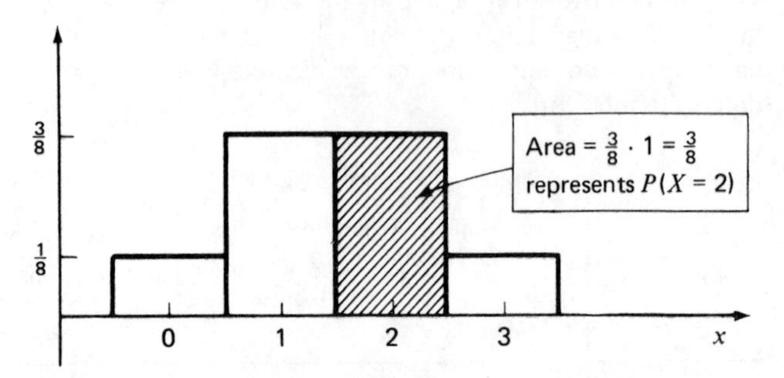

FIGURE 3.28 Frequency histogram for X: number of heads in 3 tosses of a fair coin.

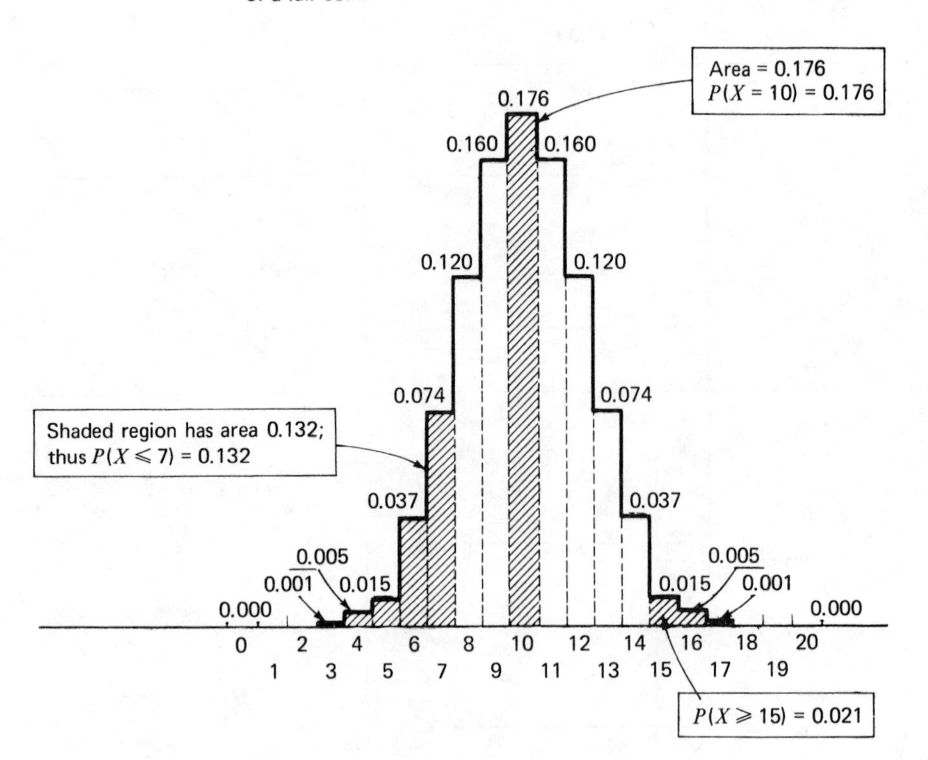

FIGURE 3.29 Histogram for $B(x; 20, \frac{1}{2})$.

The key to finding a suitable counterpart to the frequency function is to return to the histogram representation of frequency for a discrete random variable X that counts the number of heads in three tosses of a fair coin (Fig. 3.28).

The shaded rectangle has area equal to its height since the base is 1 unit wide. On the one hand it could be argued that it is a mere coincidence that the area of each rectangle has the same numerical value as the height. Certainly we could construct a frequency histogram for some discrete variable where the X values were not 1 unit apart, and consequently the component rectangles would have area different from their heights. Nevertheless, there are a great many situations where the histogram rectangles do have base 1 and where this coincidence prevails. Indeed, *all* the binomially distributed variables behave this way; so we can use the area of the rectangles in a binomial frequency histogram to find the various probabilities. In Fig. 3.29, we do just this with the variable whose frequency function is $B(x; 20, \frac{1}{2})$.

We are prepared now to introduce a function, called the probability density function, for a continuous random variable. The foregoing argument should at least suggest that our definition is not completely unmotivated. Further justification for the concept will be supplied in Chapter 5, when we return to the binomial distribution to find a suitable continuous approximation for it (namely the normal distribution).

Definition Let X be a continuous random variable. The *(probability) density function* f for X is a function whose domain consists of all possible values of X and which satisfies:

1 $\begin{pmatrix}\text{The area under the graph} \\ \text{of } f \text{ and above the } x\text{-axis}\end{pmatrix} = 1$

2 $\begin{pmatrix}\text{The area under the graph} \\ \text{of } f \text{ from } x = a \text{ to } x = b\end{pmatrix} = P(a \leq X \leq b)$

EXAMPLE 1 Round-Off Error Revisited

It is not hard to show that Fig. 3.30 represents the probability density function f for the random variable X, round-off error.

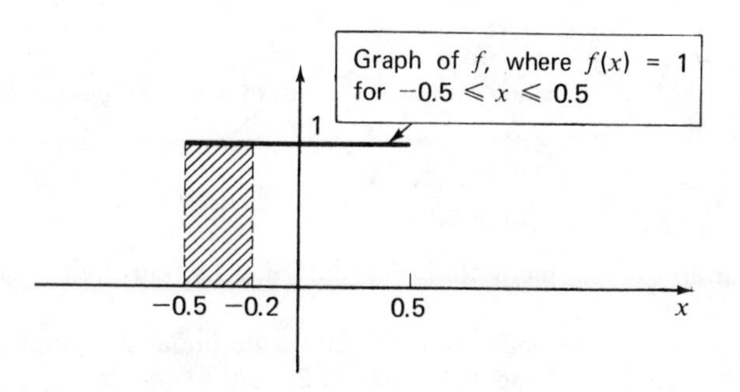

FIGURE 3.30 Probability density function for round-off error.

Observe that the total area under the graph is 1, and that a typical interval, $[-.5, -.2]$, of length .3 determines an area (the shaded region) of .3 which in turn equals $P(-.5 \leq X \leq -.2)$. Moreover it should be clear that two subintervals of $[-.5, .5]$ of equal length would determine equal areas and have equal probabilities, as should be the case.

Notice also how we can discover the distribution function F for X from its density function. We let x be in the interval $[-.5, .5]$ and we compute $P(X \leq x)$.

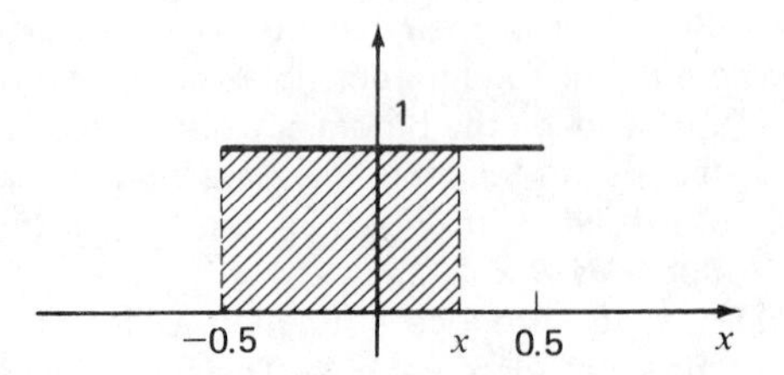

We find that the area of the shaded region determined by the interval $[-.5, x]$ is

$$\begin{aligned}(\text{base}) \cdot (\text{height}) &= (x + .5) \cdot (1) \\ &= x + .5\end{aligned}$$

But this area represents $P(-.5 \leq X \leq x)$ and, since

$$\begin{aligned}P(X \leq x) &= P(X < -.5) + P(-.5 \leq X \leq x) \\ &= 0 + P(-.5 \leq X \leq x)\end{aligned}$$

We have

$$F(x) = x + .5$$

as we discovered earlier.

Since we have such a paucity of concrete examples of continuous random variables, the remaining examples are designed to formally explore density functions and related distributions without necessarily representing random variables we might encounter in the real world.

EXAMPLE 2 A New Continous Variable

Suppose that X is a continuous random variable having density function f given by

$$f(x) = \begin{cases} 2x, & \text{if } 0 \leq x \leq 1 \\ 0, & \text{otherwise} \end{cases}$$

The graph of f is pictured in Fig. 3.31. Let us find the corresponding distribution function F.

We should first check that the total area under the graph is 1. The region is a right triangle of base 1 and height 2, so the area is

$$\tfrac{1}{2} \cdot 1 \cdot 2 = 1$$

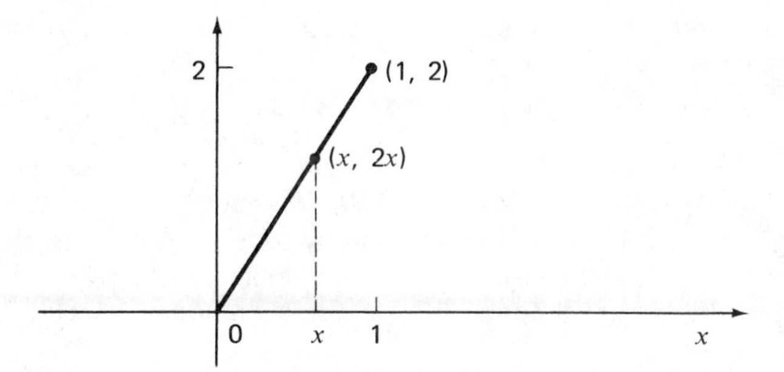

FIGURE 3.31 Graph of a density function.

The region under the graph from 0 to x is also a right triangle, its base is x and its height is $2x$; therefore its area is

$$\tfrac{1}{2} \cdot x \cdot 2x = x^2$$

or

$$P(0 \le X \le x) = x^2$$

for x in $[0, 1]$.

The entire description of F is therefore

$$F(x) = \begin{cases} 0, & \text{if } x < 0 \\ x^2, & \text{if } 0 \le x \le 1 \\ 1, & \text{if } 1 < x \end{cases}$$

and its graph is given in Fig. 3.32.

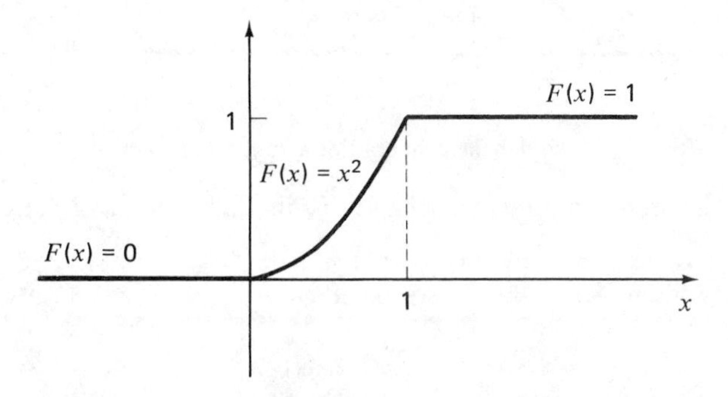

FIGURE 3.32 Graph of a distribution function.

Before leaving this example, let us examine some of the implications of the density function. For instance, let us take two subintervals of equal length

$$[0, .2] \quad \text{and} \quad [.8, 1.0]$$

and compute the probability that the value of X falls into either of these intervals

$$P(0 \leq X \leq .2) = F(.2) - F(0) = .04 - 0 = .04$$

and

$$P(.8 \leq X \leq 1) = F(1) - F(.8) = 1 - (.8)^2 = 1 - .64 = .36$$

Thus we see that X is far more likely to fall into the interval [.8, 1.0] than it is into [0, .2]. In a sense, we can draw an analogy between this density function and the frequency function for a discrete random variable:

> If $f(x_1) > f(x_2)$ then it is more likely that X will take on values *near* x_1 than it is that X will take on values *near* x_2. Likewise if f were a frequency function for a discrete random variable and if $f(x_1) > f(x_2)$ we could conclude that it is more likely that X would assume the value x_1 than it is that X would assume the value x_2.

EXAMPLE 3 A Triangular Distribution

The name triangular distribution refers to a random variable whose density function (not its distribution function) has a triangular shaped graph. One such example would be a variable X whose density function f is given by

$$f(x) = \begin{cases} 1 + x, & \text{if } -1 \leq x \leq 0 \\ 1 - x, & \text{if } 0 \leq x \leq 1 \\ 0, & \text{otherwise} \end{cases}$$

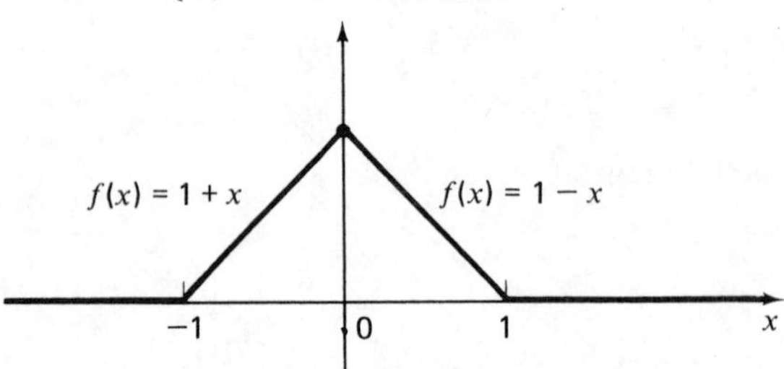

FIGURE 3.33 The density function for a continuous random variable X.

Let us determine the associated distribution function F for X.

1. When $x \leq -1$, $F(x) = P(X \leq x) = 0$.
2. When $-1 \leq x \leq 0$, we have

$$F(x) = \text{area of}$$

-1 $\quad x$

$$= \tfrac{1}{2}(\text{base})\,(\text{height})$$
$$= \tfrac{1}{2}(x - (-1))\,(1 + x)$$
$$= \tfrac{1}{2}(1 + x)^2$$

3. When $0 \leq x \leq 1$, we have

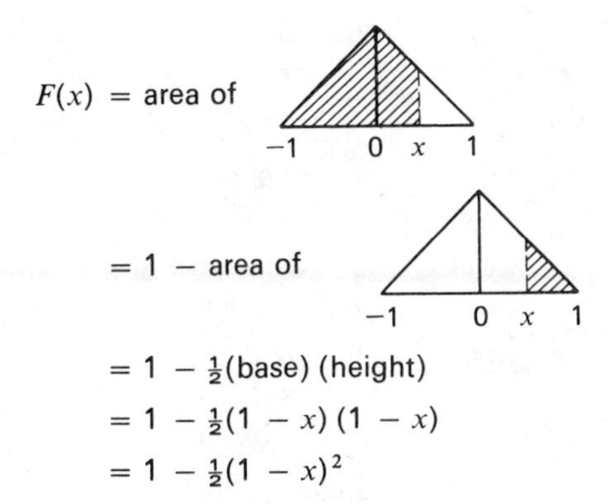

$F(x)$ = area of [figure]

= 1 − area of [figure]

$$= 1 - \tfrac{1}{2}(\text{base})(\text{height})$$
$$= 1 - \tfrac{1}{2}(1 - x)(1 - x)$$
$$= 1 - \tfrac{1}{2}(1 - x)^2$$

(Note that when $x = 1$, the above expression has value 1.)

4. When $x \geq 1$, $F(x) = 1$.
   We sketch the graph of F:

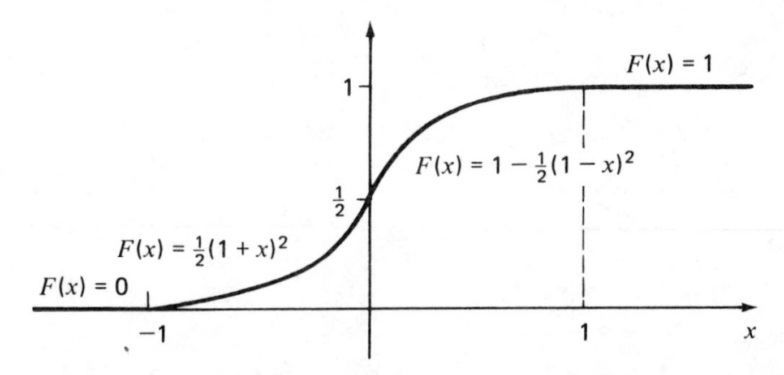

FIGURE 3.34 The distribution function for a continuous random variable X.

Finally it should be remarked that we have examined only very straightforward density functions. To do a more extensive analysis of this important concept requires that we develop some rudimentary techniques for finding area under curves in the plane. The next chapter is designed in part to meet that need.

Exercises Section 11

1 Each of the following functions is a candidate for a density function f for a random variable X. First sketch the graph of f; then check to see whether the area under the graph is 1; if it is, find the formula for the associated distribution function F.

(a) $$f(x) = \begin{cases} \frac{1}{2}, & \text{for } 1 \leq x \leq 3 \\ 0, & \text{otherwise} \end{cases}$$

(b)
$$f(x) = \begin{cases} 2, & \text{for } -1 \le x \le -\frac{1}{2} \\ 0, & \text{otherwise} \end{cases}$$

(c)
$$f(x) = \begin{cases} 1 + x, & \text{for } -1 \le x \le 0 \\ 1, & \text{for } 0 \le x \le 1 \\ 0, & \text{otherwise} \end{cases}$$

(d)
$$f(x) = \begin{cases} 2 - 2x, & \text{for } 0 \le x \le 1 \\ 0, & \text{otherwise} \end{cases}$$

(e)
$$f(x) = \begin{cases} 1, & \text{for } -\frac{1}{2} \le x \le \frac{1}{2} \\ 0, & \text{otherwise} \end{cases}$$

(f)
$$f(x) = \begin{cases} 1, & \text{for } 0 \le x \le \frac{1}{2} \\ 2, & \text{for } \frac{1}{2} < x \le \frac{3}{4} \\ 0, & \text{otherwise} \end{cases}$$

2 Each of the functions f whose graphs are sketched below is a candidate for a density function. First find the formula for f; then decide whether f can be a density function; finally, if it is, determine the formula for the associated distribution function

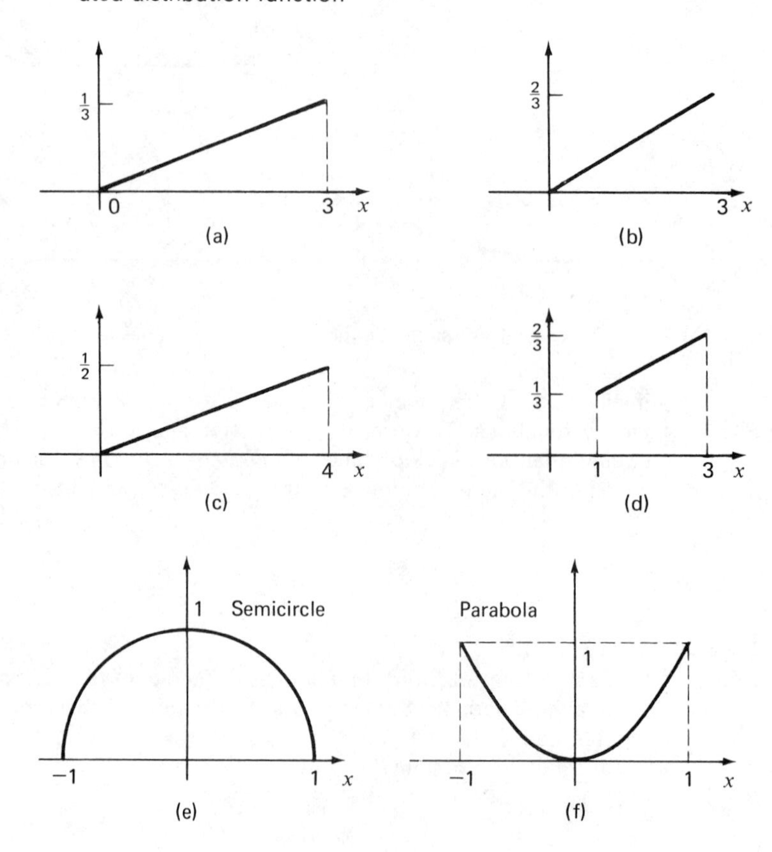

Supplementary Exercises for Chapter 3

1 A basketball player has made 80% of his free throws during the season. In his last game he has the opportunity to shoot six free throws. Find the probability that

(a) he misses them all.
(b) he makes half of them.
(c) he makes five of them.
(d) he makes at least two of them.
(e) he makes at least five of them.

2 There are n people in a room. Find the probability that they all have different birth months (assuming all months equally likely), if

(a) $n = 2$.
(b) $n = 4$.
(c) $n = 5$.
(d) $n = 6$.

3 Reformulate the preceding exercise to determine how many people must be together to give at least a 50% chance that at least two of them will have the same birth month.

4 **(The Birthday Problem)** This is a "classic" example of a rather non-intuitive probability result. The preceding exercise is a very similar problem that has the added feature that the computations are relatively easy to do by hand. The question now is, how many people need there be in a room before the chances of at least two of them having the same birthday is at least 50%? The solution requires either a lot of patience or a calculating device.

5 Which is bigger, $B(1; 3, \frac{1}{3})$ or $B(2; 6, \frac{1}{3})$?

6 An experiment consists of rolling a fair die and tossing a fair coin. If we define X by

$$X(a, H) = a^2$$

$$X(a, T) = 2a$$

Find the frequency function for X.

7 **(Counting the Number of Subsets of a Given Set)**

(a) Use the Binomial Expansion Theorem to show that

$$2^n = \sum_{x=0}^{n} C(n, x)$$

(b) Show that if S has n elements, then it has 2^n subsets.

8 An urn contains N marbles labeled 1, 2, 3, . . . , N. One marble is drawn at random and the number on it is called X. Show that $E(X) = (N + 1)/2$ and $\sigma^2_X = (N^2 - 1)/12$.

9 An instructor gives a 20 question true-false exam. He decides to award at least a D to those students whose score could be achieved only 30% of the time by straight guessing, at least a C to scores that could be achieved only 20% of the time by guessing, a B to scores that would arise only 10%

of the time through guessing and an A to scores arising only 5% of the time by guessing. Find the cut-off scores for A's, B's, C's, and D's.

10 An examination contains a section of "matching" problems where four items must be matched with four others. If you must guess at all items, which of the following strategies would yield a higher expected score?

(a) Pick one possible answer and repeat it for all four answers.

(b) Pick at random a permutation of the four answers and use it.

[*Hint:* Suppose the letters A, B, C, D are to be filled in four blanks. Strategy (a) would dictate an answer pattern like BBBB, whereas strategy (b) would dictate an answer pattern like BDAC. You may suppose that the correct answer pattern is ABCD.]

11 How much higher would you expect your score to be in the previous problem if you knew one answer for certain? Two answers? Three answers?

12 **(a)** Give definitions for the *mode* and the *median* of a discrete random variable.

(b) Using your definitions, find the mode and the median of the variable X with frequency function $B(x; 20, .4)$.

13 Two numbers are selected at random from the list 1, 2, 3, 4 and X is the *larger* of the two.

(a) If the selection is made *with* replacement, find $E(X)$.

(b) If the selection is made *without* replacement, find $E(X)$.

14 Let X be the number of times a pair of fair dice must be rolled until a "7" appears (see the exercise in Section 3 concerning tossing a fair coin until heads appears).

(a) Find the frequency function f of X.

(b) Find $E(X)$ and σ_X^2.

15 A miniature roulette wheel has positions numbered 1 through 10. A bettor pays 1 dollar to play. If his number comes up, he wins 9 dollars, otherwise he forfeits his bet.

(a) Find his expected "winnings."

(b) Find the probability that he will go broke before he first wins if he starts with 6 dollars.

(c) How much money does he have to start with to give himself a 50–50 chance of winning before he goes broke?

16 The probability that a certain type of seed will germinate is .6. Smith's Nursery wants to sell boxes with a money-back guarantee that each box contains at least 10 plants. If Smith places 20 seeds in each box, how many boxes out of 100 could he expect to be returned?

17 Repeat the preceding problem modified by the assumption that Smith wants to sell boxes guaranteed to contain at least five plants and that he places 10 seeds in each box.

18 A student is taking five courses: English, math, history, chemistry, and political science. Each course has one textbook. In his rush to an early morning class, he grabs two textbooks from his shelf at random.

(a) List the sample space for this experiment.

(b) Let X be a random variable measuring the weight of the two books. Suppose the weights of individual books are:

English	1 pound	Chemistry	$1\frac{1}{2}$ pounds
Math	$1\frac{1}{2}$ pounds	Political Science	$2\frac{1}{2}$ pounds
History	$2\frac{1}{2}$ pounds		

List the associated value of X next to each sample event in your sample space.

(c) Write down the frequency function for X.

19 An urn contains five red marbles and two blue marbles. Marbles are drawn without replacement until the first blue one is drawn. Let X denote the number of marbles drawn in a trial of the experiment. Find its frequency function.

20 A committee of three people is to be chosen from three men and four women. Let the random variable X be the number of men on the committee. Find the frequency function for X.

21 A bridge hand contains 13 cards. Let random variables X and Y be defined as follows:

X = the number of red cards in the hand
Y = the number of aces in the hand

(a) The probability that there are as many red cards as aces can be written $P(X = Y)$. Similarly, write each of the following events in terms of X and Y.

(1) The probability that there are fewer aces than red cards
(2) The probability that there are at least four red cards
(3) The probability of having all four aces
(4) The probability of having no black cards
(5) The probability that there are no aces in one hand, knowing that there are no black cards

(b) Compute the probabilities in (3), (4), and (5).

22 Four coins, three nickels and a dime, are tossed. A random variable assigns to the outcome of this experiment the value of the coins which come up heads (e.g., if two of the nickels and the dime come up heads, the value is 20).

(a) Write out the sample space for this experiment.

(b) Find the associated frequency function and graph it.

23 (A real slot machine.) One form of slot machine in Las Vegas has four wheels, each containing an assortment of 20 symbols chosen from: Cherry (C), Orange (O), Plum (P), Bell (B), Eight (8), and Jackpot (J). The following tables give the symbols on each wheel and the payoffs corresponding to betting one coin (say a nickel, dime, or a quarter):

1st: C, O, 8, O, J, O, C, O, 8, O, B, O, 8, O, C, O, P, 8, O, P
2nd: B, O, B, C, 8, C, B, C, 8, B, C, P, B, C, J, B, C, 8, B, C
3rd: B, P, O, P, 8, P, O, P, O, J, P, O, P, O, B, P, O, P, O, P
4th: P, B, P, B, O, P, B, P, B, P, B, P, J, B, P, B, P, B, P, B

	Wheel Arrangement 1st	2nd	3rd	4th	Payoff (Number of Coins)
(1)	C	—	—	—	2
(2)	C	C	—	—	5
(3)	O	O	O	—	18
(4)	O	O	O	O	23
(5)	P	P	P	—	14
(6)	P	P	P	P	23
(7)	B	B	B	—	18
(8)	B	B	B	B	23
(9)	8	8	8	—	150
(10)	J	J	J	—	150
(11)	J	J	J	J	5000

(a) Find the probabilities of each wheel arrangement (assuming, of course, that the items on a wheel are equally likely to occur). [*Hint:* Be sure to observe that arrangement (1), for example, means that (2) *does not occur*, i.e., the second wheel must not have a cherry showing.]

(b) What is the probability of no payoff on a given play?

(c) Find the expected winnings for playing this machine with one dime.

(d) If a machine is kept busy all day (24 hours) and if the players deposit on the average four dimes per minute, what is the expected revenue to the casino? [*Hint:* Determine how much money is deposited during the day and pretend that it was all played at once.]

24 A psychologist observes that one-half of the female gerbils he buys give birth within the first two months. He suspects that constant exposure to light will affect their productivity, so he buys 20 new females and subjects them to the same living conditions (same cages, same food, same availability of males) except that he leaves the laboratory lights on continuously for the first two months. Explain how he might support his conjecture at a .05 significance level.

25 The student body at Parthenon College consists of $33\frac{1}{3}\%$ women and $66\frac{2}{3}\%$ men. A certain math course has a registration of 8 women and 22 men. Can you conclude at the .10 significance level that the probability of a given man taking this course is greater than the probability of a given woman taking it?

26 A person claims to possess psychic powers that enable him to guess the correct color of cards (red or black) drawn at random from a deck. How could you support this claim at a .05 significance level?

27 If an average of three people out of four are overweight, what is the probability that

(a) less than half of the people in a group of four are overweight?

(b) less than half of the people in a group of eight are overweight?

(c) less than half of the people in a group of 12 are overweight?

28 A certain business department knows that $\frac{2}{3}$ of all students taking a particular accounting course will major in business. What is the probability that out of a class of 30 they will get no more than 10 majors.

29 An instructor feels that it is crucial that students embarking on a science project be good at addition; indeed he feels it mandatory that the probability is .9 that when they add two numbers the sum is correct. He decides to screen students by giving a test with 10 addition problems and admitting to the project only those who get all 10 right.

(a) If 100 students, who do have probability .9 of getting any individual problem right, take the test, how many will be admitted to the project?

(b) If 100 sub-par students, whose probability is only .8, take the test, how many of them will be admitted to the project?

30 If $B(10; 10, .8) = .1074$, what is $B(9; 10, .8)$?

31 Suppose that you are knowledgeable enough about a certain subject to have a probability of .95 of correctly answering any multiple choice question. If you take a test with 100 questions on it, what is the probability that you will get 90 or more correct?

32 A *circle* with center at (a, b) and radius r has the equation

$$(x - a)^2 + (y - b)^2 = r^2$$

(a) Plot several points on the circle whose equation is

$$(x - 1)^2 + (y - 2)^2 = 1$$

and then sketch the circle of radius 1 centered at $(1, 2)$.

(b) Plot several points and then sketch the graph of the equation

$$(x + 1)^2 + (y - 3)^2 = 4$$

(c) Find the equation of the circle that passes through the points $(1, 4)$, $(5, 4)$, $(3, 2)$, and $(3, 6)$.

(d) Find the equation of the circle that passes through the points $(-5, 5)$, $(-2, 8)$, and $(1, 5)$.

(e) Sketch the graph of the *semicircle* whose equation is

$$y = \sqrt{1 - (x + 2)^2}, \qquad \text{for } -3 \le x \le -1$$

33 Each of the following equations has for its graph a *hyperbola*. Plot a few points for each and sketch the resulting curve; include points for both negative and positive values of x.

(a) $y = \dfrac{1}{x}$ **(d)** $y = \dfrac{-1}{x}$

(b) $y = \dfrac{2}{x}$ **(e)** $y = \dfrac{-2}{x}$

(c) $y = \dfrac{4}{x}$

34 Each of the following functions is the distribution function for some random variable. In each case, sketch the graph and compute the indicated probabilities.

(a)
$$F(x) = \begin{cases} 0, & \text{if } x < 0 \\ x, & \text{if } 0 \le x < \frac{1}{2} \\ \frac{1}{2}, & \text{if } \frac{1}{2} \le x < 1 \\ x - \frac{1}{2}, & \text{if } 1 \le x < \frac{3}{2} \\ 1, & \text{if } x \ge \frac{3}{2} \end{cases}$$

Find:

$P(0 \le X \le 3)$
$P(0 \le X \le 1)$
$P(1 \le X \le 1.1)$
$P(1 \le X \le 1.01)$

(b)
$$F(x) = \begin{cases} 0, & \text{if } x < 0 \\ x^2, & \text{if } 0 \le x < \frac{1}{2} \\ 1 - 3(1 - x)^2, & \text{if } \frac{1}{2} \le x < 1 \\ 1, & \text{if } x \ge 1 \end{cases}$$

Find:

$P(X \ge \frac{1}{2})$
$P(0 \le X \le \frac{1}{2})$
$P(0 \le X \le \frac{1}{4})$
$P(\frac{1}{4} \le X \le \frac{3}{4})$

(c)
$$F(x) = \begin{cases} 0, & \text{if } x < -3 \\ \frac{1}{6}(x + 3) & \text{if } -3 \le x < 3 \\ 1, & \text{if } x \ge 3 \end{cases}$$

Find:

$P(X \le 2)$
$P(2 \le X \le 3)$
$P(X \ge 4)$
$P(X \le -1)$

35 At the beginning of Section 5 we discussed a poll of 100 to predict the results of an election. Using Table B in the appendix for $B(x; 100, .5)$, determine what results would lead you to predict Smith's victory

(a) at the 10% significance level.
(b) at the 5% significance level.
(c) at the 1% significance level.

36 Suppose that X is binomially distributed with mean 6 and variance 4. Find n and p.

37 **(a)** Show that

$$x(x - 1)C(n, x) = n(n - 1)C(n - 2, x - 2)$$

(b) Use the result in (a) to show that

$$E(X(X - 1)) = p^2n(n - 1)$$

where X has frequency function $B(x; n, p)$

(c) Use (b) to find σ_X^2 from

$$\sigma_X^2 = E(X(X - 1)) + \mu_X - \mu_X^2$$

CHAPTER

4

An Introduction to Calculus

1 AREA

Near the end of Chapter 3 we discussed continuous random variables in terms of their density functions and their distribution functions. In order to obtain probabilities from the density function we found it necessary to compute the area under its graph. In simple cases where this graph was a segment of a line, the computation was manageable, but in a more general (and, indeed more interesting) situation we were at a loss as to how to proceed. In this section we shall devise a method of finding successive approximations to the area under a curve. At first this technique will serve as a vehicle to perform some direct computations, and then we will discover a key that allows us in most instances to circumvent the laborious computations and to get an exact value directly.

Let us start with a concrete problem.

What is the area of the region under the graph of $y = 1 + x^2$, above the x-axis and between the lines $x = a$ and $x = b$? (See Fig. 4.1.)

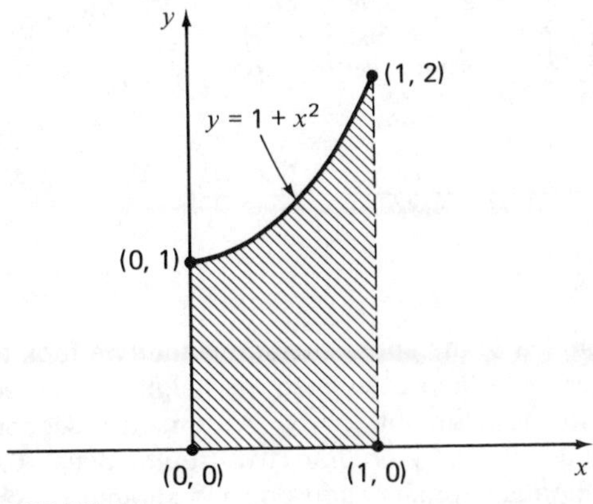

FIGURE 4.1

We can get a first approximation of this area, call the approximation R_1, from the area of the rectangle pictured below.

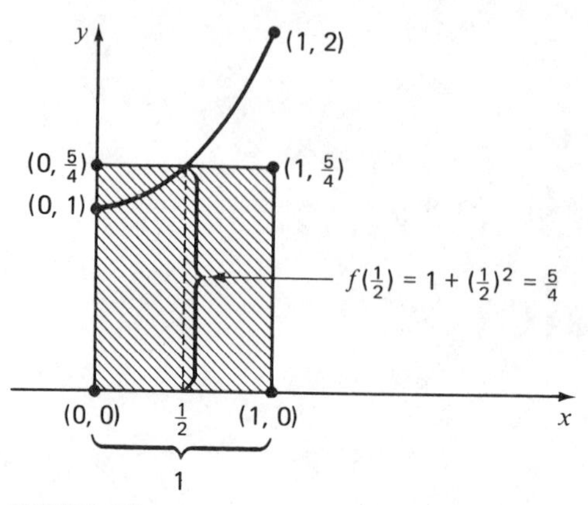

FIGURE 4.2

We obtain

$$R_1 = (\tfrac{5}{4}) \cdot 1 = \tfrac{5}{4}$$

Note (see Fig. 4.3) that we have included in this rectangle some area that isn't under the curve, and failed to include in the rectangle some area that is under the curve.

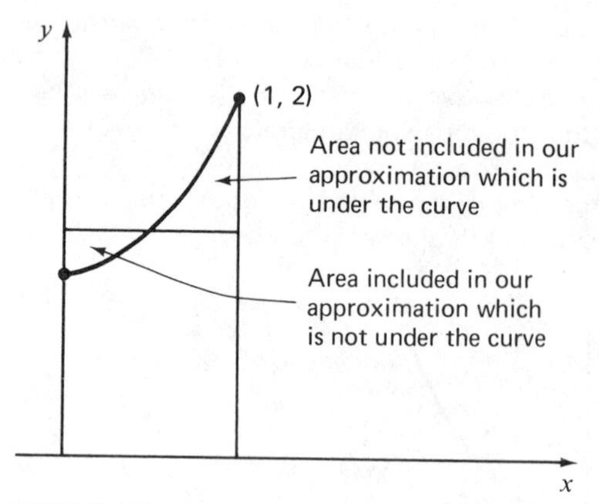

FIGURE 4.3

The essential idea of the approximation is that we took the interval [0, 1] as the base of a rectangle and calculated the height of the rectangle by applying the function f to the midpoint, $x = \frac{1}{2}$. We obtain a second approximation by breaking the interval [0, 1] up into two subintervals $[0, \frac{1}{2}]$ and $[\frac{1}{2}, 1]$ and making an analogous approximation on each subinterval (see Fig. 4.4).

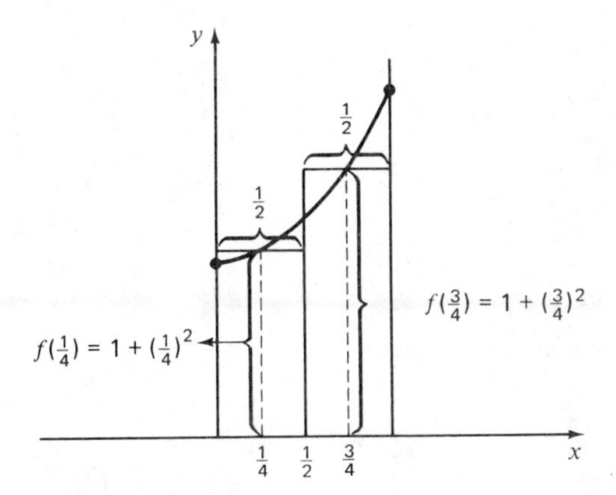

FIGURE 4.4

We obtain

$$R_2 = [1 + (\tfrac{1}{4})^2] \cdot (\tfrac{1}{2}) + [1 + (\tfrac{3}{4})^2] \cdot (\tfrac{1}{2})$$
$$= \tfrac{17}{32} + \tfrac{25}{32} = \tfrac{42}{32} = \tfrac{21}{16} \approx 1.31.$$

Once again, an examination of the picture will show that our approximation includes some area which is not under the curve and excludes some area which is.

For each positive integer n, we can calculate an approximation R_n of this type. Let us do one more, R_3.

To calculate R_3 we break the interval $[0, 1]$ up into three subintervals of equal length. They will be

$$[0, \tfrac{1}{3}], \qquad [\tfrac{1}{3}, \tfrac{2}{3}], \qquad [\tfrac{2}{3}, 1]$$

The midpoints of these subintervals are, respectively,

$$\tfrac{1}{6}, \qquad \tfrac{1}{2}, \qquad \tfrac{5}{6}$$

The heights of the rectangles to be erected on the subintervals are

$$1 + (\tfrac{1}{6})^2, \qquad 1 + (\tfrac{1}{2})^2, \qquad 1 + (\tfrac{5}{6})^2$$

or

$$\tfrac{37}{36}, \qquad \tfrac{5}{4}, \qquad \tfrac{61}{36}$$

Therefore we have (see Fig. 4.5)

$$R_3 = \tfrac{37}{36} \cdot \tfrac{1}{3} + \tfrac{5}{4} \cdot \tfrac{1}{3} + \tfrac{61}{36} \cdot \tfrac{1}{3}$$
$$= \tfrac{37}{108} + \tfrac{5}{12} + \tfrac{61}{108} = \tfrac{37}{108} + \tfrac{45}{108} + \tfrac{61}{108}$$
$$= \tfrac{143}{108} \approx 1.32$$

We could continue to calculate, obtaining R_4, R_5, etc. However as n increases, R_n is increasingly difficult to compute. The computer could be programmed to calculate R_n, and this would help, but there is a more fundamental question. Is

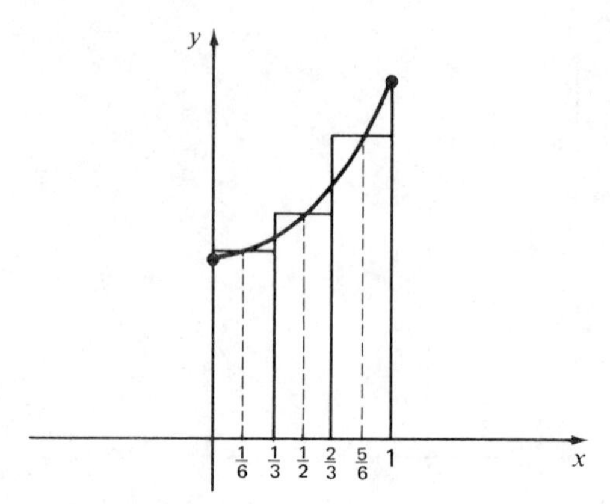

FIGURE 4.5

R_n worth calculating for large n? We said that R_n represents an approximation to the area but we have no way of knowing how good an approximation we are getting. One could, correctly, state that the Empire State Building is approximately 1 foot high. The statement is correct but the approximation leaves a good deal to be desired.

Moreover we have no guarantee that increasing the size of n makes R_n a better approximation to the area. In order to discuss this point sensibly, we need to get away from phrases such as "good approximation" or "better approximation." The usual technique, and the one which we shall adopt, is to talk about the degree of accuracy of an approximation or, alternately, the maximum error associated with an approximation. That is we might ask for an approximation of the height of the Empire State Building accurate to within 1 foot by which we would mean that the difference between the actual height and the approximation is less than 1 foot. Similarly we can talk about an approximation of the distance between the earth and the moon accurate to within 1 mile, or an approximation of the weight of a person to within 1 ounce, etc.

Thus in our area problem we shall specify some maximum allowable error, such as .005, and insist that

$$|R_n - \text{exact area}| < .005$$

[Remember that $|x - y|$ is just the (positive) distance between x and y.] in order to consider R_n to be a good approximation. Now the problem becomes, "Can we use these values R_n to obtain any such degree of accuracy, and, if so, how do we determine how large a value of n is necessary?" Well, for a large class of functions (virtually all functions we will consider), the values of R_n will zero in on the exact value of the area. As to the value of n necessary to obtain some prescribed level of accuracy, we really can't say. For most functions over most intervals this is a question which is much more difficult than the sort which we expect you to learn to answer in this course. Most problems in this book will tell you what value of n to use for the approximation. There are however certain cases where the value of n which will work is very simply determined. One such situation is the function $f(x) = 1 + x^2$ on the interval [0, 1] with which we started the discussion.

We shall go through the process of determining what value or values of n make R_n an approximation of the area accurate to within an error of .005. To do this, for each n we will define terms L_n and U_n. The L in L_n stands for lower and, for each n, L_n will be an estimate (often called the nth *lower sum*) which is always less than or equal to both R_n and the exact area. The U in U_n stands for upper and, for each n, U_n will be an estimate (often called the nth *upper sum*) which is always greater than or equal to both R_n and the exact area. We begin with $n = 1$.

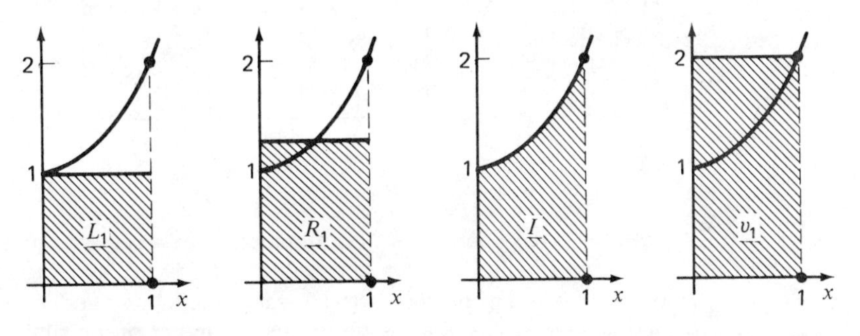

FIGURE 4.6

We let L_1 be the area of the largest rectangle completely contained in the region whose area is represented by the integral and U_1 be the area of the smallest rectangle completely containing the region. Note that on [0, 1], f attains its smallest value at $x = 0$ so

$$L_1 = f(0)(1 - 0) = 1 \cdot 1 = 1$$

The maximum value of f is attained at $x = 1$ so

$$U_1 = f(1)(L - 0) = 2 \cdot 1 = 2$$

From the picture (Fig. 4.6), we can see that both R_1 and the area (which we shall denote by I for the remainder of this discussion) are between U_1 and L_1. That is,

$$L_1 \leq R_1 \leq U_1 \quad \text{and} \quad L_1 \leq I \leq U_1$$

It follows then that the difference between R_1 and I, namely,

$$|R_1 - I|$$

is less than or equal to the difference between L_1 and U_1, so that

$$|R_1 - I| \leq U_1 - L_1$$

It follows that the error involved in using R_1 as an approximation for I is at most 1.

We now form L_2 and U_2. To do this we break $[0, 1]$ up into two subintervals $[0, \frac{1}{2}]$ and $[\frac{1}{2}, 1]$. Then, on each subinterval we erect an inner rectangle (the largest rectangle completely contained in the region) and an outer rectangle (the smallest rectangle completely containing the region). For our function f, there is a simple way to determine the area of these inner and outer rectangles. We

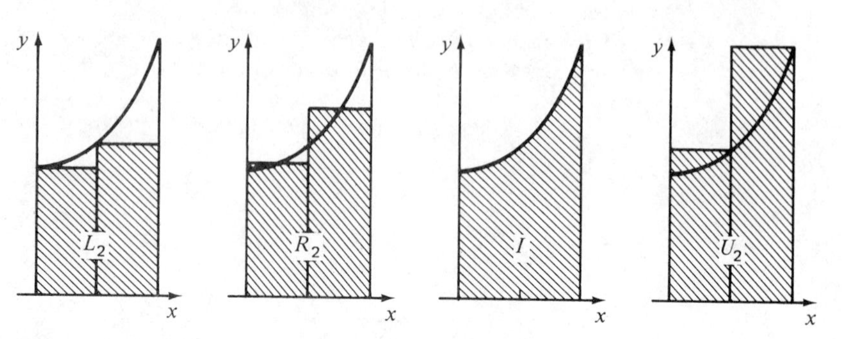

FIGURE 4.7

observe that f is an increasing function, that is, as we move from left to right along the graph, it always goes up, never down. Therefore, on any subinterval of $[0, 1]$, f attains its smallest value at the left-hand endpoint of the subinterval and its largest value at the right-hand endpoint of the subinterval.

We form L_2 by adding the areas of the inner rectangles and U_2 by adding the areas of the outer rectangles. We obtain

$$\begin{aligned} L_2 &= f(0) \cdot \tfrac{1}{2} + f(\tfrac{1}{2}) \cdot \tfrac{1}{2} \\ &= \tfrac{1}{2} + \tfrac{5}{4} \cdot \tfrac{1}{2} \\ &= \tfrac{9}{8} = 1.125 \end{aligned}$$

and

$$\begin{aligned} U_2 &= f(\tfrac{1}{2}) \cdot \tfrac{1}{2} + f(1) \cdot \tfrac{1}{2} \\ &= \tfrac{5}{4} \cdot \tfrac{1}{2} + 2 \cdot \tfrac{1}{2} \\ &= \tfrac{13}{8} = 1.625. \end{aligned}$$

Once again, both I and R_2 are between L_2 and U_2, so

$$|R_2 - I| \leq U_2 - L_2 = 1.625 - 1.125 = .5$$

Hence we know that R_2 is an approximation of I accurate to within .5.

In Fig. 4.8 we give a graphical interpretation of $U_2 - L_2$ and $U_3 - L_3$ (which we have not calculated but which you should to check your understanding). The idea behind the pictures is that over each subinterval of $[0, 1]$ there is a rectangular region representing that portion of $U_n - L_n$ which lies over the subinterval. At the right, we have piled these rectangles one on top of the other (we can do this because all the rectangles for a particular value of n have the same width) and $U_n - L_n$ is the area of the total rectangle which is formed.

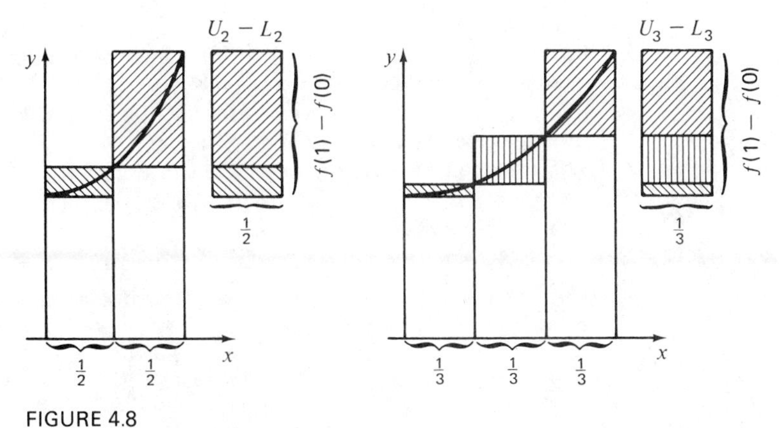

FIGURE 4.8

We shall now tackle the problem of computing R_n, U_n, and L_n for an arbitrary positive integer n and deciding what value of n will give us a prescribed accuracy.

In order to perform these calculations we must first break the interval from 0 to 1 into n equal subintervals. Clearly, they will each have length $1/n$. We shall number the intervals from left to right so that we can speak of the first subinterval, the second, etc. If you think about it for a while you will be able to come up with the following table.

Interval Number	Left-Hand Endpoint	Right-Hand Endpoint	Midpoint
1	0	$\frac{1}{n}$	$\frac{1}{2} \cdot \frac{1}{n}$
2	$\frac{1}{n}$	$\frac{2}{n}$	$\frac{3}{2} \cdot \frac{1}{n}$
3	$\frac{2}{n}$	$\frac{3}{n}$	$\frac{5}{2} \cdot \frac{1}{n}$
⋮	⋮	⋮	⋮
k	$\frac{k-1}{n}$	$\frac{k}{n}$	$\frac{2k-1}{2} \cdot \frac{1}{n}$
⋮	⋮	⋮	⋮
n	$\frac{n-1}{n}$	$\frac{n}{n}$ $(=1)$	$\frac{2n-1}{2} \cdot \frac{1}{n}$

To obtain R_n we:

1. Evaluate $f(x)$ at the midpoint of each interval to obtain the height of the rectangle erected on that subinterval.
2. Multiply each of the n answers to part (1) by $1/n$, the width of the rectangle, to obtain the areas of the n rectangles.
3. Add together the answers to part (2) to obtain R_n.

For part (1) we have

Interval Number	Midpoint	Height of Rectangle
1	$\frac{1}{2}\cdot\frac{1}{n}$	$1+\left(\frac{1}{2}\cdot\frac{1}{n}\right)^2$
2	$\frac{3}{2}\cdot\frac{1}{n}$	$1+\left(\frac{3}{2}\cdot\frac{1}{n}\right)^2$
⋮	⋮	⋮
k	$\frac{2k-1}{2}\cdot\frac{1}{n}$	$1+\left(\frac{2k-1}{2}\cdot\frac{1}{n}\right)^2$
⋮	⋮	⋮
n	$\frac{2n-1}{2}\cdot\frac{1}{n}$	$1+\left(\frac{2n-1}{2}\cdot\frac{1}{n}\right)^2$

Combining steps (2) and (3), we obtain

$$R_n = \left\{1+\left(\frac{1}{2}\cdot\frac{1}{n}\right)^2\right\}\cdot\frac{1}{n}+\left\{1+\left(\frac{3}{2}\cdot\frac{1}{n}\right)^2\right\}\cdot\frac{1}{n}+\cdots+\left\{1+\left(\frac{2n-1}{2}\cdot\frac{1}{n}\right)^2\right\}\cdot\frac{1}{n}$$

$$= \sum_{k=1}^{n}\left\{1+\left(\frac{2k-1}{2}\cdot\frac{1}{n}\right)^2\right\}\frac{1}{n}$$

To obtain L_n, the lower approximation, the procedure is much the same except (1) is replaced by

1′. Evaluate $f(x)$ at the left hand endpoint of each interval to obtain the height of the rectangle erected on that subinterval.

We obtain

$$L_n = \sum_{k=1}^{n}\left\{1+\left(\frac{k-1}{n}\right)^2\right\}\cdot\frac{1}{n}$$

Note that the only change from R_n to L_n is the value of x at which $f(x) = 1 + x^2$ is evaluated.

To obtain U_n, the upper approximation, we use

1″. Evaluate $f(x)$ at the right-hand endpoint of each interval to obtain the height of the rectangle erected on that subinterval.

followed by (2) and (3). This yields

$$U_n = \sum_{k=1}^{n} \left\{1 + \left(\frac{k}{n}\right)^2\right\} \cdot \frac{1}{n}$$

Note once again that the only difference between R_n, L_n, and U_n is the value of x where the function is evaluated.

To find an upper bound on the error involved in using R_n as an approximation to I, we calculate

$$\begin{aligned} U_n - L_n &= \sum_{k=1}^{n} \left\{1 + \left(\frac{k}{n}\right)^2\right\} \cdot \frac{1}{n} - \sum_{k=1}^{n} \left\{1 + \left(\frac{k-1}{n}\right)^2\right\} \cdot \frac{1}{n} \\ &= \left\{1 + \left(\frac{1}{n}\right)^2\right\} \cdot \frac{1}{n} + \left\{1 + \left(\frac{2}{n}\right)^2\right\} \cdot \frac{1}{n} + \cdots + \left\{1 + \left(\frac{n-1}{n}\right)^2\right\} \cdot \frac{1}{n} \\ &\quad + \left\{1 + \left(\frac{n}{n}\right)^2\right\} \cdot \frac{1}{n} - \left\{1 + \left(\frac{0}{n}\right)^2\right\} \cdot \frac{1}{n} - \left\{1 + \left(\frac{1}{n}\right)^2\right\} \cdot \frac{1}{n} \\ &\quad - \cdots - \left\{1 + \left(\frac{n-1}{n}\right)^2\right\} \cdot \frac{1}{n} \\ &= \left\{1 + \left(\frac{n}{n}\right)^2\right\} \frac{1}{n} - \left\{1 + \left(\frac{0}{n}\right)^2\right\} \cdot \frac{1}{n} \\ &= 2 \cdot \frac{1}{n} - 1 \cdot \frac{1}{n} = \frac{1}{n} \end{aligned}$$

Just as $|I - R_2| \leq U_2 - L_2$, we have $|I - R_n| \leq U_n - L_n = 1/n$. If, therefore, we wish R_n to be as close to I as .005, that is, if we wish R_n to be an approximation to I which has an error of no more than .005, we need only require

$$\frac{1}{n} \leq .005$$

or

$$n \geq \frac{1}{.005}$$

that is,

$$n \geq 200$$

We shall certainly not calculate R_{200} here. That is a task for a computer or for a person who happens to know certain algebraic tricks which would simplify the expression we obtained for R_n. We have demonstrated that

$$|R_{200} - I| \leq .005$$

and also that R_n provides as good an approximation to the exact area as we like if we take n large enough.

An expert might take issue with our error estimate and with the value of n we determined from it because the estimate is, in fact, very crude. A much smaller value of n would suffice. Furthermore, were we to use a computer to calculate R_n for such large n, the truncation error might introduce error in the calculation larger than the maximum theoretical error we have calculated. Much attention has been given, in recent years, to the problem of how to get good approximations for a relatively small value of n. We will not discuss these in any detail, but Exercise 4 shows you how to make an approximation in such a way that n can be cut in half, from 200 to 100.

Let us be sure that we have a clear understanding of the techniques involved in approximating area by considering a second example. (This will help us to see what is common to all such problems and what was specific to the problem we just did.) We shall approximate the area between the x-axis and the graph of $y = 2x$ from $x = 1$ to $x = 3$ (see Fig. 4.9). We begin by noticing that we

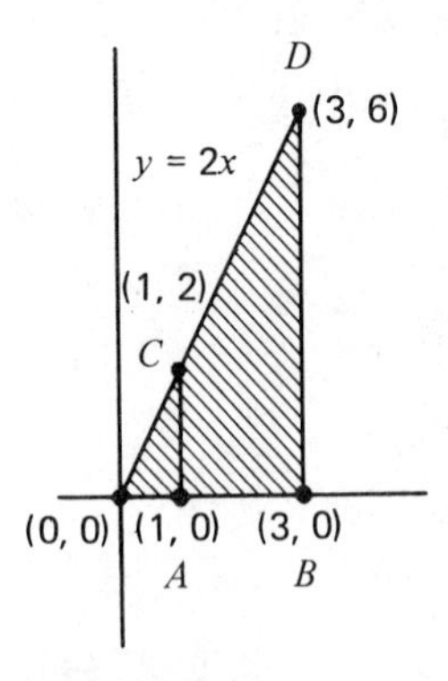

FIGURE 4.9

already know what this area is. It is simply the area of triangle OBD minus the area of triangle OAC or

$$\begin{aligned} \tfrac{1}{2} \cdot 3 \cdot 6 - \tfrac{1}{2} \cdot 1 \cdot 2 &= 9 - 1 \\ &= 8 \end{aligned}$$

This time we can check our approximations against the real answer. We begin by computing L_1, R_1, and U_1.

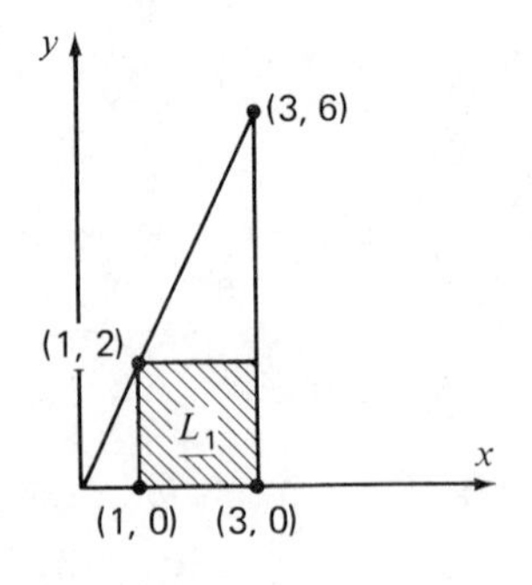

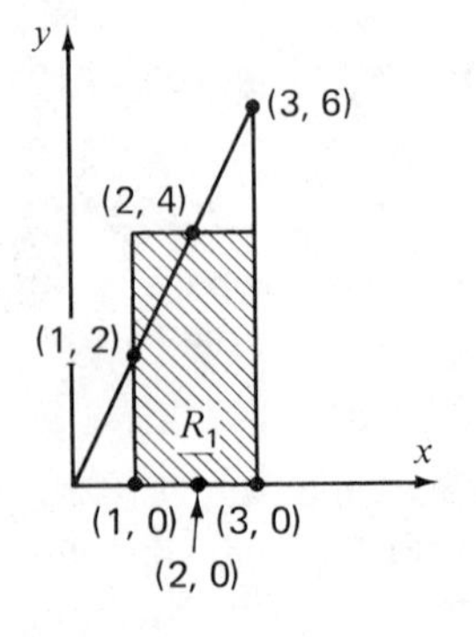

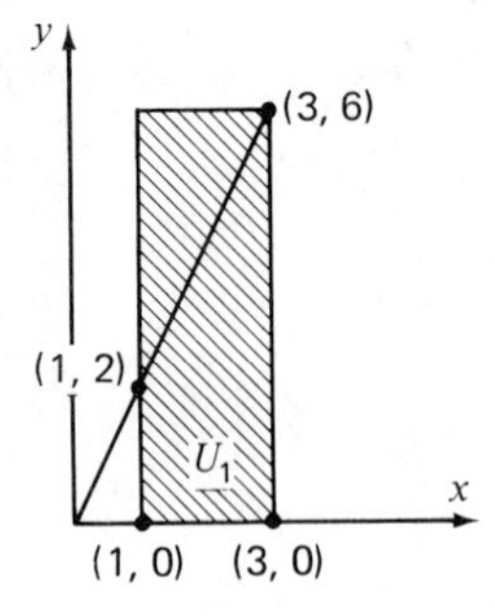

FIGURE 4.10

We obtain:

$$L_1 = 2 \cdot 2 = 4$$
$$R_1 = 4 \cdot 2 = 8$$
$$U_1 = 6 \cdot 2 = 12$$

Our first estimate, R_1, is already the correct answer to the problem, but if this were a region whose area was unknown to us we could only be sure that the answer is *between* 4 *and* 12. What happens with the second estimates, L_2, R_2, and U_2?

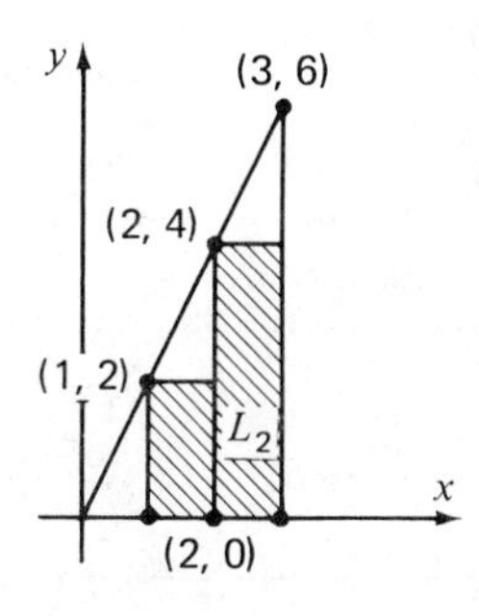

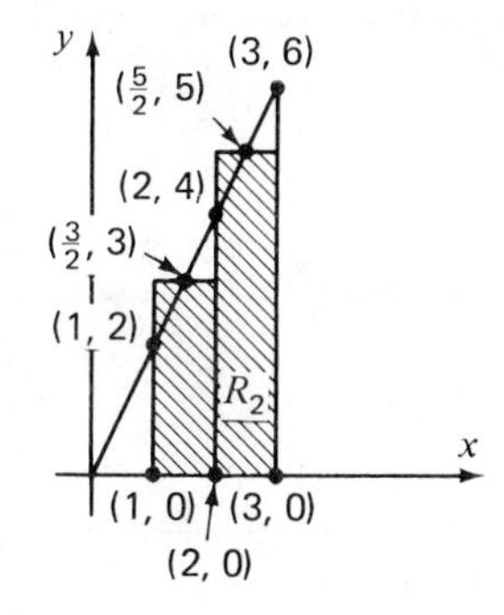

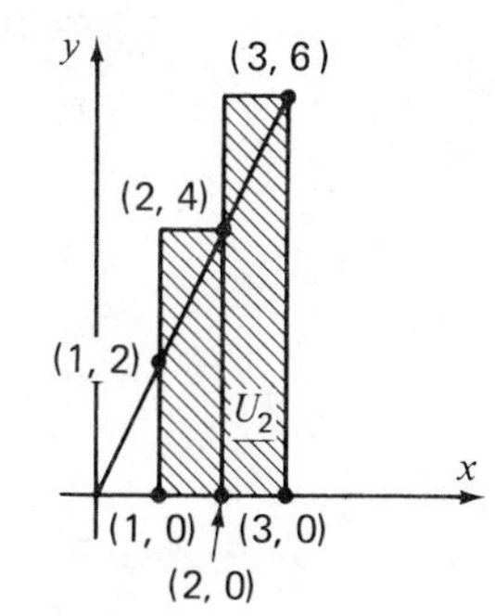

FIGURE 4.11

We have:

$$L_2 = 2 \cdot 1 + 4 \cdot 1 = 6$$
$$R_2 = 3 \cdot 1 + 5 \cdot 1 = 8$$
$$U_2 = 4 \cdot 1 + 6 \cdot 1 = 10$$

Once again R_2 is the exact answer to the problem. But, even if we did not know the area of the region, we would know that it is *between 6 and 10*.

At this point the reader should have a question: "What is all of this for? You know the area of the region. Why go to all of this trouble?" The answer is that we are not so much interested in the solution of the particular problem as we are in the *process of approximation* which is being developed.

In order to see the generality of the process let us consider the following problem. Think of the x-axis as a long straight road. We have a vehicle which goes from the origin in the positive direction in such a way that t hours after it departs its velocity is $2t$ miles per hour. How far does it travel in the time interval from one hour after departure to three hours after departure?

The simple formula

$$\text{Distance} = \text{rate} \times \text{time}$$

learned in high school doesn't work because the rate or velocity is not constant. When $t = 1$ the velocity is 2, when $t = 2$ the velocity is 4, when $t = 3$ the velocity is 6. We might decide to approximate the distance traveled by saying that in the middle of the time interval, when $t = 2$, the vehicle is going 4 miles per hour and if it went at that speed for two hours it would go

$$4 \cdot 2 = 8$$

miles. The objection might be raised that there is no reason to think this estimate is very good because sometimes the velocity is greater than 4 and sometimes less than 4. We can meet this objection by working with shorter time intervals (where there is less variation in speed). We could estimate that in the interval from $t = 1$ to $t = 2$ the vehicle travels

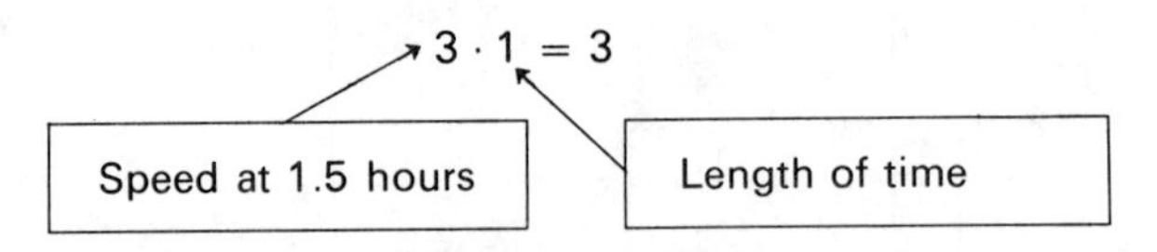

miles and from $t = 2$ to $t = 3$ it travels

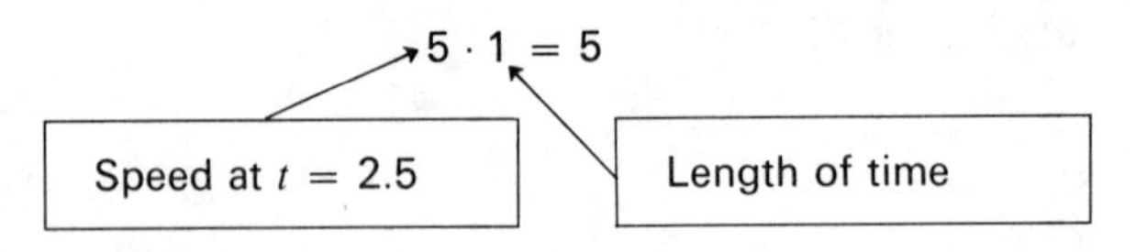

miles and we have the estimate

$$3 \cdot 1 + 5 \cdot 1 = 8$$

Compare what we have just done with the calculations of R_1 and R_2 for the function $y = 2x$ on the interval from 1 to 3. The context was very different but we had

$$R_1 = 4 \cdot 2$$

and the first estimate of distance was also $4 \cdot 2$. Moreover we had

$$R_2 = 3 \cdot 1 + 5 \cdot 1 = 8$$

and this is our second approximation to the distance traveled.

This distance example is only one of many which could be given to indicate that the approximations R_n which we are obtaining occur naturally, not only in area problems, but, in a wide variety of contexts and because of this widespread applicability it is important to understand the process.

Let us return to the area under $y = f(x) = 2x$ from $x = 1$ to $x = 3$ and compute L_n, R_n, and U_n. We break the interval from 1 to 3 (it has length 2) into n equal subintervals (each having length $2/n$). Numbering the intervals from left to right, we obtain the following chart.

Interval Number	Left-hand Endpoint	Right-hand Endpoint	Midpoint
1	1	$1 + \frac{2}{n}$	$1 + \frac{1}{n}$
2	$1 + \frac{2}{n}$	$1 + 2 \cdot \frac{2}{n}$	$1 + \frac{3}{2} \cdot \frac{2}{n}$
3	$1 + 2 \cdot \frac{2}{n}$	$1 + 3 \cdot \frac{2}{n}$	$1 + \frac{5}{2} \cdot \frac{2}{n}$
⋮	⋮	⋮	⋮
k	$1 + (k - 1)\frac{2}{n}$	$1 + k \cdot \frac{2}{n}$	$1 + \frac{2k - 1}{2} \cdot \frac{2}{n}$
⋮	⋮	⋮	⋮
n	$1 + (n - 1)\frac{2}{n}$	$1 + n \cdot \frac{2}{n}$	$1 + \frac{2n - 1}{2} \cdot \frac{2}{n}$

The areas of the approximating rectangles for R_n on each subinterval are as follows.

Interval Number	Midpoint	Height of Rectangle	Width	Area
1	$1 + \frac{1}{n}$	$2\left(1 + \frac{1}{n}\right)$	$\frac{2}{n}$	$2\left(1 + \frac{1}{n}\right) \cdot \frac{2}{n}$
2	$1 + \frac{3}{2} \cdot \frac{2}{n}$	$2\left(1 + \frac{3}{2} \cdot \frac{2}{n}\right)$	$\frac{2}{n}$	$2\left(1 + \frac{3}{2} \cdot \frac{2}{n}\right) \cdot \frac{2}{n}$
3	$1 + \frac{5}{2} \cdot \frac{2}{n}$	$2\left(1 + \frac{5}{2} \cdot \frac{2}{n}\right)$	$\frac{2}{n}$	$2\left(1 + \frac{5}{2} \cdot \frac{2}{n}\right) \cdot \frac{2}{n}$
⋮	⋮	⋮	⋮	⋮
k	$1 + \frac{2k - 1}{2} \cdot \frac{2}{n}$	$2\left(1 + \frac{2k - 1}{2} \cdot \frac{2}{n}\right)$	$\frac{2}{n}$	$2\left(1 + \frac{2k - 1}{2} \cdot \frac{2}{n}\right) \cdot \frac{2}{n}$
⋮	⋮	⋮	⋮	⋮
n	$1 + \frac{2n - 1}{2} \cdot \frac{2}{n}$	$2\left(1 + \frac{2n - 1}{2} \cdot \frac{2}{n}\right)$	$\frac{2}{n}$	$2\left(1 + \frac{2n - 1}{2} \cdot \frac{2}{n}\right) \cdot \frac{2}{n}$

Therefore

$$R_n = 2\left(1 + \frac{1}{n}\cdot\frac{2}{n}\right)\frac{2}{n} + \cdots + 2\left(1 + \frac{2k-1}{2}\cdot\frac{2}{n}\right)\frac{2}{n} + \cdots + 2\left(1 + \frac{2n-1}{2}\cdot\frac{2}{n}\right)\frac{2}{n}$$

$$= \sum_{k=1}^{n} 2\left(1 + \frac{2k-1}{2}\cdot\frac{2}{n}\right)\cdot\frac{2}{n}$$

$$= \sum_{k=1}^{n} \left(1 + \frac{2k-1}{n}\right)\cdot\frac{4}{n}$$

In a like manner we can calculate L_n,

$$L_n = \sum_{k=1}^{n} 2\left(1 + (k-1)\cdot\frac{2}{n}\right)\cdot\frac{2}{n}$$

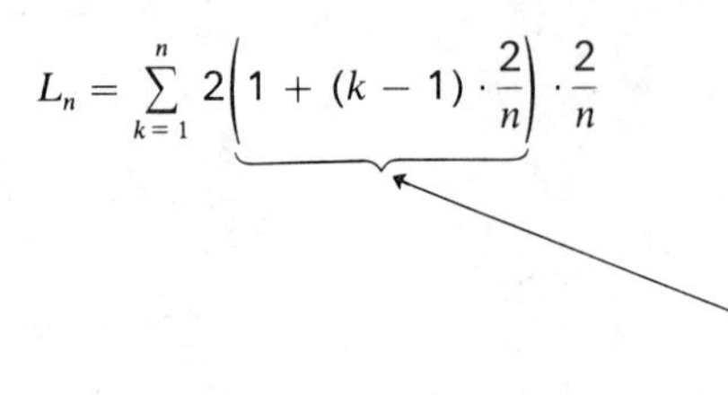

Note that this expression is the same as the expression for R_n except that the midpoint of the interval has been replaced by the left-hand endpoint.

and U_n,

$$U_n = \sum_{k=1}^{n} 2\left(1 + k\cdot\frac{2}{n}\right)\cdot\frac{2}{n}$$

Right-hand endpoint of the interval.

We now have expressions whose value is not so apparent. The interested reader is invited to try his hand at Exercise 5 which will enable him to actually compute R_n, L_n, and U_n.

The expressions L_n and U_n cannot be used, in general, in the way in which we have used them without creating considerable difficulty. We have enjoyed the fact that the low point of the graph on any subinterval was above the left-hand endpoint of the interval and the high point above the right-hand side (see Fig. 4.12). But if the function were as in Fig. 4.13, it would be very difficult indeed to find L_n and U_n.

The notions of upper sum and lower sum are important in a theoretical development of area problem for they allow us to know when our approximating procedure is working. However explicit calculation of these sums can, as Fig. 4.13 indicates, become a burdensome task. We shall therefore focus our attention on the more readily computable R_n.

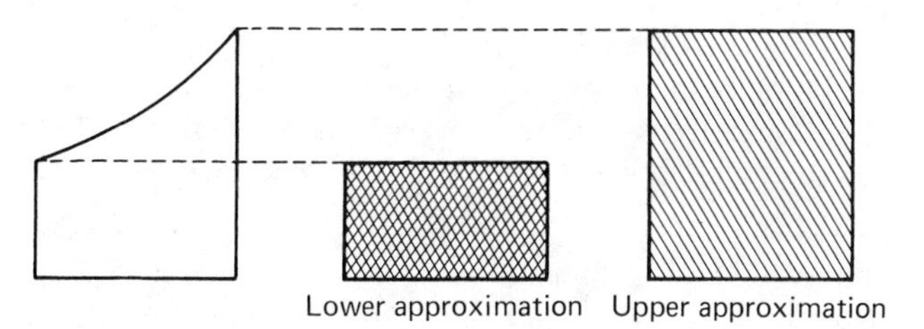

FIGURE 4.12

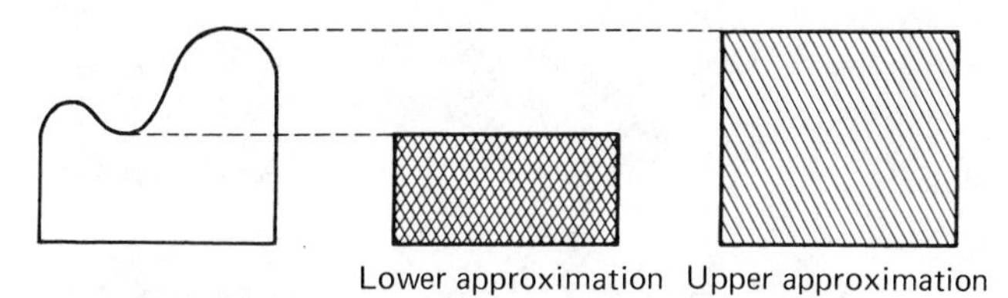

FIGURE 4.13

Let us recapitulate and generalize the discussion to this point. Each of the preceding examples has consisted of the following:

1. You are given a function $y = f(x)$ defined and positive on an interval from a to b.
2. You want to know the area under the graph and above the x-axis.

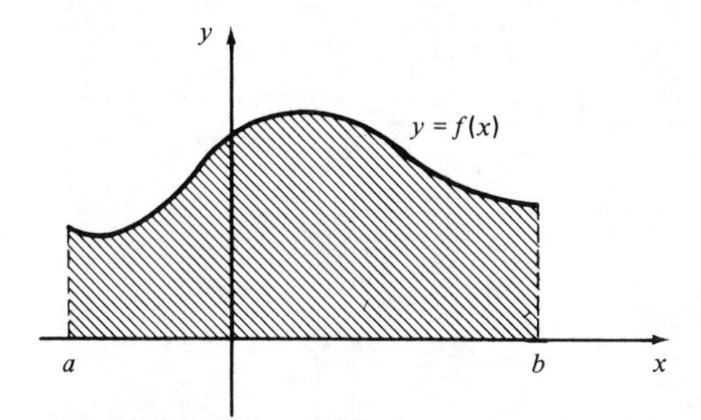

FIGURE 4.14

You cannot necessarily calculate the area exactly but you can, for each integer n, obtain an approximation.

The nth approximation (where n is a positive integer), denoted R_n, is obtained as follows.

1. The interval from a to b has length n. Break it up into n equal subintervals, each of length $(b - a)/n$.
2. Compute the midpoint of each of the intervals. This is done by numbering the subintervals from left to right and developing the following chart.

Interval Number	Left-hand Endpoint	Right-hand Endpoint	Midpoint
1	a	$a + \frac{b-a}{n}$	$a + \frac{1}{2}\frac{(b-a)}{n}$
2	$a + \frac{(b-a)}{n}$	$a + 2 \cdot \frac{(b-a)}{n}$	$a + \frac{3}{2}\frac{(b-a)}{n}$
3	$a + 2 \cdot \frac{(b-a)}{n}$	$a + 3 \cdot \frac{(b-a)}{n}$	$a + \frac{5}{2}\frac{(b-a)}{n}$
$\vdots$	$\vdots$	$\vdots$	$\vdots$
k	$a + (k-1)\frac{(b-a)}{n}$	$a + k\frac{(b-a)}{n}$	$a + \frac{(2k-1)}{2}\frac{(b-a)}{n}$
$\vdots$	$\vdots$	$\vdots$	$\vdots$
n	$a + (n-1)\frac{(b-a)}{n}$	$a + n\frac{(b-a)}{n}$	$a + \frac{(2n-1)}{2}\frac{(b-a)}{n}$

3. On each subinterval erect a rectangle whose base is the subinterval and whose height is the value of the function at the midpoint of the subinterval.

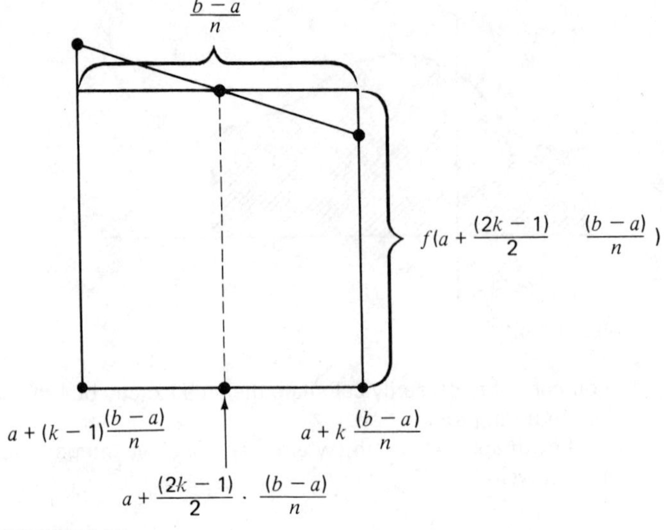

FIGURE 4.15

4. Compute the area of each of these rectangles. The area of kth rectangle is

$$f\left(a + \frac{(2k-1)}{2} \cdot \frac{(b-a)}{n}\right)\frac{(b-a)}{n}$$

5. Add these areas obtaining

$$R_n = \sum_{k=1}^{n} f\left(a + \frac{(2k-1)}{2} \cdot \frac{(b-a)}{n}\right)\frac{(b-a)}{n}$$

The resulting values

$R_1, R_2, R_3, \ldots$

are (hopefully) successively better and better approximations of I in the sense that the errors

$|R_1 - I|, |R_2 - I|, |R_3 - I|, \ldots$

are getting closer and closer to zero.

While this procedure may be enlightening, it still falls far short of being practical. Indeed, we can compute R_1, R_2, R_3, and R_4 perhaps for simple functions with a good deal of effort. But to satisfy our prescribed degree of accuracy it might be necessary to compute R_{500}, a task that would challenge the best of us. The primary value of this discussion has been to carefully formulate an important process rather than to actually solve a particular problem relating to area.

One final note, we said the process is important and we gave a simple example concerning distance traveled by an object moving at variable velocity. To fully appreciate the extent of generality, let us go further. Suppose that f is any function defined on $[a, b]$, be it positive or negative or both (that is, part or all of its graph may fall beneath the x-axis). We can still subdivide $[a, b]$ into n pieces, find the midpoint of each subinterval, evaluate f there, multiple the result by $(b-a)/n$ and add these products to obtain

$$R_n = \sum_{k=1}^{n} f\left(a + \frac{(2k-1)}{2} \frac{(b-a)}{n}\right)\frac{(b-a)}{n}$$

And now we can ask a more general question. Do these values R_n approximate something and provide better approximations as n gets larger? It is in this general context that we shall discover a practical method for eventually solving our area problem.

Summary

In this section we have developed a procedure for approximating the area of a region in the plane. We have observed that the procedure for approximating area is the same as the procedure for approximating the distance traveled by an object moving at a variable speed. Indeed the generality of the procedure we have developed leads to a wide variety of applications. We are faced with two major problems. One is the lack of a precise language to describe the approximating procedure. The development of this language is the content of Section 4. The other problem is to obtain an exact answer to the area or whatever is being approximated. The exact answer can be found in many cases and we will begin to develop the machinery for this calculation in Section 3.

Exercises Section 1

1 For each of the following functions:
(1) calculate R_1, R_2, R_3.
(2) draw three graphs of the function to show the three sets of rectangles that give R_1, R_2, and R_3.
(a) $f(x) = x^2$ on $[0, 2]$
(b) $f(x) = x$ on $[0, 1]$
(c) $f(x) = x + x^2$ on $[0, 1]$
(d) $f(x) = 1 - x^2$ on $[0, 1]$

2 Write a flow chart for computing R_n for a function f defined on an interval $[a, b]$

3 Use the flow chart developed in Exercise 2 to write a program which will print out values of R_n for all values of n which are multiples of 10 up to 200. (That is, R_n should be printed out for $n = 10, 20, 30, 40, \ldots, 180, 190, 200$.)
(a) $f(x) = x^2$ on $[0, 2]$
(b) $f(x) = \sqrt{1 - x^2}$ on $[0, 1]$
(c) $f(x) = (1 + x)^3$ on $[0, 2]$

4 For the function $f(x) = 1 + x^2$ on $[0, 1]$, set

$$S_n = \tfrac{1}{2}(U_n + L_n)$$

(notation as in discussion in text). Show that

$$|S_n - I| < .005$$

if $n > 100$.

5 Use the facts that

$$\sum_{k=1}^{n} k = \frac{n(n+1)}{2} \quad \text{and} \quad \sum_{k=1}^{n} 1 = n$$

to compute the values of R_n, U_n, and L_n for the function $f(x) = 2x$ on the interval $[1, 3]$.

2 THE DEFINITE INTEGRAL

Our effort to find the area under a curve has led us to a more general problem:

Given a function f defined on an interval $[a, b]$, we divide $[a, b]$ into n subintervals of equal length $(b - a)/n$ and form the sum

$$R_n = \sum_{k=1}^{n} f(m_k)\left(\frac{b-a}{n}\right)$$

where m_k is the midpoint of the kth subinterval (from left to right). This process produces a "sequence" of numbers

$$R_1, R_2, R_3, \ldots$$

and the question becomes, "As n increases, do these successive values of R_n tend toward some fixed value?"

Of course in the area problem we hope for an affirmative answer where R_n gets closer and closer to the exact value of the area under the graph of f from $x = a$ to $x = b$. Likewise in the distance problem we hope that R_1, R_2, R_3, and so on represent successively better approximations to the exact distance traveled (remember that it was a coincidence that each R_n gave us the exact value in that case). But more generally two things can happen:

Either, there is a number L such that the values R_n get closer and closer to L as n gets large, in which case we write

$$\lim_{n\to\infty} R_n = L$$

and say that "the limit of R_n as n grows large is L" *or*, the successive values R_1, R_2, R_3 don't settle down and approach anything, in which case we write

$$\lim_{n\to\infty} R_n \text{ does not exist}$$

We shall take a careful look at sequences and whether or not they have limits shortly, but for the time being we simply shall trust our intuition to the extent of believing that the definition that follows can be made more precise once these intricacies have been dealt with. To properly set the stage for this important definition, we must introduce a more general kind of sequence similar to the R_n's.

For a function f defined on $[a, b]$, we consider *any* subdivision of $[a, b]$ into subintervals

$$x_0 = a \quad x_1 \quad x_2 \quad \ldots \quad x_{n-1} \quad b = x_n$$

where the lengths of the subintervals may vary. Now we select *any* point x_i' in the ith subinterval, for $i = 1, 2, \ldots, n$,

$$x_{i-1} \quad x_i' \quad x_i$$

and form the sum

$$S_n = \sum_{i=1}^{n} f(x_i')(x_i - x_{i-1})$$

We shall call a sequence of the form

$$S_1, S_2, S_3, \ldots$$

a sequence of Riemann Sums if the lengths of all of the subintervals tend to 0 as n gets large. [Note that in R_n the lengths $(b - a)/n$ tend to 0.]

Definition If every sequence of Riemann Sums approaches some common value as n gets large, we call this common value *the definite integral of f from a to b* and we denote it by

$$\int_a^b f(x)\,dx$$

In particular, then, we will have

$$\lim_{n\to\infty} R_n = \int_a^b f(x)\,dx$$

provided this limit exists.

There are several important facts concerning this definition that should be made:

1. We have not restricted f to being a non-negative function on $[a, b]$. Thus any function f defined on $[a, b]$ is eligible to have a definite integral, whether its graph lies wholly above the x-axis $[f(x) > 0]$ or below the x-axis $[f(x) < 0]$ or partly above and partly below.
2. From the way in which we made the definition, it sounds very much as if we are never sure whether a given function f will have an integral or not. True, there are some functions where the necessary limit does not exist, but by and large the functions of interest to us will all have definite integrals. A crude but appropriate analogy can be seen in the statement

   If $a = bc$, then $\frac{a}{b} = c$ provided $b \neq 0$.

   Here the restriction $(b \neq 0)$ must be made to prevent dividing by 0 but in practice we are seldom bothered by that restriction.

   The real key to establishing whether the definite integral exists and to defining it, for a given function, is our earlier discussion of upper and lower sums, U_n and L_n. Recall that we defined

$$L_n = \sum_{k=1}^{n} f(x_k^*)\left(\frac{b-a}{n}\right)$$

   and

$$U_n = \sum_{k=1}^{n} f(x_k^{**})\left(\frac{b-a}{n}\right)$$

   where $f(x_k^*)$ and $f(x_k^{**})$ are, respectively, the smallest and largest values of f in the kth subinterval (provided such values exist!). Again the expressions L_n and U_n make perfectly good sense, independent of whether f is a positive function on $[a, b]$. Moreover we have

$$L_n \leq R_n \leq U_n$$

for each integer n. Thus, just as in the area problem we "trapped" the exact value of the area (along with R_n) in between L_n and R_n, now we can observe that the values of successive R_n's will tend to "settle down" or stabilize to some limit (namely the definite integral!) provided the differences

$$U_n - L_n$$

approach 0 as n gets large. Indeed, a rigorous development of the definite integral would take advantage of this fact and assert that when L_n and U_n make sense for all n, then

$$\lim_{n\to\infty} L_n \quad \text{and} \quad \lim_{n\to\infty} U_n$$

exist; moreover, if these limits are equal (say to number I), then

$$\int_a^b f(x)\,dx = \lim_{n\to\infty} R_n$$

exists and has the value I.

Whereas we cannot fully appreciate the significance of this result, it should be noted that for a large class of functions (including all functions whose graphs have no holes or breaks) it is relatively easy to show that all these limits exist.

The next three facts represent important properties of the definite integral—facts we shall use frequently for the remainder of this chapter.

3. For any functions $f(x)$ and $g(x)$ on an interval $[a, b]$ and for any constant c,

$$\int_a^b (f(x) + g(x))\,dx = \int_a^b f(x)\,dx + \int_a^b g(x)\,dx$$

and

$$\int_a^b cf(x)\,dx = c\int_a^b f(x)\,dx$$

For example,

$$\int_0^1 (x^2 + x)\,dx = \int_0^1 x^2\,dx + \int_0^1 x\,dx$$

and

$$\int_1^2 3x^3\,dx = 3\int_1^2 x^3\,dx$$

This fact is usually stated by the assertion that the integral is *linear*. It follows from the corresponding facts about limits, and Exercise 1 will give you some idea of why it is true.

4. $$\int_a^b f(x)\,dx = -\int_b^a f(x)\,dx$$

For example,

$$\int_0^2 x^3\,dx = -\int_2^0 x^3\,dx$$

Exercise 2 will give you some insight into why this is true.

5. $$\int_a^b f(x)\,dx = \int_a^c f(x)\,dx + \int_c^b f(x)\,dx$$

For example

$$\int_0^2 x^2\,dx = \int_0^1 x^2\,dx + \int_1^2 x^2\,dx$$

This will be examined in Exercise 3.

There is one question which students frequently ask and which we have ignored up to this point. What is the dx in

$$\int_a^b f(x)\,dx$$

and what is it for?

It is what is known as a differential and we shall not have occasion to explore its role. It is however useful as a marker. If you are given

$$\int_1^2 (x^2 + t^2)\,dx$$

as a problem, the dx tells you to pretend, while integrating, that t is a constant. You are integrating with respect to x.

We have not explained how one finds the *number* (and it is important to remember that it is a number) represented by

$$\int_a^b f(x)\,dx$$

That is a problem which will be tackled in the next several sections.

Finally we should say something about terminology. The function f in

$$\int_a^b f(x)\,dx$$

is called the *integrand*. The numbers a and b are referred to as the *limits of integration*.

Exercises Section 2

1 Compute R_1, R_2, R_3 for

(a) $\int_1^2 x^2\,dx$

(b) $\int_1^2 x\,dx$

(c) $\int_1^2 (x^2 + x)\,dx$

(d) $\int_1^2 3x^2\,dx$

Relate your results to fact (3) in this section.

2 Compute R_1, R_2, R_3 for $\int_a^b f(x)\,dx$, where
(a) $f(x) = x^2$, $a = 0$, $b = 2$
(b) $f(x) = x^2$, $a = 2$, $b = 0$ (Note that in this case, $(b - a)/n$ is not really the length of the subinterval.)
Relate your results to fact (4) in this section.

3 Compute R_1, R_2, R_3 for

$$\int_0^1 x^2\,dx \quad \text{and} \quad \int_1^2 x^2\,dx$$

Compute R_2, R_4, R_6 for

$$\int_0^2 x^2\,dx$$

Relate your results to fact (5) in this section.

4 Suppose

$$\int_{-2}^{3} f(x)\,dx = 12 \quad \text{and} \quad \int_{-2}^{3} g(x)\,dx = -5$$

(a) Find $\int_{-2}^{3} [f(x) + g(x)]\,dx$.
(b) Find $\int_{-2}^{3} [2f(x) - 3g(x)]\,dx$.
(c) Find $3\int_{-2}^{3} [g(x) + \frac{1}{2}f(x)]\,dx$.
(d) Find $\int_{3}^{-2} [3f(x) - 2g(x)]\,dx$.

5 A function G is called *linear* if

$$G(x + y) = G(x) + G(y)$$

and

$$G(cx) = cG(x)$$

for all cases where G is defined. Which of the following functions are linear?
(a) $G(x) = 3x$
(b) $G(x) = 3x + 2$
(c) $G(x) = x^2$
(d) $G(x) = |x|$
(e) $G(x) = x^3$

6 Compute R_1, R_2, R_3, and R_4 for $\int_0^2 f(x)\,dx$, where

$$f(x) = \begin{cases} 1, & \text{if } 0 \le x < 1 \\ 2, & \text{if } 1 \le x \le 2 \end{cases}$$

7 Compute R_1, R_2, R_3, and R_4 for $\int_0^1 f(x)\,dx$, where

$$f(x) = -3, \qquad \text{for all } x \text{ in } [0, 1]$$

8 Compute R_1 and R_2 for $\int_a^b f(x)\,dx$, where

$$f(x) = c, \qquad \text{for all } x \text{ in } [a, b]$$

9 Compute R_1, R_2, R_3, and R_4 for $\int_0^1 f(x)\,dx$, where

$$f(x) = \begin{cases} 1, & \text{if } x = \frac{1}{2} \\ 0, & \text{otherwise} \end{cases}$$

3 DERIVATIVES AND THE FUNDAMENTAL THEOREM OF CALCULUS

Now that we have defined the definite integral we need a method for evaluating it. There is a method which works for many integrals and, in order to see how it arises, let us return to the relation between the density and distribution functions which prompted the integral. We briefly review some calculations from Chapter 3. For the density function

$$f(x) = \begin{cases} 0, & \text{if } x < 0 \\ 1, & \text{if } 0 \le x \le 1 \\ 0, & \text{if } x > 1 \end{cases}$$

What is $F(z)$, its distribution function?

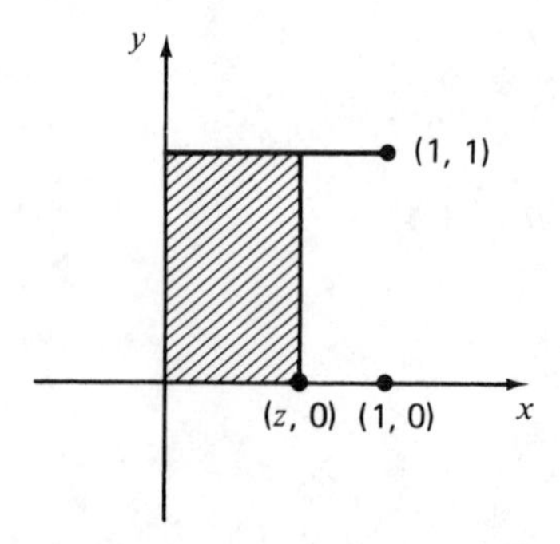

FIGURE 4.16

For $z < 0$, $F(z) = 0$. For $0 \le z \le 1$, we have

$$\begin{aligned} F(z) &= \int_0^z f(x)\,dx \qquad \text{(Using our new integral notation)} \\ &= \int_0^z 1 \cdot dx \\ &= 1 \cdot z \\ &= z \end{aligned}$$

and for $z > 1$, $F(z) = 1$. To recapitulate

$$F(z) = \begin{cases} 0, & \text{if } z < 0 \\ z, & \text{if } 0 \le z \le 1 \\ 1, & \text{if } z > 1 \end{cases}$$

A second example which we solved involves the density function

$$g(x) = \begin{cases} 0, & x < 0 \\ \frac{1}{2}x, & 0 \le x \le 2 \\ 0, & x > 2 \end{cases}$$

What is $G(z)$, the associated distribution function?

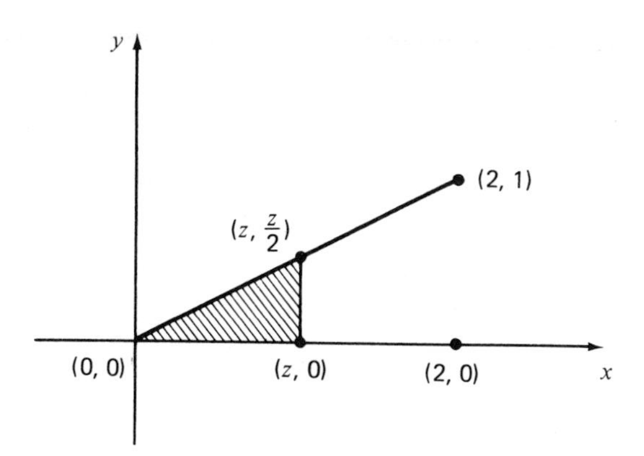

FIGURE 4.17

As in Chapter 3, $G(z) = 0$ if $z < 0$. If $0 \le z \le 2$,

$$G(z) = \int_0^z g(x)\,dx$$
$$= \frac{1}{2} \cdot z \cdot \frac{z}{2}$$
$$= \frac{z^2}{4}$$

For $z > 2$, $G(z) = 1$. Recapitulating,

$$G(z) = \begin{cases} 0, & z < 0 \\ \frac{z^2}{4}, & 0 \le z \le 2 \\ 1, & 2 < z \end{cases}$$

It was, of course, easy to compute the values of the distribution functions F and G, where the density functions f and g were 0. The challenge was to determine the values where f and g were not 0. Note that we have in this case

$$F(z) = \int_0^z f(x)\,dx$$

and

$$G(z) = \int_0^z g(x)\,dx$$

Both F and G are *functions of the right-hand endpoint* of the interval over which we integrate. What we want is a rule or formula by which we can get from f to F and from g to G. The graphs that follow give us a hint about where to look for such a formula.

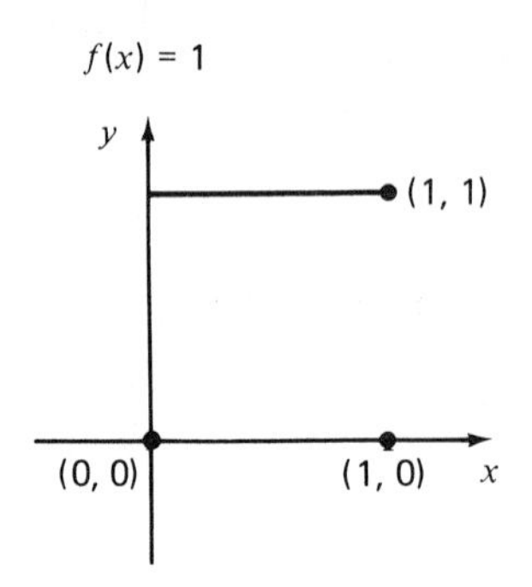

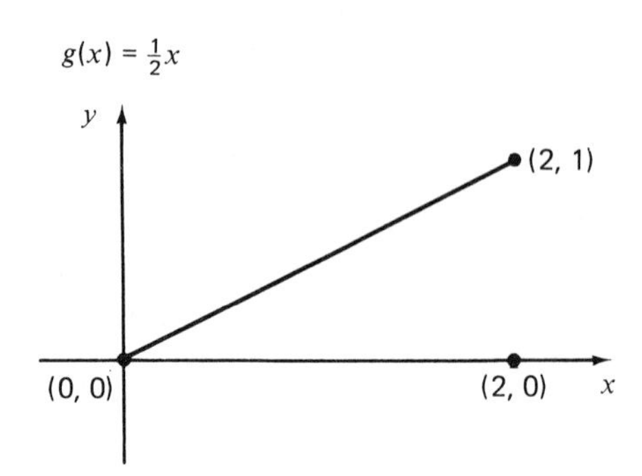

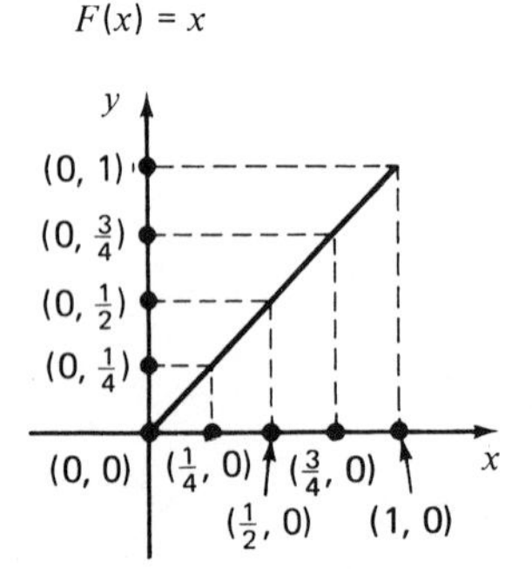

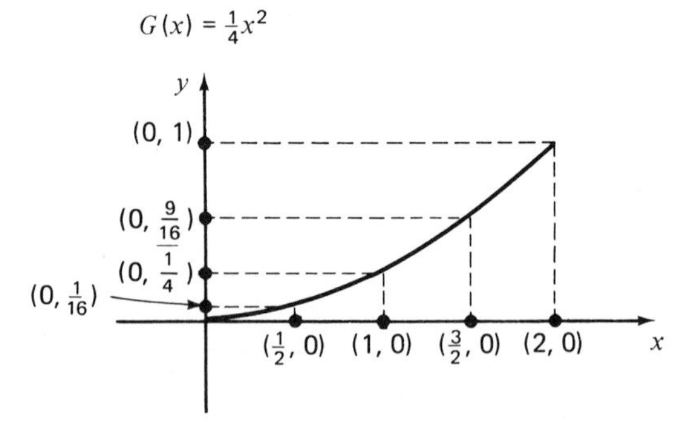

Observe that equal increments of x produce equal increments of $F(x)$. That is, if we move x to the right $\frac{1}{4}$, we move the value of $F(x)$ up $\frac{1}{4}$. The situation with G is different. As x changes from 0 to $\frac{1}{2}$, the value of G changes from 0 to $\frac{1}{16}$ but as x changes from $\frac{3}{2}$ to 2 (also an increment of $\frac{1}{2}$), G changes from $\frac{9}{16}$ to 1 (an increment of $\frac{7}{16}$). Recalling that F and G are the areas under f and g, this difference is not surprising. Since the graph of $f(x)$ is always 1 unit high the increments in $F(x)$ are constant, while as x goes from left to right $g(x)$ gets higher and so the increments in $G(x)$ become larger. Since G (or F) are functions, any change in the value of the independent variable, call it z, will produce a change in the value of the function $G(z)$ [or $F(z)$]. It is reasonable to believe that since the definition of G is so dependent upon g, this change of G is somehow related to g.

Mathematicians discovered, a long time ago, that this relation can be made precise if we look at the ratio between the change in G and the change in z. Suppose we let the independent variable change from z to $z + \Delta z$ (Δz is read "delta z" and should be viewed as the change in z). Then the value of G changes from $G(z)$ to $G(z + \Delta z)$. The ratio of these changes is

$$\frac{G(z + \Delta z) - G(z)}{z + \Delta z - z} = \frac{G(z + \Delta z) - G(z)}{\Delta z}$$

Let us compute this for the function in hand

$$G(z) = \frac{z^2}{4}$$

We have

$$\begin{aligned}
\frac{G(z + \Delta z) - G(z)}{\Delta z} &= \frac{[(z + \Delta z)^2/4] - (z^2/4)}{\Delta z} \\
&= \frac{[(z + \Delta z)^2 - z^2]/4}{\Delta z} \\
&= \frac{(z + \Delta z)^2 - z^2}{4\,\Delta z} \\
&= \frac{z^2 + 2z\,\Delta z + (\Delta z)^2 - z^2}{4\,\Delta z} \\
&= \frac{2z\,\Delta z + (\Delta z)^2}{4\,\Delta z} \\
&= \frac{2z + \Delta z}{4} \cdot \frac{\Delta z}{\Delta z} \\
&= \frac{2z}{4} + \frac{\Delta z}{4} \cdot \frac{\Delta z}{\Delta z} \\
&= g(z) + \frac{\Delta z}{4} \cdot \frac{\Delta z}{\Delta z}
\end{aligned}$$

What is this ratio? In particular, what is this ratio for small values of Δz? We compute some examples.

$$\frac{G(z + \frac{1}{2}) - G(z)}{\frac{1}{2}} = \left(g(z) + \frac{\frac{1}{2}}{4}\right) \cdot \frac{\frac{1}{2}}{\frac{1}{2}} = g(z) + \frac{1}{8}$$

$$\frac{G(z + \frac{1}{4}) - G(z)}{\frac{1}{4}} = \left(g(z) + \frac{\frac{1}{4}}{4}\right) \cdot \frac{\frac{1}{4}}{\frac{1}{4}} = g(z) + \frac{1}{16}$$

$$\frac{G(z + \frac{1}{100}) - G(z)}{\frac{1}{100}} = \left(g(z) + \frac{\frac{1}{100}}{4}\right) \cdot \frac{\frac{1}{100}}{\frac{1}{100}} = g(z) + \frac{1}{400}$$

As Δz becomes smaller, the ratio

$$\frac{G(z + \Delta z) - G(z)}{\Delta z}$$

appears to become closer and closer to $g(z)$.

Observe that, at this stage of finding the relationship between G and g, the position we are in is similar to the situation which was involved in defining

$$\int_a^b f(x)\,dx$$

There we had, for each positive integer n, a number R_n and we argued that as n becomes large, R_n becomes a better approximation of the integral. We briefly discussed the limit concept and said

$$\int_a^b f(x)\,dx = \lim_{n\to\infty} R_n$$

Now we have, for each number $\Delta z (\neq 0)$ a fraction

$$\frac{G(z + \Delta z) - G(z)}{\Delta z}$$

and we are claiming that when Δz is close to 0, the fraction is close to $g(z)$. Once again we need a limit notion to make the connection. We wish to say that

$$\lim_{\Delta z\to 0} \frac{G(z + \Delta z) - G(z)}{\Delta z} = g(z)$$

In order to avoid interrupting the main line of discussion leading to the Fundamental Theorem of Calculus, we shall postpone a formal definition and discussion to Section 5, but we will make a remark on an intuitive level.

You should have the feeling that, for each value of Δz other than 0 (the fraction isn't defined when $\Delta z = 0$) the fraction

$$\frac{G(z + \Delta z) - G(z)}{\Delta z}$$

approximates something. The closer Δz is to 0, the better the approximation. This value which the fraction approximates is called its *limit as* Δz *approaches* 0 and we write

$$\lim_{\Delta z\to 0} \frac{G(z + \Delta z) - G(z)}{\Delta z} = g(z)$$

At least two questions should occur to you. First, we have argued that if

$$g(x) = \tfrac{1}{2}x \text{ and } G(z) = \int_0^z g(x)\,dx$$

then

$$\lim_{\Delta z\to 0} \frac{G(z + \Delta z) - G(z)}{\Delta z} = g(z)$$

Is this an accident? The answer is no. We will shortly show that, for a general class of functions, if

$$F(z) = \int_a^z f(x)\,dx$$

then

$$(*) \quad \lim_{\Delta z \to 0} \frac{F(z + \Delta z) - F(z)}{\Delta z} = f(z)$$

The second question is that, even if this relation between F and f is not accidental, what good is it? The answer to this question is that it leads to a number of formulae which allow us to evaluate many integrals as well as to solve several other problems.

To motivate (*) in general let us look at the particular case $g(x) = x/2$ from a geometric point of view. What does

$$G(z + \Delta z) - G(z)$$

represent? There are two cases depending on whether Δz is positive or negative. (See Fig. 4.18 and Fig. 4.19.) When $\Delta z > 0$, $G(z + \Delta z) - G(z)$ is the area

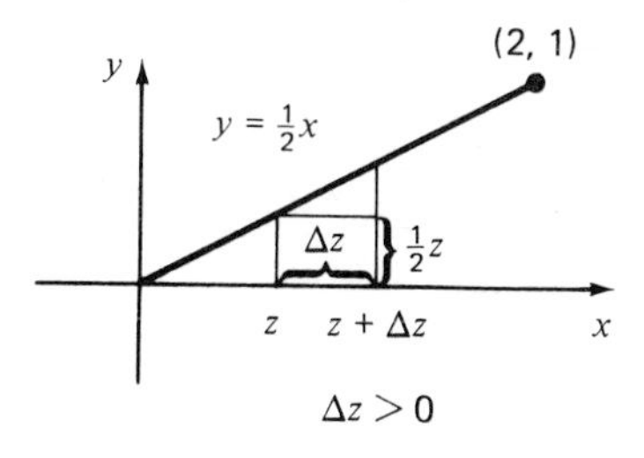

FIGURE 4.18

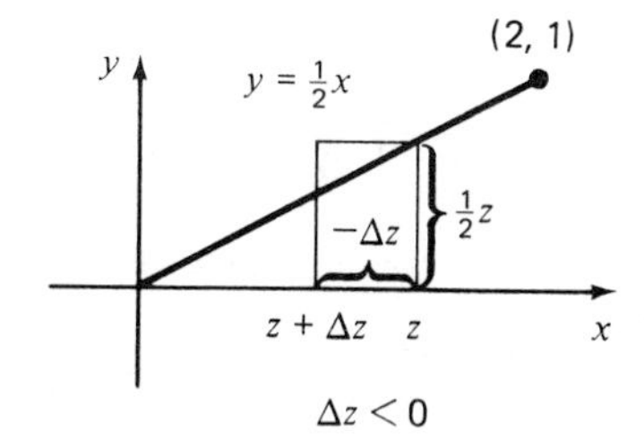

FIGURE 4.19

between the x-axis and the graph of $y = \frac{1}{2}x$ from $x = z$ to $x = z + \Delta z$. This area is approximately the area of a rectangle of height $\frac{1}{2}(z)$ and width Δz. So we have

$$G(z + \Delta z) - G(z) \approx \tfrac{1}{2}z \, \Delta z$$

and

$$(*) \quad \frac{G(z + \Delta z) - G(z)}{\Delta z} \approx \tfrac{1}{2}z$$

If $\Delta z < 0$, then

$$G(z) - G(z + \Delta z)$$

is the area under the graph of $y = \frac{1}{2}x$ and above the x-axis from $x = z + \Delta z$ to $x = z$. This is approximately equal to the area of the rectangle of height $\frac{1}{2}z$ and width $(-\Delta z)$ so

$$G(z) - G(z + \Delta z) \approx \tfrac{1}{2}z(-\Delta z)$$

$$\frac{G(z) - G(z + \Delta z)}{-\Delta z} \approx \tfrac{1}{2}z$$

$$(**) \quad \frac{G(z + \Delta z) - G(z)}{\Delta z} \approx \tfrac{1}{2}z$$

Note that (*) and (**) are the same, so we arrive at the same approximation whether Δz is positive or negative. We then have

$$\lim_{\Delta z \to 0} \frac{G(z + \Delta z) - G(z)}{\Delta z} = \tfrac{1}{2}z = g(z)$$

This method can be applied very generally. Suppose

$$F(z) = \int_a^z f(x)\, dx$$

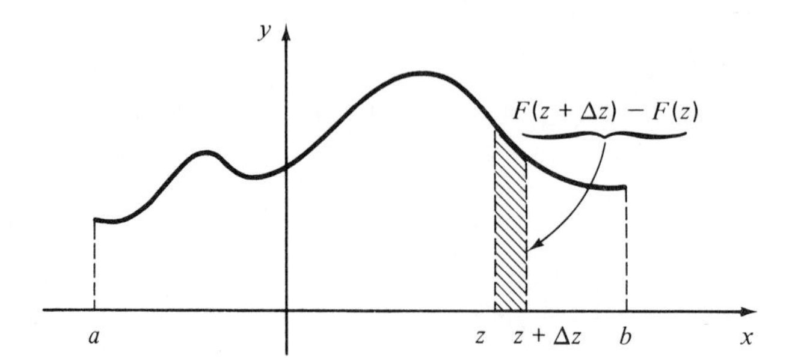

FIGURE 4.20

What is $F(z + \Delta z) - F(z)$? It is simply the area between the x-axis and the graph of $y = f(x)$ between z and $z + \Delta z$. The way we have drawn the picture is with $\Delta z > 0$. We will continue the argument for a while with $\Delta z > 0$ and then go back and fix it up for the case when $\Delta z < 0$.

Now we get an approximate value for $F(z + \Delta z) - F(z)$. We approximate this area by the area of the rectangle whose base is of length Δz and whose height is $f(z)$. We have

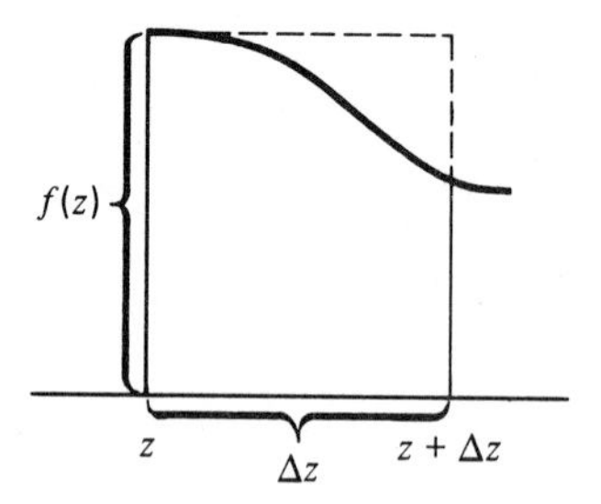

FIGURE 4.21

$$F(z + \Delta z) - F(z) \approx f(z) \cdot \Delta z$$

Dividing both sides by Δz, gives

$$(\dagger) \quad \frac{F(z + \Delta z) - F(z)}{\Delta z} \approx f(z)$$

At this point note what happens when Δz is negative. Now the area under the graph of $y = f(x)$ and the x-axis between z and $z + \Delta z$ is $F(z) - F(z + \Delta z)$. But the rectangle which we use to approximate the area has, as the length of its base, $-\Delta z$, so the approximation is

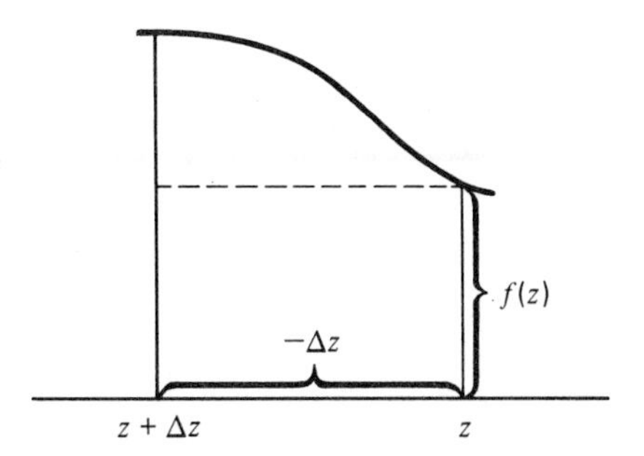

FIGURE 4.22

$$F(z) - F(z + \Delta z) \approx f(z) \cdot (-\Delta z)$$

Now, if we divide through by $(-\Delta z)$, we get (†) again so the approximation holds whether Δz is positive or negative. Now what happens at $\Delta z \to 0$? The amount of error decreases also and for the functions which will arise in this course (indeed for most functions which arise in most applications of mathematics) we have

$$\lim_{\Delta z \to 0} \frac{F(z + \Delta z) - F(z)}{\Delta z} = f(z)$$

We have seen that if

$$F(z) = \int_a^z f(x)\, dx$$

then f is related to F by being the limit of a particular fraction involving F. But, as is often the case with important concepts, this limit is important for many purposes besides evaluating integrals. We formalize it with the following definition.

Definition Let f be a function. We define a new function, denoted f', by

$$f'(x) = \lim_{\Delta x \to 0} \frac{f(x + \Delta x) - f(x)}{\Delta x}$$

The function f' is called the *derivative of* f. It should be pointed out here that, as in all cases where limits are involved, the definition is made subject to the condition that the limit actually exists.

With this definition our earlier discussion can be formalized as follows:

Fundamental Theorem of Calculus

Let $f(x)$ be continuous on $[a, b]$ and, for z on $[a, b]$, put

$$F(z) = \int_a^z f(x)\,dx$$

Then $F'(z) = f(z)$.

In addition to the notation $f'(x)$ we shall also use $D_x f$ and, if $y = f(x)$, the notation $D_x y$. Much of the importance of the derivative, aside from its relation to the evaluation of integrals, stems from its interpretation as a *rate of change*. If $y = f(x)$, then a change of the independent variable from x to $x + \Delta x$ changes the dependent variable from y to a new value $y + \Delta y$. We write

$$y = f(x)$$
$$y + \Delta y = f(x + \Delta x)$$

Therefore

$$\frac{f(x + \Delta x) - f(x)}{\Delta x} = \frac{y + \Delta y - y}{\Delta x} = \frac{\Delta y}{\Delta x}$$

The difference quotient whose limit is the derivative can be interpreted as the average ratio of the change in y to the change in x. The derivative is interpreted via this as the rate of change of y with respect to x. We shall say more about this interpretation in later sections as we apply the derivative.

EXAMPLE 1 Suppose $g(x) = 3x^2$. Let us compute $g'(x)$. We want, first, to find

$$\frac{g(x + \Delta x) - g(x)}{\Delta x}$$

We have

$$\frac{g(x + \Delta x) - g(x)}{\Delta x} = \frac{3(x + \Delta x)^2 - 3x^2}{\Delta x}$$

Note that if we set $\Delta x = 0$ in this expression, we get 0/0. This doesn't mean that the limit doesn't exist, it merely means that we have to look more closely at the problem. The technique here (and in most cases of finding a derivative from the definition) is to go through the algebra and actually perform the indicated subtraction. We have

$$\frac{g(x + \Delta x) - g(x)}{\Delta x} = \frac{3x^2 + 6x\,\Delta x + 3(\Delta x)^2 - 3x^2}{\Delta x}$$
$$= (6x + 3\,\Delta x)\frac{\Delta x}{\Delta x}$$

Therefore

$$D_x g(x) = \lim_{\Delta x \to 0} \frac{g(x + \Delta x) - g(x)}{\Delta x}$$

$$= \lim_{\Delta x \to 0} (6x + 3\,\Delta x)\frac{\Delta x}{\Delta x}$$

$$= 6x$$

This example illustrates the techniques which we commonly use in evaluating the derivative. We form the quotient

$$\frac{g(x + \Delta x) - g(x)}{\Delta x}$$

We cannot simply substitute in $\Delta x = 0$ for this would give 0/0. What we do is to perform some algebraic manipulation which factors the fraction

$$\frac{g(x + \Delta x) - g(x)}{\Delta x}$$

into some expression involving x and Δx, call it $h(x, \Delta x)$; multiplied by $\Delta x/\Delta x$, i.e.,

$$\frac{g(x + \Delta x) - g(x)}{\Delta x} = h(x, \Delta x) \cdot \frac{\Delta x}{\Delta x}$$

If either $h(x, \Delta x)$ has no denominator, or, if that denominator is not 0 when $\Delta x = 0$ is substituted in, we have

$$\lim_{\Delta x \to 0} \frac{g(x + \Delta x) - g(x)}{\Delta x} = h(x, 0)$$

There are a number of mistakes which students commonly make when they compute derivatives from the definition. Let us point out two of the more common ones.

One of them is to write

$$D_x g(x) = \frac{g(x + \Delta x) - g(x)}{\Delta x}$$

This simply isn't true. The left-hand side of the expression does not equal the right-hand side, it equals the *limit* of the right-hand side as Δx goes to 0.

A second common mistake is to write

$$(6x + 3\,\Delta x)\frac{\Delta x}{\Delta x} = 6x + 3\,\Delta x$$

This isn't so because the function on the left-hand side is not defined at $\Delta x = 0$ and the expression on the right-hand side is. What is true is that

$$\lim_{\Delta x \to 0} (6x + 3\,\Delta x)\frac{\Delta x}{\Delta x} = \lim_{\Delta x \to 0} (6x + 3\,\Delta x)$$

As another example of computing derivatives, consider the following problem:

EXAMPLE 2 $D_x(\sqrt{x}) = ?$

Let $g(x) = \sqrt{x}$. Then

$$\frac{g(x + \Delta x) - g(x)}{\Delta x} = \frac{\sqrt{x + \Delta x} + \sqrt{x}}{\Delta x}$$

How do we simplify this expression so that we can evaluate the limit? We cannot actually perform the subtraction in the numerator as we could in Example 1. The trick is to modify the fraction so that the subtraction can be performed.

$$\frac{\sqrt{x + \Delta x} - \sqrt{x}}{\Delta x} = \frac{\sqrt{x + \Delta x} - \sqrt{x}}{\Delta x} \cdot \frac{\sqrt{x + \Delta x} + \sqrt{x}}{\sqrt{x + \Delta x} + \sqrt{x}}$$

$$= \frac{x + \Delta x - x}{\Delta x(\sqrt{x + \Delta x} + \sqrt{x})}$$

$$= \frac{\Delta x}{\Delta x(\sqrt{x + \Delta x} + \sqrt{x})}$$

then

$$D_x(\sqrt{x}) = \lim_{\Delta x \to 0} \frac{\sqrt{x + \Delta x} - \sqrt{x}}{\Delta x}$$

$$= \lim_{\Delta x \to 0} \frac{\Delta x}{\Delta x(\sqrt{x + \Delta x} = \sqrt{x})}$$

$$= \lim_{\Delta x \to 0} \frac{\Delta x}{\Delta x} \cdot \frac{1}{\sqrt{x + \Delta x} + \sqrt{x}}$$

$$= 1 \cdot \frac{1}{\sqrt{x} + \sqrt{x}}$$

$$= \frac{1}{2\sqrt{x}}$$

Summary

There are two major concepts in calculus, the integral which we have discussed earlier, and the derivative defined in this section. Given any function $f(x)$, we define

$$f'(x) = \lim_{\Delta x \to 0} \frac{f(x + \Delta x) - f(x)}{\Delta x}$$

The derivative is related to the integral by the fact, known as the Fundamental Theorem of Calculus, that for a continuous function f, if

$$F(z) = \int_a^z f(x)\, dx$$

then

$$F'(z) = f(z)$$

We hinted at, and will later develop, the idea that if $y = f(x)$ then $f'(x)$ can be interpreted as the rate of change of y with respect x.

Exercises Section 3

1 Using the definition of derivative, find the derivative of each of the following functions.

(a) $g(x) = x$
(b) $g(x) = 3x$
(c) $g(x) = 1$
(d) $g(x) = 3x + 1$
(e) $g(x) = x^2$
(f) $g(x) = x^2 + 1$
(g) $g(x) = x^2 + x$
(h) $g(x) = x^2 + x + 1$
(i) $g(x) = 1/x$
(j) $g(x) = -x$
(k) $g(x) = -3x$
(l) $g(x) = -50$
(m) $g(x) = -3x - 50$

2 Using your answers to Exercise 1, find the distribution function $F(z)$ for each of the following density functions:

(a) $f(x) = \begin{Bmatrix} 3, & 0 \le x \le \frac{1}{3} \\ 0, & \text{otherwise} \end{Bmatrix}$

(b) $f(x) = \begin{Bmatrix} 2x, & \text{if } 0 \le x \le 1 \\ 0, & \text{otherwise} \end{Bmatrix}$

(c) $f(x) = \begin{Bmatrix} 2x + 1, & 0 \le x \le \dfrac{\sqrt{5} - 1}{2} \\ 0, & \text{otherwise} \end{Bmatrix}$

3 **(a)** Compute the value of $[(2 + \Delta x)^2 - 2^2]/\Delta x$ for $\Delta x = 1, -\frac{1}{2}, \frac{1}{4}, -\frac{1}{8}, \frac{1}{16}$.
(b) Compute the value of $[(0 + \Delta x)^2 - 0^2]/\Delta x$ for $\Delta x = 1, -\frac{1}{2}, \frac{1}{4}, -\frac{1}{8}, \frac{1}{16}$.

4 Write a program which will compute the value of the quotient

$$\frac{f(x + \Delta x) - f(x)}{\Delta x}$$

where $x = 2$ and for all the values $\Delta x = .5, .1, .05, .01, .005, .001$ for each of the following functions.

(a) $f(x) = \exp(x)$
(b) $f(x) = x \cdot \exp(x)$
(c) $f(x) = \dfrac{\exp(x)}{\sqrt{x}}$

4 SEQUENCES AND LIMITS

In Section 1 we discussed the area under a curve. Given a positive function $f(x)$ on an interval $[a, b]$, we defined for each integer n, an approximation R_n to

$$\int_a^b f(x)\,dx$$

We asserted, and in one case showed, that as n becomes larger, R_n becomes a better approximation to the integral, and, indeed, we can get as good an approximation as we like by letting n be large enough.

Some of you may have noticed that this discussion bears a certain resemblance to an earlier discussion when we said that as n becomes larger, the expression

$$\left(1 + \frac{1}{n}\right)^n$$

gets closer and closer to a number, denoted by e.

In this section we shall study a general situation of which these are special cases. Let us look at the common features of the approximation of e and the approximation of the integral. The first common feature is that in both cases we have a rule which assigns, to each positive number, some number (R_n in the case of the integral, $[1 + (1/n)]^n$ for e). Recalling that a function is a rule which assigns a number to each element of some set (called the domain) we have the following:

Definition A *sequence* is a function whose domain is the set of positive integers.

Note that the positive integers are those ≥ 1. On occasion we shall want to call sequences certain functions whose domains consist of all integers $\geq a$, where a is some integer other than 1. Values of a such as 0 and 2 are quite common.

Generally when f is a function and x is an element of its domain, we write $f(x)$ to denote the value of f at x. Thus if s is a sequence we might write

$$s(1), s(2), s(3), \ldots$$

for the values of s at the elements 1, 2, 3, . . . of its domain. More often than not, however, we shall write instead

$$s_1, s_2, s_3, \ldots$$

Moreover, we frequently abbreviate this list as follows:

$$\{s_n\}$$

and we speak of "the sequence $\{s_n\}$" in place of the more formal statement "the sequence s whose values are $\{s_n\}$."

The definition of a sequence was motivated by our interest in finding successive approximations of something. For example, the sequence $\{e_n\}$ defined by

$$e_n = \left(1 + \frac{1}{n}\right)^n$$

consists of successive approximations of e. What we want to do now is to define what it is that the terms of a sequence are successively approximating. One difficulty is that our definition of sequence is so general that the terms of many sequences do not successively approximate anything in any meaningful way. Consider for example the sequence $\{a_n\}$ defined by

$$a_n = (-1)^n n$$

(We will frequently refer to the sequence $\{(-1)^n n\}$ in order to avoid the tedious phrasing of the preceding sentence.) Look at a few terms of this sequence. We have

$$a_1 = -1$$
$$a_2 = 2$$
$$a_3 = -3$$
$$a_4 = 4$$
$$a_5 = -5$$
etc.

The numbers in this sequence do not appear to be getting closer and closer to anything (and indeed they are not). So we have a situation where some sequences (such as $\{[1 + (1/n)]^n\}$) serve as successive approximations of something and some sequences (such as $\{(-1)^n n\}$) do not. We introduce a definition based on our discussion in Section 1 of what constitutes a good approximation. The idea was, you recall, to introduce some maximum amount of allowable error, denoted by ε, and agree that an approximation must differ from the true value of what we are approximating by less than ε. We introduce the following.

Definition

Let $\{a_n\}$ be a sequence. We say that $\{a_n\}$ *converges to* a and we write

$$\lim_{n\to\infty} a_n = a$$

if, for any $\varepsilon > 0$, there is a positive integer N such that $n > N$ implies

$$|a_n - a| < \varepsilon$$

The number a to which the sequence $\{a_n\}$ converges is called *the limit of the sequence*. As examples, we have

$$\lim_{n\to\infty}\left(1 + \frac{1}{n}\right)^n = e$$

and

$$\lim_{n\to\infty} R_n = \int_a^b f(x)\,dx$$

A sequence which does have a limit is said to *converge* to that limit. A sequence which does not have a limit is said to *diverge*.

We do not expect you to be able to prove, from the definition, the two limits which we have stated. Indeed we shall not ask you to prove any limits from the definition. You should understand what limits are in the approximating sense we have been using and you should be able to find the limits of various sequences. We shall prove one limit directly from the definition. It is not intended to serve as an example of something you are to learn to do but we do hope that if you read through the argument you will have a better understanding of what the concept of limit means.

Let

$$a_n = \frac{1}{n}$$

We shall show that

$$\lim_{n \to \infty} a_n = 0$$

On an intuitive level, this limit should be obvious to you. Consider the following table.

n	a_n	$\lvert a_n - 0 \rvert$
1	1	1
2	$\frac{1}{2}$	$\frac{1}{2}$
3	$\frac{1}{3}$	$\frac{1}{3}$
4	$\frac{1}{4}$	$\frac{1}{4}$
5	$\frac{1}{5}$	$\frac{1}{5}$
	etc.	

Certainly, as n gets larger, the difference between a_n and 0 seems to get smaller. But how do we use the definition to prove it? Given any $\varepsilon > 0$, we must find an N such that if $n > N$ then

$$|a_n - 0| < \varepsilon$$

But

$$|a_n - 0| = \left| \frac{1}{n} - 0 \right| = \frac{1}{n}$$

If

$$\frac{1}{n} < \varepsilon$$

then

$$n > \frac{1}{\varepsilon}$$

So, if we choose an integer $N > 1/\varepsilon$, then $n > N$ implies $n > 1/\varepsilon$ and $|a_n - 0| < \varepsilon$. A formal proof would look as follows.

Let $\varepsilon > 0$ be given. Choose an integer N such that $N > 1/\varepsilon$. Then, if $n > N$, it follows that $n > 1/\varepsilon$ or $1/n < \varepsilon$, and so

$$|a_n - 0| = \left| \frac{1}{n} - 0 \right| = \frac{1}{n} < \varepsilon$$

The point to be observed is that once we have decided what the maximum amount of error is (we called it ε), we can figure out how far we must go with the approximations a_n to decide when a_n is within ε of 0.

We shall now present, without proof, a number of theorems, or rules, which are useful in calculating the limit of a sequence. In the statements of these theorems we shall use sequences $\{a_n\}$ and $\{b_n\}$ and use the symbol c to stand both for a constant (Theorems 2 and 3) and for a sequence all of whose terms equal c (Theorem 3).

Theorem 1 $\lim_{n\to\infty} \frac{1}{n^k} = 0, \qquad \text{for all } k > 0$

Theorem 2 $\lim_{n\to\infty} c \cdot a_n = c \cdot \lim_{n\to\infty} a_n$

Theorem 3 $\lim_{n\to\infty} c = c$

Theorem 4 $\lim_{n\to\infty} (a_n + b_n) = \lim_{n\to\infty} a_n + \lim_{n\to\infty} b_n$

Theorem 5 $\lim_{n\to\infty} (a_n \cdot b_n) = \lim_{n\to\infty} a_n \cdot \lim_{n\to\infty} b_n$

Theorem 6 $\lim_{n\to\infty} (a_n - b_n) = \lim_{n\to\infty} a_n - \lim_{n\to\infty} b_n$

Theorem 7 $\lim_{n\to\infty} \left(\frac{a_n}{b_n}\right) = \frac{\lim_{n\to\infty} a_n}{\lim_{n\to\infty} b_n}$

A few examples should show you how to apply these theorems to problems. We shall do these first few problems in a somewhat tedious fashion, applying only one theorem at a time, in order to clarify the procedure for you. In practice one normally would combine several of the steps into one step.

EXAMPLE 1

$$
\begin{aligned}
\lim_{n\to\infty} \left(1 + \frac{4}{n}\right) &= \lim_{n\to\infty} 1 + \lim_{n\to\infty} \frac{4}{n} && \text{(Theorem 4)} \\
&= 1 + \lim_{n\to\infty} \frac{4}{n} && \text{(Theorem 3)} \\
&= 1 + 4 \lim_{n\to\infty} \frac{1}{n} && \text{(Theorem 2)} \\
&= 1 + 4 \cdot 0 && \text{(Theorem 1)} \\
&= 1
\end{aligned}
$$

EXAMPLE 2 (We introduce an algebraic trick here which is frequently useful in cases where the terms of the sequence are fractions.)

$$\lim_{n\to\infty} \frac{n^2 + 1}{2n^2 + n} = \lim_{n\to\infty} \frac{(n^2/n^2) + (1/n^2)}{(2n^2/n^2) + (n/n^2)}$$ (We have divided numerator and denominator by the highest power of n appearing)

$$= \lim_{n\to\infty} \frac{[1 + (1/n^2)]}{[2 + (1/n)]}$$

$$= \frac{\lim_{n\to\infty} [1 + (1/n^2)]}{\lim_{n\to\infty} [2 + (1/n)]}$$ (Theorem 7)

$$= \frac{\lim_{n\to\infty} 1 + \lim_{n\to\infty} 1/n^2}{\lim_{n\to\infty} 2 + \lim_{n\to\infty} 1/n}$$ (Two applications of Theorem 4)

$$= \frac{1 + \lim_{n\to\infty} 1/n^2}{2 + \lim_{n\to\infty} 1/n}$$ (Two applications of Theorem 3)

$$= \frac{1 + 0}{2 + 0}$$ (Two applications of Theorem 1)

$$= \frac{1}{2}$$

Sometimes the limit does not exist as in the following example.

EXAMPLE 3

$$\lim_{n\to\infty} \frac{n^2 + 1}{n} = \lim_{n\to\infty} \frac{1 + (1/n^2)}{1/n}$$ (Using the trick for fractions introduced in Example 2)

$$= \frac{\lim_{n\to\infty} [1 + (1/n^2)]}{\lim_{n\to\infty} (1/n)}$$ (Theorem 7)

$$= \frac{\lim_{n\to\infty} 1 + \lim_{n\to\infty} 1/n^2}{\lim_{n\to\infty} (1/n)}$$ (Theorem 4)

$$= \frac{1 + \lim_{n\to\infty} 1/n^2}{\lim_{n\to\infty} (1/n)}$$

$$= \frac{1 + 0}{0}$$

$$= \frac{1}{0}$$

Since division by 0 is undefined, the limit does not exist. There are two similar outcomes which can arise. You might work through one of these problems and get 0/1 (or more generally 0 divided by any non-zero quantity), which means that the limit is 0. If you work a problem through and get 0/0, that means that you have to go back and use another approach to the problem because the method you are using will not give you the limit.

There is one more notion to be introduced in this section. Given a sequence $\{a_n\}$, one can form from it, the *series* (sometimes referred to as an *infinite series*) written

$$\sum_{n=1}^{\infty} a_n$$

The use of the symbol Σ should suggest to you that the terms of the sequence are to be added. The suggestion is, in an intuitive sense, the correct one. However there is a bit of a difficulty. How does one add together an infinite number of numbers? Meaning is given to this in the following way. To the infinite series we associate another sequence, the sequence of *partial sums*. The *kth partial sum* is defined by (refer to the exercises in Section 3 of Chapter 3)

$$s_k = \sum_{n=1}^{k} a_n$$

Let's see what this looks like in a concrete example. Suppose we start with the sequence $a_n = (\frac{1}{2})^n$. Then we have the series

$$\sum_{n=1}^{\infty} (\tfrac{1}{2})^n$$

$$s_1 = \sum_{n=1}^{1} (\tfrac{1}{2})^n$$

$$= \tfrac{1}{2}$$

$$s_2 = \sum_{n=1}^{2} (\tfrac{1}{2})^n$$

$$= \tfrac{1}{2} + (\tfrac{1}{2})^2 = \tfrac{1}{2} + \tfrac{1}{4} = \tfrac{3}{4}$$

$$s_3 = \sum_{n=1}^{3} (\tfrac{1}{2})^n$$

$$= \tfrac{1}{2} + (\tfrac{1}{2})^2 + (\tfrac{1}{2})^3$$

$$= \tfrac{1}{2} + \tfrac{1}{4} + \tfrac{1}{8} = \tfrac{7}{8}$$

etc.

$$\sum_{n=1}^{\infty} a_n = \lim_{n\to\infty} s_k$$

where s_k is the kth partial sum. Sometimes this limit exists, in which case we say that the series *converges* to the limit. Sometimes the limit fails to exist, in which case we say that the series *diverges*.

Occasionally the series

$$\sum_{n=1}^{\infty} a_n$$

will be written as

$$(a_1 + a_2 + \cdots + a_n + \cdots)$$

We shall close this section by calculating the limit (or *sum*) of a particularly useful convergent series. Many of you will have seen it in high school. It is called the *geometric series*. The general form is

$$\sum_{n=0}^{\infty} ar^n$$

(Note that we begin at 0, not at 1.) There is a very simple technique for finding the sum of the geometric series. We consider the kth partial sum

$$s_k = \sum_{n=0}^{k} ar^n$$

If we write this out, we have

$$s_k = a + ar + ar^2 + \cdots + ar^{k-1} + ar^k$$

We now multiply through by r, obtaining

$$rs_k = ar + ar^2 + \cdots + ar^k + ar^{k+1}$$

Subtraction yields

$$s_k - rs_k = a - ar^{k+1}$$

(Note the cancellation of terms.) Therefore

$$(1 - r)s_k = a(1 - r^{k+1})$$

and

$$s_k = a\frac{1 - r^{k+1}}{1 - r} \qquad \text{(Provided } r \neq 1\text{)}$$

We want

$$s = \lim_{k \to \infty} s_k$$

In order to obtain s, we need to evaluate

$$\lim_{k \to \infty} r^{k+1}$$

Let us see what happens for a few specific values of r. First we shall try $r = 2$. We obtain the following table.

k	2^{k+1}
1	4
2	8
3	16
4	32
etc.	

It should be clear that

$$\lim_{k \to \infty} 2^{k+1}$$

does not exist.

Now let us try $k = \frac{1}{2}$.

k	$(\frac{1}{2})^k$
1	$\frac{1}{2}$
2	$\frac{1}{4}$
3	$\frac{1}{8}$
4	$\frac{1}{16}$
5	$\frac{1}{32}$
etc.	

This table seems to indicate that

$$\lim_{k \to \infty} (\tfrac{1}{2})^k = 0$$

The indication is correct. Indeed it turns out that if $|r| < 1$, then

$$\lim_{k \to \infty} r^{k+1} = 0$$

If $r = 1$, our expression for s_k is not valid since it would involve division by 0. If $r = -1$, or $|r| > 1$, then

$$\lim_{k \to \infty} r^{k+1}$$

is not defined.

We have, therefore, for $|r| < 1$, the geometric series converges to

$$s = \lim_{k \to \infty} a \frac{1 - r^{k+1}}{1 - r}$$

$$= a \frac{1}{1 - r}$$

If $|r| \geq 1$, the geometric series diverges.

EXAMPLE 4 The Multiplier Effect

One of the common terms in economics is the *multiplier effect.* It works as follows: Suppose that by means of a tax cut the government raises each person's take-home pay by 1% and, to keep things simple, assume that everyone spends all of the increased income. How much is the new overall purchasing power of the country?

If the initial amount of purchasing power is A, you might think the answer is

$$A + (.01)A$$

But wait, for 1% more of the additional money than would otherwise be the case is available to be spent, so we have

$$A + (.01)A + (.01)(.01)A = A + (.01)A + (.01)^2A$$

But 1% more of this second increment is available to be spent than would otherwise be the case, so we have

$$A + (.01)A + (.01)^2A + (.01)^3A$$

Continuing, we find the new purchasing power to be

$$\begin{aligned}\sum_{i=0}^{\infty} (.01)^i A &= \frac{1}{1 - .01}A \\ &= \frac{1}{.99}A \\ &\approx (1.0101)A\end{aligned}$$

and the percentage increase is approximately 1.01%.

If the tax deduction increased each person's take-home pay by 2%, the *multiplier* would be

$$\begin{aligned}\sum_{i=0}^{\infty} (.02)^i &= \frac{1}{1 - .02} \\ &= \frac{1}{.98} \\ &\approx 1.0204\end{aligned}$$

and the percentage increase is approximately 2.04%.

Summary

We have introduced the definition of sequence and limit of a sequence to make precise the approximation procedure developed in Section 1 for area and earlier for the function exp (x). We also discussed series in general and the geometric series in particular as a prelude to later work. You should know the meaning of the following: *sequence, series, limit of a sequence, kth partial sum of a series, convergent series, divergent series, geometric series*. You should also know Theorems 1–7 and the formula for the sum of a geometric series.

We are now ready to return to a precise discussion of the approximation R_n but before doing so we shall, in the next section, discuss another kind of limit which we will need shortly.

Exercises Section 4

For each of the following problems compute the limit of the sequence (or the sum of the series).

1 $\lim_{n\to\infty} \left(1 + \frac{1}{n}\right)^5$

2 $\lim_{n\to\infty} \frac{n^2 + n}{n^3 + n^2 + 1}$

3 $\sum_{n=1}^{\infty} (\frac{1}{3})^n$

4 $\sum_{n=0}^{\infty} (\frac{1}{3})^n$

5 $\lim_{n \to \infty} \frac{2n}{n}$

6 $1 + \frac{2}{5} + (\frac{2}{5})^2 + \cdots + (\frac{2}{5})^n + \cdots$

7 $\lim_{n \to \infty} \frac{n^2 + 1}{n^2 + n}$

8 $\lim_{n \to \infty} (\frac{1}{2})^n$

9 $\lim_{n \to \infty} \frac{n^3 + 10n^2 + 10n + 1}{3n^3 + 1}$

10 $\lim_{n \to \infty} (-\frac{1}{2})^n$

11 $\sum_{k=0}^{\infty} (\frac{3}{5})^k$

12 $\sum_{k=2}^{\infty} (\frac{3}{5})^k$

13 Using the terminology developed in Example 4 on the multiplier effect:

(a) If the tax cut increases take-home pay by 5%, what is the percentage increase in purchasing power?

(b) If one wants to increase purchasing power by 10%, the tax cut must increase take-home pay by how much?

5 LIMITS OF FUNCTIONS

We encountered a special case of the limit of a function in Section 3. We shall now discuss the concept in some detail and will begin by looking at two simple but typical examples.

Consider the function

$$f(x) = \frac{x^2 - 1}{x - 1}$$

We can see that f is not defined at 1, but notice what happens when x is close to 1.

x	$f(x)$
2	3
1.5	2.5
1.1	2.1
1.01	2.01
.99	1.99
.9	1.9
.5	1.5
0	1

As x gets closer to 1, from either the right or left, $f(x)$ gets closer to 2. We want our definition of limit to be such that, for this function f, the limit of $f(x)$ as x approaches 1 is 2. Symbolically this is denoted

$$\lim_{x \to 1} f(x) = 2$$

Consider another function,

$$g(x) = x^2 + 2$$

What is

$$\lim_{x \to 0} g(x)?$$

We make up a table:

x	$g(x)$
1	3
.5	2.25
.1	2.01
.01	2.0001
−.01	2.0001
−.1	2.01
−.5	2.25
−1	3

After this much calculation

$$\lim_{x \to 0} g(x) = 2$$

seems reasonable.

Unlike our first example, where $f(1)$ was undefined, $g(0)$ is defined and

$$\lim_{x \to 0} g(x) = g(0)$$

The formal definition of limit follows.

Definition Let f be a function.

$$\lim_{x \to a} f(x) = L$$

means that given any $\varepsilon > 0$ there is a $\delta > 0$ such that if $0 < |x - a| < \delta$, then $|f(x) - L| < \varepsilon$.

To restate the definition in terms of approximations, it says that, given any level of accuracy ε, the number L can be approximated by $f(x)$ to that degree of accuracy provided x is sufficiently close to a. Note that the

$$0 < |x - a|$$

says that we do not care in the definition whether $f(a)$ is defined or what its value is. This is important because, as you recall, setting $\Delta x = 0$ in the definition of the derivative would give us 0/0, an undefined quantity. However it is generally true that when $f(a)$ exists,

$$\lim_{x \to a} f(x) = f(a)$$

Functions which have this property are called *continuous* at a.

Among the functions which are continuous wherever they are defined are polynomials, exp (x), and anything which can be obtained from these by the algebraic operations of addition, subtraction, multiplication, division, raising to powers, and extracting roots.

We have several limit theorems which correspond to the theorems about limits of sequences.

Theorem 1

$$\lim_{x \to a} cf(x) = c \lim_{x \to a} f(x)$$

Theorem 2

$$\lim_{x \to a} (f(x) + g(x)) = \lim_{x \to a} f(x) + \lim_{x \to a} g(x)$$

Theorem 3

$$\lim_{x \to a} (f(x) \cdot g(x)) = \lim_{x \to a} f(x) \cdot \lim_{x \to a} g(x)$$

Theorem 4

$$\lim_{x \to a} (f(x) - g(x)) = \lim_{x \to a} f(x) - \lim_{x \to a} g(x)$$

Theorem 5

$$\lim_{x \to a} \left(\frac{f(x)}{g(x)} \right) = \frac{\lim_{x \to a} f(x)}{\lim_{x \to a} g(x)}, \qquad \text{if } \lim_{x \to a} g(x) \neq 0$$

In all discussions of limits, there is a question as to whether the limit exists. For example

$$\lim_{x \to 0} \frac{1}{x}$$

does not. The theorems should all be interpreted as saying that if the limits on both sides of the equation exist, then the equality holds.

EXAMPLE 1

$$\lim_{x \to 1} \frac{x^2 + 1}{x + 1} = 1$$

Note that

$$\frac{x^2 + 1}{x + 1}$$

is continuous and defined at 1 so the limit is

$$\frac{1^2 + 1}{1 + 1} = \frac{2}{2} = 1$$

EXAMPLE 2

$$\lim_{x \to 1} \frac{x^2 - 1}{x - 1} = 2$$

Now we are taking a limit at a point where the function is undefined. The problem is handled as follows:

$$\lim_{x\to 1}\frac{x^2-1}{x-1}=\lim_{x\to 1}\frac{(x-1)(x+1)}{x-1}$$

$$=\lim_{x\to 1}\frac{x-1}{x-1}\cdot(x+1)$$

$$=\lim_{x\to 1}\frac{x-1}{x-1}\cdot\lim_{x\to 1}(x+1)$$

$$=1\cdot 2$$

There are two points which should be made here. First, in connection with

$$\lim_{x\to 1}\frac{x-1}{x-1}=1$$

note that $(x-1)/(x-1)$ is not defined at $x=1$, but it is 1 for all values of x other than 1. Our definition tells us that in evaluating

$$\lim_{x\to a}f(x)$$

we consider *values of x close to a but different from a.* The fact that $(x-1)/(x-1)$ is 1 for all values of x other than 1 gives us the limit.

Secondly, note that we have handled the problem by factoring the numerator. This is a common method. Whenever you have a fraction

$$\lim_{x\to a}\frac{f(x)}{g(x)}$$

where (1) $f(x)$ and $g(x)$ are polynomials; and (2) both $f(a)$ and $g(a)$ are 0, then $x-a$ is a factor of both $f(x)$ and $g(x)$. Taking out this common factor is a first step in finding the limit.

Now we shall look at a limit problem of the type which arose in taking derivatives.

EXAMPLE 3

$$\lim_{\Delta x\to 0}\frac{(x+\Delta x)^2-x^2}{\Delta x}=2x$$

Two things should be observed before we calculate the limit. First, there are two variables, x and Δx, in the expression but, since we are taking the limit as Δx goes to 0, we pretend that x is a constant. Secondly, the fraction is undefined at $\Delta x=0$ (it isn't defined there since the denominator is 0), so we have to modify the form of the expression. We have

$$\begin{aligned}\lim_{\Delta x\to 0}\frac{(x+\Delta x)^2-x^2}{\Delta x} &= \lim_{\Delta x\to 0}\frac{x^2+2x\,\Delta x+(\Delta x)^2-x^2}{\Delta x}\\ &= \lim_{\Delta x\to 0}\frac{2x\,\Delta x+(\Delta x)^2}{\Delta x}\\ &= \lim_{\Delta x\to 0}\frac{(2x+\Delta x)\cdot\Delta x}{\Delta x}\\ &= \lim_{\Delta x\to 0}(2x+\Delta x)\lim_{\Delta x\to 0}\frac{\Delta x}{\Delta x}\\ &= 2x\cdot 1 = 2x\end{aligned}$$

One of the most difficult limit problems we will encounter is in finding the derivative of the function exp (x). We do it as an illustrative example.

EXAMPLE 4 $D_x \exp(x) = ?$

We begin with

$$\begin{aligned}\frac{\exp(x+\Delta x)-\exp(x)}{\Delta x} &= \frac{\exp(x)\exp(\Delta x)-\exp(x)}{\Delta x}\\ &= \exp(x)\cdot\frac{\exp(\Delta x)-1}{\Delta x}\end{aligned}$$

Since Δx appears only in the quotient $[\exp(\Delta x) - 1]/\Delta x$, that is the part of the expression we must concentrate on. If we simply set $\Delta x = 0$, we have 0/0, so once again we must think of a way to modify the expression. What we need is the series which defines exp. Recall that for any number t,

$$\exp(t) = \sum_{k=0}^{\infty}\frac{t^k}{k!}$$

We have

$$\begin{aligned}\frac{\exp(\Delta x)-1}{\Delta x} &= \frac{\sum_{k=0}^{\infty}[(\Delta x)^k/k!]-1}{\Delta x}\\ &= \frac{(1+(\Delta x)+[(\Delta x)^2/2!]+[(\Delta x)^3/3!]+\cdots)-1}{\Delta x}\\ &= \frac{(\Delta x)+[(\Delta x)^2/2!]+[(\Delta x)^3/3!]+\cdots}{\Delta x}\\ &= \frac{(\Delta x)(1+[(\Delta x)/2!]+[(\Delta x)^2/3!]+\cdots)}{\Delta x}\\ &= \frac{\Delta x}{\Delta x}(1+[(\Delta x)/2!]+[(\Delta x)^2/3!]+\cdots)\end{aligned}$$

From this we can see that

$$\lim_{\Delta x \to 0} \frac{\exp(\Delta x) - 1}{\Delta x} = 1$$

Therefore

$$D_x \exp(x) = \lim_{\Delta x \to 0} \frac{\exp(x + \Delta x) - \exp(x)}{\Delta x}$$

$$= \lim_{\Delta x \to 0} \exp(x) \frac{\exp(\Delta x) - 1}{\Delta x}$$ (From our earlier computation)

$$= \exp(x) \lim_{\Delta x \to 0} \frac{\exp(\Delta x) - 1}{\Delta x}$$ [Since exp (x) can be viewed as a constant, so far as Δx is concerned]

$$= \exp(x) \cdot 1 = \exp(x)$$

We have seen that the derivative of the function exp is the function exp. We will see later that this fact that exp is its own derivative leads to a wide variety of applications in biology, economics, physics, and other fields.

In addition to

$$\lim_{x \to a} f(x)$$

we are also occasionally interested in

$$\lim_{x \to \infty} f(x)$$

Definition

$$\lim_{x \to \infty} f(x) = L$$

means, given $\varepsilon > 0$, there is a number a such that if $x \geq a$ then

$$|f(x) - L| < \varepsilon$$

The limit theorems which we have encountered for sequences and for $\lim_{x \to a} f(x)$ apply to the new limit. We conclude with some examples.

EXAMPLE 5

$$\lim_{x \to \infty} \frac{1}{x} = 0$$

EXAMPLE 6

$$\lim_{x \to \infty} \frac{x^2 + 1}{3x^2 + x} = \frac{1}{3}$$

$$\lim_{x \to \infty} \frac{x^2 + 1}{3x^2 + x} = \lim_{x \to \infty} \frac{1 + (1/x^2)}{3 + (1/x)}$$

$$= \frac{1 + 0}{3 + 0}$$

$$= \tfrac{1}{3}$$

EXAMPLE 7

$$\lim_{x \to \infty} \frac{x^2 + 1}{x + 1} \qquad \text{(Undefined)}$$

$$\lim_{x \to \infty} \frac{x^2 + 1}{x + 1} = \lim_{x \to \infty} \frac{1 + (1/x^2)}{1/x + (1/x^2)} = \frac{1 + 0}{0 + 0} \qquad \text{(Undefined)}$$

We have encountered the word "continuous" earlier in connection with random variables. Recall that a random variable X was said to be discrete if the values of X could be listed or enumerated, and X was called continuous if it could take on all values in some interval. Now that we have defined a function f being continuous at a by saying

$$\lim_{x \to a} f(x) = f(a)$$

we are in a position to relate our two usages of this word.

We shall say that a *random variable* X *is continuous* if its distribution function F is continuous at all points in $(-\infty, \infty)$. A glance back at the last two sections of Chapter 3 would reveal two kinds of graphs of distribution functions. One kind of graph is that corresponding to discrete step functions. The graphs of the distribution functions corresponding to continuous random variables were, by contrast, *connected* curves having no holes or jumps in them. Indeed, to a mathematician, the word continuous conveys the notion of connectedness and the graph of a continuous function is one that can be drawn without lifting one's pencil from the paper. In the exercises, the reader will be invited to examine some functions to determine whether they are continuous.

Summary

We have defined the limit of a function by writing

$$\lim_{x \to a} f(x) = L$$

to mean: Given any ε, there is a δ such that if $0 < |x - a| < \delta$, then $|f(x) - L| < \varepsilon$. We have also computed the derivative of exp (x). You should remember the important formula

$$\boxed{D_x \exp(x) = \exp(x)}$$

Exercises Section 5

Evaluate each of the following limits.

1 $\displaystyle \lim_{x \to 2} \frac{x^2 + 1}{x}$

2 $\displaystyle \lim_{x \to 3} \frac{x^2 - 9}{x + 3}$

3 $\displaystyle \lim_{x \to 3} \frac{x^2 - 9}{x - 3}$ [*Hint:* $x^2 - 9 = (x - 3)(x + 3)$.]

4 $\lim_{x \to 1} \frac{3x^2 - 3}{x - 1}$

5 $\lim_{x - 1} \frac{x^3 - 3x^2 + 3x - 1}{x - 1}$

6 $\lim_{x \to 0} \frac{x^2 + x}{x + 1}$

7 $\lim_{x \to 0} \frac{x^2 + x}{x}$

8 $\lim_{\Delta x \to 0} \frac{(x + \Delta x)^3 - x^3}{\Delta x}$

9 Differentiate.

(a) $g(x) = \sqrt{x + 1}$

(b) $g(x) = \sqrt{x^2 + 1}$

(c) $g(x) = \frac{1}{x}$

(d) $g(x) = \sqrt{2x}$

(e) $g(x) = \exp(2x)$

10 $\lim_{x \to 2} \frac{x^3 + 3x^2 + 3x - 26}{x - 2}$

11 $\lim_{x \to 2} \frac{3x^2 - 12}{x - 2}$

12 $\lim_{x \to 1/2} \frac{x^2 - 2x + \frac{3}{4}}{2x - 1}$

13 Sketch the graphs of each of the following functions and determine whether they are continuous at each point of their domain.

(a) $f(x) = x^2, \quad \text{for } x \in (-\infty, \infty)$

(b) $f(x) = \begin{cases} 1, & \text{if } x < 0 \\ -1, & \text{if } x \geq 0 \end{cases}$

(c) $f(x) = \begin{cases} x^3, & \text{if } x < 0 \\ x^2, & \text{if } x \geq 0 \end{cases}$

(d) $f(x) = \begin{cases} 0, & \text{if } x < 0 \\ \frac{1}{2}, & \text{if } 0 \leq x < \frac{1}{2} \\ (x - 1)^2, & \text{if } \frac{1}{2} \leq x \end{cases}$

(e) $f(x) = \begin{cases} e^{-x}, & \text{if } x < 0 \\ x + 1, & \text{if } x \geq 0 \end{cases}$

(f) $f(x) = \begin{cases} \frac{1}{x}, & \text{if } x \neq 0 \\ 0, & \text{if } x = 0 \end{cases}$

(g) $f(x) = \begin{cases} \sqrt{1 - x^2}, & \text{if } -1 \leq x \leq 1 \\ 0, & \text{otherwise} \end{cases}$

14 Evaluate each of the following limits where x becomes infinite.

(a) $\lim_{x \to \infty} \frac{1}{1 - x}$

(b) $\lim_{x \to \infty} \frac{\sqrt{1 + x^2}}{x}$

(c) $\lim_{x \to \infty} 303$

(d) $\lim_{x \to \infty} \frac{x}{\sqrt{x + 1}}$

(e) $\lim_{x \to \infty} \left(2 - \frac{1}{x}\right)$

6 ELEMENTARY PROPERTIES OF DERIVATIVES

Now that we have seen how finding the area under a curve depends upon being able to compute derivatives of functions (more precisely to compute antiderivatives) we need to develop some general formulas for handling a respectable number of routine functions. Since a derivative of a function G is a limit of another function, namely the *difference quotient*

$$\frac{G(x + \Delta x) - G(x)}{\Delta x}$$

as the variable Δx approaches 0, we will have to employ some standard limit theorems throughout our present discussion. For convenience, we now restate these theorems along with brief examples indicating how they will be used.

LIMIT THEOREMS

Theorem	Typical Example
1. The limit of a sum of (two or more) functions is the sum of their limits.	$\lim_{\Delta x \to 0} (3x + 2x\,\Delta x + (\Delta x)^2) = \lim_{\Delta x \to 0} (3x) + \lim_{\Delta x \to 0} (2x\,\Delta x) + \lim_{\Delta x \to 0} ((\Delta x)^2)$
2. The limit of a difference of functions is the difference of their limits.	$\lim_{x \to 3} (6x^3 - 5x) = \lim_{x \to 3} (6x^3) - \lim_{x \to 3} (5x)$
3. The limit of a product of (two or more) functions is the product of their limits.	$\lim_{\Delta x \to 0} (5x\,\Delta x) = \lim_{\Delta x \to 0} (5x) \cdot \lim_{\Delta x \to 0} (\Delta x)$
4. The limit of a quotient of functions is the quotient of their limits, provided the denominator is not (or does not approach) 0.	$\lim_{x \to 1} \left(\frac{x^2 - 2}{x + 3}\right) = \frac{\lim_{x \to 1} (x^2 - 2)}{\lim_{x \to 1} (x + 3)}$
5. The limit of a constant times a function is the constant times the limit of the function. (By constant we mean a factor not involved in the limiting process.)	$\lim_{x \to 3} (5x) = 5 \cdot \lim_{x \to 3} (x)$ $\lim_{\Delta x \to 0} (x\,\Delta x) = x \cdot \lim_{\Delta x \to 0} (\Delta x)$

In Theorem 5, we must emphasize that the "constant" might be like "5" in our first example, but it also might look like the "x" in our second example, where Δx is approaching 0 but x is not changing.

One of the most common errors committed by students learning calculus results from being careless about limit notation. Often we shall be forced to write a long string of expressions as we simplify the difference quotient to a point of applying limit theorems. There are basically two ways of going about this sequence of steps, and the errors are generally a result of improperly combining these ways. In our first formula we show both methods used properly.

Formula I   The derivative of a constant function is 0

We let $G(x) = c$ for all values of x.

Method 1

$$\begin{aligned} G'(x) &= \lim_{\Delta x \to 0} \frac{G(x + \Delta x) - G(x)}{\Delta x} \\ &= \lim_{\Delta x \to 0} \frac{c - c}{\Delta x} \\ &= \lim_{\Delta x \to 0} \frac{0}{\Delta x} \qquad \left(\frac{0}{\Delta x} = 0, \text{ for all } \Delta x \neq 0\right) \\ &= \lim_{\Delta x \to 0} (0) \\ &= 0 \end{aligned}$$

Method 2

$$\frac{G(x + \Delta x) - G(x)}{\Delta x} = \frac{c - c}{\Delta x} = 0 \qquad (\text{Again, for all } \Delta x \neq 0)$$

Thus

$$G'(x) = \lim_{\Delta x \to 0} (0) = 0$$

Of course the two methods are distinguished only by the fact that in Method 2, all of the simplification of the difference quotient is performed before we use the limit symbol.

> *Application of Formula I*
> Consider the distribution whose density function is given by
>
> $$f(x) = \tfrac{1}{2}, \qquad -1 \leq x \leq 1$$
>
> For any x in the interval $(-1, 1) = \{x \mid -1 < x < 1\}$, we have
>
> $$f'(x) = 0$$

Formula II   The derivative of a sum of functions is the sum of their derivatives: $D_x(G + H) = D_xG + D_xH$

We let G and H be functions, both defined on some interval containing a point x. The function $G + H$ is defined by

$$(G + H)(x) = G(x) + H(x)$$

so we have

$$\begin{aligned}
(G + H)'(x) &= \lim_{\Delta x \to 0} \frac{[(G + H)(x + \Delta x)] - [(G + H)(x)]}{\Delta x} \\
&= \lim_{\Delta x \to 0} \frac{[G(x + \Delta x) + H(x + \Delta x)] - [G(x) + H(x)]}{\Delta x} \\
&= \lim_{\Delta x \to 0} \frac{[G(x + \Delta x) - G(x)] + [H(x + \Delta x) - H(x)]}{\Delta x} \\
&= \lim_{\Delta x \to 0} \left[\frac{G(x + \Delta x) - G(x)}{\Delta x} + \frac{H(x + \Delta x) - H(x)}{\Delta x} \right]
\end{aligned}$$

and using the "limit of a sum" theorem

$$\begin{aligned}
&= \lim_{\Delta x \to 0} \frac{G(x + \Delta x) - G(x)}{\Delta x} + \lim_{\Delta x \to 0} \frac{H(x + \Delta x) - H(x)}{\Delta x} \\
&= G'(x) + H'(x) \\
&= (G' + H')(x)
\end{aligned}$$

Application of Formula II

Let $G(x) = \exp(x)$ and $H(x) = 4$, then $(G + H)(x) = \exp(x) + 4$. But $G'(x) = \exp(x)$ and $H'(x) = 0$, so

$$D_x(\exp(x) + 4) = \exp(x) + 0 = \exp(x)$$

Note that Formula II was not restricted to a sum of two functions. An obvious modification of the proof would show that the derivative of a sum of several functions is the sum of all their derivatives. In the previous section, for example, you found that

$$D_x(3x^2) = 6x$$

Thus

$$\begin{aligned}
D_x(\exp(x) + 4 + 3x^2) &= D_x(\exp(x)) + D_x(4) + D_x(3x^2) \\
&= \exp(x) + 0 + 6x \\
&= \exp(x) + 6x
\end{aligned}$$

Formula III The derivative of a difference of two functions is the difference of their derivatives: $D_x(G - H) = D_xG - D_xH$

We leave the proof of this formula to the exercises; it requires an obvious modification of the preceding argument.

Application of Formula III

$$D_x(3x^2 - \exp(x)) = 6x - \exp(x)$$

Formula IV The derivative of a product of two functions is the first function times the derivative of the second plus the derivative of the first times the second: $D_x(G \cdot H) = G \cdot D_x(H) + D_x(G) \cdot H$

By $(G \cdot H)(x)$ we mean $G(x)H(x)$; for example, if $G(x) = 3x^2$ and $H(x) = \exp(x)$, then

$$(G \cdot H)(x) = 3x^2 \exp(x)$$

The argument that follows contains a bit of trickery well-known to the mathematician: At a critical stage, the term 0 under disguise as

$$-G(x + \Delta x)H(x) + G(x + \Delta x)H(x)$$

is added to the numerator. Also we will use the fact that

$$\lim_{\Delta x \to 0} G(x + \Delta x) = G(x)$$

which should seem at least reasonable.

The difference quotient for $(G \cdot H)(x)$ is

$$\begin{aligned}
&\frac{(G \cdot H)(x + \Delta x) - (G \cdot H)(x)}{\Delta x} \\
&= \frac{G(x + \Delta x)H(x + \Delta x) - G(x)H(x)}{\Delta x} \\
&= \frac{G(x + \Delta x)H(x + \Delta x) - G(x + \Delta x)H(x) + G(x + \Delta x)H(x) - G(x)H(x)}{\Delta x} \\
&= G(x + \Delta x)\left(\frac{H(x + \Delta x) - H(x)}{\Delta x}\right) + \left(\frac{G(x + \Delta x) - G(x)}{\Delta x}\right)H(x)
\end{aligned}$$

Thus,

$$\begin{aligned}
&(G \cdot H)'(x) \\
&= \lim_{\Delta x \to 0} \left(G(x + \Delta x)\left(\frac{H(x + \Delta x) - H(x)}{\Delta x}\right) + \left(\frac{G(x + \Delta x) - G(x)}{\Delta x}\right)H(x)\right) \\
&= \lim_{\Delta x \to 0} G(x + \Delta x) \cdot \lim_{\Delta x \to 0} \left(\frac{H(x + \Delta x) - H(x)}{\Delta x}\right) \\
&\qquad + \lim_{\Delta x \to 0} \left(\frac{G(x + \Delta x) - G(x)}{\Delta x}\right) \cdot \lim_{\Delta x \to 0} H(x) \\
&= G(x) \cdot H'(x) + G'(x) \cdot H(x)
\end{aligned}$$

Application of Formula IV

Let $G(x) = 3x^2$ and $H(x) = \exp(x)$.

$$D_x(3x^2 \cdot \exp(x)) = 3x^2 D_x(\exp(x)) + D_x(3x^2) \cdot \exp(x)$$
$$= 3x^2(\exp(x)) + (6x)(\exp(x))$$
$$= (3x^2 + 6x)\exp(x)$$

Note Well: The derivative of a product of functions is NOT the product of the derivatives.

Formula V The derivative of a quotient of two functions is the denominator times the derivative of the numerator minus the numerator times the derivative of the denominator, all over the square of the denominator:

$$D_x\left(\frac{G}{H}\right) = \frac{H \cdot D_x(G) - G \cdot D_x(H)}{H^2}$$

We omit the proof of this formula.

Application of Formula V

$$D_x\left(\frac{3x^2}{\exp(x)}\right) = \frac{\exp(x) \cdot D_x(3x^2) - (3x^2) \cdot D_x(\exp(x))}{(\exp(x))^2}$$
$$= \frac{\exp(x) \cdot (6x) - (3x^2) \cdot \exp(x)}{\exp(x) \cdot \exp(x)}$$
$$= \frac{(6x - 3x^2) \cdot \exp(x)}{\exp(x) \cdot \exp(x)}$$
$$= \frac{6x - 3x^2}{\exp(x)}$$

Also

$$D_x\left(\frac{\exp(x)}{3x^2}\right) = \frac{3x^2 D_x(\exp(x)) - \exp(x) \cdot D_x(3x^2)}{(3x^2)^2}$$
$$= \frac{3x^2(\exp(x)) - \exp(x) \cdot (6x)}{9x^4}$$
$$= \frac{3x \cdot \exp(x) \cdot (x - 2)}{9x^4}$$
$$= \frac{\exp(x) \cdot (x - 2)}{3x^3}$$

Note Well: The derivative of a quotient of functions is NOT the quotient of their derivatives.

Formula VI The derivative of a constant times a function is that constant times the derivative of the function: $D_x(c \cdot H) = c \cdot D_x(H)$

The straightforward proof of this formula is also left as an exercise; it is one of the easiest of these formulas to prove. Instead we utilize two previous results and argue as follows. Let

$$G = c \cdot H$$

where c = constant. Then

$$\begin{aligned} D_x(G) &= D_x(c \cdot H) \\ &= c \cdot D_x(H) + D_x(c) \cdot H \\ &= c \cdot D_x(H) + 0 \cdot H \\ &= c \cdot D_x(H) \end{aligned}$$

as predicted.

Application of Formula VI

$$D_x(17 \cdot \exp(x)) = 17 \cdot \exp(x)$$

With these formulas at hand we can expand our repertoire of functions beyond the paltry collection of sums and products of exp (x), $3x^2$, and constants. We begin with the *derivative of an arbitrary polynomial.* Let's suppose we want the derivative of $7x^5 - 2x^2 + 6x + 5$. Using the formulas for sums and differences and for multiples, we have

$$\begin{aligned} D_x(7x^5 - 2x^2 + 6x + 5) &= D_x(7x^5) - D_x(2x^2) + D_x(6x) + D_x(5) \\ &= 7D_x(x^5) - 2D_x(x^2) + 6D_x(x) + 0 \end{aligned}$$

Now we have reduced the problem to finding the derivatives of powers of x.

Formula VII: (The Power Rule.) $D_x(x^N) = N \cdot x^{N-1}$, *for any integer* N.

The difference quotient when $N > 0$ (the case $N < 0$ is an exercise) is

$$\begin{aligned} \frac{(x + \Delta x)^N - x^N}{\Delta x} &= \frac{\sum_{k=0}^{N} C(N,k)x^k(\Delta x)^{N-k} - x^N}{\Delta x} \\ &= \frac{x^N + Nx^{N-1}(\Delta x)^1 + [\text{terms involving } (\Delta x)^2, (\Delta x)^3, \text{etc.}] - x}{\Delta x} \\ &= \frac{Nx^{N-1}(\Delta x) + [\text{terms involving } (\Delta x)^2, (\Delta x)^3, \text{etc.}]}{\Delta(x)} \\ &= Nx^{N-1} + [\text{terms involving } (\Delta x), (\Delta x)^2, \text{etc.}] \end{aligned}$$

Now taking limits as Δx approaches 0,

$$D_x(x^N) = \lim_{\Delta x \to 0} (Nx^{N-1}) + \lim_{\Delta x \to 0} [\text{terms involving } (\Delta x), (\Delta x)^2, \text{etc.}]$$
$$= Nx^{N-1}$$

Thus we complete our example,

$$\begin{aligned} D_x(7x^5 - 2x^2 + 6x + 5) &= 7D_x(x^5) - 2D_x(x^2) + 6D_x(x) + 0 \\ &= 7(5x^4) - 2(2x) + 6(1x^0) + 0 \qquad (x^0 = 1) \\ &= 35x^4 - 4x + 6 \end{aligned}$$

Using summation notation, we can write in general,

$$D_x\left(\sum_{n=0}^{N} a_n x^n\right) = \sum_{n=0}^{N} D_x(a_n x^n) = \sum_{n=1}^{N} n a_n x^{n-1}$$

We close this section with a standard example (of a function whose derivative does not exist at one point) and a standard theorem relating the derivative to the notion of continuity. Aside from the relationship to problems of finding area, the derivative is used heavily when sketching graphs of functions. To the mathematician, having a derivative at a point x says that the graph of function G in question is "smooth" at the point $(x, G(x))$. In the next section we will amplify this smoothness concept when we study tangents. For now we illustrate a non-smooth curve.

EXAMPLE A Case Where $G'(x)$ Does Not Exist

Let $G(x) = |x|$, the absolute value of x. That is, $G(2) = |2| = 2$, while $G(-3) = |-3| = 3$. Of course, $G(0) = 0$. We *try* to compute $G'(0)$. The difference quotient is

$$\frac{G(0 + \Delta x) - G(0)}{\Delta x} = \frac{|0 + \Delta x| - |0|}{\Delta x}$$
$$= \frac{|\Delta x|}{\Delta x}$$

Now when $\Delta x < 0$, we get

$$|\Delta x| = -\Delta x \quad \text{so that} \quad \frac{|\Delta x|}{\Delta x} = -1$$

But when $\Delta x > 0$, we get

$$|\Delta x| = \Delta x \quad \text{so that} \quad \frac{|\Delta x|}{\Delta x} = 1$$

Thus $G'(0)$ is supposed to be

$$\lim_{\Delta x \to 0} \frac{|\Delta x|}{\Delta x}$$

but this limit does not exist! It would have as much right being -1 as it would

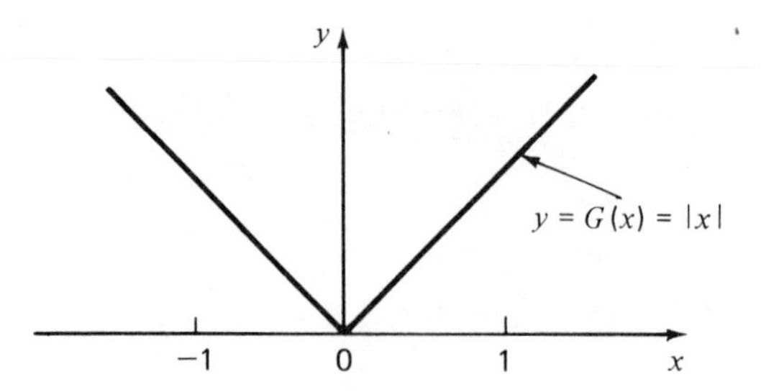

FIGURE 4.23 The graph of the absolute value function whose derivative does not exist at $x = 0$.

being $+1$, and this sort of ambiguity is not allowed in limits. We draw the graph of the function G; notice that when $x = 0$, the graph is *pointed,* not smooth. It should be mentioned that for $x \neq 0$, $G'(x)$ exists; it is $+1$ when $x > 0$ and -1 when $x < 0$.

And now for the theorem relating differentiability to continuity. We said the function G is *continuous* at the point $x = a$ if

$$\lim_{x \to a} G(x) = G(a) \quad \text{or alternately} \quad \lim_{x \to a} (G(x) - G(a)) = 0$$

Let's suppose that G has a derivative at a, i.e., $G'(x)$ exists. Now we consider the expression

$$G(x) - G(a)$$

We rewrite this, when $x \neq a$, as

$$G(x) - G(a) = \frac{G(x) - G(a)}{x - a}(x - a)$$

[Here is another piece of chicanery: multiplying by 1 disguised as $(x - a)/(x - a)$.] This factor

$$\frac{G(x) - G(a)}{x - a}$$

is really the difference quotient

$$\frac{G(a + \Delta x) - G(a)}{\Delta x}$$

in disguise, which we see by letting $x = a + \Delta x$. Then as Δx approaches 0, we see that x approaches a, and conversely. Hence

$$\begin{aligned} \lim_{x \to a} (G(x) - G(a)) &= \lim_{x \to a} \left(\frac{G(x) - G(a)}{x - a} \right) \lim_{x \to a} (x - a) \\ &= G'(a) \cdot 0 \\ &= 0 \end{aligned}$$

so G is continuous at a. We have proved

Theorem *If a function has a derivative at a point, then it is continuous at that point.* (*Note:* The example $G(x) = |x|$ at $x = 0$ shows that the converse of this theorem is false!)

Summary

After stating the basic theorems concerning limits, we established the following *rules for differentiation*:

1. $D_x(c) = 0$ for any constant c.
2. $D_x(G + H) = D_xG + D_xH$
3. $D_x(G - H) = D_xG - D_xH$
4. $D_x(G \cdot H) = G \cdot D_xH + H \cdot D_xG$
5. $D_x\left(\frac{G}{H}\right) = \frac{H \cdot D_xG - G \cdot D_xH}{H^2}$, provided $H \neq 0$.
6. $D_x(c \cdot H) = c \cdot D_xH$
7. $D_x(x^N) = N \cdot x^{N-1}$ for any integer N.

We then showed an example of a function, $f(x) = |x|$, which is continuous at a point ($x = 0$) where its derivative fails to exist, and we proved that *if a function has a derivative at a point a then it must be continuous at a.*

Exercises Section 6

1 Prove Formula III.

2 Find the derivatives of each of the following functions.

(a) $G(x) = x^2 + 7x$

(b) $G(x) = \exp(x) \cdot (7x + \exp(x))$

(c) $G(x) = \exp(2x)$ [*Hint:* Write $\exp(2x) = \exp(x) \cdot \exp(x)$.]

(d) $G(x) = \exp(3x)$ [*Hint:* Write $\exp(3x) = \exp(2x) \cdot \exp(x)$.]

(e) $G(x) = x^{-1}$ [*Hint:* $x^{-1} = 1/x$.]

(f) $G(x) = x^{-2}$

(g) $G(x) = x^{-N}$, where N is any positive integer

(h) $H(x) = \frac{x^2 + 2}{x - 1}$

(i) $H(x) = x(x^2 - 3)\exp(2x)$

(j) $H(x) = \frac{1}{G(x)}$, for any function G such that $G'(x) \neq 0$

(k) $H(x) = \exp(-x)$

3 Find a function $G(x)$ such that:

(a) $G'(x) = 2x + 3$

(b) $G'(x) = x$

(c) $G'(x) = \frac{1}{3}x^2$

(d) $G'(x) = x - \frac{1}{3}x^2$

4 Prove Formula VI using the difference quotient approach.

5 Differentiate

$$f(x) = 1 + x + \frac{x^2}{2} + \frac{x^3}{3!} + \frac{x^4}{4!} + \frac{x^5}{5!} + \frac{x^6}{6!}$$

6 Recall that if F is a distribution function for a continuous random variable X, then $F'(x) = f(x)$ is the density function for X. In each of the following cases, sketch the graphs of F and of f and indicate where, strictly speaking, f is not defined [i.e., where $F'(x)$ does not exist]:

(a) $$F(x) = \begin{cases} 0, & \text{if } x < -1 \\ \frac{1}{2}(x+1), & \text{if } -1 \le x \le 1 \\ 1, & \text{if } x \ge 1 \end{cases}$$

(b) $$F(x) = \begin{cases} 0, & \text{if } x < 0 \\ x^2, & \text{if } 0 \le x \le 1 \\ 1, & \text{if } x \ge 1 \end{cases}$$

(c) $$F(x) = \begin{cases} e^x, & \text{if } x < 0 \\ 1, & \text{if } x \ge 0 \end{cases}$$

(d) $$F(x) = \begin{cases} 0, & \text{if } x < 0 \\ 1 - e^{-x}, & \text{if } x \ge 0 \end{cases}$$

(e) $$F(x) = \begin{cases} 0, & \text{if } x < 1 \\ 1 - \dfrac{1}{x}, & \text{if } x \ge 1 \end{cases}$$

(f) $$F(x) = \begin{cases} 0, & \text{if } x < 0 \\ 1 - e^{-50x}, & \text{if } x \ge 0 \end{cases}$$

7 **(a)** Formulate a rule for the derivative of the product of three functions, $F \cdot G \cdot H$, analogous to Formula IV.

(b) Differentiate each of the following functions.

$$f(x) = (x + 3)(x^2 - 2)(x^3 - 2x)$$

$$g(x) = e^x(x^3 + x^2)(4 - 7x)$$

8 Sketch each of the following functions and determine which are differentiable at $x = 0$. Which are continuous there?

(a) $$g(x) = \begin{cases} -x^2, & \text{if } x < 0 \\ x^2, & \text{if } x \ge 0 \end{cases}$$

(b) $$h(x) = \begin{cases} -1, & \text{if } x < 0 \\ +1, & \text{if } x \ge 0 \end{cases}$$

(c) $$k(x) = \begin{cases} e^x, & \text{if } x < 0 \\ e^{-x}, & \text{if } x \ge 0 \end{cases}$$

(d) $$r(x) = \begin{cases} -x^3, & \text{if } x < 0 \\ x^3, & \text{if } x \ge 0 \end{cases}$$

(e) $$s(x) = \begin{cases} \dfrac{1}{x-1}, & \text{if } x < 0 \\ e^x, & \text{if } x \ge 0 \end{cases}$$

9 (Higher derivatives.) The function f' is often called the *first derivative* of f for the simple reason that it too is a function and is thereby eligible to have a derivative at some point x in its domain. If it exists, the value $D_x(f'(x))$ is called the *second derivative* of f at x and is denoted by

$$f''(x) \quad \text{or} \quad D_x^2 f(x)$$

For instance we have

$$f(x) = x \exp(x)$$
$$f'(x) = x \exp(x) + 1 \cdot \exp(x) = (x+1)\exp(x)$$
$$f''(x) = (x+1)\exp(x) + 1\exp(x) = (x+2)\exp(x)$$

and

$$g(x) = x^2 + 3x - 5$$
$$g'(x) = 2x + 3$$
$$g''(x) = 2$$

Similarly, the *third derivative*, the *fourth derivative*, and in general, the *nth derivative* of f at x are defined successively:

$$f'''(x) = D_x f''(x), \quad f''''(x) = D_x f'''(x), \ldots, f^{(n)}(x) = D_x f^{(n-1)}(x)$$

(a) $f(x) = x^5 + 6$; find $f'(x), f''(x)$, and $f'''(x)$.
(b) $f(x) = x^2 \exp(x)$; find $f'(x)$ and $f''(x)$.
(c) $g(x) = x^3$; find $g^{(4)}(x)$.
(d) $f(x) = ax^4 + bx^3 + cx^2 + dx + e$; find $f(0), f'(0), f''(0), f'''(0)$, and $f^{(4)}(0)$.
(e) Let

$$h(t) = 1 + t + \frac{t^2}{2} + \frac{t^3}{3!} + \frac{t^4}{4!} + \frac{t^5}{5!};$$

find $h'(t), h''(t), h'''(t)$.
(f) Let

$$H(t) = \sum_{n=0}^{100} \frac{t^n}{n!};$$

find $H'(t),\ H''(t),\ H'''(t),\ H^{(15)}(t)$.

7 THE MEAN VALUE THEOREM; TANGENTS

We are now prepared to complete our discussion of the evaluation of integrals by formula. In doing so we shall look at a new interpretation of the derivative which gives rise to many of the practical applications of this concept.

Let us try to piece together what we know about

$$\int_a^b f(x)\,dx$$

First of all, if $f(x) \geq 0$ for x in $[a, b]$, then this definite integral represents the area of the region under the graph of f and above the x-axis from $x = a$ to $x = b$.

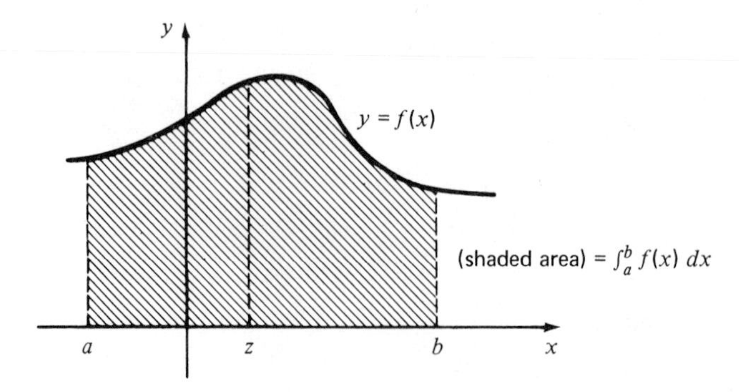

FIGURE 4.24 The area under a curve.

Moreover we know that if we define a function F by

$$F(z) = \int_a^z f(x)\,dx, \qquad \text{for } a \leq z \leq b$$

then we should have

$$F'(z) = f(z), \qquad \text{for } a < z < b$$

Now if the function f is simple enough, we can easily find *a* function F whose derivative is f. For instance, if

$$f(x) = x + 1$$

say on the interval [1, 2], then we could choose

$$F(x) = \tfrac{1}{2}x^2 + x$$

since $F'(x) = \frac{1}{2}(2x) + 1 = x + 1 = f(x)$.

However, when we examine the area determined by the graph of f from $x = 1$ to $x = 2$, we find

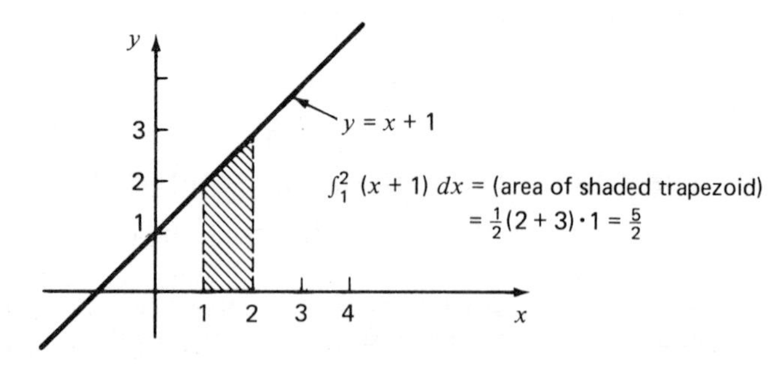

FIGURE 4.25

But of course our function F yields

$$F(2) = (\tfrac{1}{2})(2)^2 + 2 = 4$$

Since $F(2)$ does not represent the area in question, we must have overlooked something. The point is that having $F'(x) = f(x)$ for all x in [1, 2] is not enough There are, in fact, an infinitude of functions satisfying this property; any function of the form

$$F(x) = \tfrac{1}{2}x^2 + x + c$$

where c is any constant, will yield

$$F'(x) = x + 1$$

Definition

If f and F are functions such that

$$F'(x) = f(x), \qquad \text{for all } x \text{ in } [a, b]$$

We say that F is an *antiderivative of f on* $[a, b]$.

We could, of course, determine the correct choice of the value c in the present case (we would find that $c = -\tfrac{3}{2}$ works), but instead we deal with the problem in general. To do so requires an additional fact about derivatives, called the *Mean Value Theorem.*

Although the Mean Value Theorem has a variety of interpretations, and in spite of the fact that it can be stated in language we already understand, one of the most useful vehicles for describing it involves the notion of a *tangent to a curve*. Since traditionally the so-called tangent problem has been used to introduce students to calculus, it seems appropriate to sketch the problem at this juncture.

The build-up to the formal definition of tangent to a curve often seems overly pedantic at first. But the less precise definitions typically offered by students beginning to study the topic are easily seen to be erroneous. For example,

A tangent is *not* a "line that touches a curve only at one point" (as seen by Fig. 4.26).

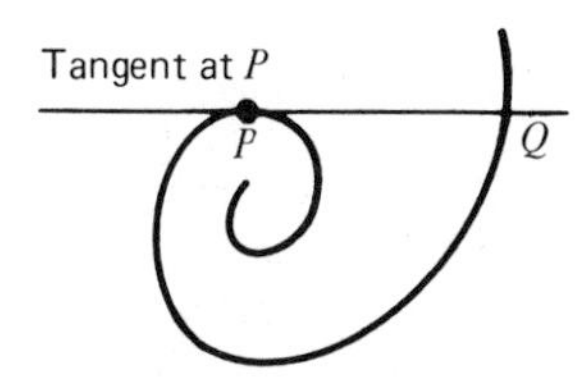

FIGURE 4.26 A tangent line (at P) that meets the curve at another point (Q).

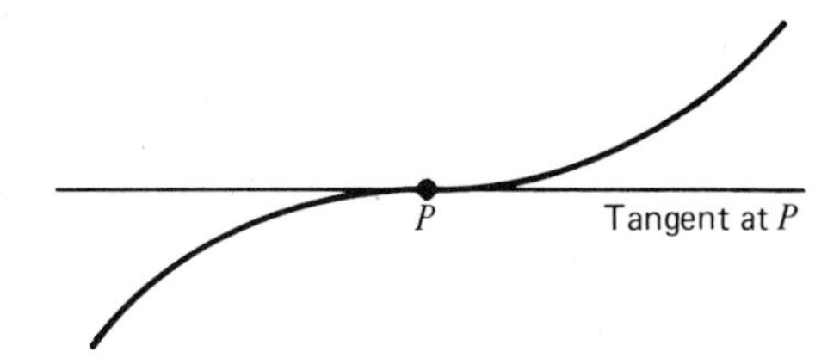

FIGURE 4.27 A tangent line (at P) that cuts through the curve.

A tangent is *not* a "line that only touches but does not cross the curve" (as seen by Fig. 4.27).

With less complicated curves (Fig. 4.28 and Fig. 4.29) most anyone would agree on what the tangent lines look like:

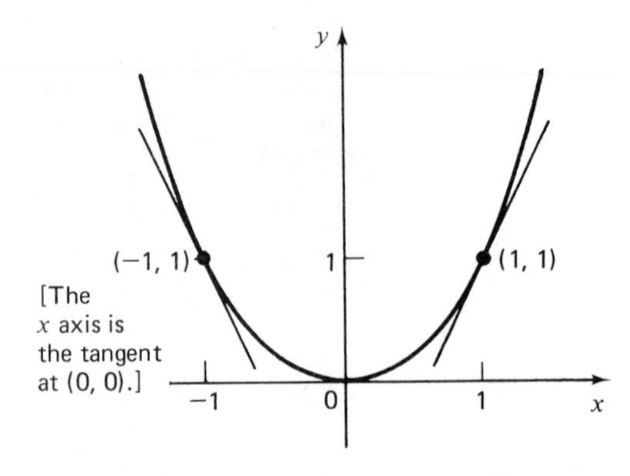

FIGURE 4.28 A circle with tangent lines at various points.

FIGURE 4.29 A parabola with tangent lines at various points.

In Fig. 4.30 we illustrate how a tangent line is constructed (and defined) in general. We select some point P at which to construct a tangent. We then select a sequence of points $Q_1, Q_2, Q_3, \ldots$ that approach P from either side.

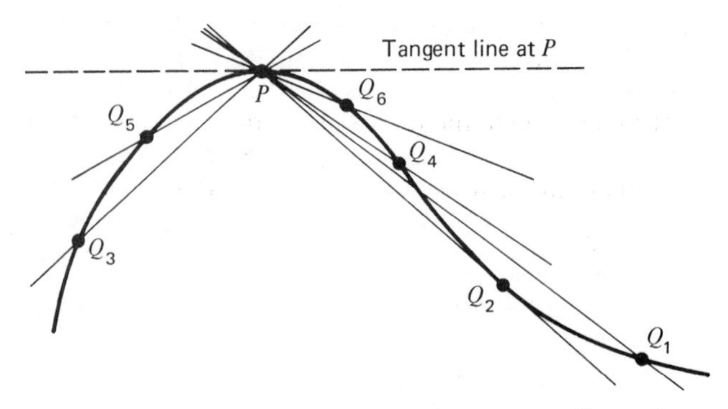

FIGURE 4.30 Several secant lines and the tangent line at P.

The lines determined by P and Q_1, P and Q_2, P and Q_3, and so on, are called *secant lines. If* it happens that these secant lines tend to settle down or stabilize toward some fixed line (through P), then we call that "limiting" line ***the tangent at*** P.

Now the curves that we are primarily interested in are graphs of functions. A point P on such a curve will have coordinates $(x, f(x))$, where f is the function. The auxiliary points $Q_1, Q_2, \ldots$ will be of the form $(x_1, f(x_1))$, $(x_2, f(x_2))$, and so on. The key to the definition of tangent line is the observation that each of these lines will have some specific *slope,* where we recall that the slope of the line joining P to Q_1 is given by the formula

$$\text{Slope of line joining } (x, f(x)) \text{ to } (x_1, f(x_1)) = \frac{f(x_1) - f(x)}{x_1 - x}$$

and to say that the secant lines tend to some specific tangent line through P is equivalent to saying that their slopes tend to a specific slope. Of course a line is determined by specifying both its slope and some point, so the tangent line at P could be specified by naming its slope (since we assume the point P lies on the line).

We are ready for the formal definition.

Definition *The tangent line to the graph of f at* $(x, f(x))$ is the line whose slope is $f'(x)$.

Interpretation: As Fig. 4.31 shows, we think of a nearby point Q as having coordinates $(x + \Delta x, f(x + \Delta x))$. The secant line joining P [i.e., $(x, f(x))$] to Q Q has slope

$$\text{Slope of secant } PQ = \frac{f(x + \Delta x) - f(x)}{(x + \Delta x) - x} = \frac{f(x + \Delta x) - f(x)}{\Delta x}$$

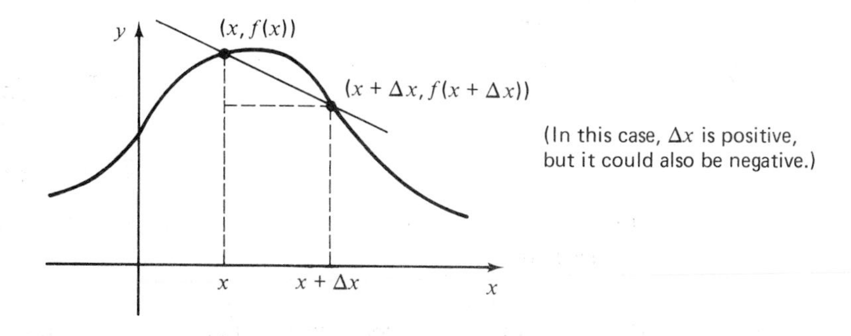

FIGURE 4.31 The secant line joining $(x, f(x))$ to $(x + \Delta x, f(x + \Delta x))$.

To say that Q approaches P (in the same sense that we had a sequence of $Q_1, Q_2, Q_3, \ldots$ approaching P) says that

Δx approaches 0

Thus the limiting slope, if it exists, must be

$$\lim_{\Delta x \to 0} \frac{f(x + \Delta x) - f(x)}{\Delta x}$$

which is, by definition, $f'(x)$.

EXAMPLE 1 Tangents to the Graph of f where $f(x) = x^2$

The graph of f is the parabola in Fig. 4.29. We see that

$$f'(x) = D_x(x^2) = 2x, \qquad \text{for all values of } x$$

In particular when $x = -1$, $f'(x) = f'(-1) = -2$, and the tangent line at $(-1, f(-1)) = (-1, 1)$ has slope -2. Similarly when $x = 0$ and $x = 1$, the tangent lines have respective slopes 0 and 2. All three of these tangent lines appear in Fig. 4.29.

A convenient interpretation of the tangent line results from imagining a tiny insect crawling along a curve, say the graph of some function f. At any given instant the insect is moving in some direction. We claim that his direction is determined by the tangent line (to the point on the curve where the insect happens to be at the given instant). Alternately, one might consider a car driving along a winding road at night. At any given instant, the headlights point in the direction of motion. Viewing the road as a curve, the headlights sweep out a tangent line at the point of the car.

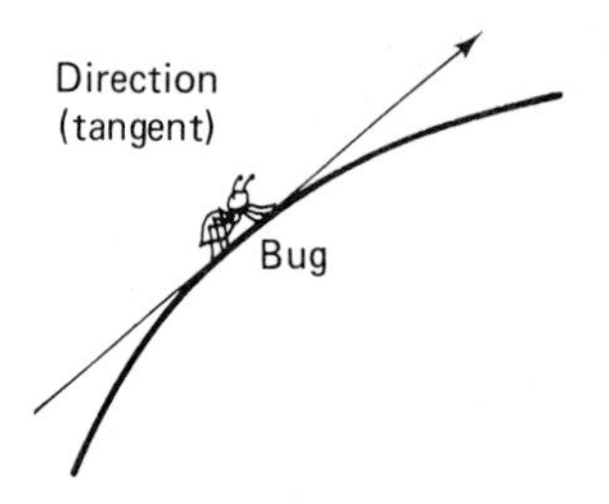

FIGURE 4.32 An insect crawling along a curve.

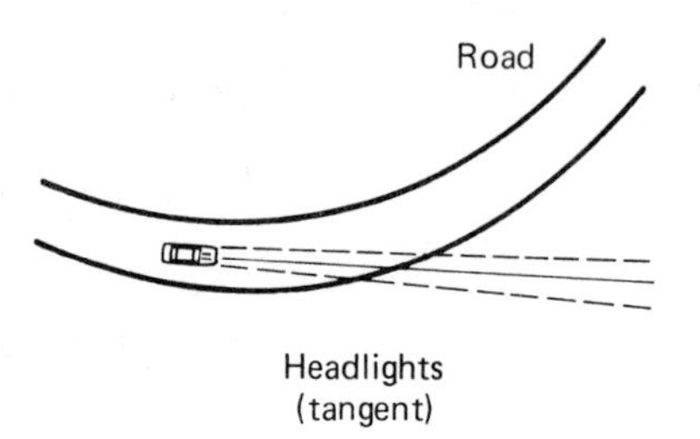

FIGURE 4.33 A car's headlights pointing in the (tangent) direction of motion.

The *Mean Value Theorem* is an assertion about a mean or average value for the derivative. We illustrate it in Fig. 4.34. The theorem says that, assuming $f'(x)$ exists for all x between a and b, that at some point $(c, f(c))$ the tangent line to the curve is *parallel* to the (secant) line joining $(a, f(a))$ to $(b, f(b))$.

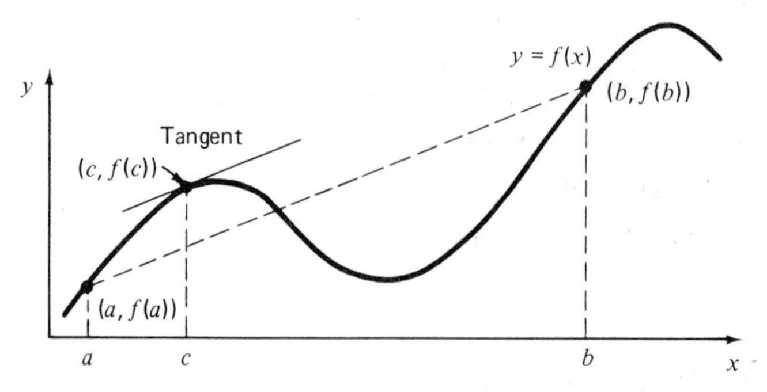

FIGURE 4.34 Mean value theorem: The tangent at $(c, f(c))$ is parallel to the line joining $(a, f(a))$ to $(b, f(b))$.

Recall that two lines are parallel if and only if they have the *same slope*. In formal language we have:

Mean Value Theorem

Let f be continuous on $[a, b]$ and let $f'(x)$ exist for $a < x < b$. Then there exists at least one point c, where $a < c < b$, such that

$$f'(c) = \frac{f(b) - f(a)}{b - a} \quad \text{or equivalently} \quad f'(c)(b - a) = f(b) - f(a)$$

(Slope of tangent at $(c, f(c))$) — (Slope of line joining $(a, f(a))$ to $(b, f(b))$)

Another interpretation of the theorem can be made in terms of the aforementioned insect: If the insect crawls along the graph of f from $(a, f(a))$ to $(b, f(b))$, then there must be some instant when he is heading in precisely the direction of the straight line joining these two points. In terms of the automobile, if the driver's ultimate goal is to travel to a destination due Northeast, say, of his origination point, then at some instant the car's headlights must point Northeast.

Aside from the meanderings of cars and insects, the Mean Value Theorem provides us with several important facts. Although we are not going to prove the Mean Value Theorem, we do sketch the proofs of some of its corollaries.

Corollary 1

If a function is constant throughout an interval, then its derivative is 0 everywhere in the interval. Conversely, if the derivative of a function is 0 throughout the interval, then the function is constant.

Proof: The first half of this theorem is a triviality (at this stage) since we know that the derivative of a constant is 0.

Conversely, let us suppose that

$$f'(x) = 0, \qquad \text{for } a < x < b$$

Then for a' and b' such that $a \leq a' < b' \leq b$ we have

$$f(b') - f(a') = f'(c')(b' - a') = 0(b' - a') = 0$$

where c' is some point between a' and b'. But then we have

$$f(b') = f(a')$$

Since a' and b' were arbitrary points in the interval, f must be constant.

Corollary 2

If two functions have the same derivative, then they differ by a constant.

Proof: Suppose $F'(x) = G'(x)$ for all F and G in some interval. Then the function $F - G$, given by

$$(F - G)(x) = F(x) - G(x)$$

has a zero derivative:

$$(F - G)'(x) = F'(x) - G'(x) = 0$$

so that

$$F(x) - G(x) = \text{constant}$$

i.e., $F(x) = G(x) + c$ for all x in the given interval.

Corollary 3 (Rolle's Theorem)

If $f(a) = f(b)$ and if f is continuous on $[a, b]$ and $f'(x)$ exists for $a < x < b$, then at some point c between a and b, $f'(c) = 0$ (see Fig. 4.35).

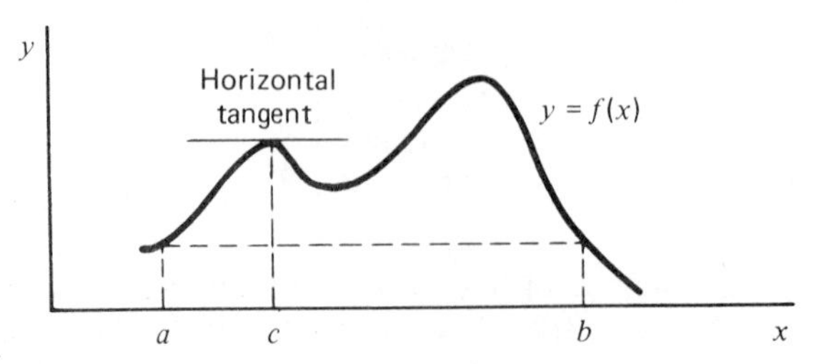

FIGURE 4.35 Rolle's theorem: If $f(a) = f(b)$, then $f'(c) = 0$ for some c.

In subsequent work we shall pay particular attention to points where the tangent line is horizontal (i.e., its slope is 0) since that behavior is characteristic of peaks and valleys in the graph of f.

EXAMPLE 2 Some Curves Where Tangent Lines Do Not Always Exist

We have already seen one example $[f(x) = |x|$, at $x = 0]$ where a derivative does not exist. Figure 4.36 illustrates two typical patterns of behavior for curves where our tangent definition does not apply. In the first, the tangent line cannot

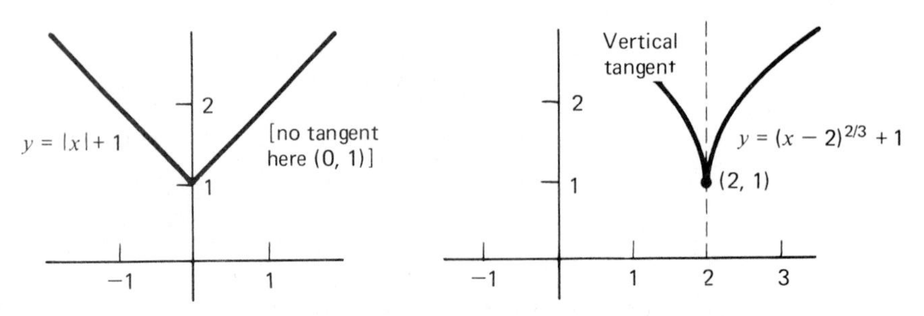

FIGURE 4.36 Examples where "$f'(x)$ = slope of tangent" does not apply.

be unambiguously defined; in the second, the tangent exists but, being vertical, it has no slope, so that $f'(x)$ would not exist.

Finally we return to the problem of computing

$$\int_a^b f(x)\,dx, \qquad \text{where } f(x) \geq 0 \text{ on } [a, b]$$

From the multitude of antiderivatives of $f(x)$ (all pairs of which differ by a constant by Corollary 2) let F be the one for which

$$F(z) = \int_a^z f(x)\,dx, \qquad \text{for all } z \text{ in } [a, b]$$

1. Then since $F(a)$ would represent the area under the curve from $x = a$ to $x = a$ (i.e., a region with 0 width), we have

$$F(a) = 0$$

2. Let G be any other antiderivative of f. Then

$$G(x) = F(x) + C, \qquad \text{for some constant } C$$

Hence

$$\begin{aligned} G(b) - G(a) &= (F(b) + C) - (F(a) + C) \\ &= F(b) - F(a) \\ &= F(b) \\ &= \int_a^b f(x)\,dx \end{aligned}$$

In words, *the definite integral of f from a to b equals $G(b) - G(a)$, where G is any antiderivative of f.*

EXAMPLE 3 Recall the area problem: $f(x) = x + 1$ from $x = 1$ to $x = 2$. We found the antiderivative

$$F(x) = \tfrac{1}{2}x^2 + x$$

$$\int_1^2 (x + 1)\,dx = (\tfrac{1}{2}(2)^2 + 2) - (\tfrac{1}{2}(1)^2 + 1) = 4 - \tfrac{3}{2} = \tfrac{5}{2}$$

(Correct value computed from picture)

Incidentally there is a convenient notation used in these definite integral computations; we write

$G(x)|_a^b$ to denote $G(b) - G(a)$

b is called the *upper limit of integration*; a the *lower limit of integration*.

Now that this technique of evaluating definite integrals is available, we hasten to point out that the restriction of

$$f(x) \geq 0, \qquad \text{for } x \text{ in } [a, b]$$

is *unnecessary*. That is if f is *any continuous function on* $[a, b]$ and if $G'(x) = f(x)$ for all x in $[a, b]$, then

$$\int_a^b f(x)\,dx = G(b) - G(a)$$

The proof of this fact can be sketched without much difficulty. Suppose f is continuous on $[a, b]$. Intuitively such a function should have a *minimum value* somewhere, say at c. In case $f(c) < 0$, we can build a new function g defined by

$$g(x) = f(x) + k$$

whose graph lies entirely above the x-axis. (See Fig. 4.37.) Now if F is an anti-

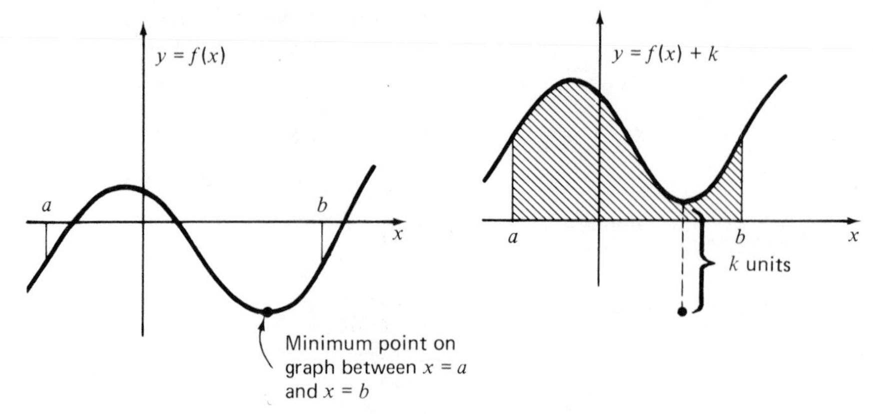

FIGURE 4.37 The graph of f (at left) and of $f + k$ (at right); raising the graph of f by k units places it entirely above the x-axis.

derivative of f, we put $G(x) = F(x) + kx$ and get

$$G'(x) = F'(x) + k = f(x) + k = g(x)$$

Thus

$$(*) \quad \int_a^b (f(x) + k)\,dx = (F(x) + kx)\Big|_a^b = F(b) - F(a) + k(b - a)$$

But by *linearity* of the integral,

$$(**) \quad \int_a^b (f(x) + k)\,dx = \int_a^b f(x)\,dx + \int_a^b k\,dx = \int_a^b f(x)\,dx + k(b - a)$$

Consequently, setting equal the right-hand expressions in (*) and (**), we have

$$\int_a^b f(x)\,dx = F(b) - F(a)$$

as desired.

Reminder: Although mentioned earlier, it is important to realize that when computing definite integrals, the following two rules (together referred to as *linearity of the integral*) apply.

$$\int_a^b k \cdot f(x)\,dx = k\int_a^b f(x)\,dx$$

$$\text{and} \quad \int_a^b (f(x) + g(x))\,dx = \int_a^b f(x)\,dx + \int_a^b g(x)\,dx$$

Summary

At the point $(x, f(x))$ on the graph of a function f, the *tangent line* is the unique line through this point whose slope is $f'(x)$, provided the derivative exists.

This geometric interpretation of derivatives has numerous useful consequences, one of which is the *Mean Value Theorem* which says that between any two points P and Q on the graph of a differentiable function there lies a third point whose tangent line is parallel to the chord joining P to Q. In turn, the Mean Value Theorem implies (1) if two functions have equal derivatives on some interval, then they differ only by a constant, and (2) if the graph of a differentiable function has two points with the same y-coordinate, then somewhere between them the tangent line to the graph is horizontal.

Finally, we were able to solve our area problem:

$$\int_a^b f(x)\,dx = F(b) - F(a)$$

where $F'(x) = f(x)$ for all x in $[a, b]$. We call F an *antiderivative* of f. While the area problem requires that $f(x) \geq 0$ for x in $[a, b]$, the above equation holds true even when $f(x)$ is negative and/or positive in the interval.

Exercises Section 7

1 Evaluate the following definite integrals.

(a) $\int_0^1 x\,dx$

(b) $\int_{-1}^0 x\,dx$

(c) $\int_{-1}^1 x\,dx$

(d) $\int_0^2 (x^2 + 3)\,dx$

(e) $\int_0^1 \exp(x)\,dx$

(f) $\int_0^1 \exp(-x)\,dx$

(g) $\int_1^2 x^{-2}\,dx$

(h) $\int_1^3 (2x^2 + x - 1)\,dx$

2 Reproduce Fig. 4.34 and Fig. 4.35 and indicate all points c in each picture whose tangent line is parallel to the chord joining $(a, f(a))$ to $(b, f(b))$.

3 Use the antiderivative $F(x) = \frac{1}{2}x^2 + x - 12$ to evaluate the integral in Example 3.

4 Sketch the graph of the function $f(x) = 4 - x^2$ by

(a) plotting the points where $x = -2, -1, 0, 1, 2$.

(b) sketching the tangent line at each of these points.

5 Repeat Exercise 4 with the function $f(x) = x^3 - 4x$, using the points where $x = -2, -1, 0, 1, 2, 3$.

6 Using the graph of $f(x) = 4 - x^2$ in Exercise 4, find the point whose tangent line is parallel to the chord joining $(-2, 0)$ to $(1, 3)$. Draw the picture with the chord and the appropriate tangent line.

7 Many of the definite integrals we wish to integrate involve functions which are defined "in pieces." To evaluate the integral, we break it up into several integrals corresponding to the intervals in which the function has a fixed formula. For example, let

$$f(x) = \begin{cases} 2 + x, & \text{if } x \text{ is in } [0, 2] \\ 4 - x, & \text{if } x \text{ is in } [2, 4] \end{cases}$$

Draw the graph of this function and find the area using

$$\int_0^4 f(x)\,dx = \int_0^2 f(x)\,dx + \int_2^4 f(x)\,dx$$

8 On the interval $[1, 2]$, let

$$f(x) = \frac{x+2}{x+1}$$

Find all points in this interval where the tangent line is parallel to the chord joining $(1, f(1))$ to $(2, f(2))$.

9 Recall that a *parabola* with vertical axis of symmetry is simply the graph of a quadratic function,

$$f(x) = cx^2 + dx + e$$

Show that for any interval $[a, b]$, the tangent line parallel to the chord joining $(a, f(a))$ to $(b, f(b))$ occurs at $(m, f(m))$ where m is the midpoint of the interval.

10 Each of the following functions is a density function for some continuous random variable. Find the specified probabilities.

(a) $f(x) = \begin{cases} 0, & \text{if } x < 0 \\ e^{-x}, & \text{if } x \geq 0 \end{cases}$

Find $P(0 \leq X \leq 3)$.

(b) $f(x) = \begin{cases} x, & \text{if } 0 \leq x \leq \frac{1}{2} \\ \frac{7}{3}x, & \text{if } \frac{1}{2} \leq x \leq 1 \\ 0, & \text{otherwise} \end{cases}$

Find $P(0 \leq X \leq \frac{1}{2})$ and $P(\frac{1}{4} \leq X \leq \frac{3}{4})$.

(c) $f(x) = \begin{cases} \dfrac{1}{2\sqrt{x}} & \text{if } 0 < x \leq 1 \\ 0, & \text{otherwise} \end{cases}$

Find $P(\frac{1}{2} \leq X \leq 1)$.

11 Find the mean and the variance for the random variables X whose density functions are given below.

(a) $f(x) = \begin{cases} 1, & \text{if } 0 \leq x \leq 1 \\ 0, & \text{otherwise} \end{cases}$

(b) $f(x) = \begin{cases} \frac{1}{2}(x+1), & \text{if } -1 \leq x \leq 1 \\ 0, & \text{otherwise} \end{cases}$

(c) $f(x) = \begin{cases} 3x^2, & \text{if } 0 \leq x \leq 1 \\ 0, & \text{otherwise} \end{cases}$

8 MIN-MAX PROBLEMS

In our day-to-day activities we are apt to encounter a variety of functional relationships. Among the common kinds of functions occurring are several where the functional value represents cost or price:

1. The postage stamp function: To each weight x (in ounces) there corresponds a price $p(x)$ (in cents) we must pay to send a letter or package, say by first-class mail.

2. The hamburger function: On a given day at a given store, there is a certain price $H(x)$ corresponding to x pounds of hamburger.
3. A profit function: A manufacturer of radios determines a certain profit $P(x)$ he can expect to make when he sells his radios at some specific cost, x each.
4. The taxi-fare function: This is similar to the postage stamp function; it costs $C(x)$ to travel x miles by cab.
5. The truck-economy function: The manager of a fleet of trucks is faced with the cost $C(v)$ for transporting goods from one station to another with average velocity v. The faster the truck goes, the less money it costs for driver time but the more it costs for fuel.

In these and other problems, it is of interest to find where the function has maximum or minimum values. At an abstract level we can put the problem as follows.

Problem 1 To find where a function f defined on an interval $[a, b]$, has *maximum* and *minimum* values.

We can solve this general problem by appealing to properties of the derivative of a function. Our discussion will be more heuristic than rigorous, and we shall appeal to the use of graphs to support much of what we assert.

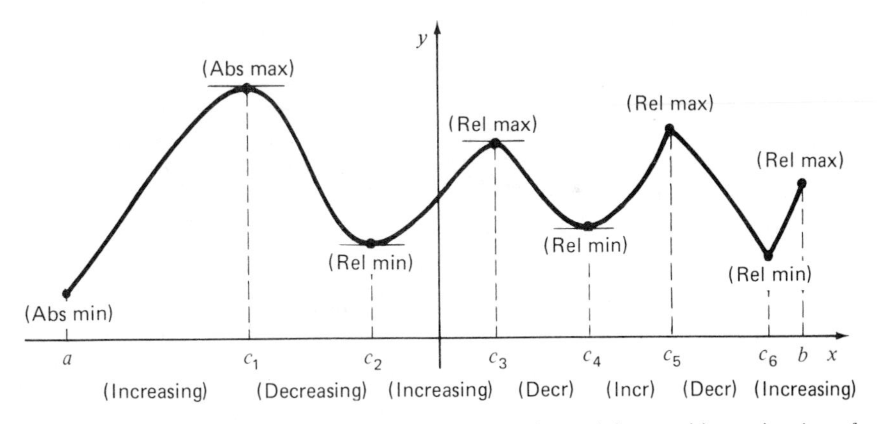

FIGURE 4.38 The graph of f illustrating various maxima, minima, and intervals where f is increasing and decreasing.

In Fig. 4.38, we indicate various points on and portions of the graph of a function where we can describe some of the following concepts.

Definitions Let f be a function defined on an interval $[a, b]$. We say that f has an *absolute maximum* value at c if

$$f(c) \geq f(x), \qquad \text{for all } x \text{ in } [a, b]$$

We say that f has an *absolute minimum* value at c if

$f(c) \leq f(x)$, for all x in $[a, b]$

We say that f has a *relative maximum* value at c if

$f(c) \geq f(x)$, for all x in some (small) interval containing c

We say that f has a *relative minimum* value at c if

$f(c) \leq f(x)$, for all x in some (small) interval containing c

We say that f is *increasing* on the (sub-) interval $[a', b']$ if

$f(c) \leq f(d)$, whenever $a' \leq c \leq d \leq b'$

We say that f is *decreasing* on the (sub-) interval $[a', b']$ if

$f(c) \geq f(d)$, whenever $a' \leq c \leq d \leq b'$

Additional Comments: $f(a)$ is the absolute minimum value of the function. $f(c_1)$ is the absolute maximum value of the function. There are *horizontal tangents* at points whose x-coordinates are c_1, c_2, c_3, c_4. There are *no tangents* at points whose x-coordinates are a, c_5, c_6, b. In each interval where f is increasing, $f'(x) \geq 0$ (tangents have positive slope). In each interval where f is decreasing, $f'(x) \leq 0$ (tangents have negative slope).

In general, we make the following assertions.

Theorem

Let f be defined on $[a, b]$.

1. If $f'(c) > 0$, f is increasing on an interval through c.
2. If $f'(c) < 0$, f is decreasing on an interval through c.
3. If $f(c)$ is a relative maximum value, then either $f'(c) = 0$ of $f'(c)$ does not exist.

Algorithm for Finding Maximum and Minimum Values

1. Compute $f'(x)$.
2. List all values $\{c_1, c_2, \ldots, c_n\}$ where $f'(c_i) = 0$ or where $f'(c_i)$ does not exist.
3. Evaluate $f(a), f(c_1), \ldots, f(c_n), f(b)$. This list contains the *absolute maximum*, the *absolute minimum* and *all relative maxima* and *minima* of f on $[a, b]$.

EXAMPLE 1

Find the maximum value of $f(x) = x^3 - 4x$ on the interval $[-3, 2]$.

Solution:

1. $f'(x) = 3x^2 - 4$.
2. $f'(x) = 0$ when $3x^2 - 4 = 0$.

$$x^2 = \tfrac{4}{3}$$

$$x = \sqrt{\tfrac{4}{3}}, \; -\sqrt{\tfrac{4}{3}}.$$

$f'(x)$ always exists!

3. $f(-3) = (-3)^3 - 4(-3) = -27 + 12 = -15.$
$f(-\sqrt{\tfrac{4}{3}}) = (-\sqrt{\tfrac{4}{3}})^3 - 4(-\sqrt{\tfrac{4}{3}}) = (-\tfrac{4}{3})(\sqrt{\tfrac{4}{3}}) + 4(\sqrt{\tfrac{4}{3}}) = (\tfrac{8}{3})(\sqrt{\tfrac{4}{3}})$
≈ 3.08
$f(\sqrt{\tfrac{4}{3}}) = (\sqrt{\tfrac{4}{3}})^3 - 4(\sqrt{\tfrac{4}{3}}) = (\tfrac{4}{3})(\sqrt{\tfrac{4}{3}}) - 4(\sqrt{\tfrac{4}{3}}) = (-\tfrac{8}{3})(\sqrt{\tfrac{4}{3}})$
$\approx -3.08.$
$f(2) = 2^3 - 4(2) = 8 - 8 = 0.$

Thus, absolute minimum $= -15$ when $x = -3$; absolute maximum ≈ 3.08 when $x = -\sqrt{\tfrac{4}{3}}$.

The graph of f is sketched in Fig. 4.39.

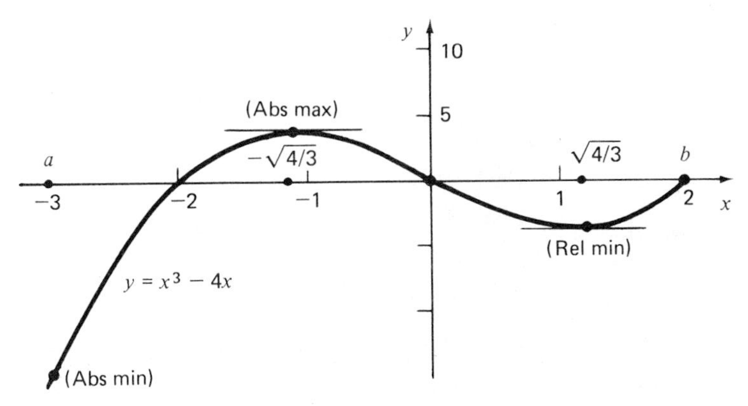

FIGURE 4.39 Graph of $x^3 - 4x$ on $[-3, 2]$.

Of course in practice, the max-min problem comes not in abstract form but usually in the guise of a "word problem" which then has to be translated into mathematics. We give two examples, the first of which is a classic to the calculus audience.

EXAMPLE 2 A farmer wished to build a rectangular pasture, one side of which is bounded by a river and the other three sides of which are formed from 1000 feet of fence. What are the dimensions of the pasture of largest area he can build?

Solution: In Fig. 4.40 we see that the rectangle is determined by specifying the number of feet, x, of fence to be used at one "end" of the pasture. For then

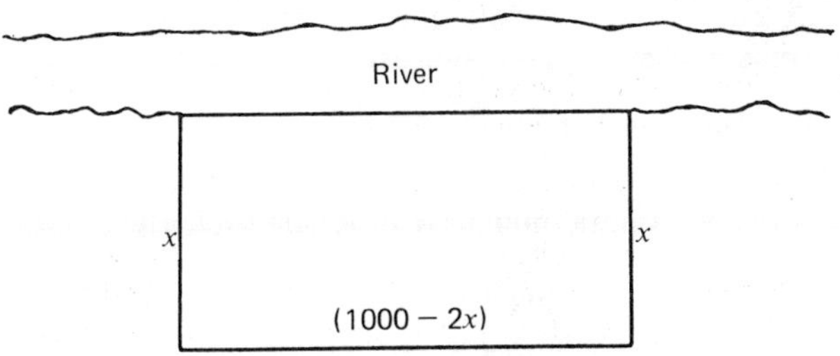

FIGURE 4.40 Rectangular pasture of area $(1000 - 2x)x$ square feet.

the other end must also measure x feet, and the side opposite the river bank must measure $(1000 - 2x)$ feet. But using the formula for the area of a rectangle, we can restate the problem mathematically as follows:

For what value of x will the function

$$A(x) = (1000 - 2x)x$$

have a maximum value, provided x is in the interval $[0, 500]$? (The extreme cases in the physical problem are $x = 0$ and $x = 500$, both of which produce "collapsed" rectangles.)

Solution:

1. Since $H(x) = (1000 - 2x)x = 1000x - 2x^2$, we have

   $$H'(x) = 1000 - 4x$$

2. $H'(x) = 0$ if $1000 - 4x = 0$, i.e.,

   $$x = 250$$

   $H'(x)$ is always defined.
3. We check

   $$H(0) = (1000 - 2 \cdot 0) \cdot 0 = 0$$
   $$H(250) = (1000 - 2 \cdot 250)(250) = (500)(250) = 125{,}000 \text{ square feet}$$
   $$H(500) = (1000 - 2 \cdot 500)(500) = 0$$

   Thus 125,000 square feet is the *maximum area*. The answer to the problem is:

   Dimensions: 250 feet × 500 feet

EXAMPLE 3 Advanced sales of a particular radio suggest to the retailer that if p is the price (in dollars) and x is the number of radios sold, then the *demand function*, as it's called, is given by

$$x = 80 - 2p$$

If the cost to the retailer of selling x radios is

$$C(x) = 8x + 140 \text{ (dollars)}$$

find the number of radios and the corresponding unit price that will yield maximum profit.

Solution: Our ultimate goal is to maximize the profit P obtained by selling x radios at p dollars per radio. Since

$$\text{Profit} = P = \text{revenue} - \text{cost}$$

we see that P depends upon the sales revenue and the cost, both of which in turn depend on the number x of radios sold. Therefore we will think of P as a function of x: $P = P(x)$, and our solution will consist of differentiating P with respect to x to seek maximal values.

First, however, let us examine some of the relationships involved in the problem. In particular, the demand function gives us x, the quantity of radios "demanded," as a function of p, the unit price. It stands to reason that x will become larger as p gets smaller and vice versa. It is also clear that we could rewrite

$$x = 80 - 2p$$

in a form to express p as a function of x:

$$p = \tfrac{1}{2}(80 - x)$$

This latter equation has the advantage of allowing is to express revenue as a function of x as follows:

$$\begin{aligned} \text{Revenue} &= (\text{unit price})(\text{quantity sold}) \\ &= p \cdot x \\ &= \tfrac{1}{2}(80 - x)x \\ &= 40x - \tfrac{1}{2}x^2 \end{aligned}$$

And finally, we can express the profit $P(x)$ explicitly as a function of x:

$$\begin{aligned} P(x) &= \text{revenue} - \text{cost} \\ &= (40x - \tfrac{1}{2}x^2) - (8x + 140) \\ &= 32x - \tfrac{1}{2}x^2 - 140 \end{aligned}$$

From this point on, our problem is straightforward:

$$\text{Maximize } P(x) = 32x - \tfrac{1}{2}x^2 - 140$$

1. $P'(x) = 32 - x$.
2. $P'(x) = 0$, if $x = 32$, and for practical purposes we must have $0 \leq x \leq 80$.
3. We check

$$\begin{aligned} P(0) &= 32(0) - \tfrac{1}{2}(0)^2 - 140 = -140 \\ P(32) &= 32(32) - \tfrac{1}{2}(32)^2 - 140 = 372 \\ P(80) &= 32(80) - \tfrac{1}{2}(80)^2 - 140 = -780 \end{aligned}$$

Hence the *maximum* profit occurs when $x = 32$ radios are sold. That *maximum profit is 372 dollars*. Figures 4.41 and 4.42 illustrate the relationship between x and p as well as the graph of the profit function $P(x)$.

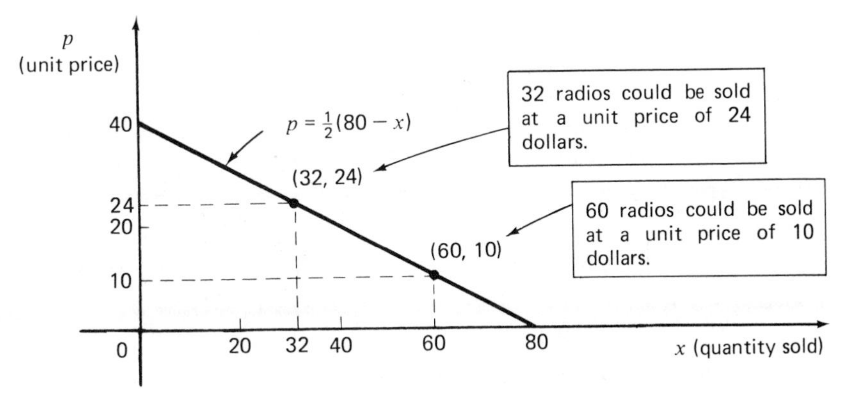

FIGURE 4.41 Graph of the demand function, $x = 80 - 2p$ or $p = \tfrac{1}{2}(80 - x)$.

Comments on Fig. 4.41: (1) The nature of the problem dictates that x should lie between 0 and 80 (inclusive), since neither x nor p should be negative, although one could argue that a negative price would correspond to *paying* someone to take a radio. (2) If we had used the horizontal axis for p and the vertical axis for x, the resulting graph [of the function $x(p) = 80 - 2p$] would also be a segment of a line with negative slope.

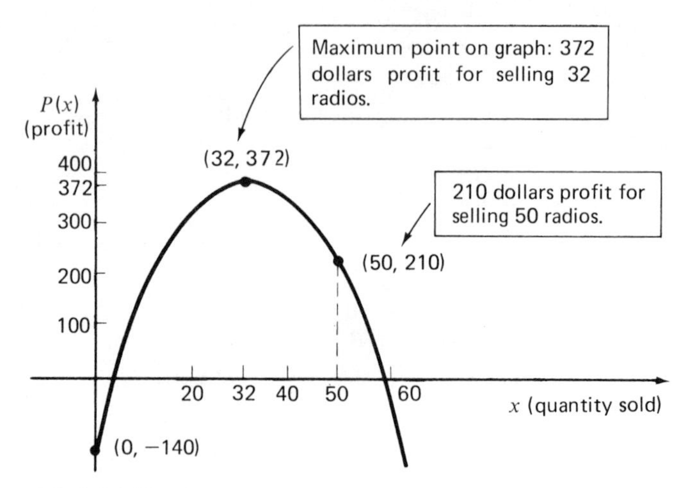

FIGURE 4.42 Graph of $P(x)$, the profit function.

Comments on Fig. 4.42: (1) Recall that the demand function was valid for all x in the interval [0, 80]. Hence we could compute $P(80)$ and get -780, which we would interpret as: "selling" 80 radios (for a unit price of 0 dollars) results in a *loss* of 780 dollars. (2) The tangent line to the graph has positive slope when $0 \leq x \leq 32$ and has negative slope when $32 \leq x \leq 80$. This slope at any value x is simply the derivative of $P(x)$ with respect to x; that is, it is the *rate of change* of profit with respect to quantity sold. In economics, this quantity

$$\frac{d}{dx}P(x) \quad \text{or} \quad P'(x)$$

is called the *marginal profit* and it is viewed as the increase in profit attributed to selling one more radio. For example, when $x = 30$, we have

$$P'(30) = 32 - 30 = 2$$

We must exercise some care in interpreting this value of $P'(30)$. It we compute the profit when $x = 29$, 30, and 31, we find

$$P(29) = 647.50$$
$$P(30) = 650.00$$
$$P(31) = 651.50$$

Thus we have a profit increase of \$2.50 for the thirtieth radio sold, and a profit increase of \$1.50 for the thirty-first radio sold. Note that neither of these increases equals the value $P'(30)$. The point is that we have ascribed a *continuous* model to a discrete situation. Whereas we have considered $P(x)$ for all values of x in the interval $[0, 80]$, in reality x can take on only integral values $0, 1, 2, 3, \ldots, 80$. The value $P'(30)$ represents an instantaneous rate of change of the (continuous) function $P(x)$ at the point 30. In practice, we can use such a value to *approximate* the *unit change* from $x = 29$ to $x = 30$ or from $x = 30$ to $x = 31$. Indeed, it is a mere coincidence that the average unit change in profit from $x = 29$ to $x = 31$ coincides with $P(30)$. Moreover, it was a coincidence that the maximal value of $P(x)$ occurred at the integral value $x = 32$. In practice we can expect "solutions" to problems of this nature to give only approximations to the actual solutions, as can be seen in some of the profit maximization problems in the exercises.

Summary

In this section we have incorporated the derivative f' of a function f to help determine points c where $f(x)$ is either a maximum or a minimum value. Our basic result is

> If $f(c)$ is either a maximum or a minimum value of f on the interval $[a, b]$, then one of two things must happen: $f'(c) = 0$ or $f'(c)$ does not exist.

Moreover, we found interpretations for intervals where $f'(x)$ is positive (the function f is increasing) and where $f'(x)$ is negative (the function f is decreasing). We distinguished *relative maxima* (*minima*) from *absolute maxima* (*minima*) by the fact that the latter are in relation to all values of f on $[a, b]$, whereas the former are simply in relation to points nearby. That is, relative extrema refer to peaks and valleys on the graph of f.

Exercises Section 8

1 For each of the following functions find the maximum and minimum values and where they occur in the prescribed intervals.
 (a) $f(x) = 3x + 2$, on $[1, 4]$
 (b) $f(x) = 9 - x^2$, on $[-2, 3]$
 (c) $f(x) = x \cdot \exp(x)$, on $[-2, 1]$
 (d) $f(x) = 1/(x^2 + 3)$, on $[-2, 2]$
 (e) $f(x) = |x| + 4$, on $[-1, 2]$ (*Hint:* When $x > 0$, $f(x) = x + 4$; when $x < 0$, $f(x) = -x + 4$.)
 (f) $f(x) = |x - 3|$, on $[0, 5]$

2 Find the largest rectangle that can be enclosed with 100 feet of rope.

3 Find the open-topped rectangular box of maximum volume that can be made from a square piece of cardboard, 8 inches per side, by cutting out squares from each corner and folding up the sides. (*Hint:* Draw a picture; let x be the side of the little squares to be cut out.)

4 Repeat Exercise 3 beginning with a rectangular piece of cardboard, 10 inches by 16 inches.

5 The cost for operating a truck at v miles per hour is

$$C(v) = 4 + \frac{v}{8} \text{ cents per mile}$$

The driver of the truck gets paid \$4.00 per hour. What is the most economical speed to suggest for a 400-mile trip?

6 Find the rectangular box of maximum volume satisfying both

(a) The bottom is square, and

(b) The sum of the height and the perimeter of the bottom is 60 inches.

7 Suppose the demand function and the cost function in Example 3 had been

$$x = 75 - 3p$$
$$C(x) = 5x + 200$$

(a) Draw the graph of p as a function of x.

(b) Find the formula for $P(x)$, the profit function, and sketch its graph.

(c) For what value of x will $P(x)$ be maximal?

(d) Use your answer to (c) to find the number of radios to sell and the corresponding unit price to maximize profits.

8 Repeat the preceding exercise with a new cost function,

$$C(x) = 15x - \tfrac{1}{6}x^2 + 200$$

9 **(a)** How would you interpret the term 200 in the cost function $C(x) = 10x + 200$?

(b) How would you interpret the term $10x$? (*Note:* The constant term is called the *fixed cost*; the remainder is called the *variable cost.*)

10 If $C(x)$ is any cost function, the value of $C'(x)$ is called the *marginal cost.*

(a) Find $C'(x)$ for the cost functions given in Exercises 7 and 8 and sketch their graphs.

(b) Compare $C'(x)$ to $[C(x + 1) - C(x)]/1$ in each of the above cases for $x = 10$, 20, and 30.

11 If the marginal cost (see Exercise 10) of selling a particular item is $45 - 6x$, can you find the cost function? What if you are told that it costs 125 dollars to sell one item?

12 (Second derivative test for determining relative maximum and minimum values) In practice, the search for extreme values of a function f on an interval $[a, b]$ can sometimes be aided by computing the second derivative f'' and evaluating this function at points c where $f'(c) = 0$. Just as the first derivative $f'(x)$ is positive when f is increasing and negative when f is decreasing, so the second derivative is positive when f' is increasing and negative when f' is decreasing. The following picture illustrates how the sign of $f''(x)$ can determine a relative maximum value for $f(x)$. We have $f'(c - h) > 0$, $f'(c) = 0$ and $f'(c + k) < 0$, where h and k are small positive numbers. Thus near c, $f'(x)$ is *decreasing*; the second derivative $f''(c)$ is *negative* and $f(c)$ is a maximum value.

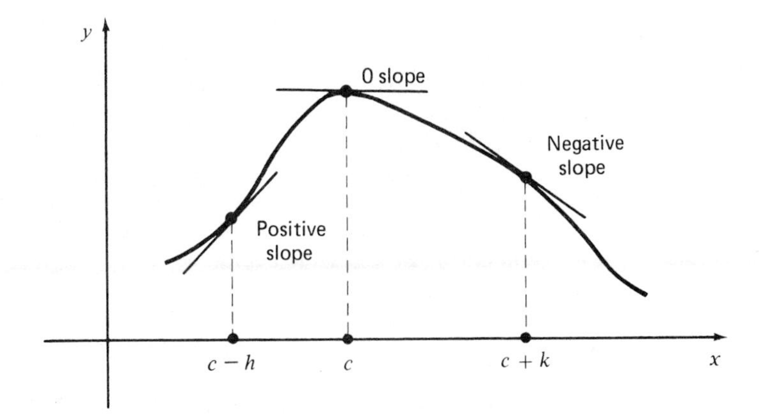

(a) Draw a corresponding picture to show how $f''(c) > 0$ and $f'(c) = 0$ would signal a relative minimum value at $f(c)$.

(b) Use the second derivative test to find the relative maximum and minimum values of the function f given by

$$f(x) = x^3 - 3x + 7, \qquad \text{on } [-4, 10]$$

(c) Explain how the second derivative test can fail to give useful information by sketching the graphs of f and g and searching for relative maxima and minima on $[-5, 5]$

$$f(x) = x^3, \qquad g(x) = x^4$$

13 In the equation

$$P(x) = R(x) - C(x)$$

where P, R, and C represent profit, revenue, and cost, respectively, we have seen that under suitable conditions $P(a)$ is a maximum when

$$P'(a) = 0$$

Note that in general we have

$$P'(x) = R'(x) - C'(x)$$

The quantities $P'(x)$ and $R'(x)$ are called *marginal profit* and *marginal revenue*. While it must be borne in mind that such a model is simplistic (for instance, it assumes no capital improvement is involved and that the duration of production is "short-term"), we can interpret the combination of the two previous equations,

$$0 = P'(a) = R'(a) - C'(a)$$

or

$$R'(a) = C'(a)$$

as saying that maximum profit occurs when the rate of increase of revenue (i.e., marginal revenue) is offset by the rate of increase of cost (i.e., marginal cost). In other words, further production (i.e., larger value of x) would result in less profit; in the layman's language, this is the so-called *Law of Diminishing Returns*.

(a) Using the demand and cost functions in Exercise 8, sketch the graphs of R' and C' on the same coordinate system to see where they meet.

(b) Using the demand and cost functions in Exercise 9, sketch the graphs of R' and C' on the same coordinate system to see where they meet.

(c) How do you interpret the condition that for $P(a)$ to be a maximum, we should have

$$P'(a) = 0 \quad \textit{and} \quad P''(a) \leq 0$$

In particular, how must R'' and C'' be related?

14 (Velocity and acceleration) If an object travels in a straight line a distance of $s(t)$ units (inches, miles, feet, etc.) in time t (hours, seconds, days, etc.) we define its *velocity* at time t by

$$s'(t) = \lim_{\Delta t \to 0} \frac{s(t + \Delta t) - s(t)}{\Delta t}$$

Note that the numerator of the difference quotient is simply the distance traveled from time t to time $t + \Delta t$, and the denominator is the time elapsed. In words, *velocity is the rate of change of distance with respect to time*. Similarly, *acceleration is the rate of change of velocity with respect to time*:

$$a(t) = v'(t) = s''(t)$$

(a) Suppose that a train travels from New York to Washington, D.C. (a distance of, say, 195 miles) in exactly 3 hours. Use the Mean Value Theorem to show that at least at one point the train was moving at exactly 65 miles per hour. (*Hint:* Let $s(t)$ be the distance from New York at t hours after departure.)

(b) A ball is thrown straight up into the air and its distance above the ground t seconds later is given by

$$s(t) = 96t - 16t^2 \text{ (feet)}$$

This formula makes sense only for $0 \leq t \leq 6$. (Why?) The velocity is positive as the ball is going up and negative (in the sense of being opposite the positive direction of upward) when the ball is coming down. When does the ball reach its highest point? What is the initial velocity?

(c) A particle moves along a horizontal line (with distance to the right called positive and to the left negative) according to the formula

$$s(t) = t^3 - 3t^2 - 9t + 6$$

When does the particle reverse direction? What is its maximal distance away from its initial position in the first 4 seconds? What is its maximal velocity in the first 4 seconds?

9 THE CHAIN RULE

So far we have seen how to compute the derivatives of such functions as

$$\exp(x),\ x^2,\ 3x^{15},\ \frac{x+1}{x^2+2},\ (x^3-6)(2x+5),\ \text{and}\ 5\cdot x^3\cdot \exp(x)$$

It is relatively easy, however, to find functions that do not yield to the differentiation formulas we have studied. For instance, consider

$$(x^2+3)^{10} \quad \text{and} \quad \exp(x^4+7x)$$

(This first function could actually be handled by expanding the binomial into a polynomial of degree 20.) These two functions have a common characteristic: Each of them can be viewed as "a function of a function of x" (more formally called a *composite function*). The purpose of this section is to introduce a very powerful differentiation formula called the Chain Rule, which tells us how to differentiate composite functions, and thereby greatly expands the class of functions which we can differentiate and integrate.

Before we state the Chain Rule, let us examine the two functions listed above. First of all to get from x to $(x^2+3)^{10}$, we can think of two steps:

1. Square x and add 3.
2. Raise the result of (1) to the tenth power.

In other words, we define two functions

$$u(x) = x^2 + 3$$
$$g(u) = u^{10}$$

We can think of u as the "inside function" and g as the "outside function."

and combine them to form the function

$$f(x) = g(u(x)) = (x^2+3)^{10}$$

Similarly, we could define

$$u(x) = x^4 + 7x$$
$$g(u) = \exp(u)$$

and combine to produce

$$f(x) = g(u(x)) = \exp(x^4+7x)$$

The Chain Rule

Suppose u is a function of x and g is a function of u. The composite function

$$f(x) = g(u(x))$$

has a derivative (with respect to x) given by

$$f'(x) = g'(u(x)) \cdot u'(x)$$

(provided both derivatives on the right-hand side of the equation exist). Note that the term $g'(u(x))$ means "the derivative of the function g with respect to u."

In words, the derivative of a function of a function of x equals the derivative of the outside function times the derivative of the inside function.

EXAMPLE 1 $f(x) = (x^2 + 3)^{10}$

Then

$$f'(x) = \underbrace{10(x^2 + 3)^9}_{} \cdot (2x)$$

The derivative of the outside function, $g(u) = u^{10}$

The derivative of the inside function, $u(x) = x^2 + 3$

Here are some similar computations

$$D_x((3x - 4)^6) = 6 \cdot (3x - 4)^5 \cdot (3)$$
$$D_x((x^2 + x + 2)^8) = 8 \cdot (x^2 + x + 2)^7 \cdot (2x + 1)$$
$$D_x((6 - 8x^2)^4) = 4(6 - 8x^2)^3 \cdot (-16x)$$

In general we have the formula

$D_x u^n = n \cdot u^{n-1} \cdot D_x(u)$	This is a generalized *Power Rule* where u is any function of x.

EXAMPLE 2 $f(x) = \exp(x^4 + x)$

The Chain Rule tells us that

$$f'(x) = \underbrace{\exp(x^4 + x)}_{} \cdot (4x^3 + 1)$$

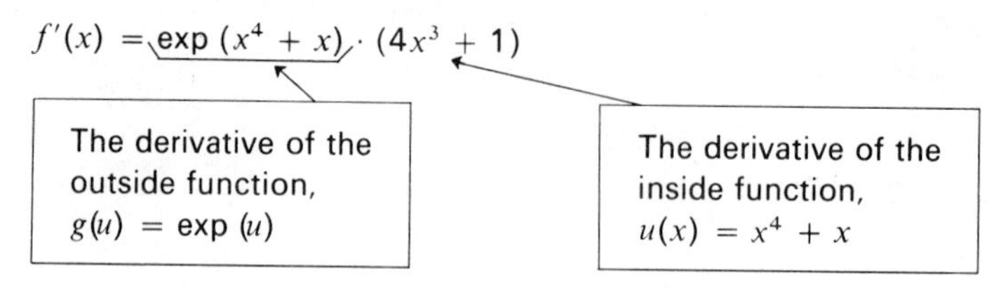

Incidentally, to avoid confusion we often rewrite this last expression in the form

$$(4x^3 + 1)\exp(x^4 + x)$$

Here are some other similar computations:

$$D_x(\exp(3x)) = \exp(3x) \cdot 3 = 3\exp(3x)$$

$$D_x(\exp(1 - 2x^2)) = \exp(1 - 2x^2) \cdot (-4x) = -4x \cdot \exp(1 - 2x^2)$$
$$D_x(\exp(\exp(x))) = \exp(\exp(x)) \cdot \exp(x) = \exp(x + \exp(x))$$

In general we have the formula

$$D_x(\exp(u)) = \exp(u) D_x u$$

The Chain Rule is also useful in evaluating definite integrals. From the Fundamental Theorem of Calculus we know that

$$\int_a^b f(x)\, dx = F(x)\Big|_a^b$$

where F is any function such that

$$F'(x) = f(x)$$

From the above examples, we can work backwards to obtain

$$\int_a^b 20(x)(x^2 + 3)^9\, dx = \int_a^b 10(x^2 + 3)^9(2x)\, dx = (x^2 + 3)^{10}\Big|_a^b$$

$$\int_a^b (4x^3 + 1) \exp(x^4 + x)\, dx = \exp(x^4 + x)\Big|_a^b$$

These are but two examples of the following general rules of integration.

$$\int_a^b u^n D_x(u)\, dx = \frac{u^{n+1}}{n+1}\Bigg|_a^b \qquad \text{where } u \text{ is any function of } x \text{ and } n \neq -1$$

$$\int_a^b \exp(u) \cdot D_x(u)\, dx = \exp(u)\Big|_a^b \quad \text{where } u \text{ is any function of } x$$

The first of these formulas could also have been stated as

$$\int_a^b nu^{n-1} D_x(u)\, dx = u^n\Big|_a^b \qquad \text{for } n \neq 0$$

In this form the illustrative example preceding the rule fits the pattern exactly. The point is that in practice we often must resort to trickery to change the given problem to one for which the rule applies. The following examples should illustrate the technique.

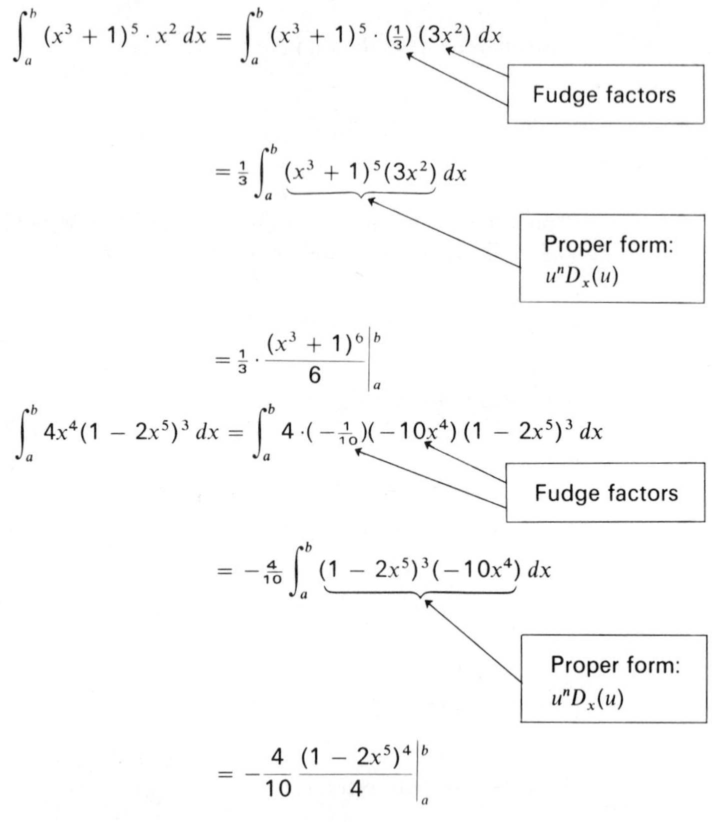

The crucial step in each case was to discover the appropriate fudge factors. But these constants were determined for us. In the first problem, we knew that

$$D_x(x^3 + 1) = 3x^2$$

and therefore we wanted to multiply x^2 by 3. At the same time we multiplied the integrand by $(\frac{1}{3})$ so as not to change its value. In the next step, we invoked the linearity of the definite integral to get

$$\int_a^b \frac{1}{3}(\qquad) = \frac{1}{3}\int_a^b (\qquad)$$

Likewise in the second problem, we observed that

$$D_x(1 - 2x^5) = -10x^4$$

and therefore we wanted to multiply x^4 by (-10). This forced us to multiply the integrand by $(-\frac{1}{10})$ as well, and we then took the product $4 \cdot (-\frac{1}{10})$ out in front of the integral sign.

In the next two problems, we perform the same kind of manipulation with functions involving exp.

$$\int_a^b x^2 \exp(x^3+2)\,dx = \int_a^b (\tfrac{1}{3})(3x^2)\exp(x^3+2)\,dx$$

$$= \tfrac{1}{3}\int_a^b \exp(x^3+2)(3x^2)\,dx$$

$$= \tfrac{1}{3}\exp(x^3+2)\Big|_a^b$$

$$\int_a^b \exp(7x+16)\,dx = \int_a^b (\tfrac{1}{7})(7)\exp(7x+16)\,dx$$

$$= \tfrac{1}{7}\int_a^b \exp(7x+16)(7)\,dx$$

$$= \tfrac{1}{7}\exp(7x+16)\Big|_a^b$$

Lest the reader get the impression that it is always possible to fix up the integrand as we have been doing, we include two problems which do not yield to this technique. Suppose you were asked to find

$$\int_a^b \exp(x^2)\,dx$$

If we try to do this as we have done the preceding problems, we have

$$\int_a^b \exp(x^2)\,dx = \int_a^b \exp(x^2)\cdot\frac{2x}{2x}\,dx$$

$$= \int_a^b \frac{1}{2x}\exp(x^2)\cdot 2x\,dx$$

The $1/2x$ *cannot* be brought through the integral sign. (The $\tfrac{1}{2}$ could, but not the x.) We cannot do this problem with the methods we have discussed, nor, indeed, with any other method.

Now that we have seen how the Chain Rule works, let us see *why* it works. We must show that

$$D_x(g(u(u))) = g'(u)\cdot u'(x)$$

By definition of the derivative,

$$D_x(g(u(x))) = \lim_{\Delta x\to 0}\frac{g(u(x+\Delta x)) - g(u(x))}{\Delta x}$$

The idea is to apply the Mean Value Theorem to the numerator. We have to worry about two cases, $u(x+\Delta x) \geq u(x)$ and $u(x+\Delta x) < u(x)$. If $u(x+\Delta x) \geq u(x)$, the Mean Value Theorem says that there is a v satisfying $u(x) < v < u(x+\Delta x)$ such that

$$g(u(x+\Delta x)) - g(u(x)) = g'(v)[u(x+\Delta x) - u(x)]$$

If $u(x+\Delta x) < u(x)$, the Mean Value Theorem says that there exists a v,

$u(x + \Delta x) < v < u(x)$ such that

$$g(u(x)) - g(u(x + \Delta x)) = g'(v)[u(x) - u(x + \Delta x)]$$

and hence, multiplying through by -1,

$$g(u(x + \Delta x)) - g(u(x)) = g'(v)[u(x + \Delta x) - u(x)]$$

In either case there is a number v between $u(x)$ and $u(x + \Delta x)$ such that

$$\frac{g(u(x + \Delta x)) - g(u(x))}{\Delta x} = \frac{g'(v)\,[u(x + \Delta x) - u(x)]}{\Delta x}$$

As Δx approaches 0, v approaches $u = u(x)$ and (provided g' is continuous at u)

$$\begin{aligned} D_x(g(u(x))) &= \lim_{\Delta x \to 0}\left[g'(v) \cdot \frac{u(x + \Delta x) - u(x)}{\Delta x}\right] \\ &= \lim_{\Delta x \to 0} g'(v) \cdot \lim_{\Delta x \to 0} \frac{u(x + \Delta x) - u(x)}{\Delta x} \\ &= g'(u) \cdot u'(x) \end{aligned}$$

The Chain Rule also allows us to extend our Power Rules

$$D_x(x^n) = nx^{n-1}$$

and

$$D_x(u^n) = nu^{n-1} \cdot u'(x)$$

to the case where n is a *rational number*. (Up till now, we had assumed that n was an integer.)

Theorem For any rational number n,

$$D_x(x^n) = nx^{n-1}$$

Proof: We write $n = p/q$, where p and q are integers, and we put

$$y = x^n = x^{p/q}$$

Our job is to find $D_x(y)$. We raise both sides of the above equation to the qth power to obtain

$$y^q = x^p$$

and now we differentiate both sides with respect to x, using the Chain Rule on the left-hand side:

$$q \cdot y^{q-1} D_x(y) = px^{p-1}$$

$$D_x(y) = \frac{p}{q}\frac{x^{p-1}}{y^{q-1}}$$

Now we substitute $x^{p/q}$ for y to get

$$D_x(y) = \frac{p}{q}\frac{x^{p-1}}{(x^{p/q})^{q-1}}$$
$$= \frac{p}{q}\frac{x^{p-1}}{x^{p-p/q}}$$
$$= \frac{p}{q}x^{(p/q)-1}$$

which is what we wanted to show since $n = p/q$.

Note: Although our proof showed only that n could be any rational number, it can be shown, by a more sophisticated argument, that

$$D_x(x^n) = nx^{n-1}, \qquad \textit{for any real number } n$$

We conclude this section with some more examples illustrating the Chain Rule in action.

$$D_x((x^2 + 2x + 1)^{-3}) = -3(x^2 + 2x + 1)^{-4}(2x + 2)$$
$$D_x((x^4 + 7)^{5/2}) = \tfrac{5}{2}(x^4 + 7)^{3/2} \cdot (4x^3)$$
$$D_x(\exp^3(1 - x^2)) = 3 \cdot \exp^2(1 - x^2) \cdot D_x(\exp(1 - x^2))$$
$$= 3 \cdot \exp^2(1 - x^2) \cdot \exp(1 - x^2) \cdot (-2x)$$
$$= -6x \cdot \exp^2(1 - x^2) \cdot \exp(1 - x^2)$$
$$= -6x \cdot \exp^3(1 - x^2)$$

[*Note:* $\exp^3(u) = [\exp(u)]^3 = \exp(u) \cdot \exp(u) \cdot \exp(u) = \exp(3u)$.]

$$D_x\left(\left(\frac{x+2}{x-1}\right)^4\right) = 4\left(\frac{x+2}{x-1}\right)^3 D_x\left(\frac{x+2}{x-1}\right)$$
$$= 4\left(\frac{x+2}{x-1}\right)^3 \cdot \left(\frac{(x-1)-(x+2)}{(x-1)^2}\right)$$
$$= 4 \cdot \frac{(x+2)^3(-3)}{(x-1)^5}$$
$$= -12\frac{(x+2)^3}{(x-1)^5}$$

$$\int_1^2 x^2(x^3 - 2)^{2/3}\,dx = \int_1^2 (\tfrac{1}{3})(3x^2)(x^3 - 2)^{2/3}\,dx$$
$$= \tfrac{1}{3} \cdot (x^3 - 2)^{5/3}\big|_1^2$$
$$= \tfrac{1}{3}(8 - 2)^{5/3} - \tfrac{1}{3}(1 - 2)^{5/3}$$
$$= \tfrac{1}{3}(6^{5/3} + 1)$$

Summary

The Chain Rule says that if y is a function of $u(y = g(u))$ and u is a function of x, then

$$D_x y = D_u y \cdot D_x u$$

On a practical level we have four consequences of the Chain Rule.

$$D_x u^n = nu^{n-1} D_x u$$

$$D_x \exp(u) = \exp(u) \cdot D_x u$$

$$\int_a^b u^n D_x u \, dx = \left. \frac{u^{n+1}}{n+1} \right|_a^b \qquad (\text{if } n \neq -1)$$

$$\int_a^b \exp(u) \cdot D_x u \, dx = \exp(u)|_a^b$$

Exercises Section 9

1 Find the derivative of each of the following functions.

(a) $f(x) = (x^3 + x)^5$

(b) $f(x) = (x^2 + 1)^{12/7}$

(c) $f(x) = (x^{3/2} + 1)^4$

(d) $f(x) = \exp(x^{4/3})$

(e) $f(x) = \exp((x^2 + x)^7)$

(f) $f(x) = x \cdot \exp(x^2)$

(g) $h(x) = \exp\left(\frac{-x^2}{2}\right)$

(h) $F(x) = (x^2 - 2)^{-3}$

(i) $G(x) = \exp^4(x^2 + 2x)$

(j) $H(x) = \left(\frac{x-3}{x^2}\right)^{3/4}$

2 Evaluate each of the following.

(a) $\int_{-1}^{0} (x^3 + 1)^{10} \cdot x^2 \, dx$

(b) $\int_0^1 (x + 1)^{5/2} \, dx$

(c) $\int_0^1 x^2 \cdot \exp(x^3) \, dx$

(d) $\int_0^1 (3x + 2)^4 \, dx$

(e) $\int_{-1}^{1} 3x(x^2 + 2)^2 \, dx$

(f) $\int_1^2 \frac{\exp((x-1)/x)}{x^2} \, dx$

(g) $\int_1^4 \frac{\exp(\sqrt{x})}{\sqrt{x}} \, dx$

(h) $\int_1^2 \frac{x+1}{x^3} \, dx$

(i) $\int_0^1 (x^2 + 1)^3 \, dx$

(j) $\int_1^2 \frac{\exp(1/x)}{x^2} \, dx$

(k) $\int_0^1 5(2x - 1)^4 \, dx$

(l) $\int_1^4 \frac{\sqrt{1 + \sqrt{x}}}{\sqrt{x}} \, dx$

(m) $\int_{-6}^{-5} 6(x + 6)^2 \, dx$

(n) $\int_0^2 2x^2(x^3 + 1)^{1/2} \, dx$

3 Find $f'(x)$ and $f''(x)$ for each of the following.

(a) $f(x) = x^n \exp(x)$

(b) $f(x) = \dfrac{a}{a - x}$

(c) $f(x) = \exp(\exp(x))$

4 Find $f'(x), f''(x), f'''(x)$ for

$$f(x) = 1 + (3x) + \frac{(3x)^2}{2!} + \frac{(3x)^3}{3!} + \frac{(3x)^4}{4!} + \frac{(3x)^5}{5!} + \frac{(3x)^6}{6!}$$

5 Let $s(x)$ and $c(x)$ be functions such that $D_x s(x) = c(x)$ and $D_x c(x) = -s(x)$. In (a) through (f), find $D_x y$.

(a) $y = s(x^2 + 1)$

(b) $y = (s(x))^2$

(c) $y = (s(x^2))^2$

(d) $y = \exp(s(x))$

(e) $y = \exp(c(x))$

(f) $y = \exp^5(s(x^2 + x))$

(g) Evaluate $\displaystyle\int_0^1 s(x)c(x)\,dx$.

(h) Evaluate $\displaystyle\int_0^1 (c(x))^5 s(x)\,dx$.

10 THE LOGARITHM FUNCTION

One of the interesting consequences of the Fundamental Theorem is that it allows us to define functions in terms of definite integrals. In other words, if we have a function f, then we can define a new function F by

$$F(x) = \int_a^x f(t)\,dt$$

(subject, of course, to appropriate restrictions on the value of x) and we know that

$$F'(x) = f(x)$$

We take advantage of this device to treat the exceptional case to our power rule. We know that

$$\int_a^b t^2\,dt = \left.\frac{t^3}{3}\right|_a^b$$

$$\int_a^b t^{-4}\,dt = \left.\frac{t^{-3}}{-3}\right|_a^b$$

and in general

$$\int_a^b t^n\,dt = \left.\frac{t^{n+1}}{n+1}\right|_a^b$$

provided $n \neq -1$. Note that the power rule fails to provide an antiderivative for t^{-1}. Indeed, no function we have studied has t^{-1} as a derivative. To fill this gap, we define a function, called Log, by

$$\text{Log}(x) = \int_1^x t^{-1}\,dt = \int_1^x \frac{1}{t}\,dt$$

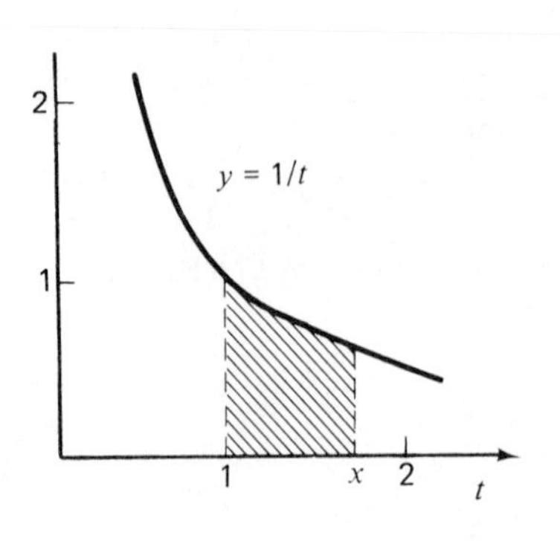

$x > 1$ and

$\int_1^x \frac{1}{t}\,dt$ = area of shaded region

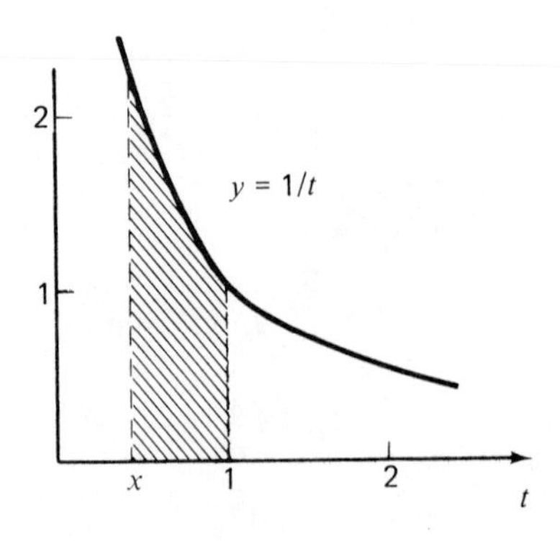

$0 < x < 1$ and

$\int_1^x \frac{1}{t}\,dt = -\int_x^1 \frac{1}{t}\,dt = -$(area of shaded region).

FIGURE 4.43

We may also observe from Fig. 4.43 that Log (1) = 0.

The use of the name Log for this function is to suggest the word logarithm. Some of you have studied logarithms in high school. While they were, in all likelihood, slightly different from the Log function, there are a number of similarities. We will explore these similar properties after an important fact is stated in the following.

Theorem

$$D_x(\text{Log}\, x) = \frac{1}{x}$$

Since

$$\text{Log}\,(x) = \int_1^x \frac{1}{t}\,dt$$

the theorem follows immediately from the Fundamental Theorem of Calculus. Application of the Chain Rule yields:

Corollary If u is a differentiable function of x, then

$$D_x(\text{Log}\, u) = \frac{1}{u} D_x u$$

As examples of typical derivative computations using this result, we offer the following:

$$D_x(\text{Log}\,(6x + 2)) = \frac{1}{6x + 2} \cdot 6 = \frac{6}{6x + 2} = \frac{3}{3x + 1}$$

$$D_x(\text{Log }(x^2 - 3x + 1)) = \frac{1}{x^2 - 3x + 1}(2x - 3) = \frac{2x - 3}{x^2 - 3x + 1}$$

$$D_x(\text{Log }(\exp(x))) = \frac{1}{\exp(x)} \cdot \exp(x) = 1$$

From the formula

$$D_x \text{ Log } u = \frac{1}{u} \cdot D_x u$$

we can deduce that

$$\int_a^b \frac{1}{u} \cdot D_x u \, dx = \text{Log }(u)\Big|_a^b$$

As examples of the use of this formula, we offer

$$\int_a^b \frac{6x + 1}{3x^2 + x + 5} dx = \text{Log }(3x^2 + x + 5)\Big|_a^b$$

$$\int_a^b \frac{5x^4 + x}{x^5 + \frac{1}{2}x^2 + 3} dx = \text{Log }(x^5 + \tfrac{1}{2}x^2 + 3)\Big|_a^b$$

$$\int_a^b \frac{x}{x^2 + 5} dx = \int_a^b \frac{\frac{1}{2} \cdot 2x}{x^2 + 5} dx = \frac{1}{2}\int_a^b \frac{2x}{x^2 + 5} dx = \tfrac{1}{2} \text{ Log }(x^2 + 5)\Big|_a^b$$

Note the use of a fudge factor in this last example.

Using the differentiation formulae, we can deduce several algebraic properties of logarithms.

Theorem For any positive numbers a and x,

$$\boxed{\text{Log }(ax) = \text{Log }(a) + \text{Log }(x)}$$

Proof: Thinking of a as a constant and x as a variable, we define

$$f(x) = \text{Log }(ax) - \text{Log }(a) - \text{Log }(x)$$

Then

$$\begin{aligned} f'(x) &= \frac{1}{ax} \cdot a - 0 - \frac{1}{x} \\ &= \frac{1}{x} - \frac{1}{x} \\ &= 0 \end{aligned}$$

Therefore $f(x)$ is a constant, i.e., there is a number k such that

$$k = f(x)$$

for all x. We determine the value of k by letting $x = 1$,

$$\begin{aligned} k &= f(1) \\ &= \text{Log}\,(a \cdot 1) - \text{Log}\,(a) - \text{Log}\,(1) \\ &= \text{Log}\,(a) - \text{Log}\,(a) - 0 \\ &= 0 \end{aligned}$$

Therefore

$$\text{Log}\,(ax) - \text{Log}\,(a) - \text{Log}\,(x) = 0$$

and

$$\text{Log}\,(ax) = \text{Log}\,(a) + \text{Log}\,(0)$$

A similar approach yields the following.

Theorem For positive x,

$$\boxed{\text{Log}\,(x^a) = a\,\text{Log}\,(x)}$$

Proof: We set

$$f(x) = \text{Log}\,(x^a) - a\,\text{Log}\,(x)$$

Then

$$\begin{aligned} f'(x) &= \frac{1}{x^a} \cdot ax^{a-1} - a \cdot \frac{1}{x} \\ &= \frac{a}{x} - \frac{a}{x} \\ &= 0 \end{aligned}$$

so $f(x)$ is a constant. Since

$$\begin{aligned} f(1) &= \text{Log}\,(1^a) - a\,\text{Log}\,(1) \\ &= \text{Log}\,(1) - a \cdot 0 \\ &= 0 - 0 \\ &= 0 \end{aligned}$$

we have

$$\text{Log}\,(x^a) - a\,\text{Log}\,(x) = 0$$

and

$$\text{Log}\,(x^a) = a\,\text{Log}\,x$$

We combine these to obtain the following.

Theorem

$$\text{Log}(u) - \text{Log}(v) = \text{Log}\left(\frac{u}{v}\right)$$

Proof:

$$\begin{aligned}
\text{Log}(u) - \text{Log}(v) &= \text{Log}(u) + (-1)\,\text{Log}(v) \\
&= \text{Log}(u) + \text{Log}(v^{-1}) \\
&= \text{Log}(u) + \text{Log}(1/v) \\
&= \text{Log}\left(u \cdot \frac{1}{v}\right) \\
&= \text{Log}\left(\frac{u}{v}\right)
\end{aligned}$$

Typical calculations using these three algebraic facts about logarithms are

$$\begin{aligned}
\text{Log}((x^2+1)\cdot\sqrt{x}) &= \text{Log}(x^2+1) + \text{Log}\sqrt{x} \\
&= \text{Log}(x^2+1) + \tfrac{1}{2}\,\text{Log}(x)
\end{aligned}$$

and

$$\begin{aligned}
\text{Log}\,4 - \text{Log}\,2 &= \text{Log}\left(\tfrac{4}{2}\right) \\
&= \text{Log}\,2
\end{aligned}$$

In addition to these algebraic facts about the Log function, there is a connection with the function exp. Let

$$f(x) = \text{Log}(\exp(x)) - x$$

Then

$$\begin{aligned}
f'(x) &= \frac{1}{\exp(x)} \cdot \exp(x) - 1 \\
&= 1 - 1 \\
&= 0
\end{aligned}$$

So $f(x)$ is a constant since

$$\begin{aligned}
f(0) &= \text{Log}(\exp(0)) - 0 \\
&= \text{Log}(1) - 0 \\
&= 0 - 0 \\
&= 0
\end{aligned}$$

We conclude that

$$\text{Log}(\exp(x)) - x = 0$$

or

$$\text{Log}(\exp(x)) = x$$

Moreover, since

$$\text{Log}(u) = \text{Log}(v)$$

implies

$$u = v \qquad \text{(See Exercise 6)}$$

we have

$$\text{Log}(\exp(\text{Log}(x))) = \text{Log}\, x$$

implying

$$\boxed{\exp(\text{Log}(x)) = x}$$

When we first encountered the function exp, it was mentioned that we sometimes write

$$e^x$$

to mean

$$\exp(x)$$

where the symbol e represents a number

$$e = \lim_{n\to\infty}\left(1 + \frac{1}{n}\right)^n = 2.71828\ldots$$

Using this exponential notation [i.e., writing e^x for $\exp(x)$], we restate the properties discussed in this section and in earlier sections:

$$e^{x+y} = e^x \cdot e^y \qquad e^0 = 1$$

$$(e^x)^y = e^{xy} \qquad \text{Log}(e^x) = x$$

$$e^{-x} = \frac{1}{e^x} \qquad e^{\text{Log}(x)} = x$$

In high school algebra, it is common to talk about "exponents" and "logarithms." We take advantage of our functions Log and exp to introduce the formal definitions and some of the properties relating to these notions.

Definition Let a be any positive number. We define

$$(1) \quad a^x = e^{x\,\text{Log}(a)}$$

and read this symbol a^x as "a raised to the x power." We also introduce the symbol

$$\log_b$$

read "log to the base b" by

$$(2) \quad \log_b x = a \quad \text{if} \quad b^a = x$$

and we note that the function Log is a special case:

(3) $\text{Log}(x) = \log_e x$

since if we put

$$y = \text{Log}(x)$$

then

$$e^y = e^{\text{Log}(x)} = x$$

In addition to the special case $\log_e$, called the *natural logarithm*, students frequently are exposed to $\log_{10}$, called the *common logarithm*. While we do not care to digress into typical usages of logarithms, it might be worth mentioning that the following computational problem

$$\frac{(6327.61) \cdot (438.5)}{(1612) \cdot (.0387)} = ?$$

becomes

$$\log_{10}(6327.61) + \text{Log}_{10}(438.5) - \log_{10}(1612) - \log_{10}(.0387) = ?$$

which is easy to solve using a table of logarithms.

We have pointed out that if

$$y = f(x)$$

then

$$D_x y = f'(x)$$

represents the rate of change of y with respect to x. There are a number of situations where the quantity of something (e.g., population or amount of radioactive material) is a function of time, say $f(t)$. In this case, $f'(t)$ represents the rate of change of the quantity in time. Such situations often lead to applications of the function Log. Three problems of this type, falling into a category called *growth and decay models* will now be considered.

PROBLEM 1 Bacterial Growth

The rate of growth of a population of bacteria in a culture is directly proportional to the size of the population. If an experiment is begun with a count of 100 bacteria in a culture and if 1 hour later the count is 150, then what is the count at the end of 24 hours?

PROBLEM 2 Population Explosion

Over a relatively short time span (say a few generations) the rate of growth of a population in a country is directly proportional to the size of the population. The population of Japan was 49 million in 1910 and 56 million in 1920. What is the predicted size for 2000?

PROBLEM 3 Radioactive Decay

Radioactive material decays at a rate proportional to the amount present. The *half-life* of a radioactive substance is the time it takes for a given quantity to decay to half its size. How long would it take 1 gram of Polonium, which has a half-life of 140 days, to reduce itself to $\frac{1}{5}$ gram?

Let us reformulate each of these "word problems" in mathematical notation. In Problem 1 we put

$N = N(t) =$ the number of bacteria present at time t

Then

$N'(t) =$ the rate of change (that is, increase) of the number of bacteria at time t

And the problem becomes:

There is some function $N(t)$ such that

$N'(t) = k \cdot N(t)$

$N(0) = 100$

$N(1) = 150$

Find

$N(24)$

Similarly, Problem 2 can be restated as follows:

There is a function $J(t)$ such that

$J'(t) = k \cdot J(t)$

$J(0) = 49$

$J(10) = 56$

Find

$J(90)$

We measure time t in years beginning with $t = 0$ in the year 1910.

The population 90 years after 1910

In like manner we restate Problem 3:

There is a function $P(t)$ such that

$P'(t) = k \cdot P(t)$

$P(0) = 1$

$P(140) = .5$

Find t_0 such that $P(t_0) = .2$

Here t is measured in days. Also, k will be *negative* because $P(t)$ is decreasing as t increases. (Recall that a function decreases if its derivative is negative.)

So in each of these problems, we have a function whose derivative is a multiple of the original function.

$$f'(t) = k \cdot f(t)$$

If we divide both sides of this equation by $f(t)$, we get

$$\frac{f'(t)}{f(t)} = k$$

The left-hand side of this equation is recognizable as the derivative of Log $(f(t))$; the right-hand side is the derivative of kt. Thus these two functions, having equal derivatives, must differ by a constant

$$\text{Log}\,(f(t)) = kt + C'$$

Now if we apply the function exp to both sides, we get

$$\exp\,(\text{Log}\,(f(t))) = \exp\,(kt + C')$$

$$f(t) = e^{kt + C'} = e^{kt} \cdot e^{C'}$$

Usually the constant $e^{C'}$ is written as C, yielding

$$f(t) = C \cdot e^{kt}$$

Returning to our three problems, we have

$$N(t) = C \cdot e^{kt}$$

$$J(t) = C \cdot e^{kt}$$

$$P(t) = C \cdot e^{kt}$$

Of course, the constants (or *parameters*) C and k will be different in each case.

What remains then is to determine the particular values of the parameters C and k for each of these three problems. Before we do, let us make some observations about this *family of curves,*

$$y = Ce^{kt}$$

In Fig. 4.44 we see some members of the family for cases where k is positive. Each is an *increasing* function. In Fig. 4.45 we see some graphs where k is negative. In all cases C is positive since the growth and decay problems we are interested in deal with positive quantities (size, population, and so on).

Upon examination of these figures, the roles of the parameters become apparent:

When $t = 0$, we get $f(0) = Ce^{k \cdot 0} = C \cdot 1 = C$

In other words, C represents the value of the function (i.e., size, population, and so on) at the initial time.

The value of k affects the graph in two ways. First, when $k > 0$, the graph increases and when $k < 0$, the graph decreases. Second, values of k near 0 cause the graph to be relatively flat; whereas values of k that are large in absolute

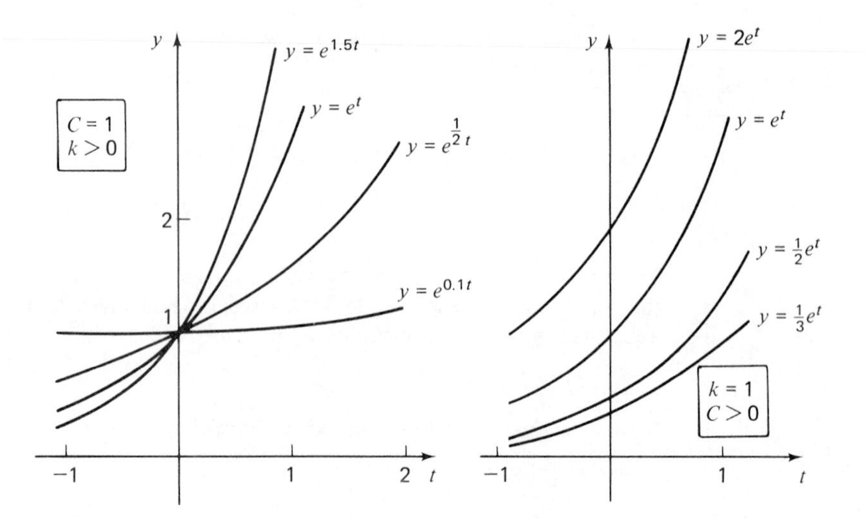

FIGURE 4.44

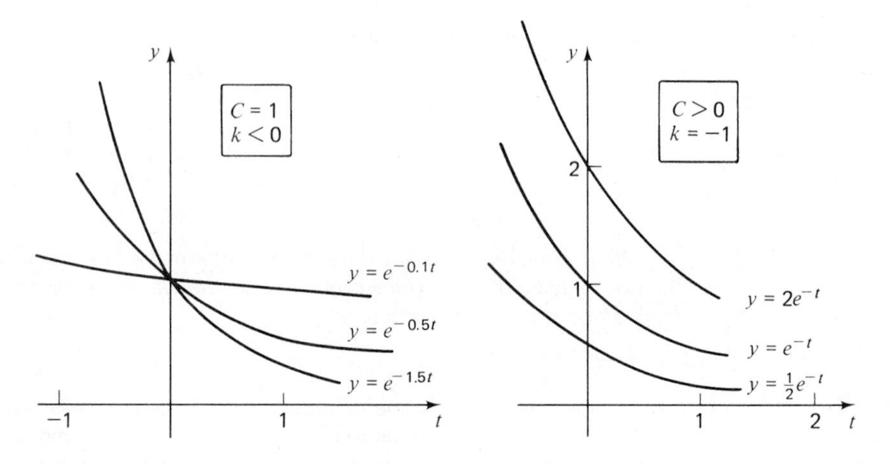

FIGURE 4.45

value cause the graph to change more dramatically, either increase sharply when $k > 0$, or decrease toward 0 rapidly when $k < 0$.

Now let us provide complete solutions to the three problems we posed earlier.

SOLUTION TO PROBLEM 1 We know that N satisfies

$$N(t) = Ce^{kt}$$

Moreover when $t = 0$, $N = 100$. Therefore

$$N(t) = 100e^{kt}$$

Now to compute k, we substitute the value $N(1) = 150$:

$$150 = 100e^{k\cdot(1)}$$
$$1.5 = e^{k}$$

and taking Log of both sides,

$$\text{Log }(1.5) = \text{Log }(e^{k}) = k$$

Checking a Log table we find that

$$\text{Log }(1.5) \approx .4055$$

Thus

$$N(t) = 100e^{(.4055)t}$$

And finally, we compute the value of N when $t = 24$ (hours).

$$N(24) = 100e^{(.4055)(24)}$$
$$= 100e^{9.722}$$

Checking a table for e^{x}, we find

$$e^{9.722} \approx 17514$$

Thus the number of bacteria present at the end of one day would be

$$N(24) \approx 1{,}751{,}400 \qquad \text{(nearly 2 million)}$$

SOLUTION TO PROBLEM 2 We know that

$$J(t) = Ce^{kt}$$

Also since when $t = 0$ (the year 1910), $J = 49$ (million),

$$J(t) = 49e^{kt}$$

and when $t = 10$ (the year 1920), $J = 56$,

$$56 = J(10) = 49e^{k\cdot(10)}$$

or

$$\tfrac{56}{49} = e^{10k}$$

Taking Log of both sides,

$$\text{Log }(\tfrac{56}{49}) = 10k$$
$$k = \tfrac{1}{10}\,(\text{Log }(56) - \text{Log }(49)) \approx .01335$$

Thus

$$J(t) \approx 49e^{(.01335)t}$$

When $t = 90$ (the year 2000) this model would predict

$$\begin{aligned} N(90) &= 49e^{(.01335)90} \\ &= 49e^{1.2015} \\ &= 49(3.32) \\ &= 162.68 \text{ (million people)} \end{aligned}$$

SOLUTION TO PROBLEM 3 We know that

$$P(t) = Ce^{kt}$$

and since $P(0) = 1$ and $P(140) = .5$, we get

$$P(t) = 1e^{kt}$$

$$.5 = 1e^{k(140)}$$

$$\text{Log } (.5) = 140k$$

$$k = (1/140) \text{ Log } (.5) \approx -.005$$

Thus

$$P(t) = e^{(-.005)t}$$

and we desire t_0 such that $P(t_0) = .2$:

$$.2 = e^{(-.005)t_0}$$

$$\text{Log } (.2) = (-.005)t_0$$

$$t_0 = \frac{\text{Log } (.2)}{-.005} \approx \frac{-1.61}{-.005} = 322$$

In other words we would expect it to take approximately 322 days for 1 gram of Polonium to decay to .2 grams. Observe that since the half-life is 140 days, it follows that there would be .25 grams after 280 days, so that our answer seems reasonable.

Exercises Section 10

1 Find the derivatives of each of the following.

(a) $f(t) = \text{Log } (t^3 + t^2 + 1)$

(b) $g(t) = t \text{ Log } (t^2 + 1)$

(c) $h(t) = \dfrac{\text{Log } (t + 3)}{t^2}$

(d) $F(x) = \text{Log } \dfrac{1}{x}$

(e) $G(x) = e^{x + \text{Log}(x)}$

(f) $H(x) = \text{Log } (\text{Log } (x))$

2 Evaluate each of the following.

(a) $\int_0^1 \frac{2x}{x^2 + 1}\,dx$

(b) $\int_0^1 \frac{x}{x^2 + 1}\,dx$

(c) $\int_0^2 \frac{x^2 + 1}{x + 1}\,dx$ (*Hint:* Try long division on the integrand.)

(d) $\int_{1.5}^2 \frac{1}{x \operatorname{Log}(x)}\,dx$

(e) $\int_1^2 \operatorname{Log}(\exp(x^2 + x + 1))\,dx$

(f) $\int_a^b \frac{5x^2}{x^3 + 17}\,dx$

3 Radioactive decay causes one type of carbon (call it type A) to change into another (call it type B). At the time a living organism dies all its carbon is type A and 1,000 years later its carbon is one-half type A and one-half type B. The remains of an animal are found and only 5 per cent of its carbon is type A. When did the animal die?

4 Given the information in Exercise 3 and given that your measurement of the percentage of type A carbon is accurate to .1 per cent, and if you want your result to be accurate to within 100 years, how old would the remains have to be before this technique would fail?

5 We have a formula for differentiating u^n, where u is a variable and n is a constant. Derive a formula for $D_x a^x$ where a is a constant.
(*Hint:* Set $y = a^x$, take logs of both sides, differentiate, and solve for $D_x y$.)

6 Show that if $\operatorname{Log}(u) = \operatorname{Log}(v)$, then $u = v$.
[*Hint:* Use the fact that $\operatorname{Log}(u/v) = \operatorname{Log}(u) - \operatorname{Log}(v)$.]

7 Why should a be positive in order to define a^x?

8 Find the derivatives of each of the following.

(a) $f(t) = 2^t$

(b) $f(x) = 2^{x^2}$

(c) $g(x) = x^x$

(d) $h(t) = \operatorname{Log}_2(t)$

(e) $F(x) = (\operatorname{Log}(x))^x$

(f) $G(x) = x^{\operatorname{Log}(x)}$

9 Solve each of the following equations for x.

(a) $\exp(xy) = \operatorname{Log}(z)$ *Answer:* $x = \frac{\operatorname{Log}(\operatorname{Log}(z))}{y}$

(b) $\exp(x + y) = \exp(a) - \operatorname{Log}(b)$

(c) $\operatorname{Log}(xyz) = \exp(t)$

11 THE POISSON DISTRIBUTION AGAIN

In Chapter 3 we pointed out that the Poisson distribution can be applied to a variety of problems other than approximating the binomial distribution. We are now in a position to justify this assertion.

Consider the following problem. A business office receives a number of calls during each day. They arrive at randomly occurring times during the day. We want a frequency function which will give us the probability that x calls are received in a time interval of t minutes. The function will be denoted

$$\mathbf{P}(x; t, \lambda)$$

where λ is a parameter to be discussed shortly.

We shall make some assumptions concerning the function $\mathbf{P}$, write them down as axioms, and, from these axioms, derive the function.

Our first assumption is that $\mathbf{P}$ is a frequency function.

Axiom I

$$\sum_{x=0}^{\infty} \mathbf{P}(x; t, \lambda) = 1$$

Our second assumption is that the number of calls received in any two non-overlapping intervals are independent.

We shall have to look closely at what this says. First, interpret $\mathbf{P}(x; t_1 + t_2, \lambda)$ as the probability of receiving x calls in the time interval from $t = 0$ to $t = t_1 + t_2$. Then break this up into two intervals $[0, t_1]$ and $[t_1, t_1 + t_2]$. The probability of getting x calls during $[0, t_1]$ is

$$\mathbf{P}(x; t_1, \lambda)$$

The probability of getting x calls over $[t_1, t_1 + t_2]$ is the same as that of getting x calls over $[0, t_2]$,

$$\mathbf{P}(x; t_2, \lambda).$$

The event of getting 0 calls over $[0, t_1 + t_2]$ is the intersection of the events of getting 0 calls over $[0, t_1]$ and 0 calls over $[t_1, t_1 + t_2]$. By our independence assumption,

$$\mathbf{P}(0; t_1 + t_2, \lambda) = \mathbf{P}(0; t_1, \lambda) \cdot \mathbf{P}(0; t_2, \lambda)$$

What about getting one call during $[0, t_1 + t_2]$? You can do this by getting one call during $[0, t_1]$ and none during $[t_1, t_1 + t_2]$ or none during $[0, t_1]$ and one during $[t_1, t_1 + t_2]$. Using the independence assumption, we have

$$\mathbf{P}(1; t_1 + t_2, \lambda) = \mathbf{P}(1; t_1, \lambda) \cdot \mathbf{P}(0; t_2, \lambda) + \mathbf{P}(0; t_1, \lambda) \cdot \mathbf{P}(1; t_2, \lambda)$$

In general we have

Axiom II

$$\mathbf{P}(x; t_1 + t_2, \lambda) = \sum_{y=0}^{x} \mathbf{P}(x - y; t_1, \lambda)\mathbf{P}(y; t_2, \lambda)$$

Next we make an assumption which clarifies the role of λ. It says that for short time intervals the probability of exactly one call is approximately λ times the length of the time interval.

Axiom III

$$\lim_{t \to 0} \frac{\mathbf{P}(1; t, \lambda)}{t} = \lambda$$

The final axiom says that for very short time intervals we can ignore the possibility of two or more calls.

Axiom IV

$$\lim_{t \to 0} \frac{\sum_{x=2}^{\infty} \mathbf{P}(x; t, \lambda)}{t} = 0$$

From the axioms stated above we can obtain an explicit formula for $\mathbf{P}(x; t, \lambda)$. We begin by finding $\mathbf{P}(0; t, \lambda)$. Using Axiom II, we have

$$\mathbf{P}(0; t + \Delta t, \lambda) = \mathbf{P}(0; t, \lambda) \cdot \mathbf{P}(0; \Delta t, \lambda)$$

If we apply Axiom I, we have

$$\mathbf{P}(0; \Delta t, \lambda) = 1 - \mathbf{P}(1; \Delta t, \lambda) - \sum_{x=2}^{\infty} \mathbf{P}(x; \Delta t, \lambda)$$

Substitution yields

$$\mathbf{P}(0; t + \Delta t, \lambda) = \mathbf{P}(0; t, \lambda) \cdot [1 - \mathbf{P}(1; \Delta t, \lambda) - \sum_{x=2}^{\infty} \mathbf{P}(x; \Delta t, \lambda)]$$

or

$$\mathbf{P}(0; t + \Delta t, \lambda) - \mathbf{P}(0; t, \lambda) = -\mathbf{P}(0; t, \lambda)[\mathbf{P}(1; \Delta t, \lambda) + \sum_{x=2}^{\infty} \mathbf{P}(x; \Delta t, \lambda)]$$

Division by Δt yields

$$\frac{\mathbf{P}(0; t + \Delta t, \lambda) - \mathbf{P}(0; t, \lambda)}{\Delta t} = -\mathbf{P}(0; t, \lambda)\left[\frac{\mathbf{P}(1; \Delta t, \lambda)}{\Delta t} + \left(\frac{1}{\Delta t}\right) \sum_{x=2}^{\infty} \mathbf{P}(x; \Delta t, \lambda)\right]$$

Taking the limit as Δt approaches 0 yields

$$D_t(\mathbf{P}(0; t, \lambda)) = -\lambda \mathbf{P}(0; t, \lambda)$$

(Axioms III and IV were used to evaluate the limit of the right-hand side of the equation.)

This is exactly the kind of situation we studied in the previous section in dealing with exponential growth. We know from the discussion there that

$$\mathbf{P}(0; t, \lambda) = a \cdot \exp(-\lambda t)$$

but a can be evaluated since $\mathbf{P}(0; 0, \lambda) = 1$ (i.e., it is certain that during an interval containing no time, nothing will happen). Therefore

$$1 = a \cdot \exp(0) = a \cdot 1 = a$$

and

$$\mathbf{P}(0; t, \lambda) = \exp(-\lambda t)$$

Next we consider the following problem. Suppose x is a positive integer ($x = 1, 2, 3, \ldots$) and we know $\mathbf{P}(x - 1; t, \lambda)$. From that, can we find $\mathbf{P}(x; t, \lambda)$? If so, setting $x = 1$ we can get from the case $x = 0$, which we have solved, to $x = 1$, then from $x = 1$ to $x = 2$, etc.

Once again, the idea is to set up an equation which will relate $D_t\mathbf{P}$ and $\mathbf{P}$. We begin by examining $\mathbf{P}(x; t + \Delta t, \lambda)$ and breaking up the time interval $[0, t + \Delta t]$ into $[0, t]$ and $[t, t + \Delta t]$, the latter of which is small enough that (by Axiom IV) either no calls or 1 call can occur there. Thus we have one of the following situations.

Case 1: x calls here (on $[0, t]$); 0 calls here (on $[t, t + \Delta t]$)

Time: $0 \qquad t \qquad t + \Delta t$

Case 2: $(x - 1)$ calls here (on $[0, t]$); 1 call here (on $[t, t + \Delta t]$)

Time: $0 \qquad t \qquad t + \Delta t$

In other words, the event "getting x calls in the interval $[0, t + \Delta t]$" is the (disjoint) union of two events, so that by Axiom II

$$\mathbf{P}(x; t + \Delta t, \lambda) = \mathbf{P}(x; t, \lambda)\mathbf{P}(0; \Delta t, \lambda) + \mathbf{P}(x - 1; t, \lambda)\mathbf{P}(1; \Delta t, \lambda)$$

As before, we substitute

$$\mathbf{P}(0; \Delta t, \lambda) = 1 - \mathbf{P}(1; \Delta t, \lambda) - \sum_{x=2}^{\infty} \mathbf{P}(x; \Delta t, \lambda)$$

into the equation and obtain

$$\mathbf{P}(x; t + \Delta t, \lambda) - \mathbf{P}(x; t, \lambda) = -\mathbf{P}(x; t, \lambda)\left[\mathbf{P}(1; \Delta t, \lambda) + \sum_{x=2}^{\infty} \mathbf{P}(x; \Delta t, \lambda)\right] + \mathbf{P}(x - 1; t, \lambda) \cdot \mathbf{P}(1; \Delta t, \lambda)$$

Dividing by Δt and taking the limit as Δt approaches 0 yields

$$D_t\mathbf{P}(x; t, \lambda) = -\lambda[\mathbf{P}(x; t, \lambda) - \mathbf{P}(x - 1; t, \lambda)]$$

This isn't quite as simple a relation between $\mathbf{P}$ and its derivative as we have had. There is a little trick which greatly simplifies the situation. We define a new function

$$R(x; t, \lambda) = \exp(\lambda t) \cdot \mathbf{P}(x; t, \lambda)$$

or

$$\mathbf{P}(x; t, \lambda) = \exp(-\lambda t)R(x; t, \lambda)$$

Then

$$D_t\mathbf{P}(x; t, \lambda) = \exp(-\lambda t)D_tR(x; t, \lambda) - \lambda \exp(-\lambda t)R(x; t, \lambda)$$

Substitution of these values yields

$$\exp(-\lambda t)D_tR(x; t, \lambda) - \lambda \exp(-\lambda t)R(x; t, \lambda) = -\lambda[\exp(-\lambda t)R(x; t, \lambda) - \exp(-\lambda t)R(x - 1; t, \lambda)]$$

From this we obtain

$$D_tR(x; t, \lambda) = \lambda R(x - 1; t, \lambda)$$

We know

$$R(0; t, \lambda) = \exp(\lambda t)\mathbf{P}(0; t, \lambda)$$
$$= \exp(\lambda t)\exp(-\lambda t) = 1$$

Therefore

$$D_t R(1; t, \lambda) = \lambda$$

from which we know

$$R(1; t, \lambda) = \lambda t + C$$

But $\mathbf{P}(1; 0, \lambda) = 0$ (you can't have something happening in a time interval of length 0) implies $R(1; 0, \lambda) = 0$ and so $C = 0$. Therefore

$$R(1; t, \lambda) = \lambda t$$

and

$$\mathbf{P}(1; t, \lambda) = (\lambda t)\exp(-\lambda t)$$

Letting $x = 2$, we obtain

$$D_t R(2; t, \lambda) = \lambda \cdot (\lambda t)$$

Therefore

$$R(2; t, \lambda) = \frac{(\lambda t)^2}{2}$$

and

$$\mathbf{P}(2; t, \lambda) = \frac{(\lambda t)^2}{2}\exp(-\lambda t)$$

In general we find that

$$R(x; t, \lambda) = \frac{(\lambda t)^x}{x!}$$

and

$$\mathbf{P}(x; t, \lambda) = \frac{(\lambda t)^x}{x!}\exp(-\lambda t)$$

The Poisson distribution as we studied it in Chapter 3 was merely this function with $t = 1$. We were, in essence, picking a unit of time and using it. We can use the Poisson tables we used before by noting that

$$\mathbf{P}(x; t, \lambda) = \mathbf{P}(x; \lambda t).$$

We can compute the theoretical mean of this distribution.

$$\begin{aligned}E[X] &= \sum_{x=0}^{\infty} x\frac{(\lambda t)^x}{x!}\exp(-\lambda t)\\ &= \lambda t \exp(-\lambda t)\sum_{x=1}^{\infty}\frac{(\lambda t)^{x-1}}{(x-1)!}\\ &= \lambda t \exp(-\lambda t)\sum_{y=0}^{\infty}\frac{(\lambda t)^y}{y!}\\ &= \lambda t \exp(-\lambda t)\exp(\lambda t)\\ &= \lambda t\end{aligned}$$

This introduces somewhat more flexibility into our use of the Poisson distribution. An example will illustrate the point.

EXAMPLE A certain realtor receives an average of eight inquiries an hour from people wishing to buy a house. What is the probability that he will receive six inquiries in a half hour?

Let us measure time in hours. We are told that $\lambda \cdot 1 = 8$, so $\lambda = 8$, and asked for

$$\mathbf{P}(6; \tfrac{1}{2}, 8) = \mathbf{P}(6; 4) = .1339$$

While our axioms were developed from observations concerning telephone calls into a business office, it turns out that they also hold in numerous other situations. Basically what we look for is a process in which discrete events occur in a continuous time interval in such a way that whether or not an event occurs in one subinterval is independent of what happens in another subinterval. We call such a phenomenon a *Poisson process*. In fact there is a nice analogy between a Poisson process and Bernoulli (or repeated) trials. Recall that if X represents the number of successes in n trials and Y represents the number of trials necessary to get the first success, then X has a binomial distribution and Y has a geometric distribution. That is, if the probability of success is p on each trial, then X has frequency function

$$f(x) = C(n, x)p^x q^{n-x} \quad \text{for} \quad x = 0, 1, \ldots, n$$

and Y has frequency function

$$g(k) = q^{k-1}p \quad \text{for} \quad k = 1, 2, 3, \ldots$$

Likewise if we have a Poisson process with an average of λ events per unit time ($t = 1$), and if X is the number of events in time t and Y is the time it takes for the first event to occur, then X, as we have just seen, has a Poisson distribution. But Y has what is known as an *exponential distribution*, since its density function is of the form

$$f(t) = \begin{cases}\lambda e^{-\lambda t}, & \text{if } t > 0\\ 0, & \text{if } t \le 0\end{cases}$$

This distribution is taken up in the Supplementary Exercises for Chapter 5, and a special case is treated in the next section.

Exercises
Section 11

1 A store sells an average of 10 watches per day. What is the probability of selling fewer than five watches in the first half of the day?

2 A door-to-door salesman sells an average of $25 worth of merchandise an hour. What is the probability that he will sell $250 worth or more in an 8-hour day?

3 What should the probability of no failure in an electronic system for one hour be, in order that the probability of no failures in 100 hours is at least .99?

4 For what value of x is $\mathbf{P}(x; t, \lambda)$ a maximum?
(*Hint:* Consider $\mathbf{P}(x; t, \lambda)/\mathbf{P}(x - 1; t, \lambda)$.)

5 If a man catches an average of one fish an hour, how many hours must he fish to have a probability of at least .6 of catching five fish?

6 Show that if λt is an integer, then

$$\mathbf{P}(\lambda t; t, \lambda) = \mathbf{P}(\lambda t - 1; t, \lambda)$$

7 In a large factory, accidents occur at a rate of one per day. Find the probability that a week goes by without an accident.

8 Verify the following identity and interpret it in terms of probabilities

$$\mathbf{P}(0; t, \lambda)^k = \mathbf{P}(0; tk, \lambda) = \mathbf{P}(0; t, \lambda k)$$

12 EXPECTATION FOR A CONTINUOUS RANDOM VARIABLE

In Chapter 3 we introduced the notion of the mean (or expected value) of a discrete random variable X:

$$\mu_X = \sum_{i=1}^{n} x_i f(x_i)$$

where $x_1, x_2, \ldots, x_n$ are the values of X and $f(x_1), \ldots, f(x_n)$ their respective probabilities. By constructing a histogram to represent the probabilities or relative frequencies of the x_i, we drew an analogy with the notion of mean value for sets of data. In this section, we extend the definition of mean, as well as that of variance, to continuous random variables. It is interesting to note that we will find parallels between the continuous and the discrete cases, both geometrically and symbolically.

Suppose then that X is a continuous random variable that ranges over some interval $[a, b]$, and has density function f. In Fig. 4.46, we sketch the graph of f together with a histogram that approximates the area under this graph. That is, the histogram is made up of n rectangles that approximate the area under the graph of f.

The histogram is formed by first subdividing $[a, b]$ into n subintervals of equal length and building a rectangle over each whose height is the value of f at the midpoint of the subinterval. That is, we have n subintervals

$$[a, x_1], [x_1, x_2], \ldots, [x_{n-1}, b]$$

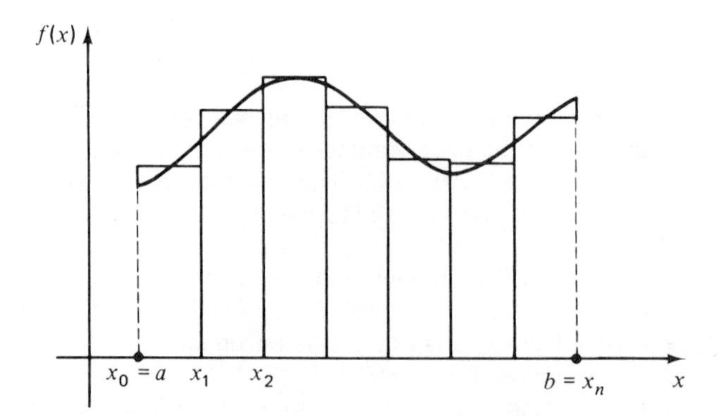

FIGURE 4.46 The graph of the density function for X and an approximating histogram.

and corresponding rectangles of height

$$f\left(\frac{a + x_1}{2}\right), f\left(\frac{x_1 + x_2}{2}\right), \ldots, f\left(\frac{x_{n-1} + b}{2}\right)$$

We know the probability

$$P(x_{i-1} \leq X \leq x_i)$$

is by definition of f equal to

$$\int_{x_{i-1}}^{x_i} f(x)\, dx$$

Area of vertical strip under graph of f between x_{i-1} and x_i

On the other hand this area is *approximately* equal to

$$f\left(\frac{x_{i-1} + x_i}{2}\right) \cdot \left(\frac{b - a}{n}\right)$$

Area of rectangle whose base is the interval $[x_{i-1}, x_i]$

Now let us imagine for a moment that the histogram represents a *discrete* random variable Y. Since the total area under the graph of f is 1, we can assume that the total area under the histogram is approximately 1. It follows that the area of each of these rectangles would represent the probability that Y takes values in the corresponding subinterval. That is,

$$P(x_{i-1} \leq Y \leq x_i) = f\left(\frac{x_{i-1} + x_i}{2}\right)\left(\frac{b - a}{n}\right)$$

Finally we assume that all values of Y occur at the midpoints of these n subintervals. Then the mean of Y is simply

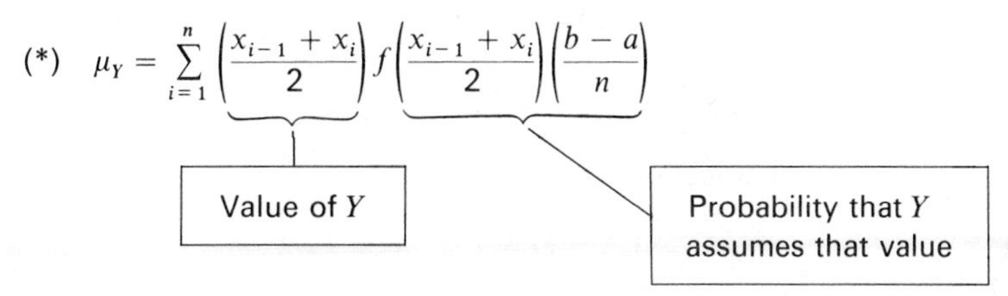

The sum on the right in (*) is, in disguise, precisely the kind of sum whose limit as $n \to \infty$ is a definite integral. Abstractly it has the form

$$\sum_{i=1}^{n} g\left(\frac{x_{i-1} + x_i}{2}\right)\left(\frac{b-a}{n}\right)$$

where

$$g(x) = xf(x)$$

Thus,

$$\lim_{n \to \infty} \sum_{i=1}^{n} \left(\frac{x_{i-1} + x_i}{2}\right) f\left(\frac{x_{i-1} + x_i}{2}\right)\left(\frac{b-a}{n}\right) = \int_a^b xf(x)\,dx$$

At the same time, the value on the left of (*),

$$\mu_Y$$

is the expectation for a discrete random variable Y that approximates the continuous random variable X. Moreover, as n increases, we get more subintervals, each of which is smaller; the resulting histogram more closely approximates the area under the graph of f; and the assumption about Y taking on only values at the midpoints of the intervals should involve less and less error within each subinterval. We conclude that the limit of these means μ_Y should be a reasonable value to assign to the mean of X, and so we make the following definition.

Definition Let X be a continuous random variable with density function f defined on $[a, b]$. The *mean* of X (also called the *expected value* or *expectation* of X) is denoted by μ_X [also by $E(X)$] and is defined by

$$E(X) = \mu_X = \int_a^b xf(x)\,dx$$

Note: The preceding discussion was intended to justify this definition. In practice, however, it is simple to remember the definition in terms of its symbolic resemblance to the discrete case:

X discrete	X continuous
$\mu_X = \sum_{i=1}^{n} x_i f(x_i)$	$\mu_X = \int_a^b xf(x)\,dx$

As with the discrete case, we immediately extend our definition to include expectation of functions of X: If g is a function of X [such as $g(X) = 3X + 5$ or $g(X) = X^2$], we define

$$E(g(X)) = \int_a^b g(x)f(x)\,dx$$

In particular, we have

$$E(X^2) = \int_a^b x^2 \cdot f(x)\,dx$$

and we can define the *variance* of X, σ_X^2, and the *standard deviation* of X (the positive square root σ_X) by

$$\text{Var}(X) = \sigma_X^2 = E(X^2) - (E(X))^2$$

Alternately,

$$\sigma_X^2 = E((X - \mu_X)^2)$$

In the exercises, the reader is invited to show that these formulas are equivalent.

EXAMPLE 1 The Uniform Distribution

Figure 4.47 illustrates two different uniformly distributed variables whose density functions are

$$f(x) = \begin{cases} 1, & \text{for } -\frac{1}{2} \le x \le \frac{1}{2} \\ 0, & \text{for all other values of } x \end{cases}$$

$$g(x) = \begin{cases} \frac{1}{4}, & \text{for } -1 \le x \le 3 \\ 0, & \text{for all other values of } x \end{cases}$$

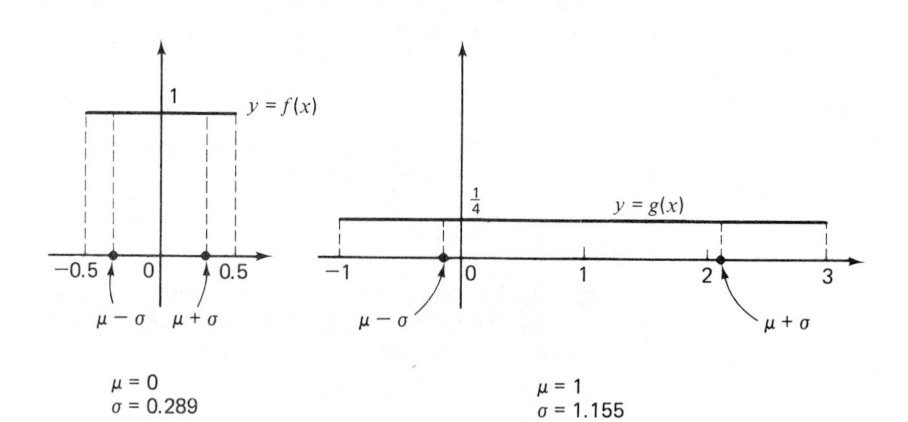

FIGURE 4.47 Two examples of a uniform distribution.

For f we have

$$\mu_X = E(X) = \int_{-1/2}^{1/2} x \cdot 1\, dx = \tfrac{1}{2}x^2\Big|_{-1/2}^{+1/2} = \tfrac{1}{2}(\tfrac{1}{2})^2 - \tfrac{1}{2}(-\tfrac{1}{2})^2 = 0$$

$$E(X^2) = \int_{-1/2}^{1/2} x^2 \cdot 1\, dx = \tfrac{1}{3}x^3\Big|_{-1/2}^{+1/2} = \tfrac{1}{3}(\tfrac{1}{2})^3 - \tfrac{1}{3}(-\tfrac{1}{2})^3 = \tfrac{1}{12}$$

Thus $\sigma_X^2 = E(X^2) - \mu_X^2 = \frac{1}{12} - 0$, and

$$\sigma_X = \sqrt{\tfrac{1}{12}} = \tfrac{1}{2}\sqrt{\tfrac{1}{3}} \approx .289$$

For g we have

$$\mu_X = E(X) = \int_{-1}^{3} x(\tfrac{1}{4})\, dx = \tfrac{1}{8}x^2\Big|_{-1}^{3} = (\tfrac{1}{8})(3)^2 - (\tfrac{1}{8})(-1)^2 = 1$$

and

$$E(X^2) = \int_{-1}^{3} x^2(\tfrac{1}{4})\, dx = \tfrac{1}{12}x^3\Big|_{-1}^{3} = \tfrac{1}{12}(3)^3 - \tfrac{1}{12}(-1)^3 = \frac{27+1}{12} \approx 2.333$$

Thus

$$\sigma_X^2 = E(X^2) - \mu_X^2 \approx 2.333 - 1 = 1.333$$

and

$$\sigma_X = \sqrt{1.333} \approx 1.155$$

As in the discrete case, these terms "mean" and "standard deviation" are intended to connote certain distinguishing characteristics of the random variable X. Although the examples of uniformly distributed variables are extremely simple, as continuous variables go, we can see that μ_X, in each case, occurs "in the middle" of the range of values of X (this interpretation is dangerous but we shall soon clarify it). Also we notice that the standard deviation of g is considerably greater than that of f, owing to the fact that the area under the g graph is more "spread out."

While we shall not offer any justification, we assert that the mean μ_X falls at precisely the "balance point" of the region under the curve. That is, if we think of the region as being a thin homogeneous solid, then it would rest in equilibrium when placed on a wedge positioned at μ_X (see Fig. 4.48).

The standard deviation is not quite so easily described. It represents the spread or dispersion of the variable X and later we shall view it as a useful kind of *unit*. We speak of a value of X being "two standard deviations below the mean" or "one standard deviation above the mean." With a given distribution we can generally say more about the standard deviation, but in general the best we can do is the continuous version of Chebychev's Theorem. We proved this result for discrete variables, and now we simply remark that it holds also for continuous variables. In the exercises we give the reader the opportunity to develop the proof.

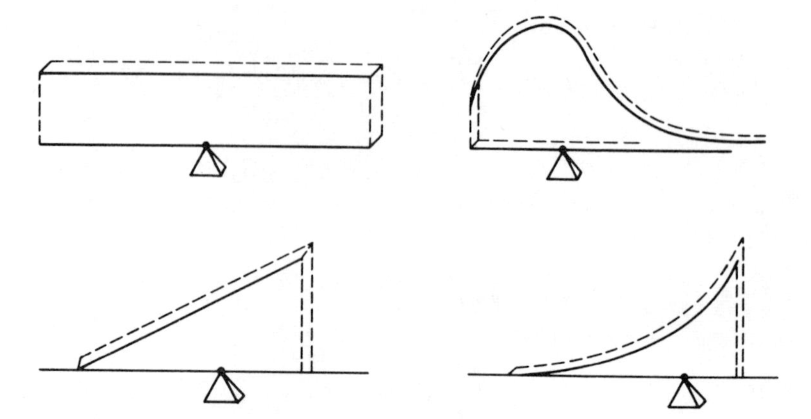

FIGURE 4.48 Some frequency functions for continuous random variables and their means.

Chebychev's Theorem

$$P(|x - \mu| \geq k\sigma) \leq \frac{1}{k^2}$$

In terms of area: The pair of regions under the graph of f and to the left of $\mu - k\sigma$ and to the right of $\mu + k\sigma$ have aggregate area less than or equal to $1/k^2$.

Incidentally, when dealing with continuous random variables, one often faces density functions that are defined on an *infinite interval* such as

$(-\infty, \infty)$,	the entire real axis
$[0, \infty)$,	the non-negative real numbers
$(-\infty, 2)$,	all real numbers less than 2

and so on. For convenience, it is customary to extend the definition of any density function to include all real numbers. For that reason, we specified, in our uniform distribution examples, that "$f(x) = 0$ for all other values of x." Moreover, it is common to find expressions such as

$$\int_{-\infty}^{\infty} xf(x)\,dx, \qquad \int_{0}^{\infty} xf(x)\,dx, \qquad \int_{-\infty}^{2} xf(x)\,dx$$

Till now, these symbols had no meaning. They are called "improper integrals" and are defined as follows:

$$\left.\begin{aligned} \int_a^{\infty} g(x)\,dx &= \lim_{r\to\infty} \int_a^r g(x)\,dx \\ \int_{-\infty}^{b} g(x)\,dx &= \lim_{r\to\infty} \int_{-r}^{b} g(x)\,dx \\ \int_{-\infty}^{\infty} g(x)\,dx &= \int_0^{\infty} g(x)\,dx + \int_{-\infty}^{0} g(x)\,dx \end{aligned}\right\}$$

Provided these limits exist!

Of course, if $g(x) = 0$ outside some interval $[a, b]$, then we have

$$\begin{aligned}\int_{-\infty}^{\infty} g(x)\,dx &= \int_{-\infty}^{a} g(x)\,dx + \int_{a}^{b} g(x)\,dx + \int_{b}^{\infty} g(x)\,dx \\ &= 0 + \int_{a}^{b} g(x)\,dx + 0 \\ &= \int_{a}^{b} g(x)\,dx\end{aligned}$$

And similarly the other improper integrals reduce to (proper) definite integrals when $g(x) = 0$ outside some $[a, b]$.

We put these ideas to use by considering a new continuous random variable whose range is infinite.

EXAMPLE 2 An Exponential Distribution

Let X be a continuous random variable whose density function is

$$f(x) = \begin{cases} e^{-x}, & \text{for } x \text{ in } [0, \infty) \\ 0, & \text{for } x \text{ in } (-\infty, 0) \end{cases} \qquad \text{(See Fig. 4.49.)}$$

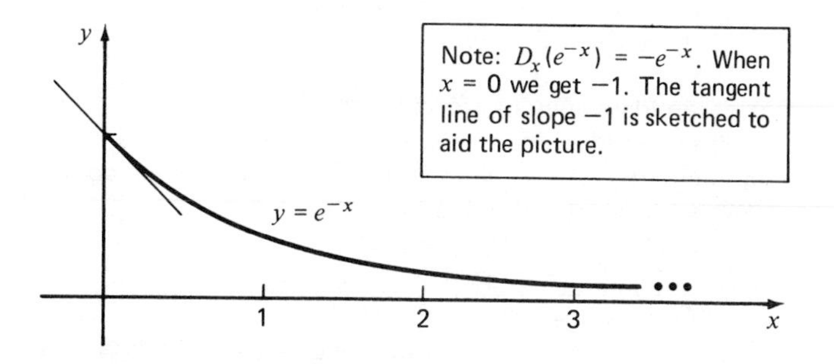

FIGURE 4.49 An exponential distribution: $f(x) = \exp(-x)$.

We first check that f *is* a density function. (This will give us practice evaluating improper integrals.)

$$\begin{aligned}\int_{-\infty}^{\infty} f(x)\,dx &= \int_{-\infty}^{0} f(x)\,dx + \int_{0}^{\infty} f(x)\,dx \\ &= 0 + \int_{0}^{\infty} e^{-x}\,dx \\ &= \lim_{r\to\infty} \int_{0}^{r} e^{-x}\,dx \\ &= \lim_{r\to\infty} \left(-e^{-x}\Big|_0^r\right) \\ &= \lim_{r\to\infty} (-e^{-r} + 1) \\ &= 1\end{aligned}$$

Now we compute the mean of X:

$$\begin{aligned} E(X) = \int_0^\infty xe^{-x}\,dx &= \lim_{r\to\infty} \int_0^r xe^{-x}\,dx \\ &= \lim_{r\to\infty} (-x-1)e^{-x}|_0^r \\ &= \lim_{r\to\infty} ((-r-1)e^{-r} - (-0-1)e^{-0}) \\ &= \lim_{r\to\infty} \left(1 - \frac{r+1}{e^r}\right) \\ &= 1 \end{aligned}$$

Note: It can be shown that $(r+1)/e^r$ tends to 0 as r gets large. Here is a suggestion:

$$\frac{r+1}{e^r} = \frac{1+r}{1+r+(r^2/2)+\ldots} = \frac{(1/r^2)+(1/r)}{(1/r^2)+(1/r)+(\frac{1}{2})+\ldots} \to \frac{0}{\frac{1}{2}+\ldots}$$

Now we compute $E(X^2)$. Already we have used the fact that $(-x-1)e^{-x}$ is an antiderivative of xe^{-x}. In the following argument we must use the fact that

$$-(x^2+2x+2)e^{-x}$$

is an antiderivative of x^2e^{-x}.

These antiderivatives are not at all obvious to the novice. However, it is easy to check (by differentiation) that the assertions are correct, and any reasonable table of integrals would contain them.

$$\begin{aligned} E(X^2) = \int_0^\infty x^2e^{-x}\,dx &= \lim_{r\to\infty} \int_0^r x^2e^{-x}\,dx \\ &= \lim_{r\to\infty} (-(x^2+2x+2)e^{-x}|_0^r) \\ &= \lim_{r\to\infty} \left(-\frac{r^2+2r+2}{e^r} + 2\right) \\ &= 2 \end{aligned}$$

Note: We use without proof the fact that

$$\lim_{r\to\infty} \left(\frac{r^2+2r+2}{e^r}\right) = 0$$

Thus,

$$\sigma_X^2 = E(X^2) - (E(X))^2 = 2 - 1 = 1$$

so that

$$\sigma_X = 1$$

EXAMPLE 3 Let X be a continuous random variable whose density function is

$$f(x) = \begin{cases} \dfrac{2x}{3}, & \text{if } 0 \le x \le 1 \\ 1 - \dfrac{x}{3}, & \text{if } 1 \le x \le 3 \\ 0, & \text{otherwise} \end{cases}$$

As seen in Fig. 4.50, the graph of f is triangular. Note that

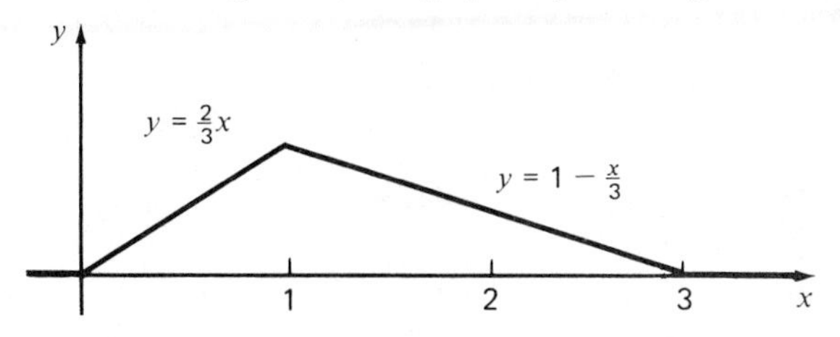

FIGURE 4.50 A triangular distribution.

$$\begin{aligned} \text{Total area under graph} &= \int_0^3 f(x)\,dx = \int_0^1 \tfrac{2}{3}x\,dx + \int_1^3 \left(1 - \frac{x}{3}\right) dx \\ &= \frac{x^2}{3}\bigg|_0^1 + \left(x - \frac{x^2}{6}\right)\bigg|_1^3 \\ &= (\tfrac{1}{3} - 0) + (3 - \tfrac{9}{6}) - (1 - \tfrac{1}{6}) \\ &= \tfrac{1}{3} + \tfrac{9}{6} - \tfrac{5}{6} \\ &= 1 \end{aligned}$$

Thus f *is* a density function.

$$\begin{aligned} E(X) &= \int_{-\infty}^{\infty} xf(x)\,dx = \int_0^1 \tfrac{2}{3}x^2\,dx + \int_1^3 \left(x - \frac{x^2}{3}\right) dx \\ &= \frac{2}{9}x^3\bigg|_0^1 + \left(\frac{x^2}{2} - \frac{x^3}{9}\right)\bigg|_1^3 \\ &= \tfrac{2}{9} + (\tfrac{9}{2} - \tfrac{27}{9}) - (\tfrac{1}{2} - \tfrac{1}{9}) \\ &= \tfrac{4}{18} + \tfrac{27}{18} - \tfrac{7}{18} \\ &= \tfrac{24}{18} \\ &= \tfrac{4}{3} \end{aligned}$$

$$\begin{aligned} E(X^2) &= \int_{-\infty}^{\infty} x^2f(x)\,dx = \int_0^1 \frac{2}{3}x^3\,dx + \int_1^3 \left(x^2 - \frac{x^3}{2}\right) dx \\ &= \frac{1}{6}x^4\bigg|_0^1 + \left(\frac{x^3}{3} - \frac{x^4}{12}\right)\bigg|_1^3 \\ &= \tfrac{1}{6} + (\tfrac{27}{3} - \tfrac{81}{12}) - (\tfrac{1}{3} - \tfrac{1}{12}) \\ &= \tfrac{1}{12} + \tfrac{27}{12} - \tfrac{3}{12} \\ &= \tfrac{25}{12} \end{aligned}$$

Thus

$$\sigma_X^2 = \tfrac{25}{12} - (\tfrac{4}{3})^2 = \tfrac{25}{12} - \tfrac{16}{9} = \tfrac{75}{36} - \tfrac{64}{36} = \tfrac{11}{36}$$

and

$$\sigma_X = \sqrt{11/36} = (\tfrac{1}{6})\sqrt{11} \approx .533$$

EXAMPLE 4 Let X be a continuous random variable with density function

$$f(x) = \begin{cases} \dfrac{1}{x^2}, & \text{for } x \geq 1 \\ 0, & \text{otherwise} \end{cases}$$

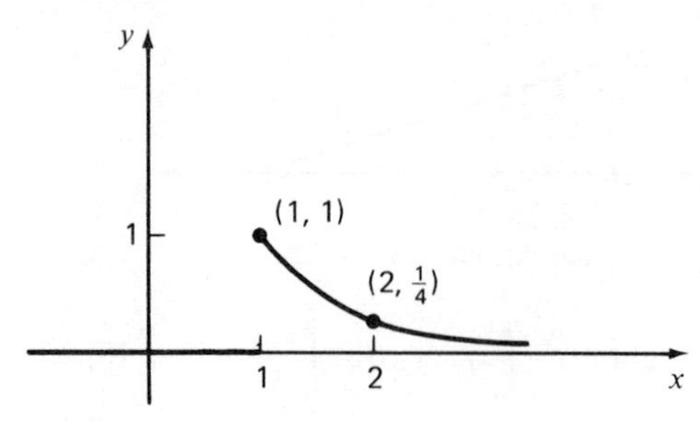

FIGURE 4.51

Since

$$\begin{aligned}\int_{-\infty}^{\infty} f(x)\,dx &= \int_1^{\infty} \frac{1}{x^2}\,dx \\ &= \lim_{r\to\infty} \int_1^r \frac{1}{x^2}\,dx \\ &= \lim_{r\to\infty} \left(-\frac{1}{x}\right)\Bigg|_1^r \\ &= \lim_{r\to\infty} \left(-\frac{1}{r} + 1\right) \\ &= 1\end{aligned}$$

$f(x)$ is a density function. But

$$\begin{aligned}E(X) &= \int_{-\infty}^{\infty} xf(x)\,dx \\ &= \int_1^{\infty} x \cdot \left(\frac{1}{x^2}\right) dx \\ &= \lim_{r\to\infty} \int_1^r \frac{1}{x}\,dx \\ & \lim_{r\to\infty} \log x|_1^r \\ &= \lim_{r\to\infty} (\log(r) - \log(1))\end{aligned}$$

which is undefined. Here we have an example of *a density function with no mean*. Whenever we deal with improper integrals, we must keep in mind that they may not exist, so it is quite reasonable to find density functions where $E(X)$ or $E(X^2)$ do not exist. The reader should calculate $E(X^2)$ for this density function and see that it too is not defined.

In the discrete case we showed that the "function" E satisfied some *linearity* properties. Likewise when X and Y are continuous random variables, it can be shown that

$$E(X+Y) = E(X) + E(Y)$$

and

$$E(cX) = cE(X), \qquad \text{for any constant } c$$

assuming of course that $E(X)$ and $E(Y)$ exist in the first place. Indeed, the proofs of these important properties depend only on the corresponding linearity properties of the integral. We present one special case.

$$\begin{aligned} E(2X+3) &= \int_{-\infty}^{\infty} (2x+3)f(x)\,dx \\ &= \int_{-\infty}^{\infty} [2xf(x) + 3f(x)]\,dx \\ &= \int_{-\infty}^{\infty} 2xf(x)\,dx + \int_{-\infty}^{\infty} 3f(x)\,dx \\ &= 2\int_{-\infty}^{\infty} xf(x)\,dx + 3\int_{-\infty}^{\infty} f(x)\,dx \\ &= 2E(X) + 3 \end{aligned}$$

(Remember that the constant random variable, $Y = 3$, has 3 for its mean, since

$$\int_{-\infty}^{\infty} 3f(x)\,dx = 3\int_{-\infty}^{\infty} f(x)\,dx = 3 \cdot 1 = 3.)$$

Summary

We have defined the *mean* and the *variance* for a continuous random variable X in a manner that closely parallels that of a discrete random variable. Using the basic definitions

$$E(X) = \int_{-\infty}^{\infty} xf(x)\,dx,$$

and more generally

$$E(g(X)) = \int_{-\infty}^{\infty} g(x)f(x)\,dx, \quad \text{provided the integral exists}$$

where g is any reasonable function. Then

$$\mu_X = E(X)$$

$$\sigma_X^2 = E(X^2) - \mu_X^2$$

We also stated the continuous version of Chebychev's Theorem, once again with the warning that it generally yields a crude estimate.

Exercises Section 12

1 Find the mean and standard deviation for the random variables X whose density functions are given below. In each case, sketch the graph of f and indicate where μ lies as well as the points $\mu + \sigma$ and $\mu - \sigma$.

(a) $f(x) = \begin{cases} \frac{1}{3}, & \text{for } x \text{ in } [-2, 1] \\ 0, & \text{otherwise} \end{cases}$

(b) $f(x) = \begin{cases} 3x^2, & \text{for } x \text{ in } [0, 1] \\ 0, & \text{otherwise} \end{cases}$

(c) $f(x) = \begin{cases} 1 + x, & \text{for } x \text{ in } [-1, 0] \\ 1 - x, & \text{for } x \text{ in } [0, 1] \\ 0. & \text{otherwise} \end{cases}$

(d) $f(x) = \begin{cases} \frac{3}{2}x^2, & \text{for } x \text{ in } [-1, 1] \\ 0, & \text{otherwise} \end{cases}$

(e) $f(x) = \begin{cases} -x, & \text{for } x \text{ in } [-1, 0] \\ x, & \text{for } x \text{ in } [0, 1] \\ 0, & \text{otherwise} \end{cases}$

2 In each of parts (a), (c), and (e) of Exercise 1, find the amount of area under the graph of f lying between $\mu - \sigma$ and $\mu + \sigma$.

3 What does Chebychev's Theorem say about the area under the graph of e^{-x} between $x = 0$ and $x = 3$? Find this area by integration.

4 Using the various properties of the integral, show that the two formulas for σ_X^2 are equivalent.

5 Prove Chebychev's Theorem for continuous random variables. (*Hint:* Mimic the proof for the discrete case using integrals instead of sums.)

6 Show that each of the following is a density function and find $E(X)$, $E(X^2)$, σ, and the distribution function.

(a) $f(x) = \begin{cases} \frac{2}{x^3}, & \text{if } x \geq 1 \\ 0, & \text{otherwise} \end{cases}$

(b) $f(x) = \begin{cases} \frac{3}{x^4}, & \text{if } x \geq 1 \\ 0, & \text{otherwise} \end{cases}$

7 (a) Sketch the graph of $y = 1/(1 + x^2)$.
(b) Show that

$$2 \leq \int_{-\infty}^{\infty} \frac{dx}{1 + x^2} \leq 4$$

Hint:

$$\int_{-\infty}^{\infty} \frac{dx}{1+x^2} = \int_{\infty}^{-1} \frac{dx}{1+x^2} + \int_{-1}^{1} \frac{dx}{1+x^2} + \int_{1}^{\infty} \frac{dx}{1+x^2}$$

Show that

$$1 \le \int_{-1}^{1} \frac{dx}{1+x^2} \le 2$$

$$\tfrac{1}{2} \le \int_{1}^{\infty} \frac{dx}{1+x^2} \le 1 \qquad (x^2 \le 1 + x^2 \le 2x^2 \text{ for } x \ge 1)$$

(c) In fact

$$\int_{-\infty}^{\infty} \frac{dx}{1+x^2} = \pi \; (\approx 3.14159)$$

so

$$f(x) = \frac{1}{\pi} \cdot \frac{1}{1+x^2}$$

is a density function. Discuss why the mean should be 0 from the graph of (a) but show that $E(X)$ is undefined.

(d) (Computer exercise.) Approximate

$$\int_{-\infty}^{\infty} \frac{dx}{1+x^2}$$

You should be able to provide an argument that it suffices to work with

$$\int_{-10}^{10} \frac{dx}{1+x^2}$$

on the computer.

8 If X is a continuous random variable with mean 10 and standard deviation 5, find

(a) $E(X^2)$
(b) $E(3X)$
(c) $E(X + 4)$
(d) $E\left(\frac{X - 10}{5}\right)$

9 Suppose X is a random variable with mean μ and standard deviation σ. Define a new random variable Y by

$$Y = \frac{X - \mu}{\sigma}$$

Find the mean and standard deviation of Y.

13 FUNCTIONS OF SEVERAL VARIABLES. PARTIAL DERIVATIVES

Earlier we argued that functional relationships abound in the real world. What we neglected to point out was that oftentimes these relationships involve more than two quantities. In a complex society such as ours, one hardly expects a typical corporation to be concerned solely with the production of one type of commodity and the resulting profit function. [Recall that we studied the profit $P(x)$ as a function of the number of items sold. In the ultimate formulation of the problem, only the two variables P and x were present.] Of course such a model, narrow as it might be, could be worthy of study. Indeed one must bear in mind that regardless of how eclectic we are in our model building we will always ignore certain factors, either because they seem insignificant or they seem difficult to quantify, or simply because we are not aware of their existence.

In this section we shall study relationships between three or more quantities. Usually such relationships will be in the form of one variable being a function of two or more other variables. This concept is by no means foreign to us; here are some well-known examples.

EXAMPLE 1 Area and Volume of Geometric Objects

(a) The area of a rectangle of length L and width W is given by

$$A = A(L, W) = L \cdot W$$

A is a function of two variables.

(b) The volume of a right circular cylinder of radius r and height h is given by

$$V = V(r, h) = \pi r^2 h$$

V is a function of two variables.

(c) The volume of a rectangular parallelepiped (a box) of length l, width w, and height h is given by

$$V = V(l, w, h) = l \cdot w \cdot h$$

V is a function of three variables.

EXAMPLE 2 The Difference Quotient

In the definition of the derivative of a function f we introduced a function of two variables

$$\theta(x, \Delta x) = \frac{f(x + \Delta x) - f(x)}{\Delta x}$$

Our next example is an extension of the earlier problems concerned with a profit function for production of a certain commodity. It represents one of the kinds of problems we wish to deal with in our applications.

EXAMPLE 3 A Two-Commodity Profit Function

Suppose a corporation manufactures two objects, say washers, W, and dryers, D. Preliminary sales figures suggest the following demand functions:

$x = 100 - p/2$ (x units of W can be sold at price p.)

$y = 120 - q/3$ (y units of D can be sold at price q.)

We suppose further that the production cost for manufacturing x units of W and y units of D is given by

$$\text{Cost} = C(x, y) = 2x^2 + 4xy + 3y^2$$

PROBLEM At what prices should these items be sold to maximize profits? Note that we are looking for values of p and q. Our demand functions can be solved for p and q once we know x and y; also we can combine our information to get a profit function.

$$\begin{aligned}\text{Profit} &= \text{sales revenue} - \text{cost}\\ &= (x \cdot p + y \cdot q) - (2x^2 + 4xy + 3y^2)\\ &= x(200 - 2x) + y(360 - 3y) - (2x^2 + 4xy + 3y^2)\\ &= 200x + 360y - 4x^2 - 6y^2 - 4xy\end{aligned}$$

Thus our problem becomes

For what values of x and y will the function

$P(x, y) = 200x + 360y - 4x^2 - 6y^2 - 4xy$ have a *maximum value*?

We shall attack this problem by reviewing our method of solving max-min problems for functions of one variable. Recall that to find the maximum and minimum values of a function $y = f(x)$ on the interval $[a, b]$, we first computed $f'(x)$ and found all points $\{c_1, c_2, \ldots, c_n\}$, where either $f'(x) = 0$ or where $f'(x)$ was undefined. Then we compiled a list $\{f(a), f(b), f(c_1), f(c_2), \ldots, f(c_n)\}$ from which we extracted the smallest and largest values. In Fig. 4.52, we illustrate graphically the reasons for looking at this kind of list.

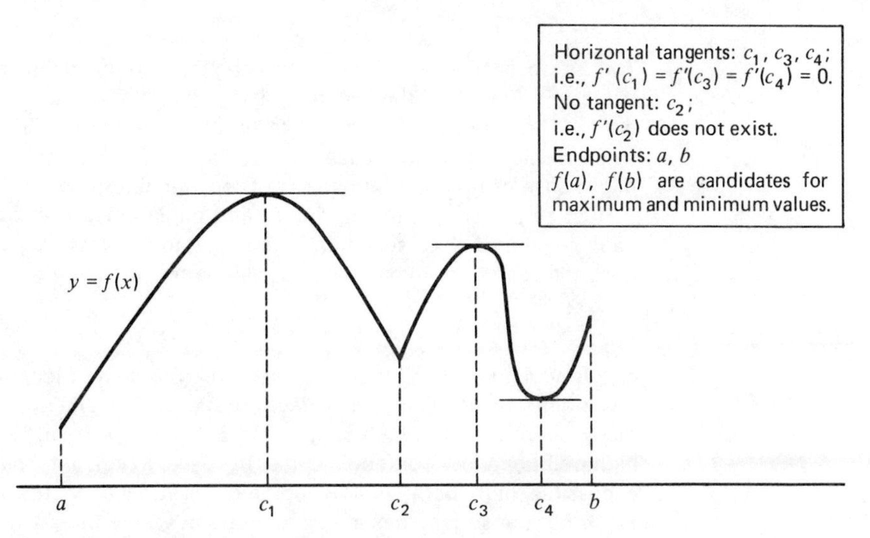

FIGURE 4.52 Some max–min considerations for a function of one variable.

So even though we look for horizontal tangents for our max-min solutions, we must be wary of endpoint solutions, and points where the tangent does not exist. Sometimes having $f'(c) = 0$ (i.e., horizontal tangent) does not even guarantee a relative max-min. We could get (Fig. 4.53) a *point of inflection*, a point where the tangent line ceases to turn in one direction and begins to turn in the other direction.

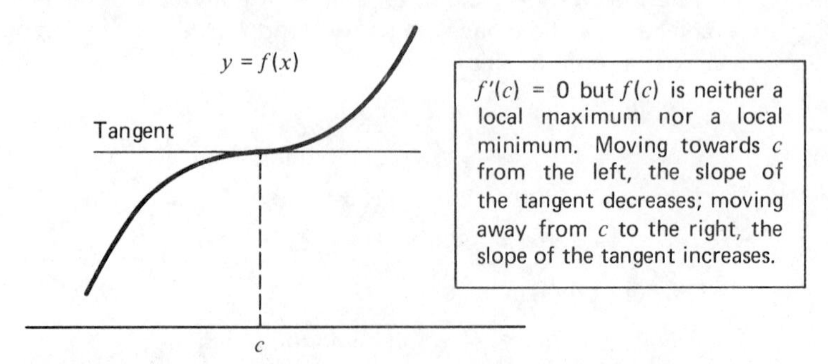

FIGURE 4.53 A point of inflection on the graph of a function of one variable.

When we deal with functions of two or more variables, the various possibilities become increasingly complex. What we are about to do is merely a sketch of a suitable approach to the max-min problem for *functions of two variables*. We begin by examining some graphs. For the first time we must draw pictures in a three-dimensional coordinate system. We label our axes x, y, and z. This context allows for graphs of functions of two variables, usually in the form

$$z = F(x, y)$$

That is, to each point (x, y) in the (horizontal) xy-plane there corresponds a value $z = F(x, y)$ of the function. The ordered triple (x, y, z) is a point on our graph. The totality of such points forms a surface. In Fig. 4.54 we illustrate some labeled points in space, and in Fig. 4.55 we illustrate two surfaces that might arise as graphs of functions of two variables.

Just as the search for maximum and minimum values of a function of one variable caused us to seek high points and low points on a curve in a plane, so does the search for maximum and minimum values of a function

$$z = F(x, y)$$

of two variables translate to the geometric problem of locating high points and low points on its graph, a surface in space. The analogy can be pursued in many situations. Geometrically a horizontal tangent line signaled a possible high point or a low point on a curve. In space, a horizontal *tangent plane* signals a possible high point or low point on a surface. In the left-hand graph of Fig. 4.55, the surface has a maximum point where the z-axis pierces it, and we could imagine a plane perpendicular to the z-axis at that point that would be a *horizontal tangent plane*. Since the surface actually extends downward in-

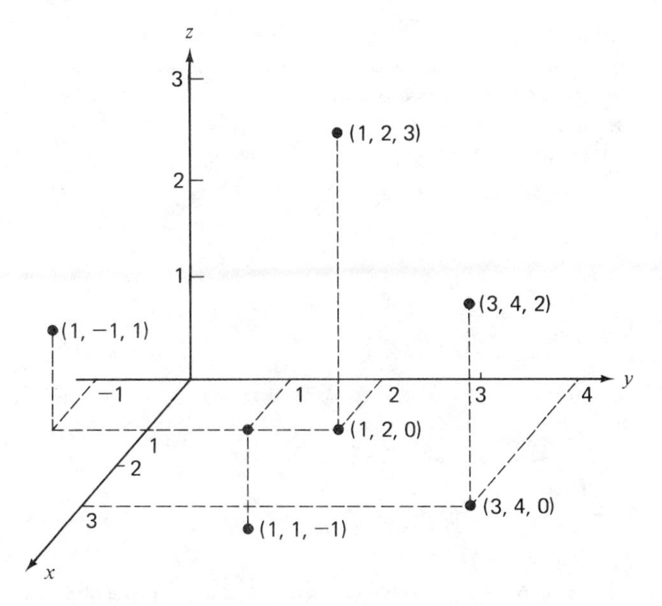

FIGURE 4.54 3-dimensional coordinate system.

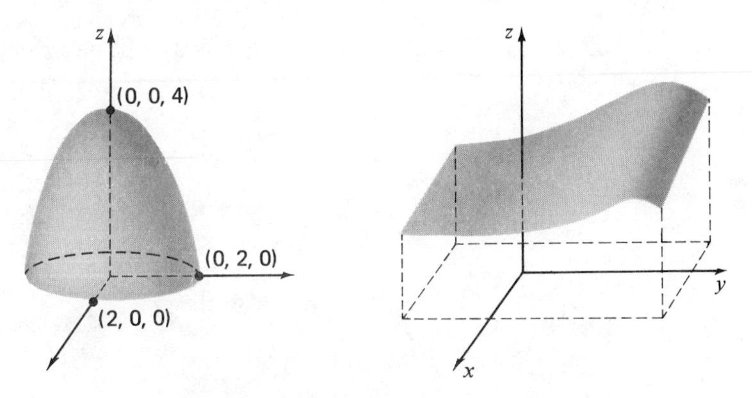

FIGURE 4.55 Examples of surfaces in space.

definitely, it has no minimum point. We would assert then that the function whose graph is the surface has a unique maximum value when $x = 0$ and $y = 0$ [i.e., above the point (0, 0, 0)] and has no minimum value.

The graph on the right in Fig. 4.55 may or may not have high points or low points, since it extends indefinitely to the right and left as well as forward and back. Even in the portion drawn, it is not clear whether we could locate a point where the tangent plane would be horizontal. More information seems to be needed.

The point is that we need some kind of analytic technique to determine where maxima and minima occur for a given function. Fortunately the analogy between functions of one variable and functions of two or more variables extends to the use of derivatives to check for horizontal tangent planes. In Fig. 4.56 we see a surface that has such a tangent plane.

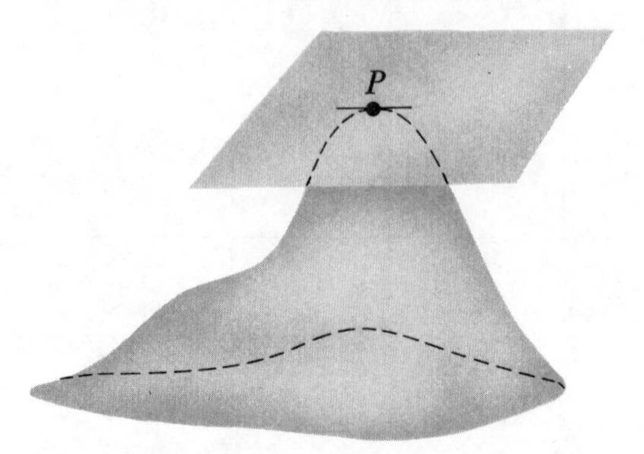

FIGURE 4.56 A surface with a horizontal tangent plane at *P*.

We note that at the point of tangency there are two key lines, one parallel to the x-axis and one parallel to the y-axis (see Fig. 4.57), called T_x and T_y. From high school geometry, we know that two intersecting lines determine a plane. Thus we can think of the horizontal (tangent) plane's being determined by T_x and T_y.

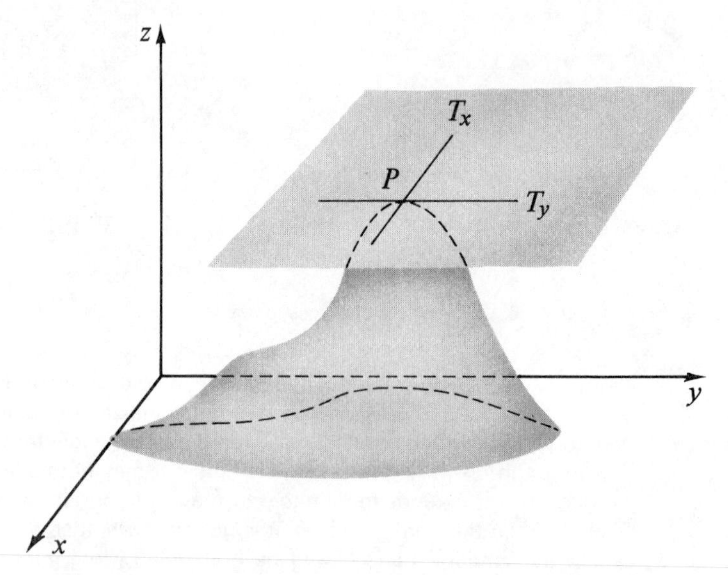

FIGURE 4.57 The tangent plane at P determined by the lines T_x and T_y.

Next we observe that T_x and T_y are really tangent lines to certain curves, called *sections*, that lie in the original surface. A section is simply a cross section or a slice of the surface. In the surface under examination, we consider two sections, one by a plane parallel to the xz-plane cutting the surface at P and the other by a plane parallel to the yz-plane cutting the surface at P.

These sections are simply curves in a plane and can be handled as graphs of functions of one variable. For instance, the section on the left in Fig. 4.58 would be the graph of

$$z = F(x, y_0)$$

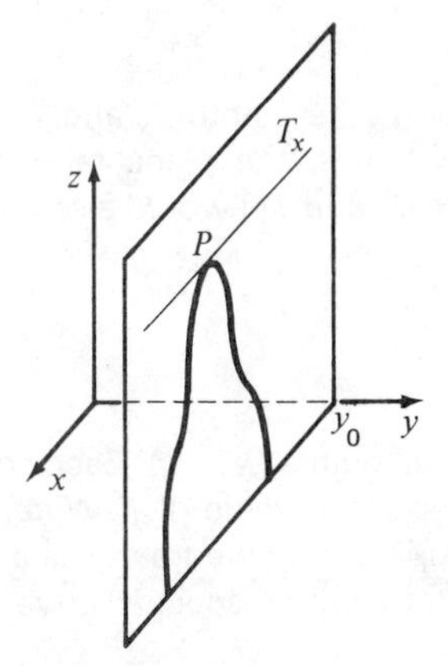

Plane parallel to xz-plane with section of surface and tangent line T_x

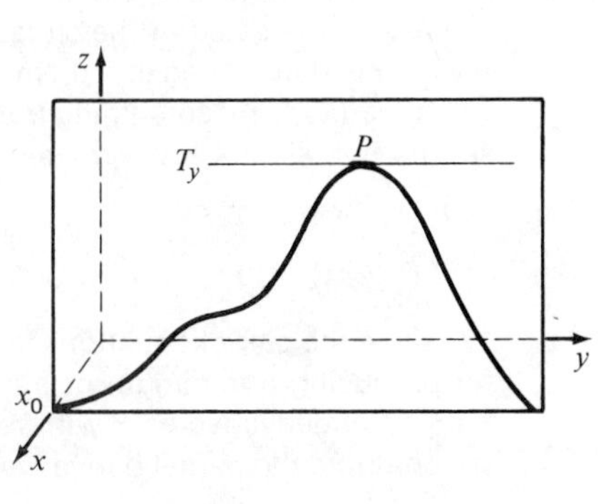

Plane parallel to yz-plane with section of surface and tangent line T_y

FIGURE 4.58

where y_0 is a constant and hence z is simply a function of x. Similarly the section on the right in Fig. 4.58 is the graph of

$$z = F(x_0, y)$$

so that z is a function of y alone.

Let us outline a procedure for finding the slopes of the tangent lines T_x and T_y, using these values to determine whether we have an extreme value of the function F at (x_0, y_0).

We consider the function of one variable

$$z = F(x, y_0)$$

Its derivative at the value x_0, namely,

$$F_x(x_0, y_0) = \lim_{\Delta x \to 0} \frac{F(x_0 + \Delta x, y_0) - F(x_0, y_0)}{\Delta x}$$

This is called the *partial derivative of F with respect to x at (x_0, y_0).*

is the slope of the tangent line T_x. Likewise, the derivative of the function

$$z = F(x_0, y)$$

at the value y_0, namely,

$$F_y(x_0, y_0) = \lim_{\Delta y \to 0} \frac{F(x_0, y_0 + \Delta y) - F(x_0, y_0)}{\Delta y}$$

This is called the *partial derivative of F with respect to y at (x_0, y_0).*

is the slope of the tangent line T_y.

We are interested in the situation where both T_x and T_y are horizontal, since the plane determined by them [that is, the plane tangent to the surface at $(x_0, y_0, F(x_0, y_0))$ containing both T_x and T_y] would also be horizontal. Thus we set the two slopes equal to zero:

$$F_x(x_0, y_0) = 0$$
$$F_y(x_0, y_0) = 0$$

We solve the pair of equations for all points (x_0, y_0). Each solution is a *candidate* for producing a minimum or a maximum value of F. *Note:* When we compute the partial derivative of F with respect to x, we treat y as a constant; and when we compute the partial derivative of F with respect to y, we treat x as a constant.

EXAMPLE 4 The function

$$z = 4 - x^2 - y^2$$

has a graph resembling the left picture in Fig. 4.55. Let us find the point (x_0, y_0) such that both partial derivatives are 0.

First we compute

$$F_x(x, y) = 0 - 2x - 0 = -2x$$

and

$$F_y(x, y) = 0 - 0 - 2y = -2y$$

Setting both partial derivatives equal to 0,

$$-2x = 0$$
$$-2y = 0$$

we get

$$x = 0$$
$$y = 0$$

and we conclude that the graph of $z = 4 - x^2 - y^2$ has a horizontal tangent plane at the point

$$(0, 0, F(0, 0)) = (0, 0, 4)$$

as we should expect from the picture.

EXAMPLE 5 We compute some partial derivatives

(a) $F(x, y) = x^2 + xy + y^2$

$$F_x(x, y) = 2x + y$$

$$F_y(x, y) = x + 2y$$

(b) $F(x, y) = \exp(x + 2y)$

$$F_x(x, y) = \exp(x + 2y) \cdot 1$$

$$F_y(x, y) = \exp(x + 2y) \cdot 2$$

(c) $F(x, y) = \dfrac{x^2 + y^2}{x + 3}$

$$F_x(x, y) = \frac{(x + 3)(2x) - (x^2 + y^2)(1)}{(x + 3)^2}$$

$$F_y(x, y) = \frac{(x + 3)(2y) - (x^2 + y^2)(0)}{(x + 3)^2}$$

Warning: Sometimes a surface will have a point with a horizontal tangent plane but that point will be neither a maximum nor a minimum point. A classic example of this phenomenon is illustrated in Fig. 4.59 where the surface

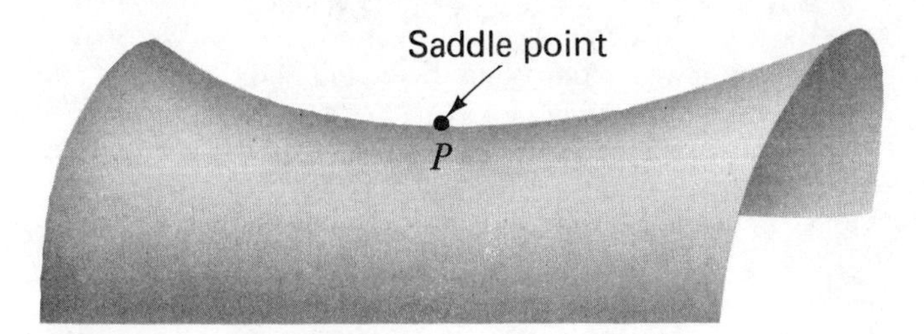

FIGURE 4.59 The saddle.

resembles the shape of a saddle and is technically called a hyperbolic paraboloid.

At the "saddle point" we note that moving from left to right the point is the lowest along the section, but moving from front to back the point is the highest along the section (see Fig. 4.60). Thus both T_y and T_x are horizontal at P, yielding a horizontal tangent plane, but neither a maximum nor a minimum point of the surface.

Second Warning: In a practical problem the domain of the function F is often restricted to certain values of x and y. Just as it sometimes happens that

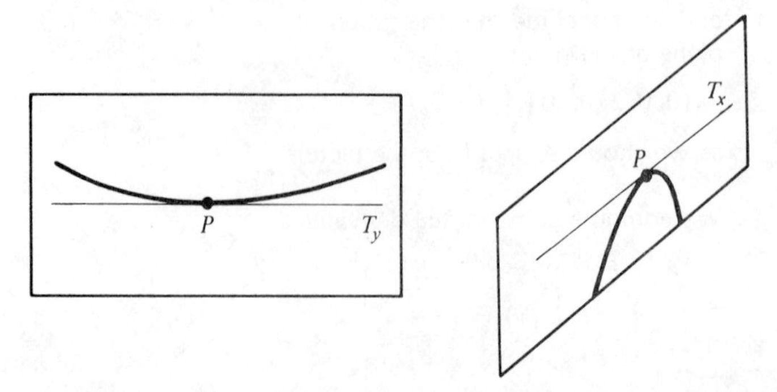

FIGURE 4.60 Two sections of the saddle illustrating horizontal tangent lines at P.

a function of one variable achieves a maximum or a minimum value at one of the endpoints of its domain, so it may happen that a function of two or more variables will achieve its extreme values at one of the points where *either x or y or both* take on their minimum or maximum values. But now instead of checking only two endpoints as candidates for extreme points, the function of several variables involves an infinitude of "endpoint" candidates. We can see the potential difficulties in the right side graph of Fig. 4.55, where we assume that the function is defined only for the rectangular domain (indicated by dotted lines) in the xy-plane. That is, we have a domain of the form

$$\{(x, y)|a \leq x \leq b \quad \text{and} \quad c \leq y \leq d\}$$

It may help to envision a canopy of a four-poster bed, where the rectangular bed represents the domain in the xy-plane and the canopy represents the graph of the function. If this were a true canopy, it would sag in the middle, guaranteeing an absolute minimum value there (Fig. 4.61). But the canopy model we wish to utilize should have more (mathematical) flexibility. We allow the four posts to be of any height, not necessarily equal. We also allow for smooth bulges

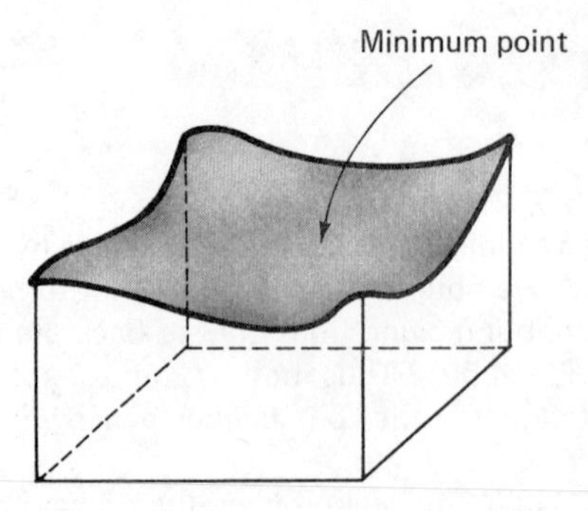

FIGURE 4.61 A canopy with a minimum point.

(imagine someone standing on the bed so that his head stretches up the canopy), or sharp peaks (imagine a tent pole pushing up the canopy in its interior), both of which would correspond to interior maximum points. And also we allow the highest or lowest point on the canopy to occur along an edge, perhaps at a corner. This deformable canopy possesses most of the common properties of graphs of functions of two variables, and it gives us some idea of the complexity of locating the maximum and minimum points. Yet, in spite of the potential abnormalities, we adopt the following test which will work in a respectable number of practical applications.

Test for Maximum and Minimum Values of a Function of Two Variables $z = F(x, y)$

Step 1: Compute the partial derivatives

$$F_x(x, y) \quad \text{and} \quad F_y(x, y)$$

Step 2: Solve the pair of equations simultaneously:

$$F_x(x, y) = 0$$
$$F_y(x, y) = 0$$

Generally there will be one solution (x_0, y_0), but there may be more, $(x_1, y_1), (x_2, y_2), \ldots$. These will correspond to points where there is a *horizontal tangent plane*.

Step 3: Locate all places where the partial derivatives do not exist. These will correspond to places where there is a sharp point or a break in the surface.

Step 4: Examine the value of F at the points in Steps 2 and 3 and compare them to a couple of values of F at the edges of the domain. That is, values such as

$$F(a, c), \quad F(a, d), \quad F(a, y), \quad \text{for} \quad c \leq y \leq d$$

In practice, one usually develops some intuition and also knows something about the behavior of the function F so that Step 4 is not terribly time-consuming. Indeed, many times we know that there will be a unique maximum or a unique minimum and we essentially stop at Step 2. Let us reconsider the profit function of Example 3. We had the function

$$z = P(x, y) = 200x + 360y - 4x^2 - 6y^2 - 4xy$$

Step 1: $P_x(x, y) = 200 - 8x - 4y$

$$P_y(x, y) = 360 - 12y - 4x$$

Step 2: $200 - 8x - 4y = 0$

$$360 - 12y - 4x = 0$$

We wish to solve this pair of equations. Rewriting,

$$8x + 4y = 200$$
$$4x + 12y = 360$$

Multiplying the second equation by 2,

$$8x + 4y = 200$$
$$8x + 24y = 720$$

Subtracting the first from the second equation,

$$20y = 520$$
$$y = 26$$

Substituting into the first equation,

$$8x + 4(26) = 200$$
$$8x + 104 = 200$$
$$8x = 96$$
$$x = 12$$

Thus the *unique* solution is

$$(x_0, y_0) = (12, 26)$$

We note that $P_x(x, y)$ and $P_y(x, y)$ are defined for all (x, y) for which P is defined, so Step 3 can be ignored. Finally we compare $P(12, 26)$ to some other values.

$$P(12, 26) = 200(12) + 360(26) - 4(12)^2 - 6(26)^2 - 4(12)(26) = 5{,}880$$
$$P(0, 0) = 200(0) + 360(0) - 4(0)^2 - 6(0)^2 - 4(0)(0) = 0$$
$$P(10, 20) = 200(10) + 360(20) - 4(10)^2 - 6(20)^2 - 4(10)(20) = 3{,}200$$
$$P(20, 30) = 200(20) + 360(30) - 4(20)^2 - 6(30)^2 - 4(20)(30) = 5{,}400$$

We conclude that

$$P(12, 26) = 5{,}880$$

is the *maximum* value of P, i.e., the maximum profit is attained.

The maximum profit is attained by selling 12 washers and 26 dryers; that maximum profit is $5,880.

EXAMPLE 6 Finding a Regression Line

The following type of problem comes up quite often in situations where one variable Y is thought to be related to another variable X by a *linear* function (whose graph is a straight line)

$$Y = mX + b$$

Typically, a researcher hypothesizes this linear relationship and proceeds to sample some values of X and Y to determine the exact line involved (i.e., to determine the slope m and the y-intercept b). We illustrate the problem graphically in Fig. 4.62. The points $(x_1, y_1), (x_2, y_2), \ldots, (x_n, y_n)$ are known, and our job is to find the equation $y = mx + b$ of the line that "comes closest" to these points. Although there are several ways to make precise this phrase "comes closest,"

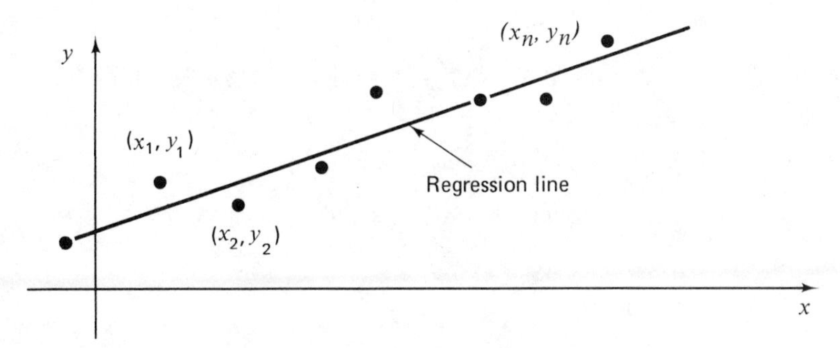

FIGURE 4.62

we adopt the value

Sum of the squares of the vertical distances from the various points to the proposed line

as a measure of closeness, and our job is to determine choices of m and b that minimize this value. Denoting the vertical distance from (x_i, y_i) by d_i, we have $d_i = |y_i - (mx_i + b)|$, so that

$$F(m, b) = [y_1 - (mx_1 + b)]^2 + [y_2 - (mx_2 + b)]^2 + \cdots + [y_n - (mx_n + b)]^2$$

Thus the problem reduces to finding a minimum value for the function $F = F(m, b)$. For the time being, we solve the problem for a special case: We are given three points (1, 1), (3, 2), and (4, 3). Our function F becomes

$$F(m, b) = (1 - (m + b))^2 + (2 - (3m + b))^2 + (3 - (4m + b))^2$$

We take partial derivatives with respect to m and b and set them equal to 0:

$$\begin{aligned} F_m(m, b) &= 2(1 - (m + b))(-1) + 2(2 - (3m + b))(-3) + 2(3 - (4m + b))(-4) \\ &= (-2)[(1 - m - b) + (6 - 9m - 3b) + (12 - 16m - 4b)] = 0 \end{aligned}$$

if

$$(1 + 6 + 12) - (1 + 9 + 16)m - (1 + 3 + 4)b = 0$$

if

$$(*) \quad 19 - 26m - 8b = 0$$

and

$$\begin{aligned} F_b(m, b) &= 2(1 - (m + b))(-1) + 2(2 - (3m + b))(-1) + 2(3 - (4m + b))(-1) \\ &= (-2)[(1 - m - b) + (2 - 3m - b) + (3 - 4m - b)] = 0 \end{aligned}$$

if

$$(1 + 2 + 3) - (1 + 3 + 4)m - (1 + 1 + 1)b = 0$$

if

$$(**) \quad 6 - 8m - 3b = 0$$

We can solve (*) and (**) simultaneously and get

$$m = \tfrac{9}{14}, \qquad b = \tfrac{2}{7}$$

The line

$$y = \tfrac{9}{14}x + \tfrac{2}{7}$$

is called the *regression line* (or *least squares line*) for the three points (1, 1), (3, 2), and (4, 3). Among other things, it is used to predict future values of y corresponding to specific values of x.

In our examples, we have restricted our attention to functions of two variables. Although max-min problems get increasingly difficult for functions of three or more variables, the notion of partial derivatives extends quite easily. In general we have: Given

$$F = F(x_1, x_2, \ldots, x_n) \qquad \text{(A function of } n \text{ variables)}$$

the partial derivative of F with respect to x_i is given by

$$F_{x_i}(x_1, x_2, \ldots, x_n) = \lim_{\Delta x_i \to 0} \frac{F(x_1, x_2, \ldots, x_i + \Delta x_i, \ldots, x_n) - F(x_1, x_2, \ldots, x_i, \ldots, x_n)}{\Delta x_i}$$

For example, if $F(x, y, z) = 2x + 3xy^2 + xz^3$, then

$$F_x(x, y, z) = 2 + 3y^2 + z^3, \qquad F_y(x, y, z) = 6xy, \qquad F_z(x, y, z) = 3xz^2$$

An alternate notation for partial derivatives involves a Greek script delta, ∂, and is used as follows.

$$\frac{\partial}{\partial x}F \text{ means } F_x$$

$$\frac{\partial}{\partial y}F \text{ means } F_y$$

$$\frac{\partial}{\partial x}F(x, y)\Big|_{(x_0, y_0)} \text{ means } F_x(x_0, y_0)$$

and so on. When we have

$$z = F(x, y)$$

it is often convenient to write

$$\frac{\partial z}{\partial x} \text{ and } \frac{\partial z}{\partial y}$$

to denote F_x and F_y, respectively.

Summary

We have considered the problem of finding extreme (i.e., maximum or minimum) values of a function F of two or more variables. Drawing an analogy with the single variable function max-min problems, we were led to define the partial derivatives of F with respect to each of its variables. The procedure for finding extreme values of F consists of four steps, which can be summarized as:

Step 1: Compute all partial derivatives of F.
Step 2: Set each partial derivative equal to 0 and solve.
Step 3: Locate points where the partial derivatives don't exist.
Step 4: Compute values of F at all points found in Steps 2 and 3 and compare these values to values of F at selected boundary points.

Exercises Section 13

1 Find the partial derivatives with respect to each possible variable for the following functions.

(a) $F(x, y) = x^2 - 5y^2 + 12$
(b) $G(x, y) = \exp(x - 3y) + xy^3$
(c) $H(x, y, z) = \log(x + 2y) - z^3(x + y)^5$
(d) $V(r, h) = \frac{1}{3}\pi r^2 h$
(e) $E(m, c) = mc^2$
(f) $S(n, a, b) = \dfrac{n}{2}(a + b)$
(g) $A(n, r, P) = P(1 + nr)$

2 Find the maximum and minimum values of

$$F(x, y) = 4x^2 - 6x - y + 2xy$$

Where the domain of F is restricted to $0 \le x \le 10$, $0 \le y \le 30$.
[*Hint:* You should make a table listing the maximum and minimum values of each of the functions

$$F(0, y), \quad F(10, y), \quad F(x, 0), \quad \text{and} \quad F(x, 30)$$

as well as values corresponding to solutions in Step 2.]

3 Find the maximum value of the product

$$xyz$$

assuming that $x + y + z = 3$. (*Hint:* Replace z by $3 - y - x$ in the product.)

4 Repeat Exercise 3 with the restriction that $x + y + z = a$.

5 Find the maximum profit and the number of washers and dryers that should be sold to yield this maximum if the demand and cost functions are

$$x = 150 - \frac{p}{2} \qquad (x \text{ washers sold at price } p)$$

$$y = 70 - \frac{q}{3} \qquad (y \text{ dryers sold at price } q)$$

$$C(x, y) = 100x + 60y + 2xy + 500$$

6 Find the regression line for the following sets of points. In each case compute the sum of the squares of the vertical distances once the line is known.
(a) $(0, -1), (0, 1), (1, -1), (1, 1)$
(b) $(-2, -1), (2, 1), (4, 2)$

7 **(Cobb/Douglas Economic Model)** This well-known model involves the following variables.

X = quantity of output (production) at price p

L = amount of labor at price w (wages)

K = amount of capital at price r (interest rate)

There are three other constants that depend on the particular firm or country under consideration: A, α_1, and α_2 (where $\alpha_1 + \alpha_2 = 1$). Finally we have the *production function*

$$X = AL^{\alpha_1}K^{\alpha_2}$$

and the *profit function*

$$\pi = pX - wL - rK$$

(a) Find the partial derivatives π_L and π_K (you must first replace X by $AL^{\alpha_1}K^{\alpha_2}$ in the profit function).
(b) Setting both partial derivatives equal to 0, rewrite these equations by isolating all factors of L and K on the left.
(c) Divide one equation into the other to get an expression for L/K in terms of r, α_1, w, and α_2. Finally write L as a multiple of K. Now interpret your work in terms of maximizing the profit function.

14 LAGRANGE MULTIPLIERS

In many problems of optimization (that is, of finding a maximum or a minimum value of a function) the real world context imposes certain restrictions on the variables of the function. This section is devoted to a method due to Lagrange that can convert such a problem to one which yields to the standard technique of setting the partial derivatives equal to zero. Before introducing the method, we consider two examples of such problems.

EXAMPLE 1 The total cost (T) to a firm for a certain product is a function of capital expenditures (C) and labor expenditures (L) and is given by

$$T = f(C, L) = 2C^2 - C - 2CL + 3L^2 + 1 \quad \text{(millions of dollars)}$$

Moreover, company policy dictates that the following relationships between C and L be maintained.

$$C = 2L - 1$$

The object is to

Minimize $f(C,L)$ subject to $g(C,L) = C - 2L + 1 = 0$

This latter condition, that $g(C,L) = 0$, is called a *constraint*. Notice that it imposes a restriction on the variables of f, the function to be minimized.

Our second example is stated in general form as a problem in analytic geometry. It should be observed, however, that all problems of optimization, once they are put into the form of mathematical equations or inequalities involving one or several variables, can be viewed as problems in analytic geometry where we seek a highest (or lowest) point on a curve or a surface, a point of intersection of various curves or surfaces, or the like.

EXAMPLE 2 The general equation for a plane in 3-space is

$$ax + by + cz + d = 0$$

where a, b, c, and d are constants, and at least one of the numbers a, b, c is not zero. It is obvious that when $d = 0$, the origin (0, 0, 0) lies on the plane since its coordinates satisfy the equation. Conversely, when $d \neq 0$, the origin cannot lie on the given plane. The problem becomes: Find the point on the given plane that is closest to the origin, and find the distance between it and the origin.

Now at first glance, this problem may not seem to represent the same type of problem as Example 1. Nevertheless, when properly translated, we see that once again we are faced with a function to be optimized subject to a constraint or restriction on the variables of that function.

First we note that the distance from the origin to any point (x, y, z) in 3-space is given by

$$h(x, y, z) = \sqrt{x^2 + y^2 + z^2}$$

[This fact is a simple extension of the Pythagorean Theorem and is left to the reader to verify by drawing a picture of a point in 3-space and constructing the line segment from the origin to that point as well as the segment from the origin to the point $(x, y, 0)$ in the xy-plane.] Thus our object is to find a point (x_0, y_0, z_0) on the plane $ax + by + cz + d = 0$ such that

$$(*) \quad \sqrt{x_0^2 + y_0^2 + z_0^2} \leq \sqrt{x^2 + y^2 + z^2}$$

for all other points (x, y, z) on the plane. Finally, we note that we may square both sides of (*) to get

$$(**) \quad x_0^2 + y_0^2 + z_0^2 \leq x^2 + y^2 + z^2$$

and we claim that if (x_0, y_0, z_0) is such that (*) holds for all (x, y, z) in the given plane, then (**) holds as well, and conversely.

In short, we may rephrase the original problem as follows:

Minimize $f(x, y, z) = x^2 + y^2 + z^2$ subject to $g(x, y, z) = ax + by + cz + d = 0$

With these two examples in mind, we introduce a method for dealing with problems of constrained optimization. As we have done in the past, we adapt an algorithmic procedure. For the sake of simplicity we specialize to functions of two variables, although as we shall see when solving Example 2, the technique easily generalizes to functions of any number of variables.

Method of Lagrange Multipliers

PROBLEM Given a function $f(x, y)$. Find (x_0, y_0) such that (1) $f(x_0, y_0)$ is optimal (either maximal or minimal) subject to the condition (2) $g(x_0, y_0) = 0$.
SOLUTION

1. Form a new function of three variables.
   $F(x, y, \lambda) = f(x, y) - \lambda g(x, y)$
2. Take the partial derivatives of F with respect to each of its variables and set them equal to zero:
   $$F_x(x, y, \lambda) = 0$$
   $$F_y(x, y, \lambda) = 0$$
   $$F_\lambda(x, y, \lambda) = 0$$
3. Find the simultaneous solutions (in some cases there will be no more than one solution) x_0, y_0, λ_0 of these equations and check the value
   $f(x_0, y_0)$
   to determine whether it is the desired optimal solution.

Comments. The value λ is called a *Lagrange multiplier*. The auxiliary function $F(x, y, \lambda)$ is called the *Lagrangean*. The function $f(x, y)$ is called the *objective function*. The method can be more tersely described as introducing a Lagrangean and optimizing it instead of the original function f. As with any general method, the practitioner must be wary of using Lagrange multipliers when a suitable direct approach is available. In our illustrations as well as in the exercises, some effort will be made to compare this technique to a more direct approach. Finally, it should be pointed out that we offer no justification for the Lagrange multiplier technique because doing so would require a fairly sophisticated excursion into properties of functions of several variables.

Solution of Example 1:

1. The Lagrangean becomes
   $$F(C, L, \lambda) = f(C, L) - \lambda g(C, L) = 2C^2 - C - 2CL + 3L^2 + 1 - \lambda(C - 2L + 1)$$

2. $$F_C(C, L, \lambda) = 4C - 1 - 2L - \lambda = 0$$
$$F_L(C, L, \lambda) = -2C + 6L + 2\lambda = 0$$
$$F_\lambda(C, L, \lambda) = -(C - 2L + 1) = 0$$

3. Solving for λ in the first two equations gives

$$\lambda = 4C - 2L - 1 \quad \text{and} \quad \lambda = C - 3L$$

Equating the two values of λ gives us

$$4C - 2L - 1 = C - 3L \quad \text{or} \quad 3C + L - 1 = 0$$

which together with the third equation from Step 2 yields

$$3C + L - 1 = 0$$
$$-C + 2L - 1 = 0$$

Solving these two equations simultaneously, we get

$$C = \tfrac{1}{7} \quad \text{and} \quad L = \tfrac{4}{7}$$

and we claim that the *minimal* value of $f(C, L)$, subject to the constraint $g(C, L) = 0$, is

$$f(\tfrac{1}{7}, \tfrac{4}{7}) = 2(\tfrac{1}{7})^2 - (\tfrac{1}{7}) - 2(\tfrac{1}{7})(\tfrac{4}{7}) + 3(\tfrac{4}{7})^2 + 1 = \tfrac{48}{49}$$

For comparison, we check the value of the objective function f at the point $(C, L) = (\tfrac{1}{3}, \tfrac{2}{3})$, noting that this point also satisfies the constraint $C - 2L + 1 = 0$:

$$f(\tfrac{1}{3}, \tfrac{2}{3}) = 2(\tfrac{1}{3})^2 - (\tfrac{1}{3}) - 2(\tfrac{1}{3})(\tfrac{2}{3})^2 + 3(\tfrac{2}{3})^2 + 1 = \tfrac{16}{9}$$

Solution of Example 2.

1. The Langrangean in this case is

$$F(x, y, z, \lambda) = x^2 + y^2 + z^2 - \lambda(ax + by + cz + d)$$

2. We have four partial derivatives to compute and to set equal to zero.

$$F_x(x, y, z, \lambda) = 2x - \lambda a = 0$$
$$F_y(x, y, z, \lambda) = 2y - \lambda b = 0$$
$$F_z(x, y, z, \lambda) = 2z - \lambda c = 0$$
$$F_\lambda(x, y, z, \lambda) = -(ax + by + cz + d) = 0$$

3. From the first three equations we get

$$\lambda = \frac{2x}{a} = \frac{2y}{b} = \frac{2z}{c}$$

from which we can solve for y and z in terms of x:

$$y = \frac{bx}{a}, \qquad z = \frac{cx}{a}$$

Now we substitute these values into the last equation in Step 2:

$$ax + b\left(\frac{bx}{a}\right) + c\left(\frac{cx}{a}\right) + d = 0$$

$$x\left(a + \frac{b^2}{a} + \frac{c^2}{a}\right) = -d$$

$$x = \frac{-ad}{a^2 + b^2 + c^2}$$

Similarly we can find

$$y = \frac{-bd}{a^2 + b^2 + c^2} \quad \text{and} \quad z = \frac{-cd}{a^2 + b^2 + c^2}$$

When we put these values, which we shall call x_0, y_0, and z_0, back into $f(x, y, z)$, we get

$$f(x_0, y_0, z_0) = \frac{a^2d^2}{(a^2 + b^2 + c^2)^2} + \frac{b^2d^2}{(a^2 + b^2 + c^2)^2} + \frac{c^2d^2}{(a^2 + b^2 + c^2)^2}$$

$$= \frac{d^2}{a^2 + b^2 + c^2}$$

Recalling that the original problem was to find the distance from this point to the origin and that $f(x_0, y_0, z_0)$ is simply the square of that distance, we see that

$$\begin{matrix}\text{Minimal distance from plane} \\ ax + by + cz + d \text{ to the origin}\end{matrix} = \sqrt{\frac{d^2}{a^2 + b^2 + c^2}} = \frac{|d|}{\sqrt{a^2 + b^2 + c^2}}$$

Alternative Methods of Solving Examples 1 and 2

In Example 1 we can take a more direct attack by first substituting the value (from the constraint equation)

$$C = 2L - 1$$

into the function $f(C, L)$ to get

$$T = f(C, L) = 2(2L - 1)^2 - (2L - 1) - 2(2L - 1)L + 3L^2 + 1$$
$$= 7L^2 - 8L + 4$$

Now we have a function of one variable, and we can set its derivative equal to 0:

$$\frac{dT}{dL} = 14L - 8 = 0$$

$$L = \tfrac{8}{14} = \tfrac{4}{7}$$

and hence

$$C = 2L - 1 = 2(\tfrac{4}{7}) - 1 = \tfrac{1}{7}$$

Clearly, this method is not only more direct, but it requires much less computation. (It is comforting to find that we arrive at the same answer.) Why, then, should we use the method of the Lagrange multiplier? The answer is simple: We *shouldn't* use it for Example 1! On the other hand, Example 2 provides us with a nice comparison; once again we attempt to solve for one of the variables, say z, in the constraint equation and substitute the result into the objective function. From

$$ax + by + cz + d = 0$$

we get

$$z = -\frac{1}{c}(ax + by + d)$$

The objective function now becomes

$$f(x, y, z) = x^2 + y^2 + \frac{1}{c^2}(ax + by + d)^2$$

We take partials with respect to x and y and set them equal to zero:

$$2x + \frac{1}{c^2}(2(ax + by + d)(a)) = 0$$

$$2y + \frac{1}{c^2}(2(ax + by + d)(b)) = 0$$

Simplifying these expressions on the left, we find

$$\left(\frac{a^2}{c^2} + 1\right)x + \frac{ab}{c^2}y + \frac{ad}{c^2} = 0$$

$$\frac{ab}{c^2}x + \left(\frac{b^2}{c^2} + 1\right)y + \frac{bd}{c^2} = 0$$

Multiplying by c^2, we have

$$(a^2 + c^2)x + aby + ad = 0$$

$$abx + (b^2 + c^2)y + bd = 0$$

Now these equations can be solved for x and y (as the reader may verify), but it should be apparent that this method is already more complicated than the Lagrange multiplier method for this problem.

Here, then, is a rule of thumb. When faced with a constrained optimization problem, first check to see whether you can solve the constraint equation for one of the variables in terms of the others (sometimes this part is quite a challenge or even impossible) and then see what kind of a problem develops when you substitute the result into the objective function, thereby decreasing the number of variables in the objective function by 1. If either of these steps is complicated, you should probably try the Lagrange multiplier method.

In certain situations, we are interested in optimizing a function subject to two or more constraint equations. In Example 3 we illustrate how to extend the method of Lagrange multipliers to handle such problems.

EXAMPLE 3 Allocation of Resources

One model for determining how to allocate labor and capital in a firm producing two commodities is given by the following extension of the Cobb–Douglas model.

$$Q_1 = L_1^{\alpha} C_1^{\beta}, \qquad Q_2 = L_2^{\gamma} C_2^{\delta} \qquad (0 \le \alpha, \beta, \gamma, \delta \le 1; \alpha + \beta = \gamma + \delta = 1)$$

where

Q_i = production level of ith commodity ($i = 1, 2$)
L_i = portion of labor costs (input) toward ith commodity
C_i = portion of capital expenditures toward ith commodity

If the unit price of the ith commodity is P_i, the problem is to maximize the *revenue*

$$R = P_1 Q_1 + P_2 Q_2$$

Subject to the budgetary constraints,

$$L_1 + L_2 = M \quad \text{and} \quad C_1 + C_2 = N$$

(where M dollars of labor and N dollars of capital are available). Although we shall leave the final computations to the reader, here is how to transform the problem. We construct

$$F(L_1, L_2, C_1, C_2, \lambda_1, \lambda_2) = P_1 Q_1 + P_2 Q_2 - \lambda_1 (L_1 + L_2 - M) - \lambda_2 (C_1 + C_2 - N)$$

and then proceed as before to differentiate this Lagrangean L with respect to each of its variables.

$$\frac{\partial F}{\partial L_1} = P_1 \alpha L_1^{\alpha - 1} C_1^{\beta} - \lambda_1$$

$$\frac{\partial F}{\partial L_2} = P_2 \gamma L_2^{\gamma - 1} C_2^{\delta} - \lambda_1$$

$$\frac{\partial F}{\partial C_1} = P_1 \beta L_1^{\alpha} C_1^{\beta - 1} - \lambda_2$$

$$\frac{\partial F}{\partial C_2} = P_2 \delta L_2^{\gamma} C_2^{\delta - 1} - \lambda_2$$

$$\frac{\partial F}{\partial \lambda_1} = -(L_1 + L_2 - M)$$

$$\frac{\partial F}{\partial \lambda_2} = -(C_1 + C_2 - N)$$

Setting each of these six partial derivatives equal to zero, we get a system of equations that, with some patience and a little algebraic juggling, give us the desired input levels for labor and capital that maximize total revenue.

Our final example shows how the Lagrangean can be used to establish a "law" of economics. It has to do with the so-called *utility function*. Suppose a consumer is interested in purchasing two products and has M dollars to spend. He wishes to determine how much money to spend on each product. In the ideal situation (i.e., ideal insofar as this problem is concerned), he would like to have as much of each product as possible, but recognizing his budgetary constraint, he desires a *utility function* $U = U(x_1, x_2)$ that measures his degree of satisfaction corresponding to the purchase of x_1 units of the first product and x_2 of the second, say purchased at respective prices of P_1 and P_2 per unit. The assumption that surfeit never sets in can be rephrased mathematically as

$$\frac{\partial U}{\partial x_1} > 0 \quad \text{and} \quad \frac{\partial U}{\partial x_2} > 0$$

which say that U increases as x_i increases for $i = 1, 2$. These partial derivatives are called *marginal utilities*, with respect to the first and the second products, respectively.

Thus we have an optimization problem:

Maximize $U = U(x_1, x_2)$

Subject to $P_1x_1 + P_2x_2 = M$

Solution: We construct the Langrangean

$$F(x_1, x_2, \lambda) = U(x_1, x_2) - \lambda(P_1x_1 + P_2x_2 - M)$$

$$\frac{\partial F}{\partial x_1} = \frac{\partial U}{\partial x_1} - P_1\lambda = 0$$

$$\frac{\partial F}{\partial x_2} = \frac{\partial U}{\partial x_2} - P_2\lambda = 0$$

$$\frac{\partial F}{\partial \lambda} = -(P_1x_1 + P_2x_2 - M) = 0$$

The first two of these equations yield

$$\lambda = \frac{\partial U/\partial x_1}{P_1} = \frac{\partial U/\partial x_2}{P_2}$$

In words, the utility function is maximized when *the ratios of marginal utilities to corresponding prices are equal.* This is sometimes called the *Second Law of Gossen*, and it generalizes rather easily to utility functions of any number of variables as can be seen in one of the exercises. Notice that the Lagrange multiplier λ in this case turns out to be (in the optimal solution) precisely this common value of the ratios. In more thorough treatments of applications of the Lagrange multiplier technique in economics, one often tries to interpret λ in the context of the problem.

Exercises Section 14

1 In each of the following problems, find the solution in two ways: (1) using the method of Lagrange multipliers and (2) by solving the constraint equa-

tion for one of the variables and substituting this value into the objective function.

(a) Minimize $f(x, y) = x^2 + 2y^2$
subject to $3x + 2y = 5$

(b) Maximize $f(x, y) = 4 - x^2 - 2y^2$
subject to $x - y - 2 = 0$

2 Use the method of Lagrange multipliers to find all optimal values (both maximal and minimal) of

$$f(x, y) = xy$$

subject to $x^2 + y^2 = 2$.

3 Minimize $f(x, y) = x^2 + xy + xy^2$, subject to $x + y = 1$.

4 Find the maximum volume of a rectangular parallelepiped inscribed in a sphere of radius a centered at the origin.

Translation: Maximize $f(x, y, z) = 8xyz$
subject to $x^2 + y^2 + z^2 = a^2$

5 Use the method of Lagrange multipliers to find the optimal solution to the single-product Cobb-Douglas model treated in the exercises of the previous section.

6 Establish the *Second Law of Gossen* in the case

$$U = U(x_1, x_2, \ldots, x_n)$$

Subject to $P_1x_1 + P_2x_2 + \cdots + P_nx_n = M$

7 Complete the solution of Example 3 in the case where

$$\alpha = \tfrac{1}{2},\ \beta = \tfrac{1}{2},\ \gamma = \tfrac{1}{3},\ \delta = \tfrac{2}{3},\ M = 10^6,\ N = 10^5$$

8 Reformulate Example 3 by optimizing the logarithm of the objective function and find the resulting partial derivatives.

15 TAYLOR SERIES

We have written the function exp as follows:

$$(1)\quad \exp(x) = \sum_{n=0}^{\infty} \frac{1}{n!}x^n = 1 + x + \frac{x^2}{2} + \frac{x^3}{3!} + \frac{x^4}{4!} + \cdots$$

where for each value of x $(-\infty < x < \infty)$ we get a series that converges to some real number. A similar kind of series expression involving the variable x results from trying to perform long division on

$$\frac{1}{1 - x}$$

We get

$$
\begin{array}{r|l}
 & 1 + x + x^2 + x^3 + \cdots \\ \hline
1 - x & 1 \\
 & \underline{1 - x} \\
 & \quad x \\
 & \quad \underline{x - x^2} \\
 & \qquad x^2 \\
 & \qquad \underline{x^2 - x^3} \\
 & \qquad\quad x^3 \\
 & \qquad\quad \underline{x^3 - x^4} \\
 & \qquad\qquad \cdot \\
 & \qquad\qquad\ \cdot \\
 & \qquad\qquad\quad \cdot
\end{array}
$$

and so it is tempting to write

$$(2) \quad \frac{1}{1-x} = \sum_{n=0}^{\infty} x^n = 1 + x + x^2 + x^3 + \cdots$$

but the expression

$$\frac{1}{1-x}$$

makes sense for all x except $x = 1$. On the other hand, the geometric series

$$1 + x + x^2 + x^3 + \cdots$$

will converge for $|x| < 1$ and diverge for $|x| \geq 1$. We conclude that (2) makes sense only when $|x| < 1$.

Even though there are some subtleties, such as the choice of values of x that guarantee convergence, it is often useful to try to express functions as infinite series containing a variable x. Without going into great detail we give a definition and some properties related to such representations.

Definition If a function f can be expressed in the form

$$f(x) = \sum_{n=0}^{\infty} a_n x^n$$

we call this expression a *Taylor series* for f; the terms of the sequence $\{a_n\}$ are called *coefficients* of the series.

Properties of Taylor Series

I. Every Taylor series has a *radius of convergence* R in the sense that the series

$$\sum_{n=0}^{\infty} a_n x^n \qquad \begin{cases} \text{converges for } |x| < R \\ \text{diverges for } |x| > R \end{cases}$$

(*Note:* There is no general rule for what happens at $x = R$ or $x = -R$; sometimes the series may converge, sometimes it may diverge at either point.)

II. Taylor series can be *differentiated term by term*. That is, if

$$f(x) = \sum_{n=0}^{\infty} a_n x^n = a_0 + a_1 x + a_2 x^2 + a_3 x^3 + \cdots$$

then

$$f'(x) = \sum_{n=1}^{\infty} n a_n x^{n-1} = 0 + a_1 + 2a_2 x + 3a_3 x^2 + \cdots$$

[Also the radius of convergence for $f'(x)$ is at least as big as for $f(x)$.]

III. Taylor series can be *integrated term by term*. If

$$f(x) = \sum_{n=0}^{\infty} a_n x^n$$

then

$$\int_c^d f(x)\,dx = \sum_{n=0}^{\infty} \left(\frac{a_n}{n+1} x^{n+1} \Big|_{x=c}^{d} \right)$$

IV. The coefficients a_n of a Taylor series are found by successive differentiation of the function f: Consider

$$f(x) = a_0 + a_1 x + a_2 x^2 + a_3 x^3 + a_4 x^4 + \cdots$$
$$f'(x) = a_1 + 2a_2 x + 3a_3 x^2 + 4a_4 x^3 + \cdots$$
$$f''(x) = 2a_2 + 3 \cdot 2a_3 x + 4 \cdot 3a_4 x^2 + \cdots$$
$$f'''(x) = 3 \cdot 2 \cdot 1a_3 + 4 \cdot 3 \cdot 2a_4 x + \cdots$$

In particular when $x = 0$ (this is one point where we are certain of convergence for all the above series),

$$f(0) = a_0$$
$$f'(0) = 1 \cdot a_1$$
$$f''(0) = (2!)a_2$$
$$f'''(0) = (3!)a_3$$

General Formula: $a_n = \dfrac{f^{(n)}(0)}{n!}$

V. If

$$f(x) = \sum_{n=0}^{\infty} a_n x^n$$

and

$$g(x) = \sum_{n=0}^{\infty} b_n x^n$$

then the functions $f \pm g$ and kf, where k is any constant, have the Taylor series

$$(f \pm g)(x) = \sum_{n=0}^{\infty} (a_n \pm b_n)x^n, \qquad (kf)(x) = \sum_{n=0}^{\infty} ka_n x^n$$

The radius of convergence for $f + g$ is at least as big as the smaller of the two radii of convergence for f and g, and the radius of convergence for kf is the same as for f, assuming $k \neq 0$.

These Taylor series representations enable us to approximate a function f by the partial sums of the series, called *Taylor polynomials*. If

$$f(x) = \sum_{n=0}^{\infty} a_n x^n$$

then we get *successive approximations*

$f(x) \approx a_0,$	Taylor polynomial of degree 0 for f
$f(x) \approx a_0 + a_1 x,$	Taylor polynomial of degree 1 for f
$f(x) \approx a_0 + a_1 x + a_2 x^2,$	Taylor polynomial of degree 2 for f
$f(x) \approx a_0 + a_1 x + \cdots + a_n x^n,$	Taylor polynomial of degree n for f

It is virtually impossible to tell in advance how good the approximations are. There are some rather complicated techniques that give formulas for the maximum error involved in the approximation

$$f(x) \approx \sum_{n=0}^{N} a_n x^n$$

But for our purposes, we shall be content to compute values for small choices of N (say 2, 3, 4, 5) and assume that the polynomial value is close to $f(x)$. A decent method to use on a computer is to compute the successive values

$$\left\{ \sum_{n=0}^{N} a_n x^n \right\}$$

for $N = 0, 1, 2, \ldots, 10$ (or more) until the difference from one term to the next is very small (say 10^{-5} or 10^{-6}). In practice this technique seems to work reasonably well for all but the most pathological functions. In other words if the difference

$$\left| \sum_{n=0}^{14} a_n x^n - \sum_{n=0}^{13} a_n x^n \right| = |a_{14} x^{14}|$$

is less than 10^{-6}, we can be pretty sure that the error

$$\left| f(x) - \sum_{n=0}^{14} a_n x^n \right|$$

is also fairly small.

EXAMPLE 1 We find the Taylor series for $f(x) = \text{Log}(1 + x)$. To find a_n, we need $f^{(n)}(0)$.

$$f'(x) = \frac{1}{1+x} = (1 + x)^{-1}$$

$$f''(x) = -(1 + x)^{-2}$$

$$f'''(x) = 2(1 + x)^{-3}$$

$$f^{(4)}(x) = -6(1 + x)^{-4} = -(3!)(1 + x)^{-4}$$

$$f^{(5)}(x) = +(4!)(1 + x)^{-5}$$

In general we will have

$$f^{(n)}(x) = (-1)^{n-1}(n-1)!(1 + x)^{-n}, \qquad \text{for } n = 1, 2, 3, \ldots$$

Thus

$$f(0) = \text{Log}(1 + 0) = 0$$

$$f'(0) = (1 + 0)^{-1} = +1$$

$$f''(0) = -1$$

$$f'''(0) = 2$$

$$f^{(4)}(0) = -(3!)$$

$$f^{(5)}(0) = +(4!)$$

$$f^{(n)}(0) = (-1)^{n-1}(n-1)!$$

Dividing $f^{(n)}(0)$ by $n!$ to get a_n, we find

$$(3) \quad f(x) = \text{Log}(1 + x) = x - \frac{x^2}{2} + \frac{x^3}{3} - \frac{x^4}{4} + \frac{x^5}{5} - \cdots$$

$$= \sum_{n=1}^{\infty} (-1)^{n-1}\left(\frac{1}{n}\right)x^n$$

Note: This series begins with $n = 1$ (i.e., the "x term"); this is quite acceptable. While it is probably not obvious what the radius of convergence for this series is, we should observe that

$$\text{Log}(1 + x) \quad \text{is not defined for} \quad x \leq -1$$

Thus we might be tempted to guess that $R = 1$, i.e., that the series converges for $-1 < x < 1$. Indeed this is the case, but the following theorem gives us a method for discovering R in general.

Theorem (Ratio Test)

The series

$$\sum_{n=0}^{\infty} a_n x^n$$

will have a radius of convergence R provided

$$\lim_{n\to\infty} \left|\frac{a_{n+1}}{a_n}\right| = \frac{1}{R}$$

EXAMPLE 2 We found that

$$\text{Log}(1+x) = \sum_{n=1}^{\infty} (-1)^{n-1}\left(\frac{1}{n}\right)x^n$$

$$\left|\frac{a_{n+1}}{a_n}\right| = \left|\frac{(-1)^n\left(\frac{1}{n+1}\right)}{(-1)^{n-1}\left(\frac{1}{n}\right)}\right| = \left|-\frac{n}{n+1}\right| = 1 - \frac{1}{n+1}$$

Thus

$$\lim_{n\to\infty}\left|\frac{a_{n+1}}{a_n}\right| = \lim_{n\to\infty}\left(1 - \frac{1}{n+1}\right) = 1$$

and $R = 1$ as we predicted.

Sometimes the process of computing enough derivatives to see a pattern for the coefficients of the Taylor series involves a good deal of work. In many instances we can save some time by appealing to the

Theorem (Substitution Rule)
If

$$f(x) = \sum_{n=0}^{\infty} a_n x^n$$

has radius of convergence R and if $|g(x)| < R$, then

$$f(g(x)) = \sum_{n=0}^{\infty} a_n (g(x))^n$$

EXAMPLE 3 Suppose we wish to find the Taylor series for

$$\frac{1}{1+x}$$

We already know that

$$\frac{1}{1-x} = \sum_{n=0}^{\infty} x^n$$

and so we put

$$f(x) = \frac{1}{1-x}$$

Thus, letting $g(x) = -x$ in the theorem, we have

$$(4) \quad \frac{1}{1+x} = f(-x) = \sum_{n=0}^{\infty} (-x)^n = \sum_{n=0}^{\infty} (-1)^n x^n$$

Actually we can (and should) carry this a step further. We know that

$$\text{Log}(1 + x) = \int_0^x \frac{1}{1+x}\,dx$$

Thus we can integrate, term by term, the series

$$\frac{1}{1+x} = 1 - x + x^2 - x^3 + x^4 - x^5 + \cdots .$$

to get

$$\text{Log}(1 + x) = x - \frac{x^2}{2} + \frac{x^3}{3} - \frac{x^4}{4} + \frac{x^5}{5} - \frac{x^6}{6} + \cdots$$

EXAMPLE 4 A function that will be useful to us in the next chapter is

$$f(x) = \frac{1}{\sqrt{2\pi}} \exp\left(-\frac{1}{2}x^2\right)$$

We take

$$\exp(x) = 1 + x + \frac{x^2}{2} + \frac{x^3}{3!} + \frac{x^4}{4!} + \cdots$$

and substitute $(-\frac{1}{2}x^2)$ for x to get the Taylor series

$$(5)\quad \exp(-\tfrac{1}{2}x^2) = 1 - \tfrac{1}{2}x^2 + \frac{x^4}{4 \cdot 2} - \frac{x^6}{8 \cdot 3!} + \frac{x^8}{16 \cdot 4!} - \cdots$$

$$= \sum_{n=0}^{\infty} (-1)^n \left(\frac{1}{2^n n!}\right) x^{2n}$$

$$(6)\quad \frac{1}{\sqrt{2\pi}} \exp(-\tfrac{1}{2}x^2) = \sum_{n=0}^{\infty} (-1)^n \frac{1}{\sqrt{2\pi}} \left(\frac{1}{2^n n!}\right) x^{2n}$$

Note that in this last series only even powers of x appear. We can think of the other coefficients as being 0 or we can admit reasonable modifications in our definitions. We have already seen the need to start a Taylor series with $n = 1$ rather than $n = 0$, and such modifications are quite common.

In practice there is a good reason to allow some point other than $x = 0$ to be the "center of convergence." For instance if we wanted to find a series representation for

$$f(x) = \text{Log}(x)$$

we could not even begin in the obvious way since

$f(x)$ is *not defined* at $x = 0$

But instead of expanding this function in powers of x, we try powers of $(x - 1)$. Since we already have the series

$$g(x) = \text{Log}(1 + x) = \sum_{n=1}^{\infty} (-1)^{n-1} \left(\frac{1}{n}\right) x^n$$

we can use our substitution rule to get

$$\text{Log}\,(x) = g(x-1) = \text{Log}\,(1 + (x-1)) = \sum_{n=1}^{\infty} (-1)^{n-1}\left(\frac{1}{n}\right)(x-1)^n \quad ,$$

that is,

$$(7) \quad \text{Log}\,(x) = (x-1) - \frac{(x-1)^2}{2} + \frac{(x-1)^3}{3} - \frac{(x-1)^4}{4} + \cdots$$

In other words, we shall refer to a representation of the form

$$f(x) = \sum_{n=0}^{\infty} a_n(x-c)^n$$

as a *Taylor series for f about the point c*. In general it will converge for all x such that $|x - c| < R$, i.e., $c - R < x < c + R$. In this case we have

$$a_n = \frac{f^{(n)}(c)}{n!}$$

Finally we remark that one important application of Taylor series is in the realm of approximating functions and approximating integrals. Often one resorts to the following approximations.

$$f(x) \approx \sum_{n=0}^{N} a_n(x-c)^n \qquad \text{(a Taylor polynomial of degree } N\text{)}$$

to give the approximation

$$\int_a^b f(x)\,dx \approx \int_a^b \sum_{n=0}^{N} a_n(x-c)^n\,dx$$

(This integral is easy to handle since the integrand is a polynomial.)

The latter is particularly useful when we do not know an antiderivative for f.

Exercises Section 15

1 Find the radius of convergence of each of the following series.

(a) $\sum_{n=1}^{\infty} \left(\frac{1}{100n}\right)x^n$

(b) $\sum_{n=0}^{\infty} \left(\frac{1}{2^n}\right)x^n$

(c) $\sum_{n=0}^{\infty} \left(\frac{1}{3^n}\right)x^n$

(d) $\sum_{n=0}^{\infty} nx^n$

(e) $\sum_{n=1}^{\infty} \left(\frac{1}{n2^n}\right)x^n$

2 Find the Taylor series for each of the following functions and determine the radius of convergence in each case.

(a) $f(x) = \exp\,(2x)$ about the point 0 (Do this directly using Property IV and then by substitution using Property V.)

(b) $f(x) = \dfrac{1}{(1+x)^2}$ about the point 0

(c) $f(x) = \sqrt{x}$ about the point 1

3 Use the first five terms of the Taylor series for

$$f(x) = \sqrt{1+x}$$

to approximate

$$\int_0^1 \sqrt{1+x}\,dx$$

Compare your answer to the true value (i.e., do the integration directly).

4 Write the Taylor series for the function

$$f(x) = x \cdot \exp(x) \quad \text{about the point} \quad x = 0$$

[*Hint:* Write the Taylor series for exp (x) and multiply by x.]

5 Write the Taylor series for the function

$$f(x) = (1-x)^{-2} \quad \text{about the point} \quad x = 0$$

and then differentiate this series term by term to get a Taylor series for

$$f(x) = (1-x)^{-3}$$

6 Differentiate the Taylor series for exp (x). What happens?

7 Use Taylor series to find the following limits

(a) $\displaystyle\lim_{x\to 0} \frac{e^x - 1}{x}$

(b) $\displaystyle\lim_{x\to 0} \frac{e^x - e^{2x}}{x}$

Supplementary Exercises for Chapter 4

1 Sketch the graphs of each of the following functions, and evaluate $\int_0^1 f(x)\,dx$ by computing $R_1, R_2, \cdots, R_6$ and guessing at $\lim_{n\to\infty} R_n$.

(a) $f(x) = \begin{cases} 3, & \text{if } x = \frac{1}{2} \\ 1, & \text{if } x = \frac{1}{3} \\ 0, & \text{otherwise} \end{cases}$

(b) $f(x) = \begin{cases} 1, & \text{if } x \text{ has the form } 1/n \text{ for some positive integer } n \\ 0, & \text{otherwise} \end{cases}$

2 Evaluate $\int_0^1 f(x)\,dx$.

(a) $f(x) = \begin{cases} 2, & \text{if } 0 \le x < \frac{1}{2} \\ -2, & \text{if } \frac{1}{2} \le x < 1 \end{cases}$

(b) $f(x) = \begin{cases} x, & \text{if } 0 \le x < \frac{1}{2} \\ 1 - x, & \text{if } \frac{1}{2} \le x \end{cases}$

3 Differentiate each of the following.

(a) $f(x) = (x^2 + 3)(x^3 - 1)(8x + 7)$

(b) $g(x) = 5\dfrac{x+1}{x-2}$

(c) $h(x) = (x^2+3)\dfrac{(x^2+2x+7)}{(2-x^2)}$

(d) $f(t) = \dfrac{1}{1-t}$

(e) $g(t) = \dfrac{50}{50-t}$

(f) $h(t) = (t+1)\exp(t)$

(g) $F(x) = (x^3+7x^2)^4$

(h) $G(x) = \exp(1-x^2)$

(i) $H(x) = \exp\left(\dfrac{x-a}{b}\right)$

(j) $F(t) = \text{Log}\left(\dfrac{1-t}{1+t}\right)$

(k) $G(t) = (\text{Log}(t+3))^2$

(l) $H(t) = 2^t \text{Log}(t)$

4 Sketch the graphs of each of the following functions along with their tangent lines at the prescribed points.

(a) $f(x) = x^3$, tangent lines where $x = -2, -1, 0, 1, 2$

(b) $g(x) = \dfrac{1}{1+x^2}$, tangent lines where $x = -2, -1, 0, 1, 2, 3$

(c) $h(x) = 4-(x-2)^2$, tangent lines where $x = 0, 1, 2, 3, 4$

(d) $h(x) = \text{Log}(4x^2+1)$, tangent lines where $x = 0, 1, 2$

5 Find the maximum and minimum values of each of the following functions on the prescribed intervals.

(a) $f(x) = x^2+2x-5$, on $[-4, 3]$

(b) $g(x) = x^4-2x^2+1$, on $[-1, 5]$

(c) $h(x) = \dfrac{3x}{x+1}$, on $[-\frac{1}{2}, 2]$

(d) $k(x) = \dfrac{x^2+2}{x+2}$, on $[-3, 0]$

(e) $m(x) = \frac{1}{3}x^3+\frac{1}{2}x^2-2x$, on $[-3, 4]$

(f) $f(t) = (t+1)e^t$, on $[-1, 3]$

6 Sketch the graph of a function f with domain $[0, 3]$ such that:

(a) f has a relative maximum at $x = 2$, an absolute minimum at $x = 1$, and an absolute maximum at $x = 3$.

(b) f is increasing on $[0, 1]$, decreasing on $[1, 2]$, constant on $[2, 3]$.

(c) f has a relative minimum at $x = 1$ but no derivative there, and f has a relative minimum at $x = 2$ where f' exists.

(d) f has no horizontal tangents on $[0, 3]$.

(e) $f(x) = 0$ at $x = 1$, $x = 2$, and $x = 3$; f is increasing on $[0, 1.5]$ and $[2, 3]$; and f has an absolute maximum at 1.5.

7 Evaluate the following definite integrals:

(a) $\int_{-1}^{2} |x|\,dx$

(b) $\int_{-3}^{3} |x-2|\,dx$

(c) $\int_{0}^{2} |3x-2|\,dx$

8 Each of the following is the density function for a continuous random variable. Find the corresponding distribution function F. Sketch the graphs of both f and F, and find the specified probabilities.

(a) $f(x) = \begin{cases} \frac{1}{2}, & \text{if } -1 \leq x \leq 1 \\ 0, & \text{otherwise} \end{cases}$ Find $P(0 \leq X \leq \frac{3}{8})$.

(b) $f(x) = \begin{cases} 2x, & \text{if } 0 \leq x \leq 1 \\ 0, & \text{otherwise} \end{cases}$ Find $P(\frac{1}{4} \leq X \leq 1)$.

(c) $f(x) = \begin{cases} e^x, & \text{if } x \leq 0 \\ 0, & \text{otherwise} \end{cases}$ Find $P(-1 \leq X \leq 0)$.

(d) $f(x) = \begin{cases} 3e^{-3x}, & \text{if } x \geq 0 \\ 0, & \text{otherwise} \end{cases}$ Find $P(X \leq 2)$.

9 Find $E(X)$, $E(X^2)$, and σ_X for the random variables whose density functions are given in the preceding exercise.

10 For each of the following distribution functions, find the corresponding density function f and sketch the graphs of f and F.

(a) $F(x) = \begin{cases} 0, & \text{if } x < 1 \\ \text{Log}(x), & \text{if } 1 \leq x \leq e \\ 1, & \text{if } x > e \end{cases}$

(b) $F(x) = \begin{cases} 0, & \text{if } x < \frac{1}{2} \\ 2x - 1, & \text{if } \frac{1}{2} \leq x \leq 1 \\ 1, & \text{if } x > 1 \end{cases}$

(c) $F(x) = \begin{cases} 0, & \text{if } x < 0 \\ \sqrt{x}, & \text{if } 0 \leq x \leq 1 \\ 1, & \text{if } x > 1 \end{cases}$

(d) $F(x) = \begin{cases} 1 - e^{-10x}, & \text{if } x \geq 0 \\ 0, & \text{otherwise} \end{cases}$

(e) $F(x) = \begin{cases} 0, & \text{if } x < 0 \\ x^2, & \text{if } 0 \leq x \leq \frac{1}{2} \\ 1 - 3(1 - x)^2, & \text{if } \frac{1}{2} < x < 1 \\ 1, & \text{if } x \geq 1. \end{cases}$

11 The sum of one number and twice another number is 40. Among all pairs of such numbers, find those whose product is the largest.

12 **(a)** Find the dimensions of the rectangle of maximal area that can be inscribed in the circle $x^2 + y^2 = 4$.

(b) Find the dimensions of the rectangle of maximal area that can be inscribed in a circle of radius R.

13 A horizontal trough is to be made from a piece of sheet metal 10 feet long and 2 feet wide by folding up equal widths on each side. What is the maximal volume this trough can be? [*Hint:* Assume that the sides are folded perpendicular to the base so that volume = (length)(width of base)(height of sides).]

14 Find the shortest line segment with ends on the positive x- and y-axes that passes through the point (2, 3).

15 Find the length of the longest ladder that can be carried horizontally around the corner in an L-shaped corridor 6 feet wide.

16 A particle moves along a straight line according to

$$s(t) = 132 + 108t - 16t^2 + 2t^3$$

where s is the distance in inches and t is the time in hours.

(a) Find the velocity and the acceleration at any time t.

(b) Is the velocity ever negative?

(c) What is the minimal velocity?

(d) When is the acceleration negative?

17 The biologist A. V. Hill published a paper in 1922 in the Physiological Reviews (Vol. II, No. 2), entitled "The Mechanism of Muscular Contraction." Hill defined "mechanical efficiency" E to be the ratio of realizable work to (heat) energy liberated, and he asserted that for the bicep of a healthy young man E is given by

$$E = \frac{.4\left(1 - \dfrac{.24}{t}\right)}{(1 + .2t)}$$

where t represents the number of seconds taken to isometrically contact the muscle. He concluded that human muscles have an optimal working speed; a higher speed (i.e., a smaller value for t in the above equation) seriously interferes with their efficiency, whereas a slower speed interferes, but not so drastically.

(a) Find the value of t that will make E maximal, and compute this optimal value of E (this value represents Hill's assumption about "optimal efficiency").

(b) Sketch a few points on the graph of E including points near the optimal value to try to interpret the effect on efficiency of a muscle contraction that is either too swift or too slow.

18 Suppose the demand function and the cost function for a particular item are both linear. That is,

$$p = a - bx \quad \text{(in dollars)}$$
$$C = c + dx \quad \text{(in dollars)}$$

where x represents the number of units to be produced, p the price per unit, and C the total production cost; a, b, c, and d are positive constants.

(a) Find the value for x, and the corresponding price p that will maximize profit.

(b) Suppose that the manufacturer receives word that he will be assessed a tax of t dollars for each item produced. Find the new cost function and show that he should, in order to maximize profits, pass on to the consumer exactly half of the tax burden.

19 Suppose that the demand function and the cost function for producing a particular item are given by

$$p = a - bx \qquad \text{(in dollars)}$$
$$C = c + dx + ex^2 \quad \text{(in dollars)}$$

(a) Find the price p (in terms of a, b, c, d, and e) to be charged in order to maximize profits.

(b) If a tax of t dollars is imposed on each unit produced, find the price that will maximize profits and compare this price to the optimal price without the tax. You may assume that a, b, c, and e are positive and that d may be positive, zero, or negative.

20 Suppose a market research program has predicted that x units of a certain type of portable color television set can be sold for a price of

$$P(x) = 50{,}000x^{-1/2} \text{ dollars}$$

The manufacturer has determined the cost of producing x units to be

$$C(x) = 2{,}500{,}000 + 100x$$

(a) Find the optimal number of sets to sell in order to maximize profits.

(b) Find the optimal unit sales price and the resulting profit.

21 Let Q be the number of items that can be sold at price p. In other words, Q and p are related in a demand function such as

$$Q = \frac{20}{p} + 10$$

We define the *elasticity of demand* $E(p)$, which is also a function of p, as follows

$$E(p) = -p\frac{Q'(p)}{Q(p)}$$

(a) Show that if $Q = c/p$ for some constants c, then $E(p) = 1$.

(b) Demand is said to be *elastic* if $E(p) > 1$. Show that if demand is elastic, then a reduction in price will increase revenue.

(c) Find the elasticity of demand for each of the following.

(i) $Q = \dfrac{4}{1 + p^2}$

(ii) $Q = ae^{-kp}$

(iii) $Q = (p - 60)^{-2}$

(iv) $Q = 100p^{-3/2}$

22 **(Average cost versus total cost.)** Suppose the total cost of producing and marketing x units of a particular item is $C(x)$. We define the *average cost* (per item) to be

$$\frac{C(x)}{x}$$

The function $C'(x)$ is called the marginal cost. It is sometimes illuminating to sketch on the same coordinate system the graphs of both of these functions.

(a) Suppose $C(x) = 100x + 250$. Sketch the graphs of $y = C'(x)$ and $y = C(x)/x$ and find where these two curves intersect.

(b) Suppose $C(x) = 6x^2 + 5x + 3$. Sketch the graphs of $y = C'(x)$ and $y = C(x)/x$ and find where these two curves intersect.

(c) Suppose $C(x) = ax^2 + bx + c$. Show that the graphs of $y = C'(x)$ and $y = C(x)/x$ intersect at the minimum point on the latter and interpret your result in words.

23 The psychologist Clark Hull used the formula

$$f(x) = 100(1 - e^{-kx})$$

as a model to describe reinforcement in learning theory. Here k is some positive constant, x is the number of repetitions performed, and $f(x)$ is the level of reinforcement or habit strength.

(a) Show that $f'(x) > 0$ for all x. What does this say about the habit strength?

(b) Sketch the graph of f in the cases where $k = 1$, $k = .5$, and $k = .1$. Pay particular attention to the places where the graph cuts the x-axis and the y-axis as well as the slopes of the graph at those points.

24 Find the area under the graph $f(x) = (1/\sqrt{2\pi}) \exp(-x^2/2)$ between $x = 0$ and $x = 1$ by:

(a) Finding the Taylor series for $f(x)$.

(b) Integrating part (a) term by term between 0 and x.

(c) Evaluating the resulting series at $x = 1$ for as many terms as are required to get error less than .0001.

25 Find $\exp(1) = e$ accurate to within .0001 using the Taylor series expansion of $\exp(x)$.

26 Let

$$f(x) = \begin{cases} 3x^2, & \text{if } -1 \le x \le 0 \\ 0, & \text{otherwise} \end{cases}$$

Find μ_x, σ_x, and $P(|x - \mu| \ge 2\sigma)$ (exactly) for the random variable whose density function is f.

27 **The Indefinite Integral.** We have used the notation $\int_a^b f(x)\,dx = F(x)|_a^b$ where $F(x)$ is an antiderivative of $f(x)$, i.e., $F'(x) = f(x)$. Another commonly used notation is $\int f(x)\,dx = F(x) + c$, where the symbol $\int - dx$ is referred to as the *indefinite integral*. The "$+ c$" acknowledges the fact that any constant can be added to an antiderivative of f and still leave us with an antiderivative. We can now state a problem such as

$$\int x^2\,dx = ?$$

and the answer is

$$\int x^2\,dx = \frac{x^3}{3} + c$$

Solve each of the following.

(a) $\int 3(2x + 1)^2\,dx = ?$

(b) $\int 2\exp(3x)\,dx = ?$

(c) $\int \frac{\text{Log}(2x)}{x}\,dx = ?$

(d) $\int \frac{\text{Log}(\text{Log}\,x)}{x\,\text{Log}\,x}\,dx = ?$

(e) $\int x^2(x^3 + 1)^5\,dx = ?$

(f) $\int 4x(3x^2 + 1)^7\,dx = ?$

(g) $\int (3x + 3)e^{(x^2+2x)}\,dx = ?$

(h) $\int xe^{\text{Log}\,x^2}\,dx = ?$

(i) $\int \frac{2\,\text{Log}\,(1/x)}{x^2}\,dx = ?$

(j) $\int \frac{3x}{x^2 + 1}\,dx = ?$

(k) $\int \frac{2x}{x^4 + 2x^2 + 1}\,dx = ?$

(l) $\int \frac{4e^{1/x}}{x^2}\,dx = ?$

28 **(Implicit Differentiation.)** There are certain occasions when, instead of being given y as a function of x and wanting $D_x y$, one is given an equation involving x and y and wants $D_x y$. For example, $x^5y^2 + xy^4 = y$. Since one cannot readily solve for y as a function of x, one simply assumes that y is a differentiable function of x and proceeds, using the Chain Rule, as follows

$$x^5y^2 + xy^4 = y$$

so

$$D_x(x^5y^2 + xy^4) = D_x y$$
$$D_x(x^5y^2) + D_x(xy^4) = D_x y$$
$$x^5D_x y^2 + y^2D_x x^5 + xD_x y^4 + y^4D_x x = D_x y$$
$$x^5 2y^2 D_x y + y^2 \cdot 5x^4 + x \cdot 4y^3 D_x y + y^4 \cdot 1 = D_x y$$
$$5x^4y^2 + y^4 = D_x y(1 - 2x^5y^2 - 4xy^3)$$
$$\frac{5x^4y^2 + y^4}{1 - 2x^5y^2 - 4xy^3} = D_x y$$

In practice, given an equation such as $\exp\,(xy) = x^2 + y^2$, one would proceed more directly to

$$\exp\,(xy)\,(xD_x y + y) = 2x + 2yD_x y$$
$$D_x y(x\,\exp\,(xy) - 2y) = 2x - y\,\exp\,(xy)$$
$$D_x y = \frac{2x - y\,\exp\,(xy)}{x\,\exp\,(xy) - 2y}$$

Try your hand at the following.

(a) $x^4y^3 + x^3y^2 + x^2y = 0$
(b) $x^2 + xy + y^2 = 1$
(c) $y\,\text{Log}\,x = x\,\text{Log}\,y$
(d) $x^3 + y^3 = xy$
(e) $x^2 + y^2 = 1$
(f) $\sqrt{2y} + \sqrt[3]{3y} = x$
(g) We can solve the equation in (e) as $y = \pm\sqrt{1 - x^2}$. It is instructive to find $D_x y$ from this and compare your answer with your answer to (e).
(h) $x^{2/3} + y^{2/3} = 1$
(i) Sketch the graph of (h).

29 **(Integration by Parts.)** A useful method of finding antiderivatives comes from an examination of the product rule for differentiation. If u and

v are differentiable functions of x, then

$$D_x(u \cdot v) = uD_xv + vD_xu$$

or

$$uD_xv = D_x(u \cdot v) - vD_xu$$

Therefore

$$\int_a^b uD_xv\,dx = \int_a^b D_x(u \cdot v)\,dx - \int_a^b vD_xu\,dx$$

$$= u \cdot v\Big|_a^b - \int_a^b vD_xu\,dx$$

As an example of the use of this formula, consider

$$\int_0^1 x \cdot e^x\,dx$$

We let

$$u = x, \qquad D_xv = e^x$$

whence

$$D_xu = 1, \qquad v = e^x$$

Substitution yields

$$\int_0^1 xe^x\,dx = xe^x\Big|_0^1 - \int_0^1 e^x \cdot 1\,dx$$

$$= xe^x\Big|_0^1 - e^x\Big|_0^1$$

$$= e - (e - 1)$$

$$= 1$$

Note that the technique consists of choosing one factor in the integrand to be u (it should be one which you can easily differentiate), the other factor to be D_xv (it should be one for which you know an antiderivative), and seeing if vD_xu is an easier integrand than uD_xv was.
Here are some problems to try.

(a) $\displaystyle\int_1^2 x \text{ Log } x\,dx$

(b) $\displaystyle\int_1^2 \text{Log } x\,dx$ (Hint: Let $D_xv = 1$.)

(c) $\displaystyle\int_0^1 x^3(x^2 + 1)^6\,dx$

(d) $\displaystyle\int_0^1 x^3e^{x^2}\,dx$

(e) $\displaystyle\int_0^1 x^2e^x\,dx$

(f) $\displaystyle\int_0^1 x^3e^x\,dx$

(g) $\displaystyle\int_0^1 \frac{x^3}{(1 + x^2)^3}\,dx$

(h) $\displaystyle\int_{-1}^0 \frac{x^2}{\sqrt{1 - x}}\,dx$

(i) $\displaystyle\int_{-1/2}^{1/2} \frac{x}{\sqrt{1 - x^2}}\,dx$

(j) $\displaystyle\int_1^2 \sqrt{x} \text{ Log } x\,dx$

(k) $\displaystyle\int_0^3 x\sqrt{x + 1}\,dx$

30 (Trigonometry.) Define functions $\sin(x)$ and $\cos(x)$ by

$$\sin x = x - \frac{x^3}{3!} + \frac{x^5}{5!} - \cdots = \sum_{k=0}^{\infty} (-1)^k \frac{x^{2k+1}}{(2k+1)!}$$

and

$$\cos x = 1 - \frac{x^2}{2!} + \frac{x^4}{4!} - \cdots = \sum_{k=0}^{\infty} (-1)^k \frac{x^{2k}}{(2k)!}$$

(a) Show that the series defining $\sin x$ and $\cos x$ converge for all numbers x.

(b) Show that $D_x \sin x$ and $D_x \cos x = -\sin x$.

(c) Show that $[\sin x]^2 + [\cos x]^2 = 1$ for all x.
(*Hint:* Show that the function $f(x) = [\sin x]^2 + [\cos x]^2$ has a 0 derivative and is therefore constant; then look at $f(0)$.)

(d) Show that $\sin(-x) = -\sin x$ and $\cos(-x) = \cos x$ for all x.

(e) In order to introduce the standard identities of trigonometry, we resort to a little trick. Let

$$f(x) = \cos(x+a) - \cos x \cos a + \sin x \sin a$$

$$g(x) = \sin(x+a) - \sin x \cos A - \cos x \sin a$$
$$h(x) = (f(x))^2 + (g(x))^2$$

Show that

(i) $f'(x) = -g(x)$
(ii) $g'(x) = f(x)$
(iii) $h'(x) = 0$
(iv) $h(x) = 0$
(v) $\cos(x+a) = \cos x \cos a - \sin x \sin a$

and

$$\sin(x+a) = \sin x \cos a + \sin a \cos x$$

(f) One can obtain, from (e), the basic identities of trigonometry. Demonstrate each of the following.

(i) $\sin(2x) = 2 \sin x \cos x$

(ii) $\cos(2x) = \cos^2 x - \sin^2 x$
$= 1 - 2\sin^2 x$
$= 2\cos^2 x - 1$

(iii) $\sin\left(\frac{x}{2}\right) = \pm\frac{1 - \cos x}{2}$

(iv) $\cos\left(\frac{x}{2}\right) = \pm\frac{1 + \cos x}{2}$

(g) Define

$$\tan x = \frac{\sin x}{\cos x}, \qquad \sec x = \frac{1}{\cos x}$$

$$\cot x = \frac{\cos x}{\sin x}, \qquad \csc x = \frac{1}{\sin x}$$

whenever the denominators are not 0.

Show that:
(i) $D_x \tan x = \sec^2 x$
(ii) $D_x \cot x = -\csc^2 x$
(iii) $D_x \sec x = \sec x \tan x$
(iv) $D_x \csc x = -\csc x \cot x$
(v) $1 + \tan^2 x = \sec^2 x$
(vi) $1 + \cot^2 x = \csc^2 x$
There are two more basic sets of facts about trigonometry. We do not offer proof but discuss them in the next two sections of this exercise.

(h) We know that $\sin(0) = 0$ and $\cos(0) = 1$. There are positive numbers of which this is true. For the very determined a proof can be based on the following facts.
(i) If $\sin x = 0$, then $\cos x = 1$ and vice versa.
(ii) $\cos(0) = 1$
(iii) If $f'(x) > 0$, then f is increasing at x and if $f'(x) < 0$, f is decreasing at x. The proof is by contradiction. For those less demanding of rigor and with access to a computer you might use the definition of $\sin x$ to approximate a root between $x = 6$ and $x = 6.5$. Do not insist upon too much accuracy or you will use a huge amount of computer time. The root is usually denoted by 2π (where $\pi \approx 3.14159$). Given that $\sin 2\pi = 0$ and $\cos 2\pi = 1$, show the following:

$$\cos(x + 2\pi) = \cos x$$
$$\sin(x + 2\pi) = \sin x$$
$$\sin \pi = 0, \quad \cos \pi = -1$$
$$\sin \frac{\pi}{2} = 1, \quad \cos \frac{\pi}{2} = 0$$
$$\sin \frac{3\pi}{2} = -1, \quad \cos \frac{3\pi}{2} = 0$$

(i) The final remark which will not be proven here is that if we draw a unit circle in the uv-plane, identify a positive number x with the angle from the positive direction of the x-axis, in a counterclockwise

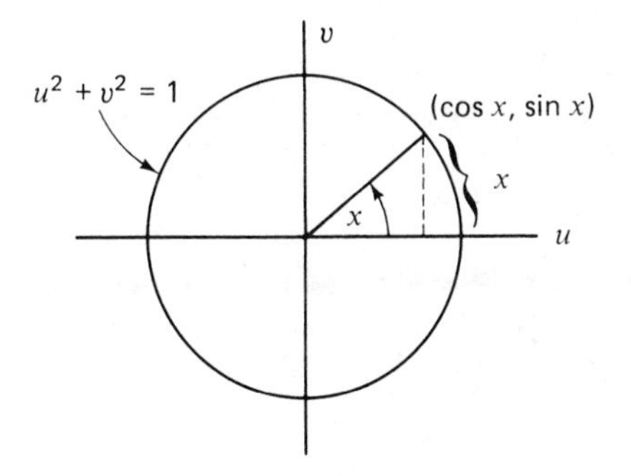

direction so that the arc length along the circle is x, then the point on the circle has coordinates $u = \cos x$, $v = \sin x$. From this, one can deduce by similar triangles the definitions often given in high school.

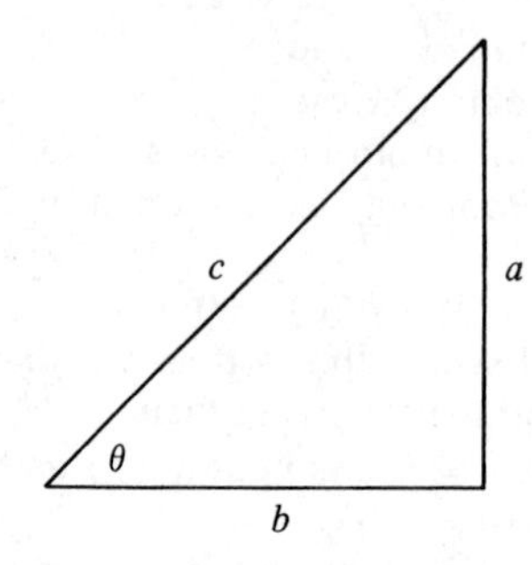

$$\sin\theta = \frac{a}{c}, \qquad \cot\theta = \frac{b}{a}$$

$$\cos\theta = \frac{b}{c}, \qquad \sec\theta = \frac{c}{b}$$

$$\tan\theta = \frac{a}{b}, \qquad \csc\theta = \frac{c}{a}$$

(j) (i) $D_x(\sin^2(2x+1)) = ?$

(ii) $D_x(x \cos x) = ?$

(iii) $D_x(\tan(3x)) = ?$

(iv) $\int_0^{\pi} \sin x \, dx = ?$

(v) $\int_0^{\pi/4} \tan x \sec^2 x \, dx = ?$

CHAPTER

5 Continuous Statistics

0 INTRODUCTION

In the first three chapters of this book we studied several basic ideas of statistics and we developed the supportive theory of probability. Where appropriate, our definitions and theorems were presented in a general context, yet our examples were essentially restricted to discrete random variables. Moreover the only contact with statistical inference we made was in a section on hypothesis testing using the binomial distribution.

Chapter 4 in large part was motivated by our desire to explore continuous random variables, and there we developed the powerful tools of calculus.

In a sense, Chapter 5 will represent a culmination of our efforts. Here we shall bring together the basic ingredients of probability, statistics, and calculus in an effort to more fully explain many of the ideas we have encountered and to answer some of the questions we have raised. The early sections are devoted to the normal distribution, which is important both as a model in its own right and as an approximation of other distributions. The remainder of the chapter is concerned with several aspects of statistical inference.

1 APPROXIMATING THE BINOMIAL DISTRIBUTION

Although the binomial distribution has a number of excellent features (e.g., it is conceptually straightforward; it applies to a vast range of situations; we were able to give a rather complete discussion of its properties, its mean, and its standard deviation without resorting to calculus), it does possess one significant drawback: There is no reasonable way to construct a table of values for $B(x; n, p)$ that would include all the choices of the parameters (n and p) that we are apt to need. Even the listing of $B(x; n, \frac{1}{2})$ for a suitable collection of n's would require a large number of pages, for we would probably want to include virtually all values of n between 1 and 30, and then perhaps all multiples of 5 from 35 to 100, and then all multiples of 10 from 110 to 200, and so on. Perhaps this seems unreasonable, but how would we deal with a problem that involves $B(x; 118, p)$ or $B(x; 249, p)$ even when $p = \frac{1}{2}$ unless we have values of n in our table at least as large as 249 and also somewhere near 118. Moreover, while it is true that $B(x; 118, \frac{1}{2})$ is very close to $B(x; 120, \frac{1}{2})$, since the former is the probability of x successes in 118 tries and the latter is the probability of x successes in 120 tries, where in each case $P(\text{success}) = \frac{1}{2}$, it is by no means obvious how far we could stretch such approximations and more importantly how *much* of an error might be involved.

Now multiply the problem of forming a table for $B(x; n, \frac{1}{2})$ by the number of different choices of p we might care to use and you get the magnitude of the problem of building a suitable binomial table. Some rather extensive tables (running several hundred pages) have been prepared in recent years with the aid of computers.

Instead of compiling a mammoth table or resorting to direct computation in every instance, we propose to find a suitable *approximation* to $B(x; n, p)$. We begin by briefly reviewing an earlier construction that should help motivate the techniques we need to develop.

When X is a continuous random variable with density function f, we find probabilities of the form

$$P(x_1 \leq X \leq x_2)$$

by evaluating the area under the graph of f,

$$\int_{x_1}^{x_2} f(x)\, dx \qquad \text{(See Figure 5.1.)}$$

Before we knew how to evaluate definite integrals, we constructed a collection of rectangles whose aggregate area approximated the area under the graph of f. In a sense, these rectangles could be viewed as forming a histogram, which in turn could be viewed as the frequency histogram for a discrete random variable Y. We might think of Y as being a discrete approximation to X. Both variables have the same range, and corresponding probabilities such as

$$P(x_1 \leq X \leq x_2) \quad \text{and} \quad P(x_1 \leq Y \leq x_2)$$

are approximately equal.

Now we wish to reverse the above procedure. With the tools of calculus available to us, we are able to treat continuous random variables with about as much ease as discrete random variables. Our objective is to start with a discrete variable X whose frequency function is of the form

$$B(x; n, p)$$

(i.e., X is binomially distributed), and to form a continuous random variable whose density function approximates the histogram of X. The particular choice of continuous approximation will be determined by the mean and the standard deviation of X; recall that

$$\mu = np \quad \text{and} \quad \sigma = \sqrt{npq}$$

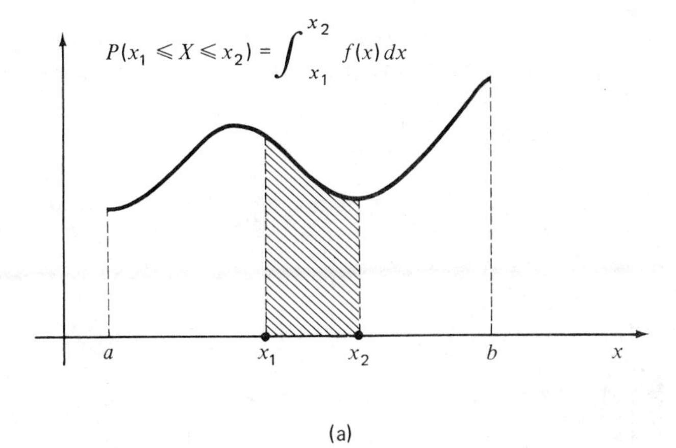

(a)

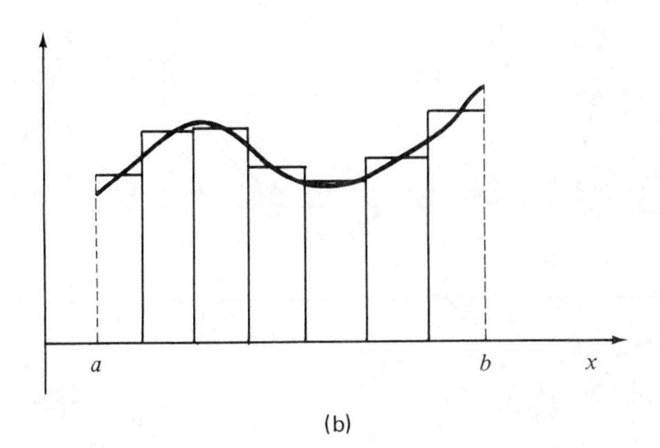

(b)

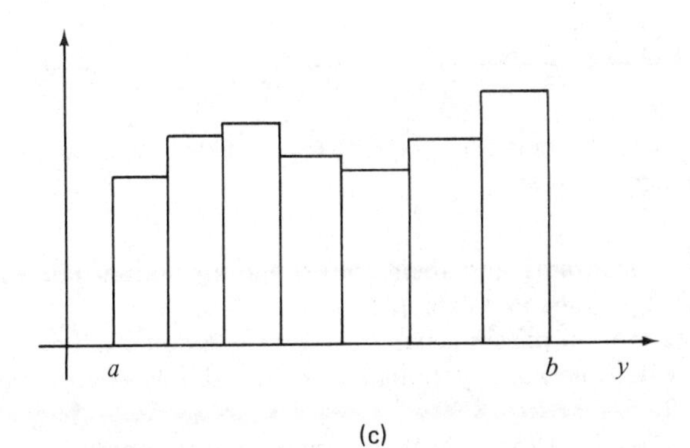

(c)

FIGURE 5.1 (a) Graph of density function f for continuous variable X. (b) Graph of f and of approximating histogram. (c) Frequency histogram for discrete random variable Y.

Our continuous variable will have the same mean and the same standard deviation. The density function's graph is called a *normal curve*. Its equation is

$$y = \frac{1}{\sigma\sqrt{2\pi}} \exp\left(-\tfrac{1}{2}\left(\frac{x-\mu}{\sigma}\right)^2\right)$$

In other words, there is one normal curve for each choice of μ and σ. We have a special notation for this function, $N(x; \mu, \sigma)$. In Fig. 5.2, we illustrate several

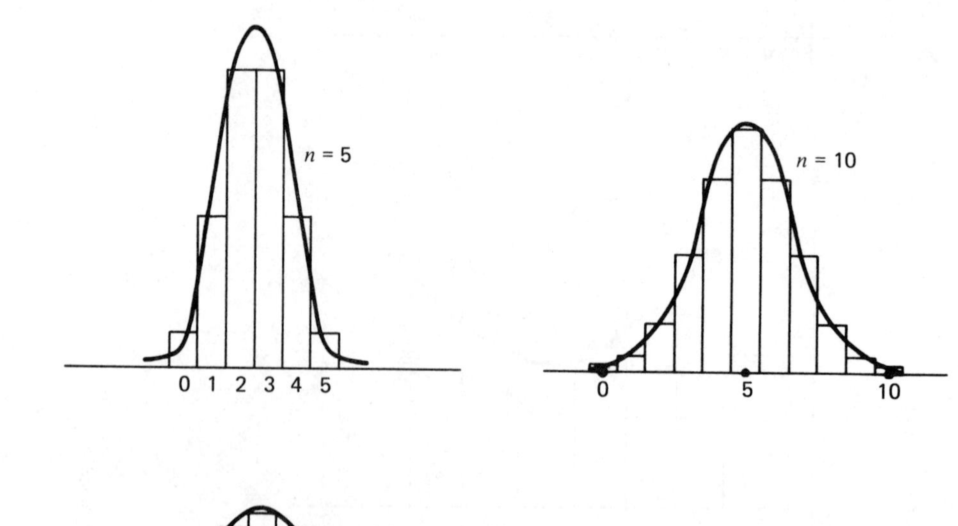

FIGURE 5.2 Histograms for $B(x; n, \frac{1}{2})$ for $n = 5, 10, 20, 30$ and superimposed normal curves.

binomial histograms and their corresponding normal curves, where $p = \frac{1}{2}$. In Fig. 5.3 we show some examples where $p = \frac{1}{3}$.

We mention in passing that $N(x; \mu, \sigma)$ is a density function in its own right, having widespread applications other than simply approximating the binomial.

QUESTION 1 Where do these normal curves come from?

ANSWER There is a theorem, whose proof is well beyond the scope of this course, that asserts that as the value of n gets large, the binomial histograms "smooth out" in such a way that the normal curves become very good approximations to them.

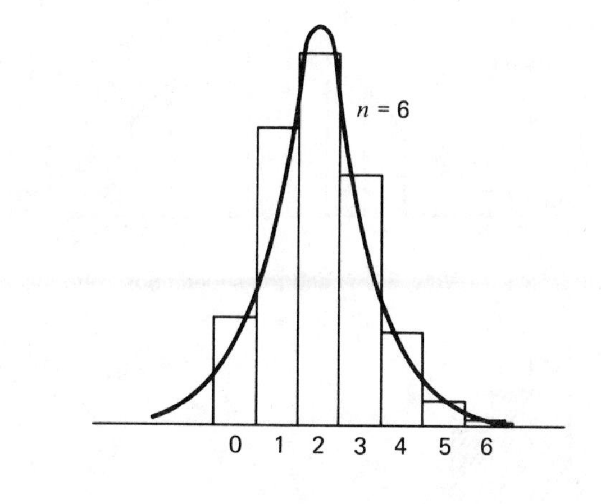

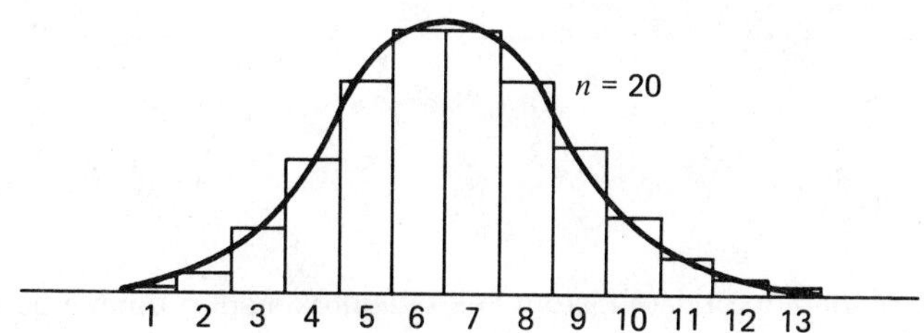

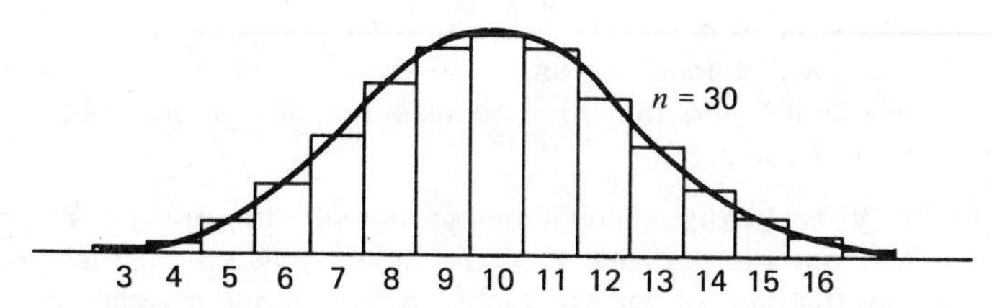

FIGURE 5.3 Histograms for $B(x; n, \frac{1}{3})$ for $n = 6, 20, 30$ and superimposed normal curves.

QUESTION 2 How does the normal curve actually help us in computing good estimates to the probabilities involving the original binomially distributed variable?

ANSWER Let's take a particular case, say a variable X whose frequency function is

$$B(x; 30, \tfrac{1}{2})$$

and let us find a good estimate for

$$P(14 \leq X \leq 20)$$

For instance, we might want to know the probability of getting between 14 and 20 heads (inclusively) if we toss a fair coin 30 times. In Fig. 5.4, we show

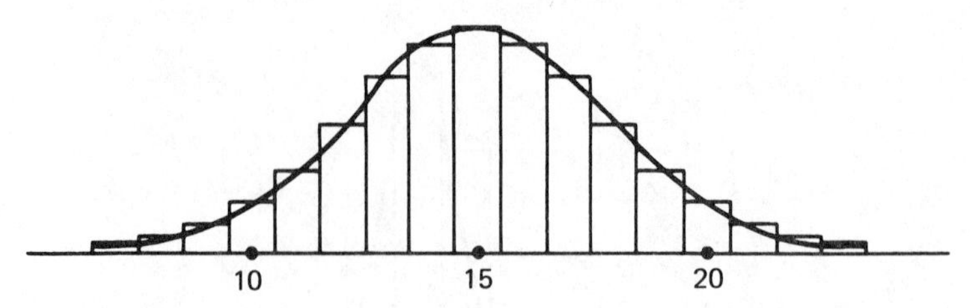

FIGURE 5.4 (a) $B(x; 30, \frac{1}{2})$ and the superimposed normal curve.

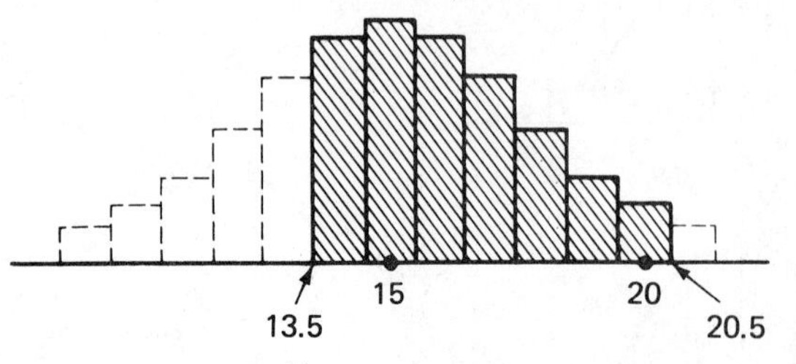

FIGURE 5.4 (b) $\sum_{x=14}^{20} B(x; 30, \frac{1}{2})$.

FIGURE 5.4 (c) $\int_{13.5}^{20.5} N(x; 15, \sqrt{\frac{30}{4}})\, dx$.

the appropriate histogram, its approximating normal curve

$$y = \frac{1}{\sqrt{\frac{30}{4}}\sqrt{2\pi}} \exp\left(-\frac{1}{2}\left(\frac{x-15}{\sqrt{\frac{30}{4}}}\right)^2\right)$$

and two shaded regions: one representing the exact probability from the histogram, and the other representing the approximate probability from the normal curve.

Note Well: Each time we convert the area of a portion of a binomial histogram to the area under a normal curve we must take into account the fact that the base of each rectangle in the histogram straddles an integer. A typical rectangle looks like

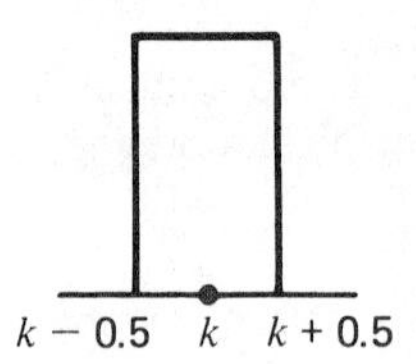

Thus the portion of the histogram ranging from $x = k$ to $x = m$ will convert to the area under a normal curve from $x = k - .5$ to $x = m + .5$. That is,

$$\sum_{x=k}^{m} B(x; n, p) \quad \text{converts to} \quad \int_{k-.5}^{m+.5} N(x; np, \sqrt{npq})\, dx$$

provided $k > 0$ and $m < n$. These special cases will be treated in the next section.

Equally important is the observation that each rectangle in a binomial histogram has a base exactly 1 unit wide. Thus the *area* of that rectangle will have the same numerical value as its *height*. But since the height represents the frequency of that value of X, we can interpret area as probability. In short,

$P(X = 14) = B(14; 30, \frac{1}{2}) =$ height of [rectangle at 14]

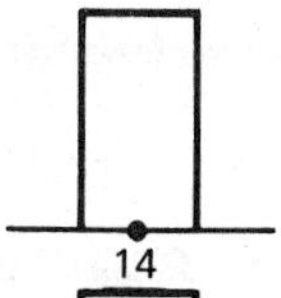

$=$ area of [rectangle at 14]

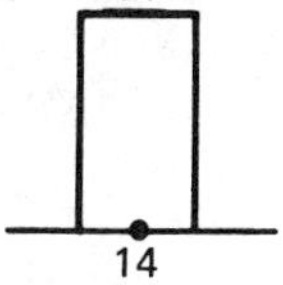

QUESTION 3 Does the fact that

$$P(14 \le X \le 20) \approx \int_{13.5}^{20.5} N(x; 15, \sqrt{\tfrac{30}{4}})\, dx$$

really help us?

ANSWER Not yet, because the definite integral does not yield to any technique currently at our disposal. Indeed, all we have accomplished is to modify the problem of computing difficult binomial probabilities to a new problem of computing difficult integrals. Of course, we wouldn't be doing all this work unless we thought we could ultimately simplify the problem. In the next section we take up the second phase of the solution, namely converting the integration problem to something more manageable.

QUESTION 4 How good are these approximations of the binomial frequency functions by the normal curves?

ANSWER It would be nice to have some convenient way to measure the error between $B(x; n, p)$ and the corresponding integral

$$\int_{x-(1/2)}^{x+(1/2)} N(x; np, \sqrt{npq})\, dx$$

but, alas, there is no such formula available. Instead, we offer some fairly standard ground rules to be observed when using this approximation.

1 The normal curve should be used to approximate $B(x; n, p)$ *only* when we cannot find a table for the exact values. Most tables include values of n up to 25 and a few values of p (see the Appendix).
2 The normal approximation works best when n is large and p is moderate. To be more specific, n should be at least 20 or 25 and p should be between

1 and .9. Moreover if p is close to .5, the approximation gets good for values of n as small as 15 or 20. One common rule of thumb is that

$$np > 5 \quad \text{and} \quad n(1-p) > 5$$

Of course when p is small (close to 0) or large (q close to 0), the Poisson distribution will often afford a better approximation.

We close this section with an example of the kind of problems to be found in the exercises. Note that we will not give a final answer to these problems since we cannot as yet evaluate the resulting definite integral.

PROBLEM A fair coin is tossed 100 times.
(a) What is the probability of getting at least 70 heads?
(b) What is the probability that the number of heads will be more than 40 but not more than 60?

SOLUTION We let X be the number of heads that occur. Clearly since the coin is fair, X has the frequency function

$$B(x; 100, \tfrac{1}{2})$$

whose mean and standard deviation are

$$\mu = 50, \qquad \sigma = \sqrt{100(\tfrac{1}{2})(\tfrac{1}{2})} = 5$$

Thus we can use the normal approximation

$$N(x; 50, 5)$$

(a) To say that X is at least 70 calls for

$$P(X \geq 70) = \sum_{x=70}^{100} B(x; 100, \tfrac{1}{2})$$

and this is approximately equal to the area

$$\int_{69.5}^{\infty} N(x; 50, 5)\, dx$$

(*Note:* To capture the rectangle in the binomial histogram centered at 70, we must start at the left end point of its base, 69.5. More will be said about the upper limit in the next section.)

(b) To say that X is more than 40 and not more than 60 calls for

$$P(40 < X \leq 60) = \sum_{x=41}^{60} B(x; 100, \tfrac{1}{2})$$

which in turn is approximated by the area

$$\int_{40.5}^{60.5} N(x; 50, 5)\, dx$$

Summary

As an alternative to compiling unwieldy tables for $B(x; n, p)$ for various values of n and p, we have introduced a class of density functions

$$N(x; \mu, \sigma) = \frac{1}{\sigma\sqrt{2\pi}} \exp\left(-\frac{1}{2}\left(\frac{x-\mu}{\sigma}\right)^2\right)$$

that will serve as good approximations via

$$\sum_{x=k}^{m} B(x; n, p) \approx \int_{k-.5}^{m+.5} N(x; np, \sqrt{npq})\, dx$$

However, the definite integrals that so arise do not yield to the Fundamental Theorem of Calculus; that is, we cannot find an antiderivative of $N(x; \mu, \sigma)$. Thus we have converted a difficult numerical computation into a seemingly equally difficult integration problem. Next we take up a second phase that actually converts the integral to a "standard form" for which we can appeal to a table.

Exercises
Section 1

1 Find the values of the parameters μ and σ in $N(x; \mu, \sigma)$ necessary to approximate each of the following binomial distributions.

(a) $B(x; 20, \frac{1}{2})$
(b) $B(x; 40, \frac{1}{2})$
(c) $B(x; 100, \frac{1}{3})$
(d) $B(x; 100, \frac{2}{3})$

2 Find the integrals that give approximations to each of the following binomial probabilities.

(a) $\sum_{x=14}^{16} B(x; 30, \frac{1}{2})$
(b) $\sum_{x=10}^{20} B(x; 30, \frac{1}{2})$
(c) $B(15; 30, \frac{1}{2})$
(d) $B(5; 36, \frac{1}{6})$

3 Express the answers to each of the following problems in terms of an approximating integral of a normal density function.

(a) A student guesses at each of 50 true-false questions on a test. What is the probability that he will get at least 40 correct?
(b) A pair of fair dice are rolled 36 times. What is the probability of getting at least ten 7's?
(c) One die is rolled 144 times. Find the probability that the combined number of 2's and 3's is at most 50.

4 Each of the following definite integrals represents an approximate binomial probability. See if you can discover what values of the parameters n and p would yield this normal approximation and then determine what probability is being approximated.

(a) $\int_{4.5}^{9.5} N(x; 8, \sqrt{\frac{20}{3}})\, dx$
(b) $\int_{60.5}^{140.5} N(x; 100, \sqrt{75})\, dx$

5 **(a)** Solve for n, p, and q in the equations

$$\mu = np \quad \text{and} \quad \sigma = \sqrt{npq}$$

(*Hint:* Remember that $p + q = 1$.)

(b) Noting that n must be an integer and that $p + q = 1$, give an example of values of μ and σ that could not arise as parameters in a normal approximation to a binomial.

(c) Is it possible for two different binomial frequency functions to yield the same approximating normal density function? Why?

6 Let

$$f(x) = \frac{1}{\sigma\sqrt{2\pi}} \exp\left(-\frac{1}{2}\left(\frac{x-\mu}{\sigma}\right)^2\right)$$

(a) Find $f(\mu)$.

(b) Find $f'(x)$.

(c) For what values of x is $f(x) = 0$?

(d) For what values of x is $f'(x) = 0$?

(e) Show that $f(\mu + a) = f(\mu - a)$ for all values of a.

2 THE STANDARD NORMAL CURVE

Among the various members of the family of normal curves, there is one we single out, namely

$$N(x; 0, 1) = \frac{1}{\sqrt{2\pi}} \exp\left(-\tfrac{1}{2}x^2\right)$$

and call this the *standard normal curve*. Since it is determined by the choice of parameters

$$\mu = 0 \quad \text{and} \quad \sigma = 1$$

this standard normal curve could not possibly be the approximation for any binomial distribution. (In any binomial distribution, np is not 0 unless $n = 0$, in which case $\sigma = 0$ also.) On the other hand this choice of parameters yields the simplest possible expression for

$$N(x; \mu, \sigma) = \frac{1}{\sigma\sqrt{2\pi}} \exp\left(-\frac{1}{2}\left(\frac{x-\mu}{\sigma}\right)^2\right)$$

More importantly we shall find that any area problem for any normal curve can easily be converted to an area problem for the standard normal curve:

Theorem (Converting to the standard normal curve)

$$\int_a^b N(x; \mu, \sigma)\, dx = \int_{(a-\mu)/\sigma}^{(b-\mu)/\sigma} N(z; 0, 1)\, dz$$

Proof: Both integrals represent areas under normal curves. As such they can be approximated (to any desired degree of accuracy) by sums of areas of

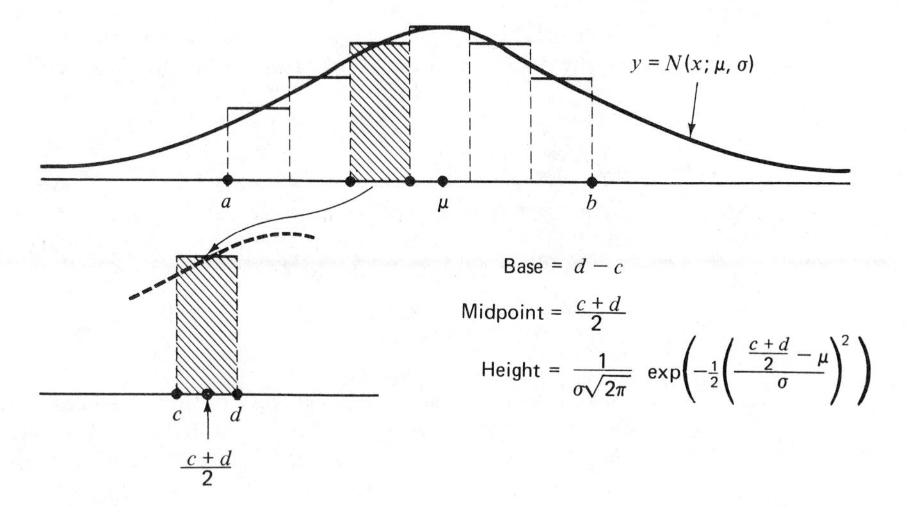

FIGURE 5.5

rectangles. In Fig. 5.5 we show a region under the graph of $N(x; \mu, \sigma)$ and in Fig. 5.6 we show a corresponding region under the graph of $N(z; 0, 1)$. By dividing the interval $[a, b]$ into n subintervals of equal length, and constructing

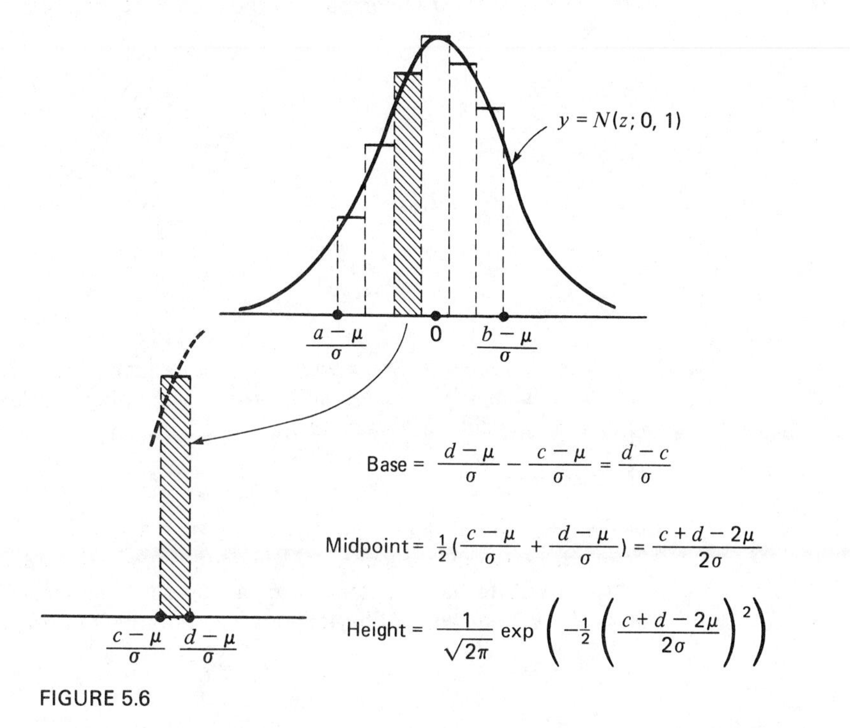

FIGURE 5.6

rectangles whose heights are the values of the function at the midpoints of the subintervals, we come up with the desired approximation to

$$\int_a^b N(x; \mu, \sigma)\, dx$$

Simultaneously we divide $[(a - \mu)/\sigma, (b - \mu)/\sigma]$ into n equal subintervals and construct the corresponding rectangles. We single out a particular (shaded) rectangle in Fig. 5.5 and its counterpart in Fig. 5.6. Multiplying the base by the height in either case results in the common area

$$\frac{d - c}{\sigma\sqrt{2\pi}} \exp\left(-\frac{1}{2}\left(\frac{c + d - 2\mu}{2\sigma}\right)^2\right)$$

It follows then that the sum of the areas of the rectangles in Fig. 5.5 equals the sum of the areas of the rectangles in Fig. 5.6. As we increase the number of rectangles in each case, the sum of their areas approaches the exact area under the curve. Thus both regions,

$$\left(\begin{array}{c}\text{Region under } N(x; \mu; \sigma) \text{ from} \\ a \text{ to } b\end{array}\right) \quad \text{and} \quad \left(\begin{array}{c}\text{region under } N(z; 0, 1) \text{ from} \\ \dfrac{a - \mu}{\sigma} \text{ to } \dfrac{b - \mu}{\sigma}\end{array}\right)$$

must have the same area, which is what we were to prove.

A convenient way to remember the theorem is to note the formal *change of variables*

$$\text{from } x \quad \text{to} \quad z = \frac{x - \mu}{\sigma}$$

This change converts

$$N(x; \mu, \sigma) \quad \text{to} \quad N(z; 0, 1)$$

and converts

$$[a, b] \quad \text{to} \quad \left[\frac{a - \mu}{\sigma}, \frac{b - \mu}{\sigma}\right]$$

We call the z-values *standard units*, and we frequently draw parallel horizontal lines to indicate the x-values and the corresponding z-values. For instance, suppose we consider a normal curve

$$N(x; 30, 4)$$

and see how we convert x-units to z-units (Fig. 5.7).

It might be worth observing that this conversion to standard units in effect gives us a way to use the standard deviation as a unit measure of dispersion. That is, if x is two standard deviations away from the mean μ, either

$$x = \mu + 2\sigma \quad \text{or} \quad x = \mu - 2\sigma$$

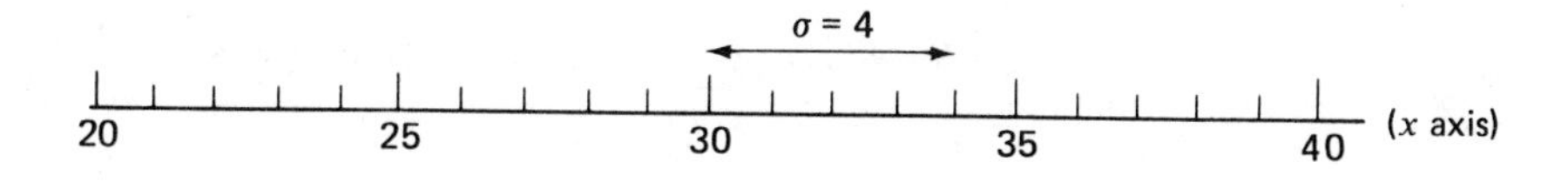

FIGURE 5.7

then the conversion produces either

$$z = 2 \quad \text{or} \quad z = -2$$

respectively. In general if x is $k\sigma$ units above (respectively, below) μ, then the corresponding z-value is k (respectively, $-k$). So, in order to find the z-value corresponding to

$$x = 38.6$$

for example, we make the substitution

$$\begin{aligned} z &= \frac{38.6 - 30}{4} \\ &= \frac{8.6}{4} \\ &= 2.15 \end{aligned}$$

Similarly, from

$$x = 25.2$$

we get

$$\begin{aligned} z &= \frac{25.2 - 30}{4} \\ &= \frac{-4.8}{4} \\ &= -1.2 \end{aligned}$$

and it would follow then that

$$\int_{25.2}^{38.6} N(x; 30, 4)\, dx = \int_{-1.2}^{2.15} N(z; 0, 1)\, dz$$

Note: Every normal curve extends indefinitely to the left and to the right. That is, the function

$$N(x; \mu, \sigma) = \frac{1}{\sigma\sqrt{2\pi}} \exp\left(-\frac{1}{2}\left(\frac{x - \mu}{\sigma}\right)^2\right)$$

is defined for all x in $(-\infty, \infty)$. Quite often we will be faced with one of the integrals

$$\int_{-\infty}^{b} N(x; \mu, \sigma)\, dx, \qquad \int_{a}^{\infty} N(x; \mu, \sigma)\, dx$$

Although we cannot treat the x-values of $-\infty$ and $+\infty$ as ordinary real numbers, we do convert them to z-values of $-\infty$ and $+\infty$, respectively, yielding

$$\int_{-\infty}^{(b-\mu)/\sigma} N(z; 0, 1)\, dz \quad \text{and} \quad \int_{(a-\mu)/\sigma}^{\infty} N(z; 0, 1)\, dz$$

A particular instance where these improper integrals arise is in connection with approximating a binomial distribution via normal curves. Suppose for example that we wanted to find the probability of getting at least 60 heads in 100 tosses of a fair coin. We use the approximation

$$\sum_{x=60}^{100} B(x; 100, \tfrac{1}{2}) \approx \int_{59.5}^{\infty} N(x; 50, 5)\, dx$$

In other words, we take the upper limit of integration to be ∞ rather than 100.5. One justification for this is that

$$\sum_{x=60}^{100} B(x; 100, \tfrac{1}{2})$$

represents "all the area under the histogram to the right of 59.5" and

$$\int_{59.5}^{\infty} N(x; 50, 5)\, dx$$

likewise represents "all the area under the corresponding normal curve to the right of 59.5."

Of course we would make the final conversion to z-units:

$$\text{When } x = 59.5,\ z = \frac{59.5 - 50}{5} = \frac{9.5}{5} = 1.9$$

and get

$$\sum_{x=60}^{100} B(x; 100, \tfrac{1}{2}) \approx \int_{1.9}^{\infty} N(z; 0, 1)\, dz$$

QUESTION How does this conversion of integrals involving normal curves into integrals involving the standard normal curve help us?

ANSWER Now that we know that every such integral can be reduced to one of the form

$$\int_{\alpha}^{\beta} N(z; 0, 1)\, dz$$

we simply construct a *table of values* for this integral. Virtually any book on statistics contains such a table, although they come in a variety of forms. Table D in the Appendix gives values for $b \geq 0$ of the distribution function

$$F(b) = \int_{-\infty}^{b} N(z; 0, 1)\, dz = \text{area of}$$

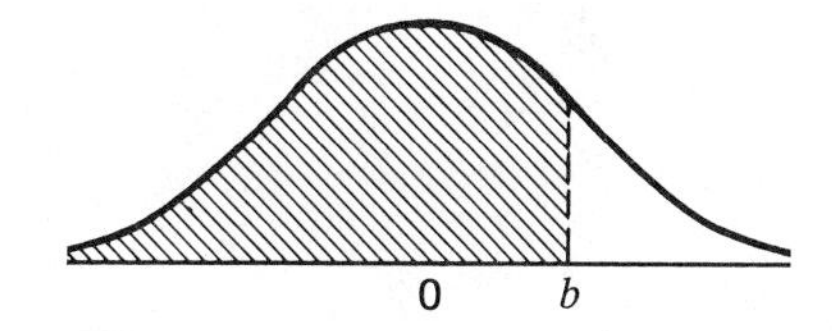

We could find, for example, that

$$\int_{-\infty}^{1.07} N(z; 0, 1)\, dz = .8577$$

The value of b can be chosen accurate to two decimal places.	The value of the area under the standard normal curve is accurate to four decimal places.

Of course, in practice we shall often be interested in finding the area of a different kind of region under the standard normal curve. Therefore we must learn a few simple methods of reducing any area problem to one of finding the area of a region extending from $-\infty$ to b (where $b \geq 0$). The following chart should reveal how to proceed in most situations.

Area Sought | *Method of Finding the Area*

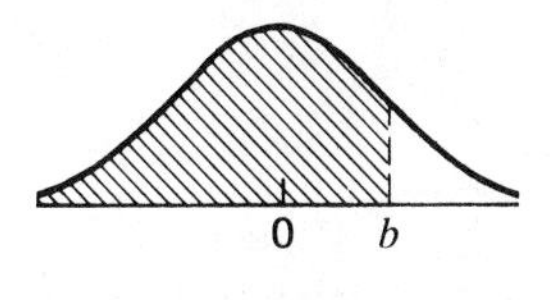

Look it up directly in the table.

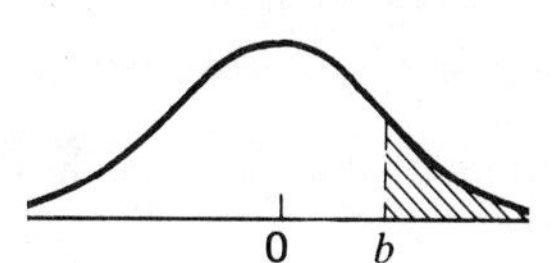

1 − area of

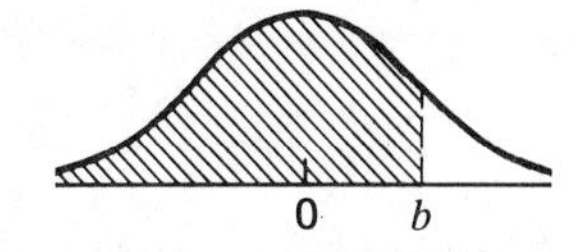

Example:

$$\int_{2}^{\infty} N(z; 0, 1)\, dz = 1 - \int_{-\infty}^{2} N(z; 0, 1)\, dz$$

$$= 1 - .9773 = .0227$$

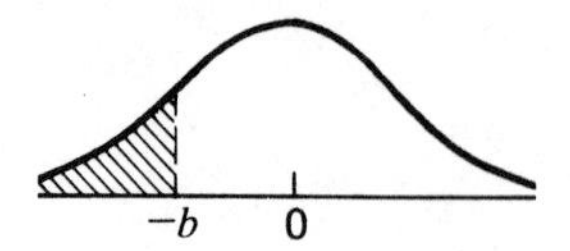

same as area of

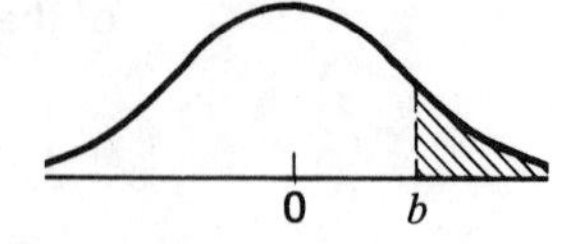

Example:

$$\int_{-\infty}^{-2} N(z; 0, 1)\, dz = \int_{2}^{\infty} N(z; 0, 1)\, dz$$

$$= 1 - \int_{-\infty}^{2} N(z; 0, 1)\, dz$$

$$= 0.227$$

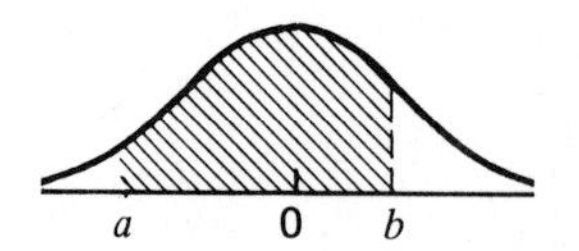

area of

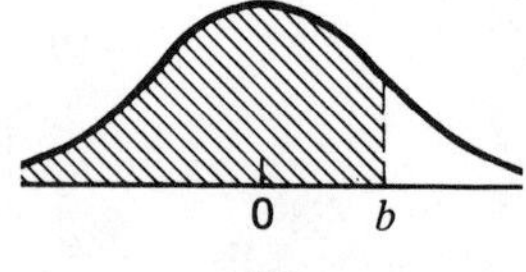

minus area of

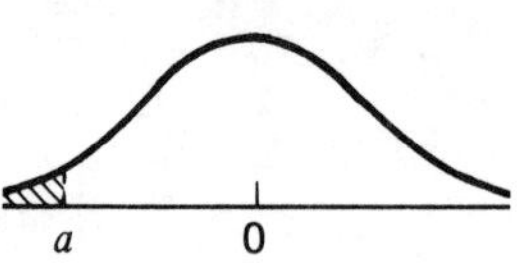

Example:

$$\int_{-2}^{1} N(z; 0, 1)\, dz = \int_{-\infty}^{1} N(z; 0, 1)\, dz - \int_{-\infty}^{-2} N(z; 0, 1)\, dz$$

$$= .8413 - .0227$$

$$= .8186$$

The facts we have used here are

1. The curve

   $$y = N(z; 0, 1)$$

   is *symmetric* about the line $z = 0$ (the y-axis).
2. The sum of the areas of two non-overlapping regions is the area of the union of those two regions.

Let us now complete the solutions of the problems in the previous section.

EXAMPLE 1 Find the probability of getting between 14 and 20 heads (inclusively) if we toss a fair coin 30 times.

Solution We had shown the approximation

$$\sum_{x=14}^{20} B(x; 30, \tfrac{1}{2}) \approx \int_{13.5}^{20.5} N(x; 15, \sqrt{\tfrac{30}{4}})\, dx$$

Now we estimate

$$\sqrt{\frac{30}{4}} = \frac{\sqrt{30}}{2} \approx \frac{5.48}{2} = 2.74$$

Converting from x-units to z-units,

$$\text{When } x = 13.5, \qquad z = \frac{13.5 - 15}{2.74} = \frac{-1.5}{2.74} \approx -.55$$

$$\text{When } x = 20.5, \qquad z = \frac{20.5 - 15}{2.74} = \frac{5.5}{2.74} \approx 2.01$$

Thus the *approximate* probability is

$$\int_{-.55}^{2.01} N(z; 0, 1)\, dz = \text{area of}$$

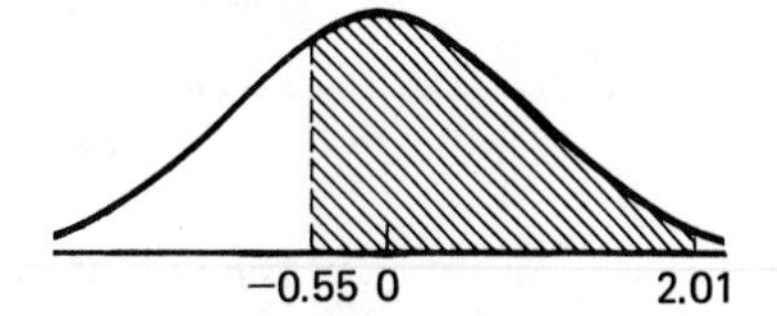

= area of

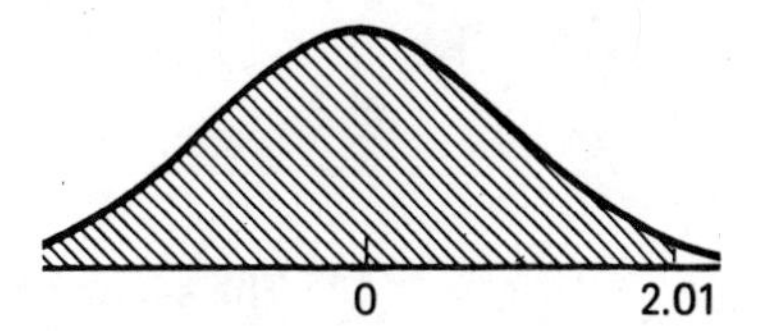

minus area of

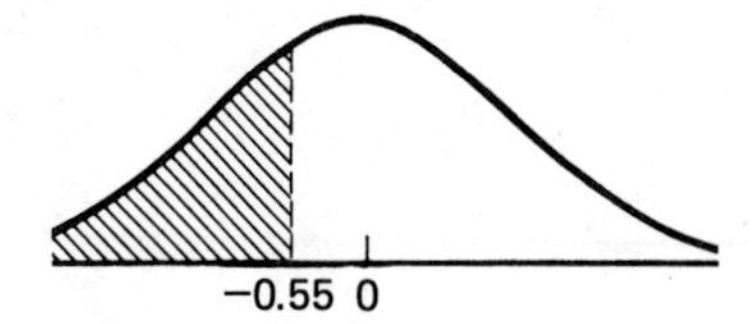

Also this latter area is equal to

1 minus area of

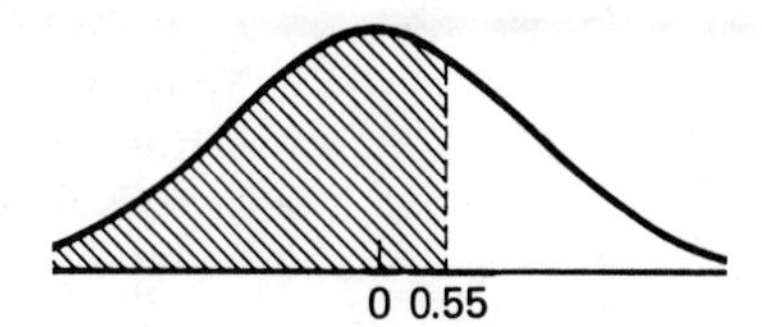

Now using Table D we find

$$\int_{-\infty}^{2.01} N(z; 0, 1)\, dz = .9778$$

$$\int_{-\infty}^{.55} N(z; 0, 1)\, dz = .7088$$

and our final answer is

$$\begin{aligned} & .9778 - (1 - .7088) \\ &= .9778 - (.2912) \\ &= .6866 \end{aligned}$$

EXAMPLE 2 A fair coin is tossed 100 times. What is the probability of

(a) getting at least 70 heads?

(b) getting more than 40 but not more than 60 heads?

We found approximations to (a) and (b), respectively, of

$$\int_{69.5}^{\infty} N(x; 50, 5)\, dx \quad \text{and} \quad \int_{40.5}^{60.5} N(x; 50, 5)\, dx$$

and we convert these to

$$\int_{3.9}^{\infty} N(z; 0, 1)\, dz \quad \text{and} \quad \int_{-1.9}^{2.1} N(z; 0, 1)\, dz$$

The first of these can be evaluated by

$$\begin{aligned} \int_{3.9}^{\infty} N(z; 0, 1)\, dz &= 1 - \int_{-\infty}^{3.9} N(z; 0, 1)\, dz \\ &\approx 1 - 1 \\ &= 0 \end{aligned}$$

The second becomes

$$\begin{aligned} \int_{-1.9}^{2.1} N(z; 0, 1)\, dz &= \int_{-\infty}^{2.1} N(z; 0, 1)\, dz - \int_{-\infty}^{-1.9} N(z; 0, 1)\, dz \\ &= \int_{-\infty}^{2.1} N(z; 0, 1)\, dz - \left(1 - \int_{-\infty}^{1.9} N(z; 0, 1)\, dz\right) \\ &\approx .9821 - (1 - .9713) \\ &= .9821 - .0287 \\ &= .9534 \end{aligned}$$

Thus the probability of getting at least 70 heads is 0. This may seem strange; nevertheless, the event is very unlikely, and to 4-decimal place accuracy, the answer is 0. Also, the probability of getting at least 41 heads and at most 60 is better than .95.

It is perhaps worth pointing out one of the subtleties of dealing with continuous random variables. Recall that when X is continuous,

$$P(X = x) = 0, \qquad \text{for all } x$$

As a result, we can be somewhat careless about distinguishing between events of the form

$$X \leq x \quad \text{and} \quad X < x$$

That is, their probabilities are equal:

$$P(X \leq x) = P(X < x)$$

The reason for this is that the events differ by the event

$$X = x$$

More formally,

$$\underbrace{\{s|X(s) < x\}}_{X < x} \cup \underbrace{\{s|X(s) = x\}}_{X = x} = \underbrace{\{s|X(s) \leq x\}}_{X \leq x}$$

A case in point where this fact is used can be found in Example 2. Converting the integral notation there, we have

$$P(-1.9 < Z < 2.1) = P(Z < 2.1) - P(Z < -1.9)$$

but the right-hand side should read

$$P(Z < 2.1) - P(Z \leq -1.9)$$

Similarly at the next step in the argument we said

$$P(Z < -1.9) = 1 - P(Z < 1.9)$$

and technically we should have written (on the right)

$$1 - P(Z \leq 1.9)$$

To sum up, we will quite often use facts such as

$$P(X \leq a) = P(X < a)$$
$$P(a \leq X < b) = P(a < X < b)$$
$$P(a \leq X \leq b) = P(a < X < b)$$
$$P(a \leq X) = P(a < X)$$

In each case there are two events under consideration; the events differ by an event of probability 0 and therefore the events have equal probability.

We close this section with two tables that offer comparisons between binomial values and the corresponding normal approximations.

x	$B(x; 20, .4)$	Normal approximation
0	.0000	.0003
1	.0005	.0015
2	.0031	.0045
3	.0123	.0142
4	.0350	.0346
5	.0746	.0723
6	.1244	.1212
7	.1659	.1607
8	.1797	.1820
9	.1597	.1607
10	.1171	.1212
11	.0710	.0723
12	.3550	.0346
13	.0146	.0142
14	.0049	.0045
15	.0013	.0015
16	.0003	.0003
17	.0000	.0000
18	.0000	.0000
19	.0000	.0000
20	.0000	.0000

x	$B(x; 100, \frac{1}{2})$	Normal approximation
50	.0796	.0796
51	.0780	.0781
52	.0735	.0736
53	.0666	.0665
54	.0580	.0579
55	.0485	.0484
56	.0390	.0389
57	.0301	.0300
58	.0223	.0221
59	.0159	.0159
60	.0108	.0109
61	.0071	.0072
62	.0045	.0045
63	.0027	.0027
64	.0016	.0015
65	.0009	.0010
66	.0005	.0005
67	.0002	.0003
68	.0001	.0001
69	.0001	.0000
70	.0000	.0000

Summary

Among the various normal curves, we singled out the graph of

$$N(z; 0, 1) = \frac{1}{\sqrt{2\pi}} \exp\left(-\tfrac{1}{2}z^2\right)$$

and called it the *standard normal curve*. In spite of the fact that we cannot easily integrate this function, we can with the aid of a computer produce a table of values for the distribution function

$$F(b) = \int_{-\infty}^{b} N(z; 0, 1)\, dz$$

Moreover we can convert any integral involving $N(x; \mu, \sigma)$ to an integral involving $N(z; 0, 1)$ via the formula

$$\int_a^b N(x; \mu, \sigma)\, dx = \int_{(a-\mu)/\sigma}^{(b-\mu)/\sigma} N(z; 0, 1)\, dz$$

where $(a - \mu)/\sigma$ and $(b - \mu)/\sigma$ are the results of converting a and b to *standard units*. This conversion, coupled with our earlier approximation to a binomial by $N(x; \mu, \sigma)$, allows us to compute, using only one table, approximations to binomial probabilities.

Exercises Section 2

1 For each of the following intervals, evaluate the definite integral

$$\int_a^b N(z; 0, 1)\, dz$$

and sketch the region under the standard normal curve whose area you are computing.

(a) $(-\infty, 0]$
(b) $(-\infty, 1]$
(c) $(-\infty, -1]$
(d) $[-1, 1]$
(e) $[-2, 2]$
(f) $[-3, 3]$
(g) $[-3, 2]$
(h) $[1.06, 2.51]$
(i) $[-.5, .5]$
(j) $[2, \infty)$

2 Convert each of the following integrals to one involving the standard normal curve and then evaluate it.

(a) $\int_{45}^{55} N(x; 100, 5)\, dx$
(b) $\int_{90}^{110} N(x; 200, 10)\, dx$
(c) $\int_{450}^{550} N(x; 500, 100)\, dx$
(d) $\int_{-\infty}^{600} N(x; 500, 100)\, dx$

3 Use the normal approximation to the binomial to evaluate

(a) $\sum_{x=18}^{30} B(x; 72, \frac{1}{3})$

(b) $\sum_{x=40}^{56} B(x; 72, \frac{2}{3})$

(c) $\sum_{x=220}^{400} B(x; 400, \frac{1}{2})$

(d) $\sum_{x=180}^{220} B(x; 400, \frac{1}{2})$

4 For each of the following inequalities, find the smallest positive number b that will work and sketch the corresponding region under the standard normal curve.

(a) $\int_{-b}^{b} N(z; 0, 1)\, dz \geq .95$
(b) $\int_{-b}^{b} N(z; 0, 1)\, dz \geq .99$
(c) $\int_{-b}^{b} N(z; 0, 1)\, dz \geq .90$
(d) $\int_{-\infty}^{b} N(z; 0, 1)\, dz \geq .95$
(e) $\int_{-b}^{b} N(x; 0, 5)\, dx \geq .95$
(f) $\int_{b}^{\infty} N(x; 10, 5)\, dx \leq .01$
(g) $\int_{-\infty}^{-b} N(x; 50, 10)\, dx \leq .05$
(h) $\int_{-\infty}^{-b} N(x; 100, 15)\, dx \leq .05$

5 Compare the exact values (from the binomial table in the Appendix) of each of the following to their normal approximations:

(a) $\sum_{x=0}^{2} B(x; 6, \frac{1}{2})$

(b) $\sum_{x=0}^{4} B(x; 12, \frac{1}{2})$

(c) $\sum_{x=0}^{6} B(x; 18, \frac{1}{2})$

6 Use the normal approximation to the binomial to approximate the probability of getting *exactly* as many heads as tails with a fair coin.

(a) In 20 tosses
(b) In 100 tosses

(c) In 500 tosses
(d) In 5,000 tosses

7 Suppose that 50% of the voters favor Smith to Jones. Find the probability that less than half of the voters polled favor Smith if
(a) 64 voters are polled.
(b) 100 voters are polled.
(c) 256 voters are polled.
(d) 1,024 voters are polled.

3 THE NORMAL DISTRIBUTION

In addition to serving as a useful approximation to a binomial distribution, each normal curve is in its own right the graph of a density function for a continuous random variable. Thus, for each choice of parameters μ and σ (with $\sigma > 0$), there is a continuous random variable X whose density function f is given by

$$f(x) = N(x; \mu, \sigma) = \frac{1}{\sigma\sqrt{2\pi}} \exp\left(-\frac{1}{2}\left(\frac{x-\mu}{\sigma}\right)^2\right)$$

Moreover, random variables having this kind of density function are quite common. Before citing some examples of such variables, let us examine some of the properties of these functions. Each of the following facts is stated without proof. Only the first one is really beyond the scope of this course, but the others involve varying degrees of manipulation with improper integrals.

Fact 1:

$$\int_{-\infty}^{\infty} N(x; \mu, \sigma)\, dx = 1$$

This tells us that the total area under each normal curve is 1 and thus $N(x; \mu, \sigma)$ can be viewed as a density function.

Fact 2:

$$\int_{-\infty}^{\infty} xN(x; \mu, \sigma)\, dx = \mu$$

This tells us that the *expected value* of the random variable with density function $N(x; \mu, \sigma)$ is μ.

Fact 3:

$$\int_{-\infty}^{\infty} (x-\mu)^2 N(x; \mu, \sigma)\, dx = \sigma^2$$

This tells us that the random variable X with density function $N(x; \mu, \sigma)$ has *variance* σ^2 and hence *standard deviation* σ.

Some other properties of normal curves are useful for drawing pictures as well as for interpreting results:

Fact 4: The *range* of the random variable X having density function $N(x; \mu, \sigma)$ is $(-\infty, \infty)$.

Fact 5: The graph of

$$y = N(x; \mu, \sigma)$$

(1) is *symmetric* about the y-axis, (2) lies entirely above the x-axis, and (3) flattens out rather rapidly toward the x-axis.

Fact 6: The function

$$f(x) = N(x; \mu, \sigma)$$

satisfies

1. $f'(x) = 0,$ only when $x = \mu$

   and thus the *highest point* on its graph is at $x = \mu$. The value of f at this point is $1/(\sigma\sqrt{2\pi})$ or about $.4/\sigma$.

2. $f''(x) = 0,$ when $x = \mu - \sigma$ and when $x = \mu + \sigma$

   This tells us that there are *points of inflection* where $x = \mu - \sigma$ and $x = \mu + \sigma$. At these points, the tangent line crosses the graph (see Fig. 5.8).

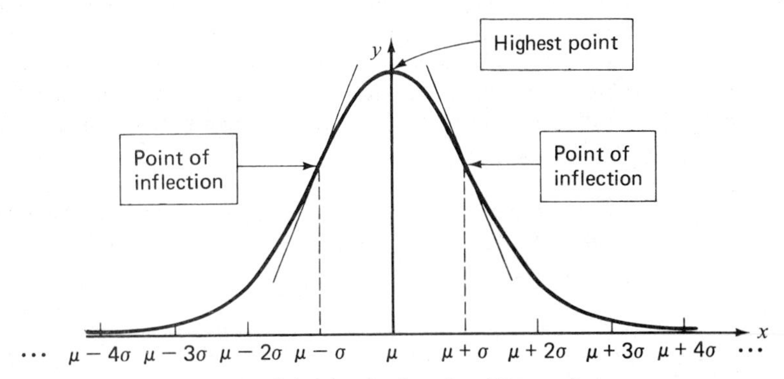

FIGURE 5.8 The graph of the density function $N(x; \mu, \sigma)$.

One final comment before we take up some examples of normally distributed variables. A glance at the table for the standard normal curve tells us that the area to the right of $z = 3$ is approximately .0013 and (by symmetry) the area to the left of $z = -3$ is approximately .0013. Using the conversion process, from x-units to z-units, we conclude that if X is a continuous random variable with density function

$$N(x; \mu, \sigma)$$

then

$$P(X < \mu - 3\sigma) \approx .0013$$

and

$$P(X > \mu + 3\sigma) \approx .0013$$

In other words, most of the action takes place *between* $\mu - 3\sigma$ and $\mu + 3\sigma$ in spite of the fact that X can theoretically take on *any* value in $(-\infty, \infty)$. In applications, we shall encounter random variables that as a practical matter cannot take on an infinite range of values. Yet the normal curve gives a very good approximation to them since within the interval $[\mu - 3\sigma, \mu + 3\sigma]$ there is suitable agreement.

EXAMPLE 1 Heights of Freshmen Males

A popular example of a random variable that is approximately normally distributed is heights of individuals in a certain group. For the approximation to be realistic, some care must be taken in suitably restricting the group. For instance, common sense would tell us that a group composed of 50 adults and 100 infants would produce a distribution (of heights) that is nowhere near normal. A frequency histogram of heights in such a case would have a mode in the neighborhood of 20 inches and another near 65 inches with virtually no values between, say, 36 inches and 60 inches. Even a group of men and women would probably not yield a height distribution that looks like a normal curve, since we could expect a cluster of values near 64 inches (for the women) and another cluster near 69 inches (for the men). Some examples of groups where the normal distribution could be sensibly used are Ohio State Highway Patrolmen, college freshmen males, college senior women, cadets at the United States Naval Academy, and so on. For sake of argument, let us consider a group of 400 freshmen males. We suppose that their average height is 70 inches with a standard deviation of 2.3 inches. Under the assumption that their heights are approximately normally distributed, we can make the following observations:

1. The probability that a randomly selected male will be at least 6 feet 4 inches becomes

$$\begin{aligned} P(X \geq 76) &= P(Z \geq 2.61) \\ &= 1 - P(Z < 2.61) \\ &= 1 - .9955 \\ &= .0045 \end{aligned}$$

> Here we found
>
> $$z = \frac{76 - 70}{2.3} \approx 2.61.$$

If we multiply this probability by 400 we would find that

$$400(.0045) = 1.8$$

or approximately two students should be 6 feet 4 inches tall or taller.

2. The probability that a randomly selected male will be between 5 feet 6 and 6 feet 2 becomes

$$\begin{aligned} P(66 < X < 74) &= P(-1.52 < Z < 1.52) \\ &= P(Z < 1.52) - P(Z < -1.52) \\ &= P(Z < 1.52) - (1 - P(Z < 1.52)) \\ &= 2P(Z < 1.52) - 1 \\ &= 2(.9357) - 1 \\ &= .8714 \end{aligned}$$

If we multiply this probability by 400 we find that

$$400(.8714) = 348.56$$

or approximately 349 students out of the 400 would be expected to be between 5 feet 6 and 6 feet 2. A related problem would be the following.

3. In (2) we found that we could expect 87.14% of the students to fall in the range

   66 to 74 inches

   How large should the range be made to include 95% of the students? By this we mean: Find a number $b > 0$ such that the range

   $70 - b$ to $70 + b$

   would include 95% of the students. Note that we have asked for an interval symmetric about the mean 70. Thus we want

$$\int_{70-b}^{70+b} N(x; 70, 2.3)\,dx = .95$$

   Converting to standard units, we have

$$\int_{-b/2.3}^{b/2.3} N(z; 0, 1)\,dz = .95$$

   But

$$\int_{-b/2.3}^{b/2.3} N(z; 0, 1)\,dz = 2\int_{-\infty}^{b/2,3} N(z; 0, 1)\,dz - 1 \quad \text{(Why?)}$$

   Thus we want

$$\int_{-\infty}^{b/2.3} N(z; 0, 1)\,dz = .975$$

Consulting our table, we find that

$$\frac{b}{2.3} = 1.96$$

or

$$b = 4.501$$

We conclude that roughly 95% of the 400 students (i.e., 380 students) would fall into the range

65.5 to 74.5 inches

EXAMPLE 2 Arrival Times

A large lecture class is scheduled to meet at 9:00 A.M. The students tend to arrive at times that are normally distributed with mean 8:55 and standard deviation two minutes. Assuming this model is accurate and assuming that the entire class attends on a given day, let us find when we would expect 99% of the class to be present.

To say that 99% are present amounts to solving:

$$P(X \leq x_0) = .99; \quad \text{find } x_0.$$

Equivalently, we have

$$\int_{-\infty}^{x_0} N(x; 55, 2)\, dx = .99$$

where $\mu = 55$ minutes (measured from 8:00 A.M.). Converting to standard units, we get

$$\int_{-\infty}^{(x_0 - 55)/2} N(z; 0, 1)\, dz = .99$$

Now using our table (backwards!) we find the value .9900 in the body of the table corresponds approximately to the standard unit

$$z_0 = 2.33$$

or

$$\frac{x_0 - 55}{2} = 2.33$$

$$x_0 = 55 + 2(2.33)$$
$$= 59.66$$

We conclude that at 8:59 + .66 minutes we should expect 99% of the class to be present.

In each of the above examples, we made the assumption that the random variable X (heights or arrival times) was *continuous*. Granted the actual height of a person is a continuous variable; as a practical matter we measure heights to the nearest half-inch or quarter-inch. Similarly, we generally measure time to the nearest second or in certain cases, such as sports events, to the nearest tenth of a second. In other words, we have used a continuous variable in a model to describe a discrete real-life situation. This is quite common and generally yields a satisfactory model of reality. Sometimes, however, we have to use a correction device similar to the one we used in connection with the normal approximation to the binomial, as the next example illustrates.

EXAMPLE 3 I.Q. Scores

A certain intelligence test is known to have its scores normally distributed with mean 100 and standard deviation 10. Out of 1,000 persons taking the test, how many would we expect to have a score of 105 or better?

Solution: We first find the probability that a randomly selected score is 105 or better. That is, we want

$$P(X \geq 105)$$

Now the random variable in this case is really *discrete*: A student cannot get a score of 104.7215 . . . . He either gets 104 or 105. Thus we take this into account by finding

$$\int_{104.5}^{\infty} N(x; 100, 10)\, dx$$

The justification for choosing 104.5 as the left endpoint for the region under the normal curve is that if the scores were continuous, then since 104 and 105 are nearly equally likely, about half of the scores between 104 and 105 would fall to the left of 104.5 and the other half to the right. Hence, rounding off the scores to the nearest whole number, we would lump into the 105 category all scores of at least 104.5. We compute

$$\begin{aligned}\int_{104.5}^{\infty} N(x; 100, 10)\, dx &= \int_{.45}^{\infty} N(z; 0, 1)\, dz \\ &= 1 - \int_{-\infty}^{4.5} N(z; 0, 1)\, dz \\ &= 1 - .6736 \\ &= .3264\end{aligned}$$

Finally, out of 1,000 students taking the test we would expect

$$1{,}000(.3264) = 326.4$$

or approximately 326 students to have a score of at least 105.

These three examples are somewhat typical of normal distributions that occur in practice. To help spot candidates for normally distributed random variables, we list some general criteria for a variable X:

1. Values of X to the left of the mean should be equally likely as corresponding values to the right of the mean. That is,

   $$P(X < \mu - \alpha) \approx P(X > \mu + \alpha)$$

2. Probabilities of events (i.e., intervals for X) should decrease as the values get farther away from the mean. That is, values of X tend to cluster about the mean.
3. Subject to the previous comments, X should be a continuous variable with range $(-\infty, \infty)$.

Summary

In addition to serving as an approximation to the binomial, the function $N(x; \mu, \sigma)$, for any μ and any $\sigma > 0$ is a density function in its own right. A random variable X having $N(x; \mu, \sigma)$ for its density function has the properties:

The range of X is $(-\infty, \infty)$.

$\mu_X = \mu, \qquad \sigma_X = \sigma$

$P(X < a) = P(X > -a)$ (Symmetry)

Exercises
Section 3

1 Assume that math SAT scores are normally distributed with mean 500 and standard deviation 100. Find the probability that a randomly selected student has a score
   (a) greater than 675.
   (b) greater than 700.
   (c) greater than 750.
   (d) between 450 and 550.
   (e) between 400 and 600.

2 Assume that the weights of yellow perch in Lake Mendota are normally distributed with mean 15 ounces and standard deviation 4 ounces. Find the probability of catching a fish that weighs
   (a) at least 2 pounds.
   (b) less than 6 ounces.
   (c) between 8 ounces and 22 ounces.

3 Which of the following random variables would you think might be approximately normally distributed? In each case make a guess for the mean and the standard deviation.
   (a) Ages of people in the United States (to the nearest year)
   (b) Ages of college freshmen (to the nearest month)
   (c) Percentages of butterfat in quarts of whole milk produced by a dairy
   (d) Lengths of home runs hit during the major league baseball season
   (e) Diameters of gum balls in a vending machine
   (f) Weights of 1-inch number 6 machine screws
   (g) Heights of college freshmen

4 A tire manufacturer claims his premium tire has an average life of 40,000 miles. Assuming a normal distribution with standard deviation 5,000 miles, find the probability of buying a tire that lasts only 28,000 miles or less.

5 Using the approximation

$$\frac{1}{\sqrt{2\pi}} \approx .4$$

Sketch each of the following normal curves by plotting their maximum points and making the graph get very close to the x-axis near $x = \mu \pm 3\sigma$.
   (a) $N(x; 0, 2)$
   (b) $N(x; 0, \frac{1}{2})$
   (c) $N(x; 10, \frac{1}{2})$
   (d) $N(x; -5, 5)$

6 The results of a campus survey show that students study on the average 12 hours per week with standard deviation 2 hours.
   (a) If a student is selected at random, what is the probability that he studies more than 15 hours per week?
   (b) What percentage of students study less than 8 hours per week?

7 Assuming that weights of children born in City Hospital are normally distributed with mean 7 pounds 3 ounces and standard deviation 13 ounces,

find the probability that a newborn infant will weigh

(a) more than 10 pounds.
(b) between 6 and 8 pounds.
(c) less than 5 pounds.

8 A commuter observes that on the average his morning bus tends to arrive about 5 minutes later than scheduled. Assuming that arrival times are normally distributed with variance 9 minutes, find the probability that

(a) the bus arrives more than 10 minutes late.
(b) the bus arrives early.
(c) the bus arrives no more than 6 minutes late.

9 A certain blood test is designed so that 80% of the patients tested should have a count between 80 and 120. Assuming that these test counts are normally distributed with mean 100, find

(a) the standard deviation.
(b) the probability that a randomly selected patient will have a count greater than 130.

10 Suppose you are told that X is a normally distributed random variable with mean 100 and that $P(90 \leq X \leq 120) = .80$. Can you find σ_X?

4 BINOMIAL HYPOTHESIS TESTING REVISITED

With the normal approximation to the binomial distribution at our disposal, let us review our earlier efforts at hypothesis testing and somewhat extend the range of applications.

Recall that a *statistical hypothesis* is a statement about one or more of the parameters of a random variable X. For example, if X is known to be normally distributed with unknown mean μ, we could form one of the following statistical hypotheses about μ.

$$\mu = 12$$
$$\mu \neq 12$$
$$\mu < 12$$
$$\mu \geq 12$$

Of if X is known to be a binomially distributed variable we might posit one of the following:

$$p \leq .5$$
$$p = .5$$
$$p \neq .5$$
$$p > .5$$

We have been rather careful thus far to restrict our attention only to hypotheses about binomially distributed random variables, even though we have used the Poisson distribution to approximate the binomial in this connection. Thus we have been concerned with the following kind of procedure:

We wish to conduct n repeated trials and let X be the number of successes. The true probability of success p on a single trial is unknown to us, but we are led to believe that certain values of p are not reasonable. We formulate a *null hypothesis* H_0 and an *alternative hypothesis* H_1, where generally we try to reject H_0 in favor of H_1. We establish a significance level α, usually $\alpha = .05$ or $\alpha = .01$, and then we determine a critical region R as follows: R is a subset of the set of numbers from 0 through n that would cause us to reject H_0. That is, R consists of those values of X that would be considered "extreme" if H_0 were indeed true. We are guided by $P(R)$.

Typical null hypotheses and the corresponding types of critical regions are:

1. $H_0: p = \frac{1}{2}$, $R = \{0, 1, \ldots, k, m, \ldots, n\}$ (two-tail)
2. $H_0: p < \frac{1}{2}$, $R = \{m, \ldots, n\}$ (right tail)
3. $H_0: p > \frac{1}{2}$, $R = \{0, 1, \ldots, k\}$ (left tail)

In (1) for example, values of X too far away from the mean np *in either direction* would cause us to be suspicious of H_0, whereas in (3) only relatively small values of X would cause us to doubt H_0.

With this decision rule (reject H_0 if X falls in R; accept H_0 otherwise) we can make two kinds of mistakes: rejecting a true H_0 (called a type I error and having probability α) or accepting a false H_0 (called a type II error).

The object of this section is to take advantage of the normal approximation to the binomial to convert each type of critical region to a subset of the range of the standard normal random variable. For example, if X has frequency function $B(x; n, p)$, then a critical region of the type

$$\{m, \ldots, n\}$$

can be converted to an interval of the form

$$\left[\frac{m - .5 - np}{\sqrt{npq}}, \infty\right)$$

in the sense that

$$P(X \geq m) \approx P\left(Z \geq \frac{m - .5 - np}{\sqrt{npq}}\right)$$

Similarly the other two kinds of critical regions can be transformed to subsets of z-values. The following set of graphs indicate the particular z-values that determine common critical regions:

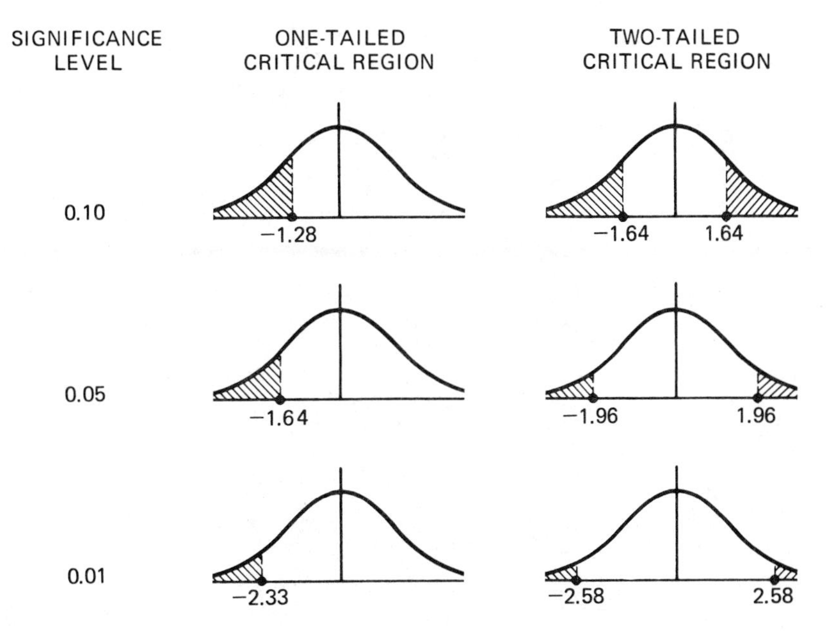

EXAMPLE 1 A coin is to be tossed 100 times. Can you conclude, at a .05 significance level, that it is not a fair coin if it comes up heads 60 times?

Our null hypothesis is

$$H_0 : p = .5$$

and the alternative hypothesis is

$$H_1 : p \neq .5$$

Since $n = 100$, we will have

$$\begin{aligned} \mu &= np \\ &= 100(.5) \\ &= 50 \end{aligned}$$

and

$$\begin{aligned} \sigma &= \sqrt{npq} \\ &= \sqrt{100(.5)(.5)} \\ &= \sqrt{25} \\ &= 5 \end{aligned}$$

To determine k, the upper bound of the left-hand tail, we convert $k + \frac{1}{2}$ to standard units

$$\frac{k + \frac{1}{2} - 50}{5} = \frac{k - 49.5}{5}$$

and require the result to be at most -1.96.

$$\frac{k - 49.5}{5} \leq -1.96$$

$$k - 49.5 \leq -9.80$$

$$k \leq 49.5 - 9.8$$

$$k \leq 39.7$$

The largest integer k satisfying this inequality is 39.

To determine m, the lower bound of the right-hand tail, we take $m - \frac{1}{2}$, convert it to standard units

$$\frac{m - \frac{1}{2} - 50}{5} = \frac{m - 50.5}{5}$$

and require the result to be at least 1.96.

$$\frac{m - 50.5}{5} \geq 1.96$$

$$m - 50.5 \geq 9.8$$

$$m \geq 50.5 + 9.8$$

$$m \geq 60.3$$

The smallest integer m satisfying this condition is 61, so the critical region is

$$\{0, 1, \ldots, 39, 61, 62, \ldots, 100\}$$

Since 60 is not in the critical region, we cannot reject the fairness of the coin (the null hypothesis) at the .05 significance level.

It is worthwhile to make a comparison between this critical region and the critical region obtained in Section 4 Chapter 3 when the coin was tossed 20 times. When the coin was tossed 20 times it had to come up heads (or tails) 75% of the time before we could conclude, at a .05 significance level, that it was unfair. But with 100 tosses, the 75% is reduced to 61%. This means that we have cut the size of the type II error (accepting a false hypothesis), since we stand a better chance of getting at least 61% heads (or tails) than at least 75%, even with an unfair coin. When we study the Central Limit Theorem, we will learn to determine the sample size to make the type II error as small as we like.

EXAMPLE 2 A thumbtack can land either point up or point down. Find the critical region for a test at the .10 significance level that the probability of landing up is $\frac{1}{3}$ where the test is to consist of 288 tosses of the tack.

Our null hypothesis is

$$H_0 : p = \tfrac{1}{3}$$

We have

$$\begin{aligned}\mu &= np\\ &= 288(\tfrac{1}{3})\\ &= 96\end{aligned}$$

and

$$\begin{aligned}\sigma &= \sqrt{npq}\\ &= \sqrt{288(\tfrac{1}{3})(\tfrac{2}{3})}\\ &= \sqrt{64}\\ &= 8\end{aligned}$$

The problem calls for a two-tailed test. The integer k, the upper bound of the left-hand tail, is the largest integer such that

$$\frac{k + \frac{1}{2} - 96}{8} \leq -1.64$$

$$\begin{aligned}k - 95.5 &\leq -13.12\\ k &\leq 95.50 - 13.12\\ k &\leq 82.38\end{aligned}$$

The largest integer satisfying this inequality is 82. Similarly for m, the lower bound of the right-hand tail, we convert $m - \frac{1}{2}$ to standard units

$$\frac{m - \frac{1}{2} - 96}{8} = \frac{m - 96.5}{8}$$

and require

$$\frac{m - 96.5}{8} \geq 1.64$$

$$\begin{aligned}m - 96.5 &\geq 13.12\\ m &\geq 96.5 + 13.12\\ m &\geq 109.62\end{aligned}$$

The smallest integer satisfying this inequality is 110 and the critical region is

$$\{0, 1, \ldots, 82, 110, 111, \ldots, 288\}$$

EXAMPLE 3 The election campaign between Smith and Jones is still on. Armed with the normal approximation to the binomial, you are going to sample 100 voters. How many people must say that they will vote for Jones so that you can predict his victory at the .05 significance level?

As before, let p be the probability that a randomly selected voter will vote for Jones. Our null hypothesis is that Jones won't win, i.e.,

$$H_0 : p \leq .5$$

Our critical region is everything to the right of 1.64 in standard units so we want the smallest integer m such that when $m - \frac{1}{2}$ is translated into standard units

$$\frac{m - \frac{1}{2} - 50}{5} = \frac{m - 50.5}{5}$$

the result is greater than or equal to 1.64.

$$\frac{m - 50.5}{5} \geq 1.64$$

$$m - 50.5 \geq 8.20$$

$$m \geq 50.5 + 8.2$$

$$m \geq 58.7$$

The smallest integer satisfying this inequality is 59 so our critical region is

$$\{59, 60, \ldots, 100\}$$

and if 59 or more people say they will vote for him we can predict his victory at a significance level of .05.

Exercises Section 4

1 Find the critical regions corresponding to .10, .05 and .01 significance levels for each of the following histograms.
 (a) $B(x; 100, .5)$
 (b) $B(x; 500, .5)$
 (c) $B(x; 1000, .5)$
 (d) $B(x; 100, \frac{1}{3})$
 (e) $B(x; 500, \frac{1}{3})$
 (f) $B(x; 1000, \frac{1}{3})$

2 In a true-false test consisting of 100 questions, how many must a student get right before the professor can predict at a .10 significance level that he is not guessing? Explain your choice of null hypothesis.

3 A new drug claims to substantially reduce the number of deaths from a disease. In the past, one-third of all people with the disease died. In a group of 300 treated with the drug, only 90 died. What can be said about the effectiveness of the drug. Explain your answer in terms of accepting or rejecting a suitable null hypothesis?

4 Referring to Exercise 3, another manufacturer comes out with a second drug and only 180 out of 600 die. Compare the second drug's effectiveness with that of the first.

5 Under normal conditions $\frac{4}{5}$ of the students taking Math 789 pass. Last year with the same instructor, text, etc., only 34 out of 200 failed. Can you conclude that it was a better class than usual?

6 Suppose we wish to test the hypothesis $H_0 : p \leq .5$ and we adopt the decision rule

"Reject H_0 if $\frac{x}{n} \geq .55$."

Find the corresponding significance level in the following cases.
(a) $n = 20$
(b) $n = 50$
(c) $n = 100$
(d) $n = 1{,}000$
(e) $n = 10{,}000$

5 POINT ESTIMATION OF PARAMETERS, MAXIMUM LIKELIHOOD

Statistical inference may be viewed as making educated decisions or guesses under uncertainty. Hypothesis testing falls into this category as does the topic of this section, estimation of parameters. The context once again is that we are given one (or sometimes more than one) random variable X, some of whose characteristics are unknown to us. Whereas in hypothesis testing we first make a conjecture (i.e., a hypothesis) about some characteristic (i.e., some parameter) and proceed to support or discredit that conjecture by obtaining some sample information, the process is somewhat reversed when one does estimation of parameters: First some sample information is obtained and then an educated guess is made about some characteristic of X. We begin by establishing some basic terminology couched in the framework of a familiar example.

Suppose that the big election between Smith and Jones has 1 million eligible voters. To each of these voters we assign a number: 0 if the voter is going to vote for Smith, 1 if he is going to vote for Jones. In effect we have specified a random variable X on a sample space (the 1 million voters). The resulting collection of 1 million numbers (0's and 1's) is called the *population* for the random variable X. In some books, the term "population" would refer to the voters, not the numbers assigned. When we take a poll, we simply select a subset of these 0's and 1's. This subset is called a *random sample* if the selection process (i.e., the poll) is conducted in such a way that all possible subsets of the same size are equally likely to be selected. The number of elements in the subset is called the *sample size*.

Generally in a sampling problem, we are in quest of an unknown characteristic of the population. In the voter problem we are interested in predicting the proportion of 1's in the population, and this proportion, p, is an unknown *parameter* of X. A piece of relevant information is the proportion of 1's in the sample, and in general we shall use the term "statistic" to refer to a measurable characteristic of the sample. In short, then, a parameter is to a population as a statistic is to a random sample. While in this problem we are interested in the proportion of 1's in the sample, statistics come in all shapes and sizes—they are simply functions of the sample. Here, for example are some typical statistics that would correspond to a sample of the form $\{x_1, x_2, \ldots, x_n\}$

x_3

$\frac{1}{n}\sum_{i=1}^{n} x_i$ (Called the *sample mean* and denoted $\bar{x}$)

$\sum_{i=1}^{n} x_i^2$

$\frac{1}{(n-1)}\sum_{i=1}^{n} (x_i - \bar{x})^2$ (Called the *sample variance* and denoted $\hat{s}^2$)

$\frac{1}{n}\sum_{i=1}^{i=n} (x_i - \bar{x})^2$ (The usual variance, s^2 of the numbers $x_1, \ldots, x_n$)

The largest value of $\{x_1, x_2, \ldots, x_n\}$

The smallest value of $\{x_1, x_2, \ldots, x_n\}$

The median of the set of numbers $\{x_1, x_2, \ldots, x_n\}$ (called the *sample median*)

Remarks: It is important to note that properly speaking *a statistic is a random variable,* since it changes from one sample to another. It is also significant that the sample variance $\hat{s}^2$ is slightly different from the variance s^2 of a set of n numbers in that the sum of the squares of the deviations from the mean is divided by $n - 1$ rather than n. Reasons for this modification will be given later.

Armed with the above definitions we can now describe a process called estimation of parameters.

1. X is a random variable known to satisfy a particular kind of distribution (e.g., binomial, Poisson, normal) but the parameters of the density function are unknown (e.g., we don't know p, λ, μ, and σ).
2. A random sample of size n is drawn from the values of X and one or more sample statistics are computed.
3. An estimate is made for the values of the unknown parameters on the basis of the sample statistics.

In essence, there are two kinds of estimates that we can make. We could say for example that

$\mu = 12$

where μ is the unknown mean of a normally distributed variable X, or we could say

$11.4 \leq \mu \leq 12.6$

In the first case we would be making a *point estimate,* that is, a single value estimate of μ. In the second case we would be making an *interval estimate,* providing a range of possible values for μ. In this section we illustrate one method for determining point estimates and in the next two sections we shall describe techniques for finding interval estimates.

The process we are about to describe is but one of several methods for finding point estimates. In contrast to interval estimates, where a range of possible values are postulated (along with a probability that the true value of the parameter falls into that interval), a point estimate is somehow a "best possible" choice for the unknown parameter.

Let us suppose that X is a random variable (either discrete or continuous) and that f is its density or frequency function. Suppose further that $\{x_1, x_2, \ldots, x_n\}$ is a random sample of values of X. We define the *likelihood function* L as follows:

$$L = f(x_1)f(x_2)\cdots f(x_n)$$

Not only does L depend upon the sample values $x_1, x_2, \ldots, x_n$, but it also depends upon the various parameters involved in f. For example if X were normally distributed, we would have

$$L = N(x_1; \mu, \sigma)N(x_2; \mu, \sigma)\cdots N(x_n; \mu, \sigma)$$

and L would depend upon the values of the parameters μ and σ. Similarly, if X were a discrete variable satisfying a Poisson distribution, we would have

$$L = P(x_1; \lambda)P(x_2; \lambda)\cdots P(x_n; \lambda)$$

and hence L would be a function of λ. The next point to note is that this product

$$L = f(x_1)f(x_2)\cdots f(x_n)$$

does somehow represent the "likelihood" of getting exactly the values $x_1, x_2, \ldots, x_n$ out of a random sample of n values of X. Indeed, *in the discrete case* L is precisely the probability of drawing that sample as a sequence of independent events. In case X is continuous, L is no longer a probability of any event, but it should seem reasonable that values of X that make $f(x)$ larger are in intervals that are more likely to be occupied by randomly selected values of X. [To be more concrete, if X is a normally distributed variable with density function $f(x) = N(x; \mu, \sigma)$, then we know that $f(x)$ is largest when x is near μ and that the probability of a randomly chosen value of X being in an interval near μ is higher than being in an interval of the same width far away from μ.] The problem then becomes:

> Find the values of the parameters that will maximize L. In other words, once the sample is determined, we seek the particular frequency function or density function that would make this sample selection most likely.

Certainly, if we knew that X were normally distributed, and we took a sample and found $\{30, 28, 31, 27, 29, 27\}$, we would strongly suspect the mean value of X to be around 28 or 29; we would be surprised to find the true mean to be 12 or 87 since that would make our sample very unlikely.

Now, how do we maximize the function L? Granted it depends on the particular choice of distribution, but we follow the same pattern in general. First we treat $x_1, x_2, \ldots, x_n$ as constants, and the parameters as variables. Then we utilize our max-min techniques from calculus: Take all partial derivatives and set them equal to 0. Fortunately in these problems we have to go no further. There is always a unique solution for the parameters that can be found by solving these equations. We illustrate the procedure with the normal distribution.

EXAMPLE 1 Maximum Likelihood Estimates for the Normal Distribution

Let $f(x) = N(x; \mu, \sigma)$ be the density function for a random variable X and suppose that $\{x_1, x_2, \ldots, x_n\}$ is a random sample of values of X. The likelihood function is

$$L(x_1, \ldots, x_n; \mu, \sigma) = N(x_1; \mu, \sigma)N(x_2; \mu, \sigma)\cdots N(x_n; \mu, \sigma)$$

$$= \left(\frac{1}{\sqrt{2\pi}}\right)^n \frac{1}{\sigma^n} \exp\left(-\frac{1}{2}\left(\frac{x_1 - \mu}{\sigma}\right)^2\right) \cdots \exp\left(-\frac{1}{2}\left(\frac{x_n - \mu}{\sigma}\right)^2\right)$$

$$= \left(\frac{1}{\sqrt{2\pi}}\right)^n \sigma^{-n} \exp\left(-\frac{1}{2\sigma^2} \sum_{k=1}^{n} (x_k - \mu)^2\right)$$

The partial derivative with respect to μ is

$$L_\mu(x_1, \ldots, x_n; \mu, \sigma) = \left(\frac{1}{\sqrt{2\pi}}\right)^n \sigma^{-n}\left[-\frac{1}{\sigma^2} \sum_{k=1}^{n} (x_k - \mu)\right] \exp\left(-\frac{1}{2\sigma^2} \sum_{k=1}^{n} (x_k - \mu)^2\right)$$

When we set $L_\mu = 0$, the only factor that could equal 0 is

$$\sum_{k=1}^{n} (x_k - \mu) = 0$$

and solving this for μ yields

$$\mu = \frac{1}{n} \sum_{k=1}^{n} x_k = \bar{x} \quad \text{(the sample mean)}$$

Similarly, it can be shown that with $\mu = \bar{x}$, the only way to make $L_\sigma = 0$ is to put

$$\sigma^2 = \frac{1}{n} \sum_{i=1}^{n} (x_i - \mu)^2 = s^2$$

Thus, we have shown that the only way L can achieve a maximum value is if

$$\mu = \bar{x} \quad \text{and} \quad \sigma = s$$

On the basis of these results we make a definition. The choices of values for the parameters that maximize the likelihood function for any density function f are called *maximum likelihood estimates* for those parameters.

EXAMPLE 2 Test Scores

The following set of 100 numbers represent scores on a mathematics examination. Under the assumption that these scores (or, if you like, a larger set of scores from which these were taken as a random sample) are normally distributed, let us find the maximum likelihood estimates for μ and σ that should produce the best-fitting normal curve.

40	85	35	66	93	26	98	61	98	29
17	31	76	44	51	78	36	12	78	84
40	51	70	6	36	32	76	87	95	57
70	91	33	53	72	73	96	38	30	71
64	29	72	85	25	43	44	53	45	36
31	44	51	30	64	38	18	79	64	36
77	79	21	59	75	67	35	62	21	47
81	66	57	37	33	56	72	28	72	69
28	15	57	78	79	68	89	55	91	20
73	94	91	28	85	30	94	33	49	74

Some rather laborious computation yields

$$\bar{x} = 56.21 \quad \text{and} \quad s = 24.018$$

Rounding off these values, we see that

$$N(x; 56.2, 24)$$

should be the best-fitting normal curve. We compute a couple of values to see how good the approximation is. Remember since X (the original random variable of scores) is discrete, we should use the approximation

$$P(a \le X \le b) \approx \int_{a-.5}^{b+.5} N(x; 56.2, 24)\, dx$$

where a and b are integers. For instance,

$$\begin{aligned} P(40 \le X \le 60) &\approx \int_{39.5}^{60.5} N(x; 56.2, 24)\, dx \\ &= \int_{-.7}^{.17} N(z; 0, 1)\, dz \\ &\approx .3255 \end{aligned}$$

Multiplying this probability by 100 gives

32.55

or approximately 33 scores should be in the range [40, 60]. Glancing at the list of 100 scores, we count

20

This kind of discrepancy between the theoretical model (the normal curve) and fact (the raw data) should be greeted with veiled skepticism at best. Let us try another interval

$$\begin{aligned} P(30 \le X \le 49) &\approx \int_{29.5}^{49.5} N(x; 56.2, 24)\, dx \\ &\approx .258 \end{aligned}$$

Thus the model predicts approximately 26 scores between 30 and 49, whereas in actuality there are 27 such scores. By most standards this approximation is sufficiently accurate. So on the one hand we have an interval where the curve fits the data nicely; and yet on another interval the fit is poor. The matter of "goodness of fit" will be dealt with later in this chapter when we shall devise some quantitative method of measuring how good the curve fits the data. For the time being, let's consider another likelihood function.

EXAMPLE 3 Prussian Cavalry

In Section 6 of Chapter 3 there is an exercise about deaths per corps-year in the Prussian Cavalry due to being kicked by horses. The data is

Number of deaths per corps-year	Frequency (absolute)
0	109
1	65
2	22
3	3
4	1

Under the assumption that X, the number of deaths, is a random variable whose frequency function is Poisson, we want the best choice for the parameter λ in

$$f(x) = \mathbf{P}(x; \lambda) = \frac{e^{-\lambda}\lambda^x}{x!}$$

The likelihood function in this case would be

$$\begin{aligned} L(x_1, x_2, \cdots, x_n; \lambda) &= \mathbf{P}(x_1; \lambda)\mathbf{P}(x_2; \lambda)\cdots\mathbf{P}(x_n; \lambda) \\ &= \frac{(e^{-\lambda})^n \lambda^{x_1+x_2+\cdots+x_n}}{(x_1!)(x_2!)\cdots(x_n!)} \end{aligned}$$

To simplify this expression, we put

$$a = x_1 + x_2 + \cdots + x_n$$
$$p = (x_1!)(x_2!)\cdots(x_n!)$$

and we get

$$L = \frac{e^{-n\lambda}\lambda^a}{p}$$

which we view as a function of one variable (λ), since the values of x_i and n are determined by the sample. Hence to maximize L, we take its derivative with respect to λ and set it equal to 0.

$$D_\lambda(L) = \frac{1}{p}\{e^{-n\lambda}(a\lambda^{a-1}) + \lambda^a(-ne^{-n\lambda})\}$$

Setting this expression equal to 0, we get

$$e^{-n\lambda}(a\lambda^{a-1}) + \lambda^a(-ne^{-n\lambda}) = 0$$
$$e^{-n\lambda}\lambda^{a-1}(a - n\lambda) = 0$$
$$a - n\lambda = 0$$

$$\boxed{\lambda = \frac{a}{n}}$$

But a/n is simply the sample mean (recall that $a = x_1 + x_2 + \cdots + x_n$). Hence our maximum likelihood estimate for λ is simply $\bar{x}$, the sample mean.

Back to the Prussian Cavalry. We compute

$$\begin{aligned}\bar{x} &= \tfrac{1}{200}(0 \cdot 109 + 1 \cdot 65 + 2 \cdot 22 + 3 \cdot 3 + 4 \cdot 1)\\ &= \tfrac{1}{200}(65 + 44 + 9 + 4)\\ &= \tfrac{122}{200}\\ &= .61\end{aligned}$$

Thus the best fitting Poisson "curve" (the graph of a Poisson frequency function is a collection of disconnected points) is

$$\mathbf{P}(x; .61)$$

Finally, since we use a table for the Poisson distribution, letting $\lambda = .6$ would give a suitable approximation.

Recall that we said that a sample statistic is a random variable. The estimates we have made in our examples are real numbers and when we wish to state in general what function of the sample we use (i.e., what statistic) to get our estimate, we use the term *estimator*. Thus we would say that the maximum likelihood estimators of μ and σ in a normal distribution are

$$\frac{1}{n}\sum_{i=1}^{n} x_i \quad \text{and} \quad \frac{1}{n}\sum_{i=1}^{n} (x_i - \bar{x})^2$$

respectively.

One final comment about our examples. In each instance we have found that the "best" estimate of the population mean has been the sample mean. Intuitively this should seem perfectly reasonable. We hasten to point out, however, that this is not always the case. In the uniform distribution, for instance, the maximum likelihood estimate of the population mean is *not* the sample mean but rather it is the average of the largest and the smallest values. The proof of this fact is left as an exercise (see Exercises 7 and 8).

Summary

When a random variable X is known to have a particular type of distribution, but its parameters are unknown, we can find *point estimates* for these parameters by means of taking a random *sample* of values of X from the *population* of all values of X. A function of the random sample is called a *statistic* and certain statistics are called *maximum likelihood estimators* of the unknown parameters, since their values represent the hypothetical value of the population parameters that maximize the probability of drawing the specified sample.

Exercises Section 5

1 Find the maximum likelihood estimate for the parameter p in the binomial distribution. In other words, let X be a discrete random variable with frequency function $B(x; n, p)$ and let $\{x_1, x_2, \ldots, x_k\}$ be a random sample of values of X. Write down the likelihood function, differentiate it with respect to p, and set it equal to 0, and solve for p.

2 If X is the random variable that gives the incomes of all persons (in some specified population) whose income is at least I (some fixed minimum income), then X satisfies (in theory) the Pareto distribution with density function

$$PA(x; \theta) = \frac{\theta I^{\theta}}{x^{\theta+1}}$$

Find the maximum likelihood estimate for the parameter θ.

3 Find the maximum likelihood estimate for the parameter θ in the density function

$$f(x; \theta) = \theta e^{-\theta x}, \qquad x \geq 0$$

4 Using the approximation $\lambda = .6$, compute the estimates for the frequencies of the number of deaths per corps-year in Example 3 and compare them to the actual figures.

5 Assume that the data given below on gum ball diameters is normally distributed with mean 3.55 and standard deviation .094. Find estimates for each of the following and compare these to the actual count.

(a) The number of diameters at least 3.55

(b) The number of diameters at least 3.45 but at most 3.65

(c) The number of diameters less than 3.70

48 Gum Ball Diameters

3.63	3.56	3.44	3.47	3.75	3.64	3.46	3.47
3.70	3.64	3.39	3.73	3.40	3.36	3.48	3.57
3.76	3.65	3.46	3.54	3.47	3.55	3.52	3.56
3.50	3.57	3.49	3.53	3.54	3.46	3.61	3.72
3.46	3.56	3.55	3.54	3.51	3.58	3.62	3.39
3.47	3.56	3.57	3.54	3.55	3.60	3.60	3.68

6 Suppose the weight of a shipment of 1,000 Winesap apples are suspected to be normally distributed with unknown mean μ and standard deviation σ. The produce manager decides to select some of the largest ones to package

separately at a higher price. The weights of the 9 randomly selected apples are (in ounces) 5.1, 3.8, 4.6, 7.3, 4.8, 6.9, 5.6, 5.3, 5.7. Find a particular weight for him to choose to separate the 100 largest apples in the shipment.

7 A *uniformly* distributed random variable Y has density function

$$f(y) = \begin{cases} \dfrac{1}{c}, & \text{if } 0 < y < c \\ 0, & \text{otherwise} \end{cases}$$

Set up the likelihood function for a sample of size n and find what value of c would maximize this function. (*Hint:* You will observe that the derivative of the likelihood function is never 0, but that a certain choice of c will still yield a minimum value; remember that the sample elements must be between 0 and c.)

8 Find the maximum likelihood estimator for the uniformly distributed random variable X whose density function is

$$f(x) = \begin{cases} \dfrac{1}{(b-a)}, & \text{if } a < x < b \\ 0, & \text{otherwise} \end{cases}$$

6 INTERVAL ESTIMATES. THE CENTRAL LIMIT THEOREM

We are now going to make a substantial change in our point of view. Given a sample from a population, and under the assumption that the population can be described by some distribution (e.g., binomial, normal, Poisson) we have attempted to determine the value of the parameter or parameters involved. But there is an implicit difficulty with the procedure. Suppose we are trying to estimate the value of the parameter p of the binomial distribution from a sample and, using the maximum likelihood technique, arrive at the estimate $p = .432$. How do we interpret this? We certainly are not very confident that p actually is .432. After all, p could take on any of the infinite number of values between 0 and 1, so the probability that any given sample will lead to the exact value of p is 0. We do mean that in terms of the criteria of the maximum likelihood technique, .432 is the best estimate we have. Moreover, on an intuitive level, our estimate is probably fairly close to the true value. But how close is fairly close?

What we shall do in this section is to develop a technique for finding an *interval* of values in which we can be fairly sure the true value of the parameter lies and where the degree of certainty can be quantified. To do this we need information relating the properties of a population with the properties of samples drawn from it.

We begin with a concrete example. For the population we take five balls numbered 1–5. We can identify this population with the sample space associated with the experiment of choosing one of the balls. The random variable X which assigns to each outcome of this experiment the number of the ball which is drawn can be regarded as generating the population.

We shall frequently speak of a population P and a random variable X on it. This is a convenient abuse of language. More formally, X is a random variable associated with the experiment of choosing an element of the population.

When we calculate $\mu = \mu_x$, we obtain

$$\begin{aligned}\mu &= \tfrac{1}{5}\cdot 1 + \tfrac{1}{5}\cdot 2 + \tfrac{1}{5}\cdot 3 + \tfrac{1}{5}\cdot 4 + \tfrac{1}{5}\cdot 5 \\ &= \tfrac{1}{5}(1 + 2 + 3 + 4 + 5) \\ &= \tfrac{1}{5}\cdot 15 \\ &= 3\end{aligned}$$

Then $\sigma = \sigma_X$ is given by

$$\begin{aligned}\sigma^2 &= \tfrac{1}{5}(1^2 + 2^2 + 3^2 + 4^2 + 5^2) - 3^2 \\ &= \tfrac{1}{5}(55) - 9 \\ &= 11 - 9 \\ &= 2\end{aligned}$$

and

$$\sigma = \sqrt{2}$$

Suppose we take samples of size 2 by drawing (with replacement) two of the balls. The possible samples are

1, 1	2, 1	3, 1	4, 1	5, 1
1, 2	2, 2	3, 2	4, 2	5, 2
1, 3	2, 3	3, 3	4, 3	5, 3
1, 4	2, 4	3, 4	4, 4	5, 4
1, 5	2, 5	3, 5	4, 5	5, 5

We define a new random variable X_2 which assigns to each sample of size 2 the mean of the sample. The values of X_2 for each of the above samples of size 2 are given in the following table.

1.0	1.5	2.0	2.5	3.0
1.5	2.0	2.5	3.0	3.5
2.0	2.5	3.0	3.5	4.0
2.5	3.0	3.5	4.0	4.5
3.0	3.5	4.0	4.5	5.0

We shall now calculate $\mu_2 = \mu_{X_2}$ and $\sigma_2 = \sigma_{X_2}$ and compare them with μ and σ. Using the fact that all samples are equally likely, we obtain

$$\begin{aligned}\mu_2 &= \tfrac{1}{25}(1.0 + 1.5 + 2.0 + 2.5 + 3.0 + 1.5 + \cdots + 4.0 + 4.5 + 5.0) \\ &= \tfrac{1}{25}(75) \\ &= 3\end{aligned}$$

Also

$$\sigma_2^2 = \tfrac{1}{25}((1.0)^2 + (1.5)^2 + \cdots + (4.5)^2 + (5.0)^2) - 3^2$$
$$= \tfrac{1}{25}(250) - 9$$
$$= 10 - 9$$
$$= 1$$

and

$$\sigma_2 = 1$$

Observe that

$$\mu_2 = \mu$$

and

$$\sigma_2 = \frac{\sigma}{\sqrt{2}}$$

As an exercise you will be asked to calculate μ_3 and σ_3, the mean and standard deviation of the set of all means of samples of size 3 from our population. You will find that

$$\mu_3 = \mu$$

and

$$\sigma_3 = \frac{\sigma}{\sqrt{3}}$$

Now consider the sample space consisting of all possible samples of size n. We define a new random variable, X_n, which assigns to each sample of size n its sample mean. We calculate $\mu_n = \mu_{X_n}$ and $\sigma_n = \sigma_{X_n}$ under the assumption that all samples are equally likely. (This assumption is what is meant by random sampling.)

In the few cases we have looked at we obtained

$$\mu_n = \mu$$

and

$$\sigma_n = \frac{\sigma}{\sqrt{n}}$$

Part of the Central Limit Theorem which we will state later says that these equations hold true for the kinds of random variables we will encounter in this course and in most applications.

What is all of this telling us? It is telling us that the average mean of a sample of size n is the mean of the original population ($\mu_n = \mu$) *and* that the sample means vary much less than the value of the original random variable, less, in fact, by a factor of $1/\sqrt{n}$ (that is, $\sigma_n = \sigma/\sqrt{n}$).

Look at the implications of this in a concrete case. Suppose that X is the height of college freshmen males in the United States, and we know that $\mu = \mu_X = 70$ (inches) and $\sigma = \sigma_X = 2.3$ (inches). If X is normally distributed, then, as we observed in Section 3, approximately 95% of all men will be between $70 - 1.96(2.3) = 65.5$ inches and $70 + 1.96(2.3) = 74.5$ inches in height.

Now suppose we take samples of size 100. What about the means of these samples. The average sample mean will be 70 again, but the standard deviation of these sample means will be

$$\frac{2.3}{\sqrt{100}} = \frac{2.3}{10} = .23$$

Therefore, the sample mean of 95% of all samples of size 100 will be between $70 - 1.96(.23) = 69.55$ and $70 + 1.96(.23) = 70.45$.

One can conclude from this that while you will find individuals whose height is 71 inches, you are much less likely to find a random sample of 100 whose average height is 71 inches. It is still less likely that you would find a random sample of 400 whose average height is 71.

The careful reader may have noted the assertion that 95% of the means of samples of size 100 were between 69.55 and 70.45 implies that these sample means are normally distributed. Part of the content of the Central Limit Theorem which we will soon state is that, for reasonably large n, this is approximately true and the assumption will not get us into trouble. In particular the fact that X_n is approximately normally distributed for large n does not depend upon X being normally distributed. It holds for a very general class of random variables. We shall not give an explicit description but will comment that it works for most random variables encountered in applied work.

We summarize the preceding discussion in the statement of the Central Limit Theorem. The reader who looks at other books may find a more general statement under the same label. What we are stating is actually a special case, in which we focus our attention on those aspects of the theorem generally needed for applications.

Central Limit Theorem

Let X be a random variable (not necessarily normally distributed) with mean μ and standard deviation σ. Let X_n be the random variable denoting the means of samples of size n drawn at random from the population which X defines. Then, for reasonably large n, X_n is approximately normally distributed with mean μ_n and standard deviation σ_n given by

$$\mu_n = \mu$$

$$\sigma_n = \frac{\sigma}{\sqrt{n}}$$

Applications of the Central Limit Theorem generally require a slight restatement of the result. One of the implications of the fact that X_n is (approximately) normally distributed with mean μ and standard deviation $\sigma/\sqrt{n}$ is that the variable

$$\frac{X_n - \mu}{\sigma/\sqrt{n}}$$

had mean 0 and standard deviation 1. Therefore, given a probability c, and using the table for $N(z; 0, 1)$, we can find a value for z such that

$$P\left(-z \leq \frac{X_n - \mu}{\sigma/\sqrt{n}} \leq z\right) = c$$

The inequalities

$$z \leq \frac{X_n - \mu}{\sigma/\sqrt{n}} \leq z$$

can be interpreted in two ways. We can note that

$$\frac{-z\sigma}{\sqrt{n}} \leq X_n - \mu \leq \frac{z\sigma}{\sqrt{n}}$$

or

$$\mu - \frac{z\sigma}{\sqrt{n}} \leq X_n \leq \mu + \frac{z\sigma}{\sqrt{n}}$$

This says that X_n is in an interval of length $2z\sigma/\sqrt{n}$ centered at μ.

$\mu - (z\sigma/\sqrt{n})$ $\qquad$ μ $\qquad$ $\mu + (z\sigma/\sqrt{n})$

But we can also first multiply the inequality

$$-z \leq \frac{X_n - \mu}{\sigma/\sqrt{n}} \leq z$$

by -1 obtaining

$$z \geq \frac{\mu - X_n}{\sigma/\sqrt{n}} \geq -z$$

or

$$-z \leq \frac{\mu - X_n}{\sigma/\sqrt{n}} \leq z$$

whence

$$\frac{-z\sigma}{\sqrt{n}} \leq \mu - X_n \leq \frac{z\sigma}{\sqrt{n}}$$

or

$$X_n - \frac{z\sigma}{\sqrt{n}} \leq \mu \leq X_n + \frac{z\sigma}{\sqrt{n}}$$

This inequality says that μ is in an interval of length $2z\sigma/\sqrt{n}$ centered at X_n.

$$X_n - (z\sigma/\sqrt{n}) \qquad\qquad X_n \qquad\qquad X_n + (z\sigma/\sqrt{n})$$

It is this second interpretation which allows us, knowing σ and having a value of X_n from a sample, to assert with probability c that μ is in a certain interval. Let us return to the problem of the average height of college freshmen males in the United States. This time assume that we don't know what the average height is but we will pretend we know that the standard deviation is 2.3. Suppose we take our random sample of 100 men and the average height is 70.0 inches. If we choose the probability $c = .95$, then $z = 1.96$ and the interval is

$$(70 - 1.96\sigma_{100}, 70 + 1.96\sigma_{100})$$

Since

$$\begin{aligned}\sigma_{100} &= \frac{\sigma}{\sqrt{n}}\\ &= \frac{2.3}{\sqrt{100}}\\ &= .23\end{aligned}$$

and since

$$(1.96)(.23) \approx .45$$

our interval is

$$(69.55, 70.45)$$

How are we to interpret this? In particular, what is it that .95 is the probability of? A naive statement, all too frequently heard, is that the probability is .95 that the true population mean μ is in the interval. But that makes no sense. There is a true value of μ, whether we know what it is or not, and it is either in the interval (69.55, 70.45) or it is not. What is meant is that if you use the above procedure to find an interval, then the probability is .95 that the interval contains the true value of μ. It is the interval, which could change from sample to sample, to which the probability is attached, not μ.

The interval which we have found is called a *confidence interval*. Since we choose $c = .95$, we would refer to the interval

$$(69.55, 70.45)$$

as a *.95 confidence interval* or a *95% confidence interval*. If we chose $c = .99$, we would have a *.99 confidence interval*, etc.

We have avoided the issue of what to do when the standard deviation of the population is unknown (as it usually is in practical problems). Your natural inclination is, probably, to use the standard deviation of the sample as an estimate, but this creates difficulties. To see what goes wrong, let us return to the sampling problem discussed earlier in the section when we drew, with replacement, samples of size 2 from a population of five balls numbered 1, 2, 3, 4, and 5. If we compute the variance of the 25 equally likely samples, we obtain

0.00	0.25	1.00	2.25	4.00
0.25	0.00	0.25	1.00	2.25
1.00	0.25	0.00	0.25	1.00
2.25	1.00	0.25	0.00	0.25
4.00	2.25	1.00	0.25	0.00

The average of these variances is 1 while the variance of the population is 2. We would, therefore, on the average make a mistake by using the variance of the sample as an estimate of the variance of the population.

The difficulty can be precisely stated in terms of expected value. If we let X_n be the random variable which assigns to each sample of size n its mean, then

$$E(X_n) = \mu \quad \text{(The true population mean)}$$

But if we let S_n^2 be the random variable which assigns to each sample of size n its variance, then

$$E(S_n^2) \neq \sigma^2 \quad \text{(The true population variance)}$$

It can be shown that

$$E(S_n^2) = \frac{n}{n-1}\sigma^2$$

To deal with this we shall define a new statistic $\hat{S}_n$ on samples of size n defined as follows. For a sample $\{x_1, x_2, \ldots, x_n\}$ with mean $\bar{x}$,

$$\hat{S}_n^2 = \frac{1}{n-1}\sum_{i=1}^{n}(x_i - \bar{x})^2$$

This statistic $\hat{S}_n^2$ is called the *sample variance* and

$$E(\hat{S}_n^2) = \sigma^2$$

Formally, this last equation asserts that $\hat{S}_n^2$ is an *unbiased estimate* of σ^2. We shall take $\hat{S}_n$, the *sample standard deviation*, rather than S_n as the estimate for σ.

EXAMPLE 1 If, on a sample, the random variable assumes the values $\{1, 5, 4, 2, 8\}$, then the mean is given by

$$\bar{x} = \frac{1 + 5 + 4 + 2 + 8}{5}$$

$$= \frac{20}{5} = 4$$

and the sample variance $\hat{S}^2$ would be

$$\hat{S}^2 = \tfrac{1}{4}\{(1-4)^2 + (5-4)^2 + (4-4)^2 + (2-4)^2 + (8-4)^2\}$$

$$= \tfrac{1}{4}\{9 + 1 + 0 + 4 + 16\}$$

$$= \tfrac{1}{4}(30) = 7.5$$

EXAMPLE 2 A random sample of 100 new cars of a given make were driven 100 miles. They used an average of 6.7 gallons of gasoline with a sample standard deviation of .51. What can be said about the average amount of gasoline used by all new care of this make in traveling 100 miles?

The 95% confidence interval is

$$\left(6.7 - (1.96)\frac{.51}{\sqrt{100}}, 6.7 + (1.96)\frac{.51}{\sqrt{100}}\right)$$

or

$$(6.7 - .09996, 6.7 + .09996) = (6.60004, 6.79996)$$

If we wanted a higher level of confidence, we could obtain the 99% confidence interval, which is

$$\left(6.7 - (2.58)\frac{.51}{\sqrt{100}}, 6.7 + (2.58)\frac{.51}{\sqrt{100}}\right)$$

or

$$(6.56842, 6.81358)$$

One final comment. The statement of the Central Limit Theorem used the phrase "reasonably large n." We should ask what is reasonably large. The answer depends upon the degree of accuracy desired. We shall, somewhat arbitrarily, say $n = 50$. It should be understood that this is a choice made for the sake of convenience and that people who demand less accuracy require n to be 25 or 50; in some cases where more accuracy is desired n must be even larger. In the next section we will learn what to do with smaller samples.

Summary

In this section we have studied the notion of an *interval* estimate of a parameter as opposed to the point estimates developed earlier. If we take a sample of size n from a population with standard deviation σ and if the mean of the sample is $\bar{x}$, then the 95% confidence interval for μ, the mean of the population, is

$$\left(\bar{x} - 1.96\frac{\sigma}{\sqrt{n}}, \bar{x} + 1.96\frac{\sigma}{\sqrt{n}}\right)$$

The 95% refers to the fact that if intervals are formed by the above formula, then 95% of the time they will contain the mean of the population. To deal with the cases where the standard deviation of the population is not known we introduced $\hat{S}$, the sample standard deviation, as an estimate. The formula for $\hat{S}^2$ is

$$\hat{S}^2 = \frac{1}{n-1}\sum_{i=1}^{n}(x_i - \bar{x})^2$$

where $\{x_1, x_2, \ldots, x_n\}$ is the sample and $\bar{x}$ is its mean.

Exercises
Section 6

1 A sample of 81 dentists were found to have a mean income of $40,000 with a sample standard deviation of $1,400. Find the 90%, 95%, and 99% confidence intervals for the mean income of all dentists.

2 A sample of 70 tires of a particular grade were found to have an average life of 31,000 miles with a sample standard deviation of 2,100 miles. Find the 90%, 95%, and 99% confidence interval for the average life of this type of tire.

3 Redo Exercises 1 and 2 with the same mean and sample standard deviation but with the sample sizes changed to

(a) 100 (c) 900
(b) 400 (d) 2,500

4 Let us suppose for the sake of simplicity that the bacteria count of a certain kind of bacteria in hamburger patties is such that when the count is 10 the average person will be slightly affected by eating the hamburgers, though he will not be ill, and at 11 the average person will become ill. A busy store sells several thousand hamburger patties per day.

(a) A random sample of 100 shows a mean count of 8.6 with a sample standard deviation of 1.3. Find the 95% confidence interval for the average bacteria count.

(b) Using the data from (a), in what percentage of hamburgers is the count 10 and in what percentage is the count 11?

(c) Assuming that the estimate 1.3 of the standard deviation holds, what would the average bacteria count have to be before the percentage of hamburgers with a bacteria count of 10 is as small as .1%?

(d) For a sample of 100, what range of sample means would convince you that only .1% of the hamburgers have a bacteria count of 10?

5 Exercise 4 should indicate to you that there are circumstances where one might want a one-sided confidence interval since we would not be disturbed by having too few bacteria.

(a) Describe a procedure for finding a one-sided 95% confidence interval [i.e., one of the form $(-\infty, a]$ or $[b, \infty)$] from a sample of size n with mean $\bar{x}$ and sample standard deviation $\hat{S}$.

(b) Find a one-sided confidence interval for the average number of soot particles per cubic foot in the air if a sample of size 100 had a mean of 32 and a standard deviation of 12.

6 You have 100 real numbers to add. Instead of adding them, you round each off to the nearest integer before adding. Assume that the round-off errors are uniformly distributed between $-.5$ and $+.5$. If the sum you obtain is 427, find a 95% confidence interval for the true sum.

7 THE t-DISTRIBUTION

The t-distribution was invented by William Gosset (who wrote under the name Student and hence the distribution is often called the Student t-distribution) to overcome two limitations of the Central Limit Theorem.

First, in order to apply the Central Limit Theorem, the value of n must be large—at least 50, perhaps more. In practice the sample may not be that large and there may be no convenient way of enlarging it. Secondly, one should know the value of the standard deviation of the population to apply the Central Limit Theorem, but in practice one has only the sample standard deviation and, for small n, there may be a considerable difference between the population standard deviation and the sample standard deviation.

The t-distribution is designed to be used for samples too small for the Central Limit Theorem and it takes into account the fact that the standard deviation of the sample is being used instead of the standard deviation of the population. As you might expect, you don't get something for nothing and there is a restriction on the applicability of the distribution which isn't present in the Central Limit Theorem. The Central Limit Theorem can be applied whether or not the random variable X on the population P is normal. But, *the t-distribution only applies when X is normally distributed.*

In order to introduce the random variable t and compare it with the central Limit Theorem, we reintroduce and extend the notation of the previous section. We have a population defined by a random variable X. We assume X has mean μ and standard deviation σ. We let X_n denote the mean of a sample of size n $\hat{S}_n^2$ its estimated variance. The Central Limit Theorem asserts that for large n, the variable

$$\frac{X_n - \mu}{\sigma/\sqrt{n}}$$

is normally distributed with mean 0 and standard deviation 1.

The t-variable is defined by replacing σ by its estimate $\hat{S}_n$, yielding

$$t = \frac{X_n - \mu}{\hat{S}_n/\sqrt{n}}$$

Like the standard normal distribution, the t-distribution is symmetric about the origin. However, while there is only one standard normal curve, there are

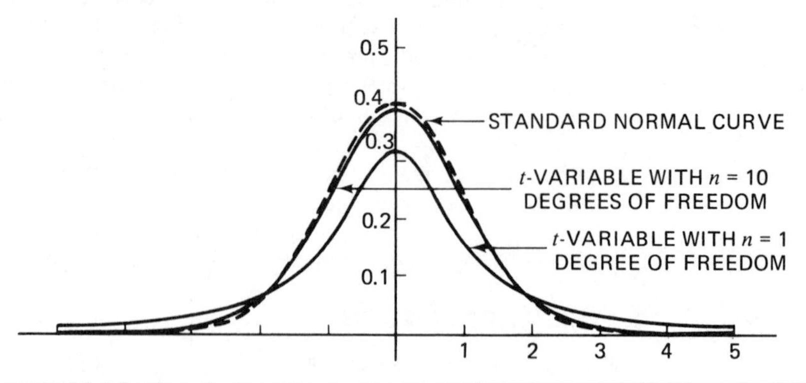

FIGURE 5.9 Density functions for the t-variable with $n = 1$ and with $n = 10$ degrees of freedom and the standard normal curve.

many for t, one for each integer n. For small values of n, the t curve is more spread out than the normal curve and as n increases, the t curve gets closer to the normal curve as indicated in Fig. 5.9.

Instead of a single table, we need a set of tables, one for each value of n. However, custom has it that in setting up the tables we do not use the sample size n, but the number of degrees of freedom, which is $n - 1$.

EXAMPLE 1 In a large introductory history class a random sample of four students had grades on the first hour exam of 63, 83, 71, 75. Find a 95% confidence interval for the class average on the test.

Solution: The mean, $\bar{x}$, of the sample is

$$\bar{x} = \tfrac{1}{4}(63 + 83 + 71 + 75) = \tfrac{1}{4}(292) = 73$$

and the estimated standard deviation $\hat{S}_4$ is

$$\begin{aligned} S_x &= \sqrt{\tfrac{1}{3}[(63 - 73)^2 + (83 - 73)^2 + (71 - 73)^2 + (75 - 73)^2]} \\ &= \sqrt{\tfrac{1}{3}(208)} \\ &\approx \sqrt{69.33} \\ &\approx 8.33 \end{aligned}$$

Since the sample size is 4, there are $4 - 1 = 3$ degrees of freedom. Consulting Table E in the Appendix, we find that

$$P(t \leq 3.182) = .025$$

and, since t is symmetric about 0,

$$P(-3.182 \leq t \leq 3.182) = .05$$

Since

$$t = \frac{73 - \mu}{8.33/\sqrt{4}}$$

(where μ is the actual class average), we have

$$-3.182 \leq \frac{73 - \mu}{8.33/\sqrt{4}} \leq 3.182$$

or

$$3.182 \geq \frac{\mu - 73}{(8.33)/\sqrt{4}} \geq -3.182$$

Therefore, a 95% confidence interval is

$$73 - \frac{(3.182)(8.33)}{2} \leq \mu \leq 73 + \frac{(3.182)(8.33)}{2}$$

$$73 - 13.25 \leq \mu \leq 73 + 13.25$$

$$59.75 \leq \mu \leq 86.25$$

The interval here is quite large. This agrees with your intuition which should tell you that the mean of a small sample can differ from the mean of a population by quite a bit. A glance at the table will show you that as the sample size, and hence the number of degrees of freedom, grows, the interval becomes smaller since the factor 3.182 is replaced by a smaller number (for example by 2,571 for a sample of size 6 with, therefore, 5 degrees of freedom), and the denominator also becomes larger. Once again your intuition should tell you that a larger sample should be close to the true mean.

It might be noted that our numerical calculations in this exercise should be carried out, in general, to give the following fact. Once the t-distribution is applicable, and we have found, for some number p, $0 \leq p \leq 1$, the value t_p such that

$$P(-t_p \leq t \leq t_p) = p$$

then the $100p\%$ confidence interval for μ is given by

$$\bar{x} - \frac{t_p\hat{S}}{\sqrt{n}} \leq \mu \leq \bar{x} + \frac{t_p\hat{S}}{\sqrt{n}}$$

EXAMPLE 2 A tire manufacturer advertises that his tires should last for 25,000 miles. A company bought 16 of them and put them on company cars. They were found to last an average of 24,300 miles with a sample deviation of 800 miles. What does this say about the claim of the manufacturer?

The 95% confidence interval for μ, the average tire life, is from

$$24{,}300 - \frac{(2.131)800}{\sqrt{16}} \quad \text{to} \quad 24{,}300 + \frac{(2.131)800}{\sqrt{16}}$$

or from 24,300 − 426 to 24,300 + 426, i.e., 23,874 to 24,726.

We see that the manufacturer's claim is outside the 95% confidence interval. We are therefore led to doubt his claim. Indeed, this problem seems to bring us back to the hypothesis testing which we did earlier. We shall explore this in the next section.

Summary

Given a normally distributed random variable X with mean μ, we have defined a random variable t on samples of size n as follows. If the sample has mean X_n and sample standard deviation $\hat{S}_n$,

$$t = \frac{X_n - \mu}{\hat{S}_n/\sqrt{n}}$$

We use this variable, rather than the normally distributed

$$\frac{X_n - \mu}{\sigma/\sqrt{n}}$$

to find a confidence interval for μ, where (1) n is small (50 or less), and (2) σ is unknown.

In point of fact the above distinction, while operationally accurate, is open to some theoretical discussion. The real difference between the t-variable and the standard normal variable is not so much the sample size as it is whether or not σ is known. In the previous section, where we used the normal distribution for large samples and replaced σ by $\hat{S}_n$, we were really using the fact that for large n the distribution of t is approximately normal.

Exercises Section 7

1. A random sample of seven lawyers had IQ's of 127, 120, 142, 131, 128, 148, 119. Construct a 95% confidence interval for the average IQ of lawyers.
2. A random sample of 11 seniors at a college had an average grade point average of 2.45 with variance .37. Construct a 90% confidence interval for the average senior grade point average.
3. Three different sources assert that the average income of accountants is $16,700, $15,000, and $16,200. Confused by the discrepancy, you take your survey of 11 members of the profession and find that they have an average income of $15,800 with standard deviation $500. Which of the three asserted averages is
   (a) in the 95% confidence interval?
   (b) in the 99% confidence interval?
4. **(a)** A random sample of 10 workers in a city were found to spend an average of 42 minutes getting to work with a standard deviation of 8 minutes. Find the 95% confidence interval for the average time it takes a worker to get to work.
   (b) Suppose the random sample is of 100 workers with the average still 42 minutes and the standard deviation 8 minutes. Find the 95% confidence interval and compare with your answer to (a).
5. A manufacturer of flashlight batteries claims that his batteries have an average life of 12 hours. You test a sample of 15 of them and they have an average life of 11 hours and 20 minutes with a standard deviation of 30 minutes. What can you say in terms of confidence intervals about the manufacturer's claim?
6. A golfer posts scores of 86, 94, 82, 97, 85, 89, and 90. Find a 90% confidence interval for his scores.

8 HYPOTHESIS TESTING AND CONFIDENCE INTERVALS

The material which we have just covered on the Central Limit Theorem and the t-test allows us to handle a number of hypothesis testing problems which were previously beyond our capacity. In this section we will consider several types of hypothesis testing problems.

One of the most common hypotheses to be tested is an hypothesis about the value of the mean μ of a random variable X. There are, effectively, two types of null hypotheses about μ. One type is

$$H_0: \mu = a$$

and the other is

$$H_0: \mu < a$$

(or $\mu > a$). This distinction is the distinction between the two-tailed test and the one-tailed test, which we have discussed previously.

But there is another factor which must be considered in deciding how to test any hypothesis concerning μ. This second factor is a decision as to whether we will invoke the Central Limit Theorem (and hence use the normal distribution) or use the t-distribution. We decide as follows.

1. If σ, the standard deviation of the population, is known and if n is large (>50), we use the Central Limit Theorem.
2. If σ is unknown and n is large (>50), we use the sample standard deviation s as an estimate for σ and, technically, the t-test. However for such values of n the t-distribution is approximately normal and so we use the normal distribution.
3. Finally, if $n < 50$, we use the t-distribution.

EXAMPLE 1

Les and Bob run competing hamburger stands. Les advertises a 10-ounce hamburger, but Bob doesn't believe it. He gets a sample of 25, weighs them and finds the mean to be 9.91 ounces with a sample standard deviation of .16 ounce. At the 5% significance level can he reject Les' claim?

Our hypothesis is

$$H_0: \mu = 10$$

Since $n = 25$ and we do not have the standard deviation of the population (which consists of the weights of all of Les' hamburgers) and must use the estimate .16, we should employ the t-test. Our critical region should be a range of sample means that will contain only 5% of the sample means if the hypothesis is true. Consulting our table for the t-distribution, we see that if the mean is 10 and if $\sigma = .16$, then 95% of all samples of size 25 will fall in the interval

$$\left(10 - 2.064\frac{.16}{\sqrt{25}}, \qquad 10 + 2.064\,\frac{.16}{\sqrt{25}}\right)$$

or (9.93395, 10.06605).

Our critical region at the 5% significance level is those sample means X_{25} such that

$$X_{25} < 9.93395 \quad \text{or} \quad X_{25} > 10.06605$$

Bob's sample mean of 9.91 is in the critical region and so he can challenge Les at the 5% significance level.

There are two observations to be made here. First, we chose to use a two-tailed test but one might be inclined to try a one-tailed test. Now how would we proceed? The null hypothesis is

$$H_0: \mu \geq 10$$

The alternate hypothesis is

$$H_1 : \mu < 10$$

We assume the same significance level, 5%. Consulting the tables, we see that, if H_0 is true, then 95% of the means of samples of size 25 will be greater than

$$10 - 1.711 \frac{.16}{\sqrt{25}} = 9.94525$$

The critical region is then all sample means less than 9.94525.

It should be pointed out that this can well be an unsound procedure. Ideally the steps in hypothesis testing should come in a certain order;

1. Form the hypothesis.
2. Decide the significance level.
3. Determine sample size.
4. Calculate the critical region, and take the sample.

In practice one can rarely be that pure but you should be very cautious about any change in the order. The nature of one-tailed as opposed to two-tailed tests allows Bob to reject Les' claim in the example even if Bob's sample mean were closer to 10. So if you decide on a one-tailed test instead of a two-tailed test after the data is collected, you are making it easier to reject hypotheses and it is not entirely clear that the level of significance is unchanged. Exercise 13 goes into detail on this problem.

Secondly we should note the relationship between hypothesis tests and interval estimates. In the two-tailed test, the critical region for a test at the .05 significance level consisted of those sample means outside the interval.

$$(*) \quad \left(10 - 2.064 \frac{.16}{\sqrt{25}},\ 10 + 2.064 \frac{.16}{\sqrt{25}}\right)$$

The 95% confidence interval for the true mean, based on the same sample, is

$$(**) \quad \left(9.91 - 2.064 \frac{.16}{\sqrt{25}},\ 9.91 + 2.064 \frac{.16}{\sqrt{25}}\right)$$

Hypothesis testing for the mean can be done from an interval estimate because 9.91 is outside interval (*) (i.e., in the critical region) precisely when 10 is outside of interval (**) (i.e., the value of μ asserted in the null hypothesis is not in the confidence interval).

The Central Limit Theorem can also be used to determine sample size as in the following example.

EXAMPLE 2

We return again to the election between Smith and Jones. We want to have a .05 level of significance for our prediction, and we want to be able to make a prediction if as few as 51% of our sample choose a particular man.

This last phrase deserves discussion. The point is that if the number of people preferring Smith (or Jones) is between 49 and 51% we will say the race is too close to predict but if either of them gets 51% or more, we will predict his victory at a .05 significance level.

We set the problem up as in the past. The population is the set of eligible voters, the random variable X assigns a 0 to those who are for Smith and a 1 to those who are for Jones. Then $\mu = p$, the probability that a randomly selected voter will vote for Jones. If $p > .5$, Jones will win; if $p < .5$, Smith will win. We know, as observed earlier, that

$$\sigma \leq \tfrac{1}{2}$$

We take as our null hypothesis

$$H_0: p \leq .5$$

i.e., Jones will not win, which requires a one-tailed test.

Suppose we take a sample of size n. The mean μ_n of the means of these samples is μ and the standard deviation σ_n is given by

$$\sigma_n = \frac{\sigma}{\sqrt{n}}$$

But

$$\sigma \leq \tfrac{1}{2} \text{ (See Exercise 4.)}$$

so

$$\sigma_n \leq \frac{1}{2\sqrt{n}}$$

Let μ' be the mean of the sample we actually take. For a one-tailed test at the .05 significance level, we should have

$$\mu' - 1.65\sigma_n > .5$$

or

$$.5 + 1.65\sigma_n < \mu'$$

Since

$$\sigma_n \leq \frac{1}{2\sqrt{n}}$$

this will hold if

$$.5 + \frac{1.65}{2\sqrt{n}} < \mu'$$

We want to be able to reject H_0 if μ' is as small as .51. Therefore, we must have

$$.5 + \frac{1.65}{2\sqrt{n}} < .51$$

$$\frac{1.65}{2\sqrt{n}} < .01$$

$$\frac{1.65}{.02} < \sqrt{n}$$

$$82.5 < \sqrt{n}$$

$$6806.25 < n$$

We deduce that a sample of 6,807 or more will meet our requirements.

We now move to another type of hypothesis test. Consider the following problem. The manager of a large fleet of cars, all of the same make and model, has had them using various brands of gasoline. One group of 14 cars using brand X last week averaged 17.43 miles per gallon with a sample standard deviation of 3.21. A second group of 11 averaged 16.12 miles per gallon with a sample standard deviation of 2.71. Do these differences represent the normal fluctuations due to sampling or do they indicate a real difference in the mileage one should expect to obtain from the gasoline?

This problem is typical of a large number of problems which arise in business and in the social and biological sciences. They center about determining whether or not two populations have the same mean. In the problem mentioned above, we want to know whether cars of a particular make using brand X gasoline have the same average mileage as those cars using brand Y.

In other problems we might want to know whether sighted and blind people have the same hearing capacity or whether people who have been taught reading by the phonic method have the same reading skills as those taught by the sight method. The two populations may represent the same people at different times. For example, are the attitudes of voters in October the same as they were in August? In this section we will introduce a method which will allow us to answer problems of this type.

In all of these problems we have two random variables X_1 and X_2 with means μ_1 and μ_2 and standard deviations σ_1 and σ_2. Our null hypothesis for these problems is always

$$H_0 : \mu_1 = \mu_2$$

The alternative hypothesis is always

$$H_1 : \mu_1 \neq \mu_2$$

How do we test these hypotheses? Suppose we have a sample of size n_1 from X_1 with mean $\bar{x}_1$ and sample variance $\hat{S}_1^2$ and an independently chosen sample of size n_2 from X_2 with mean $\bar{x}_2$ and sample variance $\hat{S}_2^2$. The variance which we use to test the hypothesis depends upon the size of n_1 and n_2.

Case 1

$$n_1 + n_2 - 2 > 50$$

In this case the variable defined by

$$Z = \frac{\bar{x}_1 - \bar{x}_2}{\sqrt{\frac{\hat{S}_1^2}{n_1} + \frac{\hat{S}_2^2}{n_2}}}$$

is a *normal variable* with mean 0 and standard deviation 1.

Case 2

$$n_1 + n_2 - 2 \leq 50$$

In this case the variable

$$t = \frac{\bar{x}_1 - \bar{x}_2}{\sqrt{\frac{(n-1)\hat{S}_1^2(n_2-1)\hat{S}_2^2}{n_1+n_2-2}}\sqrt{\frac{1}{n_1}+\frac{1}{n_2}}}$$

is a *t-variable* with $n_1 + n_2 - 2$ degrees of freedom.

In the example concerning gasoline mileage we have $n_1 = 14$ and $n_2 = 11$, so

$$n_1 + n_2 - 2 = 14 + 11 - 2 = 23 \leq 50$$

and we have Case 2. Since $\bar{x}_1 = 17.43$ and $\bar{x}_2 = 16.12$, and $\hat{S}_1 = 3.21$ and $\hat{S}_2 = 2.71$, we have

$$\begin{aligned} t &= \frac{\bar{x}_1 - \bar{x}_2}{\sqrt{\frac{(n_1-1)\hat{S}_1^2 + (n_2-1)\hat{S}_2^2}{n_1+n_2-2}}\sqrt{\frac{1}{n_1}+\frac{1}{n_2}}} \\ &= \frac{17.43 - 16.12}{\sqrt{\frac{13(3.21)^2 + 10(2.71)^2}{23}}\sqrt{\frac{1}{14}+\frac{1}{11}}} \\ &\approx \frac{1.31}{\sqrt{9.017}\sqrt{.162}} \\ &\approx \frac{1.31}{\sqrt{1.461}} \\ &\approx 1.08 \end{aligned}$$

In Table E we see that for a t-variable with 23 degrees of freedom

$$P(-2.069 \leq t \leq 2.069) = .95$$

so we cannot reject the hypothesis at a .05 significance level. Indeed

$$P(-1.714 \leq t \leq 1.714) = .90$$

so we cannot reject H_0 at a .10 significance level. Therefore we cannot conclude that there is a real difference in average gas mileage between the two brands of gasoline.

EXAMPLE 3 The 254 freshmen entering Tedium College have an average math SAT (Scholastic Aptitude Test) of 502 with sample variance 630, while the 312 freshmen entering Lethargy U. have an average math SAT of 518 with sample variance 682. Can one conclude at the .05 significance level that, on the basis of SAT scores, the freshmen at Lethargy U. are better math students than the freshmen at Tedium?

In this case

$$254 + 312 - 2 = 564 > 50$$

so we use the normal variable. We have

$$\begin{aligned} z &= \frac{\bar{x}_1 - \bar{x}_2}{\sqrt{\dfrac{\hat{S}_1^2}{n_1} + \dfrac{\hat{S}_2^2}{n_2}}} \\ &= \frac{502 - 518}{\sqrt{\dfrac{630}{254} + \dfrac{682}{312}}} \\ &\approx \frac{-16}{\sqrt{4.67}} \\ &\approx -7.4 \end{aligned}$$

For a standard normal variable

$$P(-1.96 \leq Z \leq 1.96) = .95$$

and $-7.4 < -1.96$, so we can reject at a .05 significance level the null hypothesis that the two groups of students are the same in ability.

Summary

We have discussed once again the subject of hypothesis testing. While later parts of the book contain some hypothesis tests, this is the last general discussion of the subject we will have and it seems appropriate to talk about certain features of hypothesis testing which may get lost in the midst of the technical development.

One of the points we made a long time ago is that there is no certainty in these tests. It is always possible to get 100 heads in a row tossing a fair coin or 50 heads in 100 tosses of an unfair coin. What hypothesis testing accomplishes is that it gives you a method for making decisions concerning the value of a parameter while knowing how much chance of error you are running. If you test hypotheses at the .05 significance level, then you will only reject a true hypothesis five times in one hundred on the average (but you *will* be wrong those five times). If you are willing to risk more frequent error, you can test at the .10 significance level (wrong ten times out of one hundred) or, if you prefer more caution, you can test at the .01 significance level (only wrong one time in a hundred).

There is another type of error. The error involved in accepting the null hypothesis when it is false. We have said very little about this. We did say a little about it in our last look at the Smith-Jones election campaign when we discussed sample size. You can find it discussed in more advanced books under the *power* of a test. (See also Supplementary Exercise 14.)

Exercises Section 8

1 Redo the example concerning political forecasting first for a significance level of .01 and then for a significance level of .1.

2 Determine the sample size for testing the fairness of a coin at the .01, .05, and .1 significance levels.

3 Determine the sample size for testing the hypothesis that the probability of a thumbtack landing point up is $\frac{1}{3}$ at the .01, .05, and .1 significance levels.

4 Let $f(p) = p(1 - p)$ for $0 \leq p \leq 1$. (Note that this is simply the variance of the binomial distribution when $n = 1$.) Show that f has a maximum at $p = \frac{1}{2}$, and hence the standard deviation in the case $n = 1$ of the binomial distribution is always less than or equal to $\frac{1}{2}$.

5 It is known that the distribution of heights among male college students is normal with mean 71.4 inches and standard deviation 1.1 inch. A sample of 100 students at Proust University has a mean of 72.1 inches. At what level of significance can one assert that Proust draws tall students?

6 Brand X soap is used by 10 per cent of all soap users. After playing a new commercial for a week in a particular city, they conduct a poll of 100 soap purchasers and find that 13 have purchased Brand X soap. At what level of significance can the company conclude that its new commercial is effective in selling its soap?

7 A task designed to measure hand and eye coordination was given to a sample of 100 college professors who took an average of 10 minutes to complete it with a sample standard deviation of 3. When the same task was given to a sample of 60 college students, they took an average of 8 minutes to complete it with a sample standard deviation of 2 minutes. Can you conclude at a .05 significance level that students have better hand and eye coordination than professors?

8 A sample of eight Brand A 100-watt bulbs had an average life of 112 hours with a sample variance of 72 while a sample of six Brand B 100-watt bulbs had an average life of 107 hours with a sample variance of 65. Can you conclude at a .05 significance level, that Brand A bulbs are longer lasting than Brand B?

9 While observing the posted golf scores at the Duns Scotus Country Club, you note that Joe Shank has recently had scores of 96, 87, 99, 86, and 101, while Ralph Bird has had scores of 94, 96, 101, 90, and 104. Can you reject at a .10 significance level that they are equally good golfers?

10 A question arose as to the comparability of grades assigned by two professors. Ten students took an essay exam, xerox copies were made, and each professor was given the set to grade. The results follow:

	1	2	3	4	5	6	7	8	9	10
Professor Lax	73	80	82	57	78	90	75	83	65	84
Professor Stern	70	75	79	40	83	89	78	81	50	96

What information can you deduce about the differences in this grading?

11 Two statisticians got into an argument over the relationship between income and the amount of time spent commuting each day. They took a random sample of 22 people. Five of them earned over $15,000 a year and spent an average of 1 hour 30 minutes commuting with a sample variance of 140 minutes. The 17 with an income under $15,000 spent an average of 1 hour 15 minutes commuting with a sample variance of 120 minutes. What does this say about the argument?

12 In a sample of 200 married couples, the men were found to drive an average of 135 miles per week with a sample standard deviation of 20, while the women drove an average of 130 miles per week with a sample standard deviation of 15. What does this say about the hypothesis that men and women do about the same amount of driving?

13 If you conduct a two-tailed hypothesis test at the .05 significance level, you will reject a true hypothesis 5% of the time. Suppose you adopt the following procedure for a test of

$$H_0: \mu = a$$

Look at the sample mean. If it is less than a, conduct a one-tailed test at the 5% significance level on the hypothesis $\mu \geq a$. If it is greater than a, conduct a one-tailed test at the 5% significance level on the hypothesis $\mu \leq a$.

Calculate the percentage of rejection of a true hypothesis under this procedure.

14 For a group of 100 students, their mean grade point average as freshmen was 1.4000 (on a 3.0 scale) with sample standard deviation .4485. As sophomores their mean grade point average was 1.439 with sample standard deviation .4294.

(a) Can you reject, at the 10% significance level, the hypothesis that the grade point average of the entire class as freshmen is 1.439?

(b) Can you reject at the 10% significance level the hypothesis that the grade point average of the entire class as sophomores is 1.4000?

(c) Can you reject the hypothesis that the class's average grade point average was the same both years?

9 THE F-TEST

In the previous section we learned to decide whether two samples are from the same population or different ones. We now want a technique which will allow us to compare three or more samples. Consider the following problem.

EXAMPLE 1 Over a 5-week period, a record was kept of the weekly sales volume of four real estate salesmen. Salesman A sold an average (in thousands of dollars) of 72 with a sample variance of 35. Salesman B averaged 66 with a sample variance of 42. Salesman C averaged 79 with a sample variance of 49 and Salesman D averaged 68 with a sample variance of 30. Can one conclude at a .05 significance level that there is no difference in the true average weekly sales of these salesmen?

The method for solving this problem and similar ones involves two assumptions. The first is that the populations from which we are sampling are normally distributed. In our example this means that, for any one salesman, the distribution of weekly sales is normal. The second assumption is that all of the populations have the same variance.

If we have random variables $X_1, X_1, \ldots, X_n$ satisfying these assumptions and let $\mu_1, \mu_2, \ldots, \mu_n$ be their means, we let the null hypothesis be

$$H_0: \mu_1 = \mu_2 = \cdots = \mu_n$$

The null hypothesis is that for some i and j,

$$\mu_i \neq \mu_j$$

If the null hypothesis is true, the populations are really the same, since normal distributions are uniquely determined by their mean and standard deviation.

In order to test this null hypothesis we introduce a variable F which is the ratio of two estimates of σ^2. We may give an informal definition of F as

$$F = \frac{\text{variance between samples}}{\text{variance within samples}}$$

Let us calculate F for the data of Example 1. The estimate of σ^2 based on the variance between samples is calculated as follows. Our four samples have means 72, 66, 79, and 68, so the average sample mean is

$$\frac{72 + 66 + 79 + 68}{4} = \frac{285}{4} = 71.25$$

The sample variance of these sample means, denoted by $\hat{S}_{\bar{x}}^2$, is

$$\hat{S}_{\bar{x}}^2 = \frac{(72 - 71.25)^2 + (66 - 71.25)^2 + (79 - 71.25)^2 + (68 - 71.25)^2}{3}$$

$$= 30.42$$

The Central Limit Theorem says that for samples of size n

$$\sigma_n^2 = \frac{\sigma^2}{n}$$

or

$$\sigma^2 = n\sigma_n^2$$

Since $\hat{S}_{\bar{x}}^2$ is an estimate of σ_5^2, we have

$$5(30.42) = 152.10$$

as an estimate of σ^2 based on the *variance between samples*.

The estimate of σ^2 based on the *variance within samples* is simply the average sample variance of the samples. We have

$$\frac{35 + 42 + 49 + 30}{4} = \frac{156}{4}$$
$$= 39$$

as this estimate. Therefore,

$$F = \frac{152.1}{39}$$
$$= 3.9$$

How do we interpret F? Intuitively, if the samples are all from the same population, i.e., the null hypothesis is true, then the variance between samples should not be much larger than the variance within samples, and F should be not too much larger than 1. However, if the null hypothesis is false, then the variance between samples should be greater than the variance within samples; that is, F should be greater than 1. Exactly how much larger F should be than 1 has been calculated and you will find that value of F for various significance levels in Table F in the Appendix.

As was the case in the t-distribution, there are various distribution functions of F depending on degrees of freedom. For the variable F, there are two degrees of freedom, one known as the *vertical degree of freedom*, the other as the *horizontal degree of freedom*. If one has n samples of size k, then

Vertical degrees of freedom $= n - 1$

Horizontal degrees of freedom $= n(k - 1)$

In our example we have $4 - 1 = 3$ vertical degrees of freedom and $4(5 - 1) = 16$ horizontal degrees of freedom. Then referring to our table, we obtain

$$P(F \leq 3.24) = .95$$

and our value of F, $F = 3.9$, is large enough to allow us to reject the null hypothesis and conclude that there is a difference in true average weekly sales among the salesmen.

EXAMPLE 2

The owner of a fleet of trucks has kept records on the mileage of four sets of five brands of tires. The results are given in the following table.

Brand	A	B	C	D	E
Average life (in thousands of miles)	17.2	16.7	15.9	16.3	16.8
Sample variance (in thousands of miles)	4.7	5.1	4.9	4.2	3.8

Can one conclude, at a .05 significance level, that there is a difference in the average life of the brands of tires? The average tire life is

$$\frac{17.2 + 16.7 + 15.9 + 16.3 + 16.8}{5} = \frac{82.9}{5}$$
$$= 16.58$$

and the sample variance of the averages is

$$\frac{(17.2 - 16.58)^2 + (16.7 - 16.58)^2 + (15.9 - 16.58)^2 + (16.3 - 16.58)^2 + (16.8 - 16.58)^2}{4}$$
$$= \frac{(.62)^2 + (.12)^2 + (.68)^2 + (.28)^2 + (.22)^2}{4}$$
$$= \frac{1.3748}{4}$$
$$= .3437$$

Therefore, the estimate of σ^2 based on the *variance between samples* is

$$4(.3437) = 1.3748$$

The estimate of σ^2 based on the *variance within samples* is

$$\frac{4.7 + 5.1 + 4.9 + 4.2 + 3.8}{5} = \frac{22.7}{5}$$
$$= 4.54$$

So

$$F = \frac{1.3748}{4.54} < 1$$

and we cannot deduce any difference in average tire life among the five brands.

Summary

The F-test is used to test a particular statistical hypothesis. We assume that we have several normally distributed random variables $X_1, X_2, \ldots, X_n$ with means $\mu_1, \mu_2, \ldots, \mu_n$ and all having standard deviation σ. The null hypothesis is

$$H_0 : \mu_1 = \mu_2 = \cdots = \mu_n$$

For each of the n variables, assume we have a sample of size k and that the sample mean for X_i is $\bar{x}_i$ and the sample standard deviation is $\hat{S}_i$. Then F is computed as follows: Let

$$m = \frac{1}{n}\sum_{i=1}^{n} \bar{x}_i$$

Then

$$F = \frac{[k/(n-1)] \sum_{i=1}^{n} (\bar{x}_i - m)^2}{(1/n) \sum_{i=1}^{n} \hat{S}_i^2}$$

The variable F has degrees of freedom in two directions. The vertical degree of freedom is $n - 1$ and the horizontal degree of freedom is $n(k - 1)$.

Exercises Section 9

1 A college has five large sections of sociology. From each section a sample of six students were given an examination with the following results.

Class	Average Grade	Standard Deviation
I	72	10
II	78	13
III	69	15
IV	74	9
V	81	12

Test at a .05 significance level, the hypothesis that the classes are equally good.

2 Over a period of six work days, three shifts of widget makers had the following production totals.

	1	2	3	4	5	6
Shift I	7	9	8	8	9	10
Shift II	6	10	7	8	5	12
Shift III	12	11	8	6	9	8

Are the shifts equally productive?

3 Annoyed by the rapid rate at which bulbs burnt out in a teaching machine, an instructor decided to see if there was a difference between brands. He located five different brands, bought six bulbs of each kind, and kept a record of how long (in hours) each bulb lasted. The mean and sample standard deviation, by brand, is in the following table.

	BRAND				
	A	B	C	D	E
MEAN	4.2	4.6	3.9	4.1	4.5
Standard Deviation	.4	.7	.7	.5	.3

Can he conclude that there is a difference between brands? Give the level of significance of your response.

4 The following table purports to be the mean and sample standard deviation of four samples of size 6.

	A	B	C	D
MEAN	21	18	19	24
Standard Deviation	3	4	4	x

Are there any values of x for which one can claim, at a .05 significance level, that the four true means are not identical? If there are such values of x, indicate all values of x which will allow the claim and all values of x which won't.

10 THE CHI SQUARE DISTRIBUTION

In the section on the maximum likelihood estimate we discussed the problem of fitting data to a distribution. We gave a method for choosing the parameters in the "best" possible way to make the distribution "fit" the data. However it is quite possible that this best fit isn't very good. We may have made an unwise choice of distribution. The *chi square distribution* gives us a method for deciding how good a fit we have. It will also allow us to detect whether two random variables are related.

What is involved in deciding whether or not data fits some theoretical model? To gain insight into this question we refer to the data on page 48 concerning a die which has been rolled 72 times. We reproduce parts of Tables 2.1 and 2.2 on that page concerning the observed and the theoretical results of this experiment.

Result:	1	2	3	4	5	6
Observed Frequency:	12	8	15	14	6	17
Expected Frequency:	12	12	12	12	12	12

As you look at this table trying to see whether or not the data fits the model, i.e., whether or not the die is fair, you look first at the difference between the observed and expected frequencies in each column. By now you should know that you can't expect exact agreement but nevertheless there should not be "too much" difference. The problem is, how much is too much? We shall answer this by means of the variable χ^2 (chi squared) defined below in several steps.

1. We always begin with a random variable X and a frequency function f.

X assigns to each roll of the die the number of spots showing. The frequency function f is defined by

$$f(1) = f(2) = \cdots = f(6) = \tfrac{1}{6}$$

2. Break the range of X up into k disjoint subsets; $I_1, I_2, \ldots, I_k$.

In our example we broke the range of X up into six subsets, $\{1\}$, $\{2\}$, $\{3\}$, $\{4\}$, $\{5\}$, and $\{6\}$. We could do things differently. We could have broken it up into three subsets $\{1, 2\}$, $\{3, 4\}$, $\{5, 6\}$, or the three subsets $\{1, 3\}$, $\{2, 4\}$, $\{5, 6\}$. There are some rules governing this choice of subsets which we will discuss later but it is, to a considerable degree, an arbitrary procedure.

3. Suppose there have been n trials of the experiment. Let o_i denote the number of trials where the value of X fell into I_i and let e_i be the number of trials which f predicts X will fall into I_i.

To tie this in with our example we have $I_1 = \{1\}, I_2 = \{2\}, \ldots, I_6 = \{6\}$. We have $o_1 = 12, o_2 = 8, \ldots, o_6 = 17$ and $e_1 = e_2 = \cdots = e_6 = 12$.

4. Set

$$\chi^2 = \sum_{i=1}^{k} \frac{(e_i - o_i)^2}{e_i}$$

In the example we have

$$\chi^2 = \frac{(12-12)^2}{12} + \frac{(12-8)^2}{12} + \frac{(12-15)^2}{12} + \frac{(12-14)^2}{12} + \frac{(12-6)^2}{12} + \frac{(12-17)^2}{12}$$

$$= \tfrac{1}{12}(0 + 16 + 9 + 4 + 36 + 25)$$

$$= \tfrac{1}{12}(90)$$

$$= \tfrac{15}{2} = 7.5$$

However the value of χ^2 depends not only on the data you have and the model or distribution you have chosen. It also depends on the subsets $I_1, \ldots, I_k$ into which the range of X has been decomposed. Note what happens in our example when we change the subsets.

Result:	$\{1, 2\}$	$\{3, 4\}$	$\{5, 6\}$
Observed frequency:	20	29	23
Predicted frequency:	24	24	24

$$\chi^2 = \frac{(24-20)^2}{24} + \frac{(24-29)^2}{24} + \frac{(24-23)^2}{24}$$

$$= \tfrac{1}{24}(16 + 25 + 1)$$

$$= \tfrac{42}{24} = \tfrac{7}{4} = 1.75$$

Changing again, we have

Result:	$\{1, 3\}$	$\{2, 4\}$	$\{5, 6\}$
Observed frequency	27	22	23
Predicted frequency:	24	24	24

and

$$\chi^2 = \frac{(24 - 27)^2}{24} + \frac{(24 - 22)^2}{24} + \frac{(24 - 23)^2}{24}$$
$$= \tfrac{1}{24}(9 + 4 + 1)$$
$$= \tfrac{14}{24} = \tfrac{7}{12} \approx .583$$

These examples illustrate the wide range of values of χ^2 depending upon our choice of subsets of the range of X. The greatest fluctuation was caused by changing the number of subsets from six to three, though the two examples with three subsets give vastly different values of χ^2.

> 5. In a problem where χ^2 is used with a distribution function f which has no parameters which were estimated from the data the *number of degrees of freedom is* $k - 1$, where k is the number of subintervals into which the range of X was divided.

The reader might well ask why we bother with degrees of freedom when it seems that the only thing involved is the number of subintervals. However we shall soon see an example where other things are involved. In our example we have six subsets and, hence, five degrees of freedom. But we have observed that there is more to the value of χ^2 than the number of degrees of freedom (or subsets of the range of X). That is cleared up next.

> 6. If the number of subsets of the range of X (*not* the number of degrees of freedom) is at least five and if f predicts at least five for each subset, then, the value χ^2 which we have defined is *approximately* distributed like a theoretically defined variable, also denoted χ^2, with the appropriate number of degrees of freedom.

The exact definition of this theoretical χ^2 variable is beyond the scope of this course. What we have done is to provide you with Table G in the Appendix.

In our example we have made the null hypothesis that the die is fair. What would lead us to reject this hypothesis? Note that as the difference between observed and predicted results increases, the value of χ^2 increases. So if our model is good (i.e., the die is fair), χ^2 should be small. If the model isn't very good (i.e., the die is not fair), χ^2 should be larger. We have calculated in our example

$$\chi^2 = 7.5$$

with five degrees of freedom. Referring to the table, we see that for five degrees of freedom

$$P(\chi^2 < 7.78) = .90$$

and

$$P(\chi^2 < 5.39) = .75$$

Therefore we *cannot* reject the fairness of the die at the 10% significance level but *can* do so at the 25% significance level.

We turn now to a different kind of example. In Section 5 we took 100 test scores and fitted them to the normal distribution. We can now discuss the question as to how well the normal distribution fits this data. The problem resembles very closely the one we just solved but there is one major procedural difference. The calculation of the number of degrees of freedom has to be modified.

The frequency function for the experiment with the die was, as we emphasized at the time, based on an idea and we did not use the data in developing the model. However we used the test data in fitting it to the normal distribution. In particular, we used the mean ($m = 56.21$) of the data as the mean μ of the density function and the standard deviation ($s = 24.018$) of the data as the standard deviation σ of the density function. The rule for computing the number of degrees of freedom is as follows

> 5′. Number of degrees of freedom = number of subsets − 1 − (number of parameters estimated from sample data).

For our problem with the test scores this means that the number of degrees of freedom will be 3 less than the number of subsets of the range of X, since we will subtract the 1 which is always subtracted, 1 for the estimation of μ, and 1 for the estimation of σ.

We turn now to determining the subsets. Since one frequently thinks in terms of intervals of 10 for test scores, we try [0, 9,5], [9.5, 19.5], [19.5, 29.5], . . . , [89.5, 99.5]. We then use the normal distribution

$N(x; 56.21, 24.018)$

to predict how many of the 100 test scores would fall into each interval and count the number which actually do. The results are summarized in the following table.

Interval	Predicted	Counted
[0, 9.5]	2.6	1
[9.5, 19.5]	3.7	4
[19.5, 29.5]	6.8	10
[29.5, 39.5]	11.0	18
[39.5, 49.5]	14.8	9
[49.5, 59.5]	14.8	11
[59.5, 69.5]	15.3	10
[69.5, 79.5]	12.5	20
[79.5, 89.5]	8.4	7
[89.5, 99.5]	8.3	10

Recall that in rule 6 we said that the density function should predict at least 5 in each subset of the range of X. But the intervals [0, 9.5] and [9.5, 19.5] do not satisfy this requirement. The difficulty is resolved by combining them.

Our new subsets are [0, 19.5], [19.5, 29.5], . . . , [89.5, 99.5]. The predicted number of scores in [0, 19.5] is

$$2.6 + 3.7 = 6.3$$

and the number of scores counted there is

$$1 + 4 = 5$$

Now we have to do some arithmetic.

$$\begin{aligned}\chi^2 &= \frac{1.3^2}{6.3} + \frac{3.2^2}{6.8} + \frac{9^2}{11} + \frac{5.8^2}{14.8} + \frac{5.6^2}{16.6} + \frac{5.3^2}{15.3} + \frac{7.5^2}{12.5} + \frac{1.4^2}{8.4} + \frac{1.7^2}{8.3} \\ &= .268 + 1.506 + 7.364 + 2.273 + 1.889 + 1.836 + 4.500 + \\ &\qquad + .233 + .340 \\ &= 20.7\end{aligned}$$

We have nine subintervals and hence

$$9 - 3 = 6$$

degrees of freedom. Consulting the table, we see that

$$P(\chi^2 < 18.5) = .995$$

and so the hypothesis that the test scores are normally distributed can be rejected at the .005 significance level.

At the beginning of this section we said that, in addition to testing the hypothesis that a set of data is distributed in a certain way, we can use the chi square distribution to test the dependence, or independence of two random variables. This is done by means of a contingency table. Let us consider a concrete example.

Suppose we consider two random variables, one giving us a person's annual income and the other giving us the highest grade in school completed by that person. We might suspect that there is a relationship between these two variables but how do we verify or refute this suspicion?

A reasonable first step is to go out and take a random sample of people and ask them what their annual income is and what was the highest grade in school they completed. Next we must analyze the data, and the chi square distribution gives us a reasonable way to do so.

We first break the range of the random variables up into intervals. For the random variable representing the highest completed grade in school let A be people who have graduated from college, B those who have graduated from high school but not from college, and C those who have not graduated from high school. For the random variable representing annual income we let D be less than \$5,000, E between \$5,000 and \$10,000, and F more than \$10,000. We then use our sample of people to construct a *contingency table*. One is given below.

	A	B	C
F	40	40	39
E	33	47	50
D	7	15	28

The 40 in column A and row F means that in our sample there were 40 people who had graduated from college and earned more than $10,000 per year (i.e., fell in both A and F). The 33 in column A and row E means that 33 people in our sample had graduated from college and earned between $5,000 and $10,000 a year (i.e., fell in both A and E). You should be able to interpret the rest of the table.

Suppose we take as our hypothesis that the two variables are independent. Adding the numbers in our table,

$$40 + 33 + 7 + 40 + 47 + 15 + 39 + 50 + 28 = 299$$

we determine the total number of people sampled. Adding the number of people in column A.

$$40 + 33 + 7 = 80$$

we obtain the number of people in the sample who fell in A, i.e., had graduated from college. Therefore $P(A)$, the probability that a randomly selected member of our sample is in A, is given by

$$P(A) = \frac{80}{299}$$

Similar calculations yield

$$P(F) = \frac{40 + 40 + 39}{299} = \frac{119}{299}$$

If our hypothesis, that the two random variables are independent, is correct, then

$$P(A \cap F) = P(A) \cdot P(F) = \frac{80}{299} \cdot \frac{119}{299}$$

and the number of people, out of the sample, in $A \cap F$ would be

$$299 \cdot \frac{80}{299} \cdot \frac{119}{299} = 31.7$$

We actually have 40 in $A \cap F$. Similarly we have 47 people in $B \cap E$, but

$$P(B) = \frac{40 + 47 + 15}{299} = \frac{102}{299}$$

and

$$P(E) = \frac{33 + 47 + 50}{299} = \frac{130}{299}$$

so our independence hypothesis predicts

$$\begin{aligned} 299 \cdot P(B \cap E) &= 299 \cdot P(B) \cdot P(E) \\ &= 299 \cdot \tfrac{102}{299} \cdot \tfrac{130}{299} \\ &= 44.6 \end{aligned}$$

in $B \cap E$.

Continuing this process, we obtain the following table of predictions (based on the hypothesis that the variables are independent).

	A	B	C
F	31.7	40.4	46.7
E	35.0	44.6	51.5
D	13.4	17.0	19.7

Using the predicted and observed values, as in earlier calculations of χ^2, we have

$$\begin{aligned} \chi^2 &= \frac{(31.7 - 40)^2}{31.7} + \frac{(40.4 - 40)^2}{40.4} + \cdots + \frac{(17.0 - 15)^2}{17.0} + \frac{(19.7 - 28)^2}{19.7} \\ &= 2.173 + .004 + 1.270 + .114 + .004 + .004 + 3.057 \\ &\qquad + .235 + 3.487 \\ &= 10.388 \end{aligned}$$

We need a rule for obtaining the number of degrees of freedom associated with a contingency table.

> 5.″ In an n by m contingency table there are $(n - 1) \cdot (m - 1)$ degrees of freedom.

Therefore we have

$$(3 - 1) \cdot (3 - 1) = 4$$

degrees of freedom. Consulting the tables, we see

$$P(\chi^2 < 9.49) = .95$$

and, since our value of χ^2 is 10.388, we can reject, at a .05 significance level, the hypothesis that the variables are independent.

Once we know that there is a relation between two variables we might ask what this relationship is. Chapter 6 will tell us how to go about finding the relation.

Summary

In this section we learned to use χ^2 for two purposes: to decide how well a certain model fits a set of data and to decide if two random variables are independent. In both cases we assert a null hypothesis (that the model does fit the data or that the variables are independent) and we attach a level of significance to any rejection of the hypothesis. It should be observed that in neither case are we testing a *statistical hypothesis* (a conjecture about the value of a parameter). However we are testing what might be called a *scientific hypothesis*.

Exercises Section 10

1 In Section 5 you fitted the normal distribution to data consisting of 48 gum ball diameters. Using the intervals [3.36, 3.40], [3.41, 3.45], [3.46, 3.50] . . . , [3.66, 3.70], [3.71, 3.76] as in Chapter 1, Section 1, apply the χ^2 test to

(a) the normal distribution model of the data, and

(b) the uniform distribution model of the data.

Which do you think is the better model?

2 Below you will find a set of data on 100 students. As an aid to answering the questions which follow we give you the following information. The grade point averages (GPA) are based on a 3-point scale: A = 3, B = 2, C = 1, D = F = 0.

	Mean	Standard Deviation
Freshman GPA	1.4000	.4485
Sophomore GPA	1.4390	.4294
Verbal SAT	527.45	81.58
Math SAT	553.88	85.50

Answer the following questions concerning the data. Each answer should indicate your significance level.

(a) Are the math SAT's normally distributed?

(b) Are the sophomore GPA's normally distributed?

(c) Are the verbal and math SAT's independent?

(d) Is the sophomore GPA more dependent on (or less independent of) the freshman GPA or the verbal SAT?

	Freshman GPA	Sophomore GPA	Verbal SAT	Math SAT	High School Rank in Class	Number of Students in High School Class
1	1.666700	1.318700	414	599	42	180
2	1.757600	1.393900	539	685	24	195
3	1.641800	1.818200	449	527	34	174
4	1.225800	1.344300	513	521	114	330
5	0.871000	1.033300	430	524	137	537
6	0.727300	0.887100	456	459	110	161
7	1.294100	1.469700	634	550	154	591
8	2.382400	2.439400	716	523	33	264

	Freshman GPA	Sophomore GPA	Verbal SAT	Math SAT	High School Rank in Class	Number of Students in High School Class
9	1.818200	1.317500	498	561	141	736
10	1.424200	1.593800	541	519	112	914
11	1.032300	1.069000	587	577	85	160
12	1.687500	1.548400	464	543	30	154
13	0.970600	1.343300	525	536	58	176
14	0.900000	0.610200	489	580	130	222
15	1.166700	1.338700	505	537	40	162
16	1.064500	1.206300	555	604	31	221
17	1.718800	1.571400	554	591	10	54
18	1.933300	1.322000	546	487	81	641
19	0.937500	1.460300	496	552	84	205
20	0.970600	1.058800	470	515	83	336
21	1.166700	1.000000	443	533	63	455
22	1.527800	1.623200	546	624	35	148
23	0.760000	0.862700	458	381	71	93
24	1.264700	1.102900	641	661	149	623
25	0.812500	0.934400	509	541	93	155
26	1.470600	1.772700	639	645	24	186
27	1.903200	1.682500	580	533	15	237
28	1.117600	1.106100	402	524	96	339
29	1.916700	2.085700	634	766	23	257
30	1.666700	1.691200	690	708	63	1,140
31	1.181800	1.096800	473	478	50	202
32	1.189200	1.450700	719	737	23	321
33	3.000000	2.952400	737	691	10	155
34	1.277800	1.000000	525	598	221	395
35	1.214300	1.311500	450	505	133	307
36	0.818200	0.806500	518	515	115	240
37	1.035700	1.183300	374	384	229	605
38	2.111100	1.929600	559	580	10	374
39	0.892900	0.842100	444	510	11	53
40	0.705900	1.033300	484	408	117	321
41	1.400000	1.666700	559	549	28	625
42	1.533300	1.444400	498	595	163	414
43	0.781300	1.016900	573	568	114	316
44	1.371400	0.970100	707	682	47	63
45	2.096800	1.892300	552	680	17	210
46	1.411800	1.253700	450	617	19	371
47	1.000000	1.406300	573	523	48	60
48	1.647100	1.718800	540	539	19	176
49	2.406300	2.625000	599	573	25	591
50	0.931000	0.966100	469	452	138	454
51	2.064500	2.063500	593	652	198	736
52	1.322600	1.460300	506	539	24	305
53	1.172400	1.250000	533	522	27	412
54	1.121200	1.287900	539	402	22	123

	Freshman GPA	Sophomore GPA	Verbal SAT	Math SAT	High School Rank in Class	Number of Students in High School Class
55	1.757600	1.907700	588	580	24	373
56	1.000000	1.100000	447	518	84	510
57	1.333300	1.303000	395	552	116	321
58	1.093800	1.265600	511	440	93	321
59	1.333300	1.261500	546	582	13	74
60	1.571400	2.016900	356	454	15	260
61	1.705900	1.626900	491	483	180	478
62	0.862100	0.786900	614	608	24	85
63	1.555600	1.304300	573	671	29	252
64	1.655200	1.508200	464	440	176	229
65	1.636400	1.796900	532	539	81	349
66	1.200000	1.311500	539	533	101	270
67	2.264700	2.264700	614	505	86	305
68	1.133300	1.133300	429	716	100	736
69	2.147100	2.161800	599	633	17	977
70	1.916700	1.913000	714	497	4	295
71	1.032300	1.333300	507	515	48	217
72	0.964300	1.216700	464	366	103	313
73	1.733300	1.876900	546	626	6	197
74	1.444400	1.584600	586	492	6	209
75	1.852900	1.835800	477	617	14	39
76	1.823500	1.597000	443	532	45	205
77	1.500000	1.333300	628	699	13	459
78	1.187500	1.050000	526	645	15	45
79	1.031300	0.950800	450	440	34	305
80	1.965500	2.149300	562	575	135	537
81	0.875000	1.032300	486	605	107	349
82	1.766700	1.950000	535	595	95	206
83	1.121200	1.393900	508	681	16	150
84	1.542900	1.764700	587	543	23	267
85	1.633300	1.854800	470	455	33	750
86	1.269200	0.933300	443	444	29	135
87	0.923100	1.145500	580	621	143	321
88	2.093800	1.951600	395	468	108	430
89	1.500000	1.852900	632	614	36	321
90	0.777800	1.017200	375	338	10	97
91	1.468800	1.689700	595	543	17	89
92	1.600000	1.447800	546	650	57	283
93	2.193500	2.229500	641	654	2	114
94	1.733300	1.790300	402	487	131	1,035
95	1.031300	1.218800	467	496	115	328
96	0.909100	1.147500	566	617	101	323
97	1.030300	1.061500	484	598	141	461
98	0.903200	1.140400	450	552	33	131
99	1.241400	1.278700	580	561	16	305
100	1.161300	1.296900	505	403	221	410

3 In Chapter 2, Section 1, we gave some data on rolling a pair of dice 144 times. Use a χ^2 test to decide whether the dice were fair. State your conclusion in terms of a significance level.

4 The following contingency table tabulates the results of asking people two questions: (1) Do they approve or disapprove of the anti-poverty programs? (2) Are they primarily interested in local government or in state or national government?

Major Governmental Interest	Appraisal of Anti-Poverty Program	
	Approve	Disapprove
National or State	59	115
Local	67	49

Do you think this data shows a connection between people's attitudes on these issues? Include a level of significance in your answer.

5 The following data is from a survey of members of the Mathematics Association of America concerning the work of the Committee on the Undergraduate Program in Mathematics. The letters across the top of each contingency table mean the following:
A = total number of responses
B = responses from universities offering the Ph.D. degree
C = responses from universities and four-year colleges not offering the Ph.D. degree
D = responses from two-year colleges
Check the phrase which best expresses your opinion of the effect which the work of the Committee on the Undergraduate Program in Mathematics (CUPM) has had on collegiate mathematics.

	A	*B*	*C*	*D*
Strongly beneficial	1,726	343	886	144
Moderately beneficial	2,903	806	1,165	250
Negligible	355	122	106	30
Adverse	173	66	60	12
No opinion	1,337	252	161	40
No response	254	58	45	11

How has the work of CUPM affected you in your professional life?

Has helped a great deal	1,082	172	636	123
Has been moderately helpful	2,365	587	1,143	229
Has had little effect	1,546	508	450	96
Has had an adverse effect	68	18	28	7
Has been largely unrelated to my professional life	1,359	276	112	22
No response	328	86	54	10

Is there, at the .05 significance level, a relationship between the answers to the above questions and the kind of institution where the respondents work?

Supplementary Exercises for Chapter 5

1 For each of the following binomial probabilities, find the exact value from the table and also find the normal approximation.

(a) $B(5; 10, \frac{1}{2})$
(b) $B(10; 20, \frac{1}{2})$
(c) $B(25; 50, \frac{1}{2})$
(d) $\sum_{x=0}^{4} B(x; 10, \frac{1}{2})$
(e) $\sum_{x=3}^{10} B(x; 50, .1)$
(f) $\sum_{x=40}^{60} B(x; 100, \frac{1}{2})$
(g) $B(8; 20, .4)$
(h) $B(3; 9, \frac{1}{3})$
(i) $B(1; 10, .1)$
(j) $B(2; 10, .2)$
(k) $B(3; 10, .3)$
(l) $B(4; 10, .4)$

2 A certain type of projector bulb is advertised to have an expected life of 9.4 hours with standard deviation 1.1 hours. Find the probability, assuming the life is normally distributed, that a randomly selected bulb has a life of

(a) more than 11 hours.
(b) between 8 and 10 hours.

3 If the yearly salaries of 300 junior engineers in a large corporation are approximately normally distributed with mean $13,000 and standard deviation $1,500, find

(a) how many engineers earn more than $15,000 per year.
(b) what percentage of engineers earn between $10,000 and $12,000 per year.

4 **Error in Approximation.** It is revealing to compare some of the binomial probabilities with their normal approximations to develop some feeling for the degree of accuracy involved. One measurement that can be used is the *percentage of error*, defined in general by

$$\text{Percentage of error} = \frac{(\text{approximate value} - \text{true value})}{\text{true value}} \cdot 100$$

Find the percentage of error between each of the following binomial probabilities and their corresponding normal approximations.

(a) $B(1; 5, \frac{1}{2})$
(b) $B(1; 10, \frac{1}{2})$
(c) $B(5; 10, \frac{1}{2})$
(d) $B(5; 20, \frac{1}{2})$
(e) $B(10; 20, \frac{1}{2})$
(f) $B(1; 10, .1)$
(g) $B(5; 10, .1)$
(h) $\sum_{x=0}^{5} B(x; 20, \frac{1}{2})$
(i) $\sum_{x=0}^{3} B(x; 12, \frac{1}{2})$
(j) $\sum_{x=0}^{2} B(x; 10, .1)$

5 The heights of male students at a large state university are approximately normally distributed with mean 69.5 (inches) and variance 3.24.

(a) If two students are selected at random, what is the probability that they are both over 6 feet tall?

(b) If four students are selected at random, what is the probability that all their heights are between 67 and 72 inches?

(c) What is the smallest size of a randomly selected group that would give you a 90% chance of having at least one member over 6 feet 4 inches tall?

6 **(a)** Sometimes the *sample median* $\tilde{x}$ is used as an estimator of the population mean. In particular, that is how swim meets are timed, with three timers per lane. Compute $E(\tilde{x})$ and $\sigma_{\tilde{x}}^2$ for samples of size 3 drawn (without replacement) from the random variable that takes on values 1, 2, 3, 4 with equal probability. (*Hint:* There are 64 samples of size 3. Rather than trying to list them, simply complete the following chart.)

Possible *Ordered* Sequences

$\tilde{x} = 1$, frequency		$\tilde{x} = 2$, frequency		$\tilde{x} = 3$, frequency	$\tilde{x} = 4$, frequency
111	(1)	122	(3)		
112	(3)	123	(6)		
113	(3)	124	(6)		
114	(3)	222	(1)		
		223	(3)		
		224	(3)		

(b) Compare the values in (a) with the mean and variance of $\bar{x}$, the sample means of samples of size 3.

7 The following two contingency tables, one for white citizens, the other for black, records their response to two questions. One question concerned their fear of being physically attacked (in the street, in their home, etc.). The other question asked was their opinion about the amount of support (financial and otherwise) the police get. The answers are classified as to whether the respondent felt the police should receive more support, are getting enough, or the support they are getting should be curtailed.

White

	Fear of Attack			
Support Police	Very Likely	Somewhat Likely	Somewhat Unlikely	Very Unlikely
More	67%	57%	50%	53%
Enough	30%	37%	46%	42%
Curtail	3%	6%	4%	5%
Number of Respondents	63	175	227	154

Black

	Fear of Attack			
Support Police	Very Likely	Somewhat Likely	Somewhat Unlikely	Very Unlikely
More	42%	21%	28%	24%
Enough	41%	65%	58%	53%
Curtail	17%	14%	14%	23%
Number of Respondents	52	57	50	17

Does this data indicate a relationship between people's attitudes on the two questions among either whites or blacks? Give the significance level of any assertion you make. Do you think, on the basis of the data, that there is a difference between the two groups?

Note that the above tables are in percentages and some conversion is needed. The data is from an article "Fear of Crime and Fear of Police" by Richard L. Block in the magazine *Social Problems*, Summer 1971, Vol. 19, No. 1.

8 Four neighborhoods are picked in a large city. In each of these neighborhoods five grocery stores are chosen at random and items on a typical grocery list are priced. The total bill at each of the stores is presented in the following table.

	Neighborhood			
Store	*A*	*B*	*C*	*D*
1	16.22	16.74	17.12	16.10
2	16.87	17.01	17.50	16.35
3	15.80	16.10	16.90	16.05
4	19.50	16.18	18.10	16.80
5	16.10	17.03	17.14	15.95

Can one conclude at a 10% significance level that grocery stores in different neighborhoods follow different pricing policies?

9 Due to a dispute over teaching methods, the following experiment in the teaching of introductory probability was conducted. Out of 60 students wishing to take a course, 20 were put into group A, which was taught by traditional means, 20 were put into group B, which was taught from programmed textual material, and 20 were put into group C, which was taught by CAI (computer assisted instruction). At the end of the course all 60 students took the same final examination. Students in group A had a mean score of 75.5 with standard deviation 4.1, those in group B had a mean score of 76.2 with a standard deviation of 6.3, and those in group C had a mean score of 72.6 with a standard deviation of 5.2. Do these results indicate a difference in the effectiveness of the teaching methods at the .05 significance level?

10 Four alumni of four different high schools were debating the relative merits of their football teams during the 1960's. Each team had played eight games each year with the number of wins as indicated below.

	Jefferson	Washington	Bismark	Lincoln
1960	7	5	7	6
1961	4	4	8	6
1962	8	4	7	5
1963	5	5	4	6
1964	7	6	3	7
1965	6	7	2	6
1966	6	8	6	5
1967	7	8	8	7
1968	5	4	8	6
1969	8	5	7	6

Is there a difference in overall success at the .01, .05, or .10 level?

11 The mean weight of a tablet of a certain drug is to be 50 milligrams. A sample of 64 tablets shows a mean weight of 50.15 milligrams with a sample standard deviation of .4 milligrams. Using a .01 significance level, can we conclude that the desired weight of the tablet is not being maintained?

12 A random sample of 30 members was selected from the faculty of a large university in order to estimate the average length of teaching experience. The teaching experience (measured in years) for the 30 members are:

3, 4, 4, 6, 2, 3, 4, 6, 2, 4

6, 4, 3, 3, 3, 7, 3, 4, 5, 6

1, 6, 4, 5, 4, 3, 2, 4, 3, 4

(a) Using the above data, establish a .99 confidence interval for the average length of teaching experience of a faculty member.

(b) Use the χ^2 test to decide whether the number of years of teaching experience is a uniformly distributed variable or a normally distributed variable.

13 A high school genetics student hypothesizes that the offspring of a certain pair of guinea pigs will be distributed in the ratio 9:3:3:1 with respect to color and texture of coat. He predicts that 6.25% will have smooth white coats, 18.75% smooth black coats, 18.75% rough white coats, and 56.25% rough black coats. A total of 96 progeny are raised and they include 4 with smooth white coats, 14 smooth black, 23 rough white, and 55 rough black. Test the hypothesis at the .05 level.

14 **(Type II error.)** Recall that a type II error is committed when a false null hypothesis is accepted. For an illustration of how one measures type II error, consider the following situation. You are given a normally distributed random variable X with unknown mean μ and known standard deviation, $\sigma = 2$. You test the null hypothesis

$$H_0: \mu = 10$$

against the two-sided alternative

$$H_1: \mu \neq 10$$

(a) Find the critical region corresponding to a .05 significance level.
(b) Suppose that the true value of μ is 9. Find the probability that X take on a value in the "acceptance region," i.e., not in the critical region from (a). This is the probability of making a type II error.
(c) Repeat (b) assuming that $\mu = 8$. Notice that the probability of making a type II error depends very much on the true value of μ.
(d) Repeat (a), (b), and (c) using a .01 level of significance (that is, with a smaller chance of making a type I error).
(e) Repeat (a), (b), and (c) using a .10 significance level (that is, with a larger chance of making a type I error).

15 **(Percentiles.)** Let X be a random variable with distribution function F. For any number k, $0 \leq k \leq 100$, we define the *kth percentile*, which we denote x_k, to be the smallest value of x such that

$$F(x) \geq \frac{k}{100}$$

That is, $F(x_k) \geq (k/100)$ and $F(x) < (k/100)$ for each $x < x_k$. For example, the 20th percentile would satisfy

$$F(x_{20}) \geq .20$$

In case X is continuous, we always have $F(x_k) = (k/100)$, but the inequality $F(x_k) > (k/100)$ may hold when X is discrete. Note that the 50th percentile is precisely the *median*.

(a) Find the 20th, 40th, and 75th percentiles for a random variable X whose density function is $N(x; 100, 5)$.
(b) Find the 25th and 90th percentiles for a discrete random variable whose frequency function is $B(x; 15, \frac{1}{3})$.
(c) Find the 60th, 70th, 80th, and 90th percentiles for the standard normal variable.
(d) Find the 50th, 60th, 70th, and 80th percentiles for the discrete variable X that takes on values 1, 2, 3, 4, and 5 with respective frequencies of .1, .2, .2, .3, .2.
(e) Explain how the three types of critical regions (left-tailed, right-tailed, and two-tailed) corresponding to a significance level α can be defined in terms of percentiles.

16 **(Moment Generating Functions.)** For any random variable X we define its *moment generating function*, $m_X(t)$, as follows

$$m_X(t) = E(e^{tX})$$

$$= \begin{cases} \int_{-\infty}^{\infty} e^{tx} f(x)\,dx, & \text{if } X \text{ is continuous} \\ \sum e^{tx} f(x), & \text{if } X \text{ is discrete} \end{cases}$$

provided, of course, that these expectations exist. The function m_X gets its name from the fact that:

$m_X'(0) = E(X)$, the *1st moment* of X

$m_X''(0) = E(X^2)$, the *2nd moment* of X

$\vdots$

$m_X^{(k)}(0) = E(X^k)$, the *kth moment* of X

Compute $m_X(t)$ and use it to find μ_X and σ_X^2 in each of the following situations.

(a) X is discrete with frequency function $B(x; 2, \frac{1}{2})$

(b) X is discrete with frequency function $B(x; 3, \frac{1}{3})$

(c) X is discrete, having values 1, 2, 3 with equal probability

(d) X is continuous with density function f given by

$$f(x) = \begin{cases} 0, & \text{if } x < 0 \\ e^{-x}, & \text{if } x \geq 0 \end{cases}$$

(*Hint:*

$$\int_0^{\infty} e^{-ax}\,dx = 1/a \text{ if } a > 0.)$$

(e) X is continuous with density function f given by

$$f(x) = \begin{cases} 0, & \text{if } x < 0 \\ 10e^{-10x} & \text{if } x \geq 0 \end{cases}$$

17 **(a)** The geometric distribution (see Section 6 of Chapter 3) is used when X represents the number of repeated trials necessary to yield the first success, where $P(S) = p$ and $P(F) = q = 1 - p$. In this case X is infinite but discrete and it can be shown that its moment generating function (see the preceding exercise) is

$$m_X(t) = \frac{pe^t}{1 - qe^t}$$

Find μ_X and σ_X^2.

(b) The moment generating function for a normally distributed random variable with density function $N(x; \mu, \sigma)$ is given by

$$m_X(t) = \exp\left(t\mu + \frac{\sigma^2 t^2}{2}\right)$$

Verify that $m'_X(0) = \mu$ and that $m''_X(0) = \sigma^2 + \mu^2$.

(c) The moment generating function of the Poisson variable X with frequency function $P(x; \lambda)$ is given by

$$m_X(t) = \exp(\lambda(\exp(t) - 1))$$

Verify that $\mu = \sigma^2 = \lambda$.

(d) The Bernoulli distribution has for its moment generating function

$$m_X(t) = pe^t + q$$

Find μ_X and σ_X^2.

18 **(The Exponential Distribution.)** Suppose a certain Poisson process (see Section 11 of Chapter 4) has parameter λ. If T represents the time necessary for the first event to occur, then T is a continuous random variable with density function

$$f(t) = \begin{cases} \lambda e^{-\lambda t}, & \text{if } t > 0 \\ 0, & \text{if } t \leq 0 \end{cases}$$

(a) Show that the area under the graph of f equals 1.

(b) Find the distribution function F for T.

(c) Use the fact that the moment generating function for F is

$$m_T(t) = \frac{\lambda}{\lambda - t}, \qquad \text{for } |t| < \lambda$$

to find μ_T and σ_T^2.

(d) Show that for any positive numbers a and b,

$$P(T > a + b \mid T > a) = P(T > b)$$

and interpret this result in words.

(e) Suppose the average life of an overhead projector bulb is 10 hours and suppose that failures (i.e., burn-outs) in a batch of 100 such bulbs is approximately a Poisson process. How many bulbs would you expect to burn out during the first 5 hours of use?

(f) If in a large factory the accident rate is one per day, find the probability that there will be no accidents for three consecutive days.

(g) Using Table A in the Appendix, sketch the graphs of the density functions for exponential random variables with $\lambda = 1$ and $\lambda = 5$ using the same set of coordinate axes.

CHAPTER

6

Correlation and Regression

0 INTRODUCTION

In this chapter we will be concerned with finding and evaluating relations between two or more random variables. You have all seen examples of the kind of thing we are after. The tables found in doctors' offices and in magazines relating height, sex, and age to weight is an example. The table tells you that if you are a male of a particular height and particular age you should weigh a certain amount. A second example is "the predicted grade point average" used by colleges to try to predict, from the student's rank in his high school class and his SAT scores, how well he will do in college.

Such predictions are not as precise as those the physicist attains through the formula

distance = rate · time

There is always error attached to the prediction of weight (though not as much as certain overweight people believe) or to the prediction of grade point average. Because of this we must not only find a relation between random variables, we must also estimate the error involved.

We have already encountered two useful tools. The regression lines studied on page 354, along with certain generalizations, will be our major method for finding relations between random variables. The χ^2 test applied to contingency tables, as studied in Chapter 5, Section 10, will often be useful in telling us whether or not there is a relation between variables and hence whether or not the computationally complicated methods of this chapter should be employed. We shall begin by reviewing the earlier discussion of regression and extending it in various ways.

1 REGRESSION

In Section 13 of Chapter 4, we looked at the problem of finding a line (called the *regression line*) which comes closest to containing some given points in the plane, $(x_1, y_1), (x_2, y_2), \ldots, (x_n, y_n)$

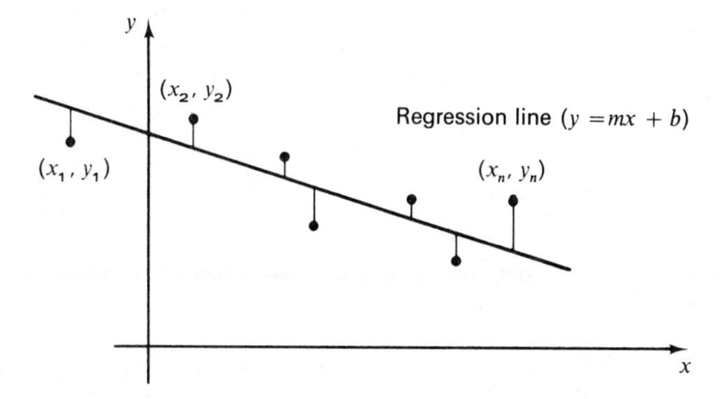

FIGURE 6.1

We "solved" the problem by letting d_i^2 be the square of the vertical distance of (x_i, y_i) from a line $y = mx + b$ and minimizing the function

$$f(m, b) = \sum_{i=1}^{n} d_i^2 = \sum_{i=1}^{n} (mx_i + b - y_i)^2$$

The result of setting

$$\frac{\partial f}{\partial m} = 0 \quad \text{and} \quad \frac{\partial f}{\partial b} = 0$$

was a pair of equations (called *normal equations*):

$$\left(\sum x_i^2\right)m + \left(\sum x_i\right)b = \sum x_i y_i$$
$$\left(\sum x_i\right)m + nb = \sum y_i$$

Normal equations for a regression line

These two linear equations in m and b yield the simultaneous solution

$$m = \frac{n\sum x_i y_i - \sum x_i \sum y_i}{n\sum x_i^2 - \left(\sum x_i\right)^2}$$
$$b = \frac{\left(\sum x_i^2\right)\sum y_i - \sum x_i \sum x_i y_i}{n\sum x_i^2 - \left(\sum x_i\right)^2}$$

In many ways this solution is satisfying: Although the resulting formulas appear messy, they accommodate data consisting of any (finite) number of points; the algebraic manipulation necessary involves only addition, subtraction, multiplication, and division and thus can be easily handled by a variety of calculating devices. Finally the solution is *unique*, or at least seems to be within our context, since minimal values of f should produce partial derivatives that are 0.

On the other hand, it is important to keep in mind the initial assumptions we made and to know their limitations. As Fig. 6.2 illustrates, we have several

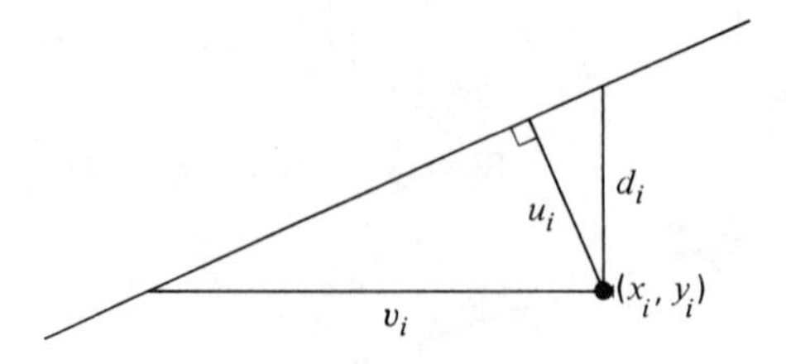

FIGURE 6.2

options for measuring the distances from the points (x_i, y_i) to a line. We selected d_i, the *vertical* distance to the line, though we could have chosen u_i (the *perpendicular* distance) or v_i (the *horizontal* distance). Moreover, we did not try to minimize the sum of the absolute values

$$g(m, b) = |d_1| + |d_2| + \cdots + |d_n|$$

but rather the sum of their squares

$$f(m, b) = d_1^2 + d_2^2 + \cdots + d_n^2$$

The reader may recall that we ran into the same kind of problem with standard deviation. The function $g(m, b)$ is really unsuitable for minimizing since $d_i = |mx_i + b - y_i|$ could have either $+x_i$ or $-x_i$ for its partial derivative with respect to b, and there is no way of knowing in advance (while m and b are still undetermined) which values to choose. The function $f(m, b)$, on the other hand can be differentiated with relative ease, so we accept it as a measure of the aggregate vertical discrepancy between the points (x_i, y_i) and the line $y = mx + b$.

The decision to use d_i rather than u_i or v_i is based on the assumption that we wish to use the resulting regression line, $y = mx + b$, to *predict* values of y corresponding to given values of x. Thus d_i measures the discrepancy between a "predicted" y value $(mx_i + b)$ and an actual y value (y_i). It should be noted that if the ultimate goal is to use the line to predict values of x from given values of y, then the *co-regression* line should be sought, i.e., the line $x = cy + d$ that minimizes the sum of the squares of the horizontal distances from the given points to a line.

When you have found a regression line $y = mx + b$ which allows you, given a value for x, to predict a value of y, we often say that *y has been regressed on x*. The line $x = cy + d$ which, from a value of y, predicts a value of x is said to *regress x on y*.

It should be pointed out that there are, in a sense, two kinds of regression problems. In one kind we have a random and a non-random variable. For example we might be conducting polls at various times preceding a referendum vote and we would have the proportion of the electorate intending to vote yes as the random variable and the time as the non-random variable. We will regress a random variable on a non-random variable but never a non-random variable on a random variable.

The second kind of regression problem involves two random variables, for example, the mathematical and verbal SAT scores. Here we can regress either variable on the other. We acknowledge the distinction between random variables and non-random ones by the use of capital letters for random variables and lower case letters in the non-random case.

The next three examples should clarify the computational problems in finding a regression line.

EXAMPLE 1 Find the regression line for the points (1, 1), (2, 3), (4, 3).

Solution.

$$m = \frac{3(1 + 6 + 12) - (7)(7)}{3(1 + 4 + 16) - 7^2} = \frac{57 - 49}{63 - 49} = \frac{8}{14} = \frac{4}{7}$$

$$b = \frac{(1 + 4 + 16)(7) - (7)(1 + 6 + 12)}{3(1 + 4 + 16) - 7^2} = \frac{147 - 133}{63 - 49} = \frac{14}{14} = 1$$

Therefore the equation of the regression line is

$$y = \tfrac{4}{7}x + 1$$

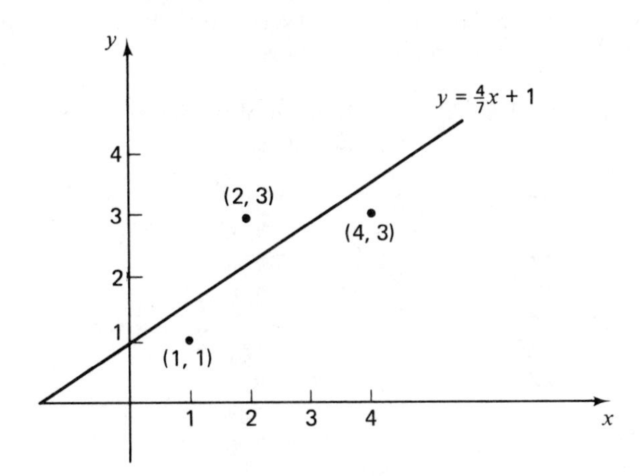

FIGURE 6.3

EXAMPLE 2 Find the regression line for the points (1, 0), (2, 1), (4, 3).

Solution.

$$m = \frac{3(0 + 2 + 12) - (7)(4)}{3(1 + 4 + 16) - 7^2} = \frac{42 - 28}{63 - 49} = \frac{14}{14} = 1$$

$$b = \frac{(1 + 4 + 16)(4) - (7)(0 + 2 + 12)}{63 - 49} = \frac{84 - 98}{14} = \frac{-14}{14} = -1$$

Therefore the regression line has the equation $y = x - 1$; moreover the three given points actually *lie on* the regression line.

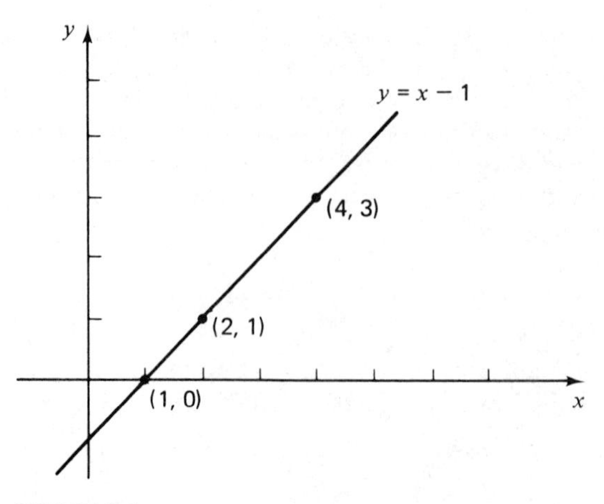

FIGURE 6.4

EXAMPLE 3 Let us do a more realistic problem. Referring to the data on 100 college students found on page 459, we might expect to see some relation between a student's rank in class and the size of his class. We shall check this for the first 20 students in the list. The raw data is as follows.

High School Rank in Class	Number of Students in High School Class
42	180
24	195
34	174
114	330
137	537
110	161
154	591
33	264
141	736
112	914
85	160
30	154
58	176
130	222
40	162
31	221
10	54
81	641
84	205
83	336

We shall not produce the arithmetic but, after some calculations, letting Y be high school rank and X be class size, we obtain

$$Y = .124589X + 34.79564$$

If we had a student from the college where this data was compiled and if we were told that he graduated in a class of 200, the best prediction we could make is that his class standing was

$$\begin{aligned}(.124589)200 + 34.79564 &= 24.9178 + 34.79564 \\ &= 59.7\end{aligned}$$

or, that he was about 60th. As you can see from trying the equations out on several people in the data set, the prediction is not precise. But is it good enough to use for some purposes? We shall return to that question in the next section.

Forgetting momentarily the practical applications of the regression line, let us view the problem geometrically. Suppose we are given n points in the plane whose coordinates are

$$(x_1, y_1), (x_2, y_2), \ldots, (x_n, y_n)$$

Only in a very restricted situation will these points appear to be nearly collinear. In the next section we shall look into the problem of determining the extent to which the points are collinear, but for the time being we can treat the matter informally. Figures 6.5, 6.6, 6.7, and 6.8 portray some simple-minded examples.

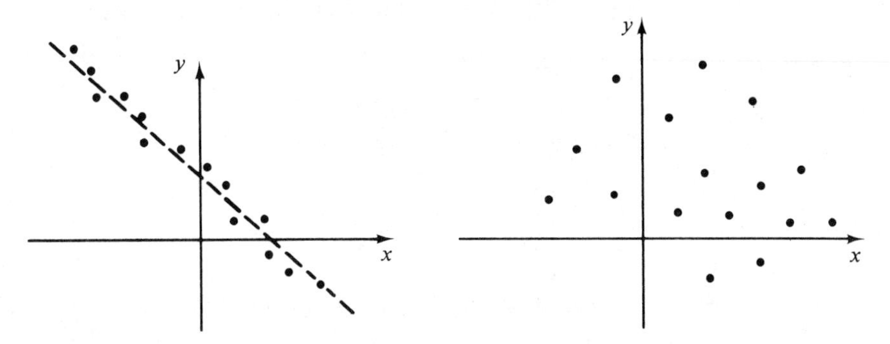

FIGURE 6.5

FIGURE 6.6

Roughly speaking, Fig. 6.5 represents a set of data points where *linear* regression is most appropriate. We would expect the regression line to have negative slope and positive y-intercept. Figure 6.6, on the other hand, represents data for which linear regression would serve no useful purpose; the data points appear to be more or less haphazardly distributed. Figures 6.7 and 6.8 illustrate (very) regular patterns, although the points are by no means collinear. Indeed these examples beg for a non-linear curve to serve as a model, and we shall presently devise suitable descriptions in a more general setting of a regression curve. These graphs (Figs. 6.5, 6.7, and 6.8) are commonly called *scatter diagrams*.

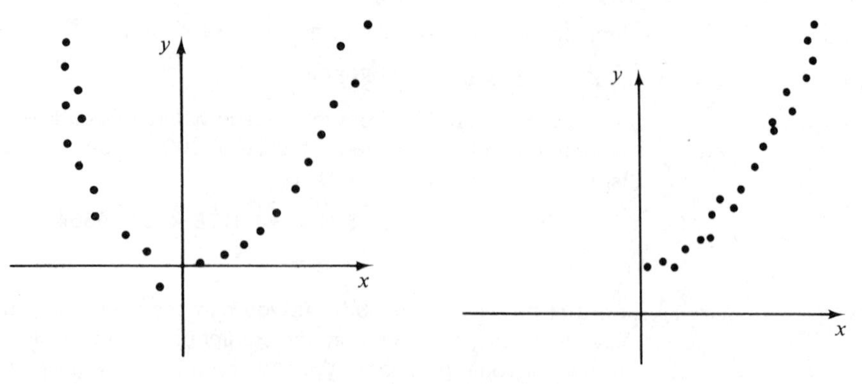

FIGURE 6.7

FIGURE 6.8

Let us formulate the notion of regression in a general context. We begin with a set of variables (two or more) that appear to bear some relationship to each other. We conjecture the nature of the relationship by prescribing a type of function or curve that "fits" some given values (data) of the variables. The functional relationship we propose involves parameters and our task is to find the values of these parameters that yield a particular function or curve which best fits our data. The problem at this stage is not unlike that of finding the value of a parameter for a distribution that yields a density function or a frequency function which best fits given sample data. The following examples illustrate the variety of problems that fall into the general category of regression.

EXAMPLE 4

Variables: X, Y.
Conjecture: $Y = aX^2 + bX + c$
Data:

X	-1	0	1	2	3	4
Y	5	1	.9	3.5	6	9

Method of attack: Minimize the sums of the squares of the differences between given Y values and the predicted Y values by finding suitable values for the parameters a, b, and c. Specifically: Put

$$f(a, b, c) = \sum_{i=1}^{6} (ax_i^2 + bx_i + c - y_i)^2$$

and set

$$\frac{\partial f}{\partial a} = \frac{\partial f}{\partial b} = \frac{\partial f}{\partial c} = 0$$

then solve for a, b, c.
Note: The reader might check the data to see that $Y = X^2 - 2X + 2$ is a pretty good estimate.

EXAMPLE 5

Variables: X, U, Y
Conjecture: $Y = aX + bU + c$
Data:

X	1	1	2	2	3	3	3
U	1	2	1	2	1	2	3
Y	.8	−.8	3	1.9	5.4	4	2.9

Method of attack: Minimize the sums of the squares of the differences between given Y values and predicted Y values by finding suitable values for a, b, and c. Specifically: Put

$$f(a, b, c) = \sum_{i=1}^{7} (ax_i + bu_i + c - y_i)^2$$

and set

$$\frac{\partial f}{\partial a} = \frac{\partial f}{\partial b} = \frac{\partial f}{\partial c} = 0$$

then solve for a, b, and c.
Note: The reader might check that $Y = 2X - U$ is a good guess.

Let us complete this solution in detail. From the data we can compute the following sums.

$$\sum x_i = 15, \quad \sum y_i = 17.2, \quad \sum u_i = 12, \quad \sum x_i^2 = 37$$
$$\sum u_i x_i = 27, \quad \sum x_i y_i = 46.7, \quad \sum u_i^2 = 24, \quad \sum u_i y_i = 28.1$$

Taking partial derivatives of f, we find

$$\frac{\partial f}{\partial a} = \sum 2(ax_i + bu_i + c - y_i)(x_i) = 0, \text{ if } (\sum x_i^2)a + (\sum x_i u_i)b + (\sum x_i)c = \sum x_i y_i$$

$$\frac{\partial f}{\partial b} = \sum 2(ax_i + bu_i + c - y_i)(u_i) = 0, \text{ if } (\sum x_i u_i)a + (\sum u_i^2)b + (\sum u_i)c = \sum u_i y_i$$

$$\frac{\partial f}{\partial c} = \sum 2(ax_i + bu_i + c - y_i) = 0, \quad \text{ if } (\sum x_i)a + (\sum u_i)b + 7c = \sum y_i$$

These three normal equations become

(1) $37a + 27b + 15c = 46.7$

(2) $27a + 24b + 12c = 28.1$

(3) $15a + 12b + 7c = 17.2$

We proceed to eliminate "a" from the second and third equations. First we *divide the first equation by 37* to get

(1′) $a + \frac{27}{37}b + \frac{15}{37}c = \frac{46.7}{37}$

Now we subtract 27 times (1′) from (2) to get

(2′) $4.3b + 1.05c = -5.98$

and we subtract 15 times (1′) from (3) to get

(3′) $1.05b + .9c = -1.73$

Solving (2′) and (3′) simultaneously, either by formulas or by eliminating one of the variables, yields

$b = -1.3$

$c = -.37$

$a = 2.37$

The regression equation is therefore

$$Y = 2.37X - 1.3U - .37$$

The following table compares the given values of y_i with the values $\tilde{y}_i$ computed from this regression formula

x_i	u_i	y_i	$\tilde{y}_i$
1	1	.8	.7
1	2	−.1	−.6
2	1	3	3.07
2	2	1.9	1.77
3	1	5.1	5.44
3	2	4	4.14
3	3	2.9	2.84

EXAMPLE 6

Variables: X, U, Y
Conjecture: $Y = aXU^b$
Data:

X	0	1	1	2	2	3	3
U	1	1	2	1	2	1	2
Y	0	.5	2.8	1.1	8.0	1.7	11.5

Method of attack: Here it pays to transform the problem using properties of logarithms: From $Y \Rightarrow aXU^b$ it follows that

$$\text{Log } Y = \text{Log } a + \text{Log } X + b \text{ Log } U$$

which can be thought of as

$$y = a' + x + bu$$

The data is transformed accordingly and the problem reduced to much the same thing as Example 4.

It should be mentioned at this point that the level of difficulty of solving the (normal) equations that arise by setting partial derivatives equal to zero tends to increase with the number of variables involved. Typically the equations are in the form

$$(*) \quad \begin{cases} p_1 a + q_1 b = r_1, \\ p_2 a + q_2 b = r_2, \end{cases} \qquad \text{corresponding to } Y = aX + b$$

$$(\dagger) \quad \begin{cases} p_1 a + q_1 b + r_1 c = d_1, \\ p_2 a + q_2 b + r_2 c = d_2, \\ p_3 a + q_3 b + r_3 c = d_3, \end{cases} \qquad \text{corresponding to } Y = aX + bU + c \text{ or } Y = aX^2 + bX + c$$

and so forth, where the p_i, q_i, r_i, and d_i are expressions such as $\sum x_i$, $\sum x_i y_i$, $\sum x_i^2$, etc. Standard techniques of linear algebra are applicable. Bearing in mind that a, b, and c are the unknowns, the solutions to (*) and (†) are

$$(**) \quad \begin{cases} a = \dfrac{r_1 q_2 - r_2 q_1}{p_1 q_2 - p_2 q_1} \\[2ex] b = \dfrac{p_1 r_2 - p_2 r_1}{p_1 q_2 - p_2 q_1} \end{cases}$$

$$(\dagger\dagger) \quad \begin{cases} a = \dfrac{d_1(q_2 r_3 - q_3 r_2) - d_2(q_1 r_3 - q_3 r_1) + d_3(q_1 r_2 - q_2 r_1)}{p_1(q_2 r_3 - q_3 r_2) - p_2(q_1 r_3 - q_3 r_1) + p_3(q_1 r_2 - q_2 r_1)} \\[2ex] b = \dfrac{p_1(d_2 r_3 - d_3 r_2) - p_2(d_1 r_3 - d_3 r_1) + p_3(d_1 r_2 - d_2 r_1)}{p_1(q_2 r_3 - q_3 r_2) - p_2(q_1 r_3 - q_3 r_1) + p_3(q_1 r_2 - q_2 r_1)} \\[2ex] c = \dfrac{p_1(q_2 d_3 - q_3 d_2) - p_2(q_1 d_3 - q_3 d_1) + p_3(q_1 d_2 - q_2 d_1)}{p_1(q_2 r_3 - q_3 r_2) - p_2(q_1 r_3 - q_3 r_1) + p_3(q_1 r_2 - q_2 r_1)} \end{cases}$$

While these formulas are not the sort of thing one tries to memorize, they can be reproduced by careful bookkeeping. Note that all the denominators in (††) are the same. Also the numerators resemble the denominators; in fact to get the numerator for a, we simply take the denominator and replace p_i by d_i. Similarly, for b we replace q_i by d_i and for c we replace r_i by d_i. It is obvious how the same procedure applies in (**). Of course just about the time the reader gets all of this in order, he is apt to encounter a system of normal equations with four or more parameters (unknowns) to solve for and he must start over again discovering new formulas. There are techniques for solving such systems either by hand or using a computer and anyone needing to solve systems with several unknowns shouldn't have any trouble finding them.

A more significant observation to make is that Examples 4 and 5 above are really identical. For if we view the equation $Y = aX^2 + bX + c$ as an equation relating the three variables X, X^2, and Y, we see that the problem of finding suitable a, b, and c is precisely as in $Y = aX + bU + c$. Geometrically, $Y = aX^2 + bX + c$ is the equation of a parabola in the plane, hopefully approximating the data points

$$(x_1, y_1), (x_2, y_2), \ldots, (x_n, y_n)$$

whereas the equation $Y = aX + bU + c$ represents a *plane* in space, hopefully approximating the data points

$$(x_1, u_1, y_1), (x_2, u_2, y_2), \ldots, (x_n, u_n, y_n)$$

One final comment on the geometric interpretation of regression is in order. For any sets of n points $(x_1, y_1), \ldots, (x_n, y_n)$ in the plane, we define the *centroid* of that set to be the point $(\bar{x}, \bar{y})$ where

$$\bar{x} = \frac{1}{n}\sum_{i=1}^{n} x_i \quad \text{and} \quad \bar{y} = \frac{1}{n}\sum_{i=1}^{n} y_i$$

That is, the centroid is the point whose x-coordinate is the *average* of all the x-coordinates and whose y-coordinate is the average of all the y-coordinates of points in the set. In Fig. 6.9, the centroid of (1, 2), (1, 4), (2, 0), (5, 1), and

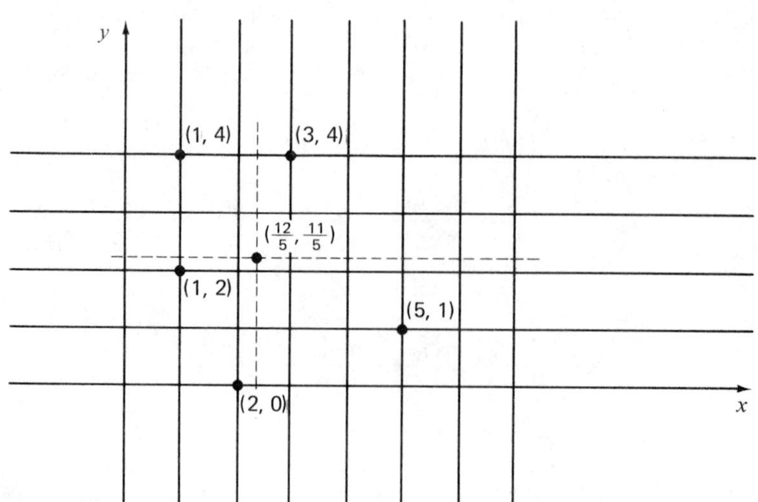

FIGURE 6.9

(3, 4) is $(\frac{12}{5}, \frac{11}{5})$. Centroids are useful in physical problems. If Fig. 6.9 were sketched on a lightweight plastic square and if one-pound weights were placed at each of the given points, then the resulting object could be balanced on a rod that touched the square at the centroid.

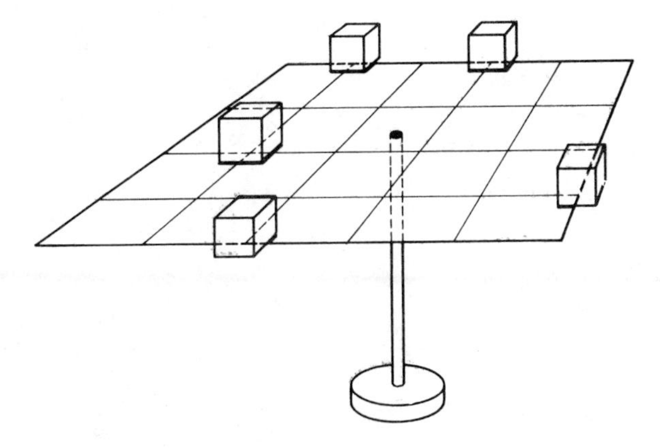

FIGURE 6.10

Anyone who has played on a teeter-totter has first-hand experience of some one-dimensional centroid properties.

We can show that regression lines always pass through the centroid of the scatter diagram. To see this we compute the value of $y = mx + b$ corresponding to $x = \bar{x}$. From the original solution to this two-variable problem we see

$$
\begin{aligned}
y &= m\bar{x} + b \\
&= \frac{n\sum x_i y_i - \sum x_i \sum y_i}{n\sum x_i^2 - (\sum x_i)^2}\frac{1}{n}\sum x_i + \frac{(\sum x_i^2)(\sum y_i) - (\sum x_i)(\sum x_i y_i)}{n\sum x_i^2 - (\sum x_i)^2} \\
&= \frac{(\sum x_i y_i)(\sum x_i) - (1/n)(\sum x_i)^2 \sum y_i + (\sum x_i^2)(\sum y_i) - (\sum x_i)(\sum x_i y_i)}{n\sum x_i^2 - (\sum x_i)^2} \\
&= \frac{1}{n}\sum y_i \\
&= \bar{y}
\end{aligned}
$$

As an immediate consequence of this result, we can observe that in practice it suffices to find the value of m, the slope of the regression line, since one point $(\bar{x}, \bar{y})$ is always known. In some instances it is helpful to take advantage of the definitions of the means and standard deviations to rewrite the formulas for m and b. Recalling that $\bar{x} = (1/n)\sum x_i$, $\bar{y} = (1/n)\sum y_i$, and

$$S_x^2 = \frac{1}{n}\sum x_i^2 - \left(\frac{1}{n}\sum x_i\right)^2$$

We can rewrite m as follows:

$$m = \frac{n\sum x_i y_i - \sum x_i \sum y_i}{n\sum x_i^2 - (\sum x_i)^2} = \frac{n\sum x_i y_i - n^2\bar{x}\bar{y}}{n^2 S_x^2}$$

or simply

$$m = \frac{\sum x_i y_i - n\bar{x}\bar{y}}{nS_x^2}$$

Also the above argument shows that

$$b = \bar{y} - m\bar{x}$$

Summary

Given a set of points $(x_1, y_1), (x_2, y_2), \ldots, (x_n, y_n)$, we have found a line, called the *regression line* of y on x, which is, given x, the best *linear* predictor of y. If the line is

$$y = mx + b$$

then we know

$$m = \frac{\sum_{i=1}^{n} x_i y_i - n\bar{x}\bar{y}}{nS_x^2}$$

$$b = \bar{y} - m\bar{x}$$

Exercises
Section 1

1 Find the regression line for the points:
  (a) (1, 1), (3, 1), (5, 4)
  (b) (0, 0), (1, 0), (0, 1), (1, 1)

2 Find the regression parabola for each of the sets of points in Exercise 1.

3 Use the least squares technique to fit the following data on the population of Japan to a curve of the form

$$x = c \cdot \exp(at)$$

Year	Population (x) (in millions)
1900	44
1910	49
1920	56
1930	64
1940	72
1950	83
1960	93

(*Hints:* (1) See Example 6. (2) Let $t = 0$ in the year 1900.)

4 Consider the data on 100 freshmen found on page 459. Given variables X and Y, the regression of Y on X means the regression line

$$Y = mX + b$$

Given variables X, Y, and U, the regression of Y on X and U means the regression plane

$$Y = aX + bU + c$$

Do each of the following, first with only students 1–20 on the list and then with all 100.

(a) Find the regression of the freshman GPA on the verbal SAT.

(b) Find the regression of the freshman GPA on the math SAT.

(c) Find the regression of the freshman GPA on the high school rank in class.

(d) Find the regression of the freshman GPA on the math SAT and verbal SAT.

(e) Find the regression of the freshman GPA on the math SAT and verbal SAT and high school rank in class.

Remark: For part (e) you need to be able to solve the simultaneous equations

$$\begin{aligned} p_1a + q_1b + r_1c + s_1d &= e_1 \\ p_2a + q_2b + r_2c + s_2d &= e_2 \\ p_3a + q_3b + r_3c + s_3d &= e_3 \\ p_4a + q_4b + r_4c + s_4d &= e_4 \end{aligned}$$

corresponding to

$$Y = aX + bU + cV + d$$

The solution is as follows. The denominator of each of a, b, c, and d is

$$(*)\quad \begin{cases} p_1q_2(r_3s_4 - r_4s_3) - p_1q_3(r_2s_4 - r_4s_2) + p_1q_4(r_2s_3 - r_3s_2) \\ -p_2q_1(r_3s_4 - r_4s_3) + p_2q_3(r_1s_4 - r_4s_1) - p_2q_4(r_1s_3 - r_3s_1) \\ +p_3q_1(r_2s_4 - r_4s_2) - p_3q_2(r_1s_4 - r_4s_1) + p_3q_4(r_1s_2 - r_2s_1) \\ -p_4q_1(r_2s_3 - r_3s_2) + p_4q_2(r_1s_3 - r_3s_1) - p_4q_3(r_1s_2 - r_2s_1) \end{cases}$$

The expression for the numerator of a is obtained from (*) by replacing p_i by e_i for $e = 1, 2, 3, 4$. The expression for the numerator of b is obtained from (*) by replacing q_i by e_i for $i = 1, 2, 3, 4$, etc.

5 **(a)** Generate a set of points (x, y) by the following experiment. Take six coins and toss them. Let x be the number of heads which appear. Toss the x coins which landed heads and let y be the number of heads obtained this time. Repeat the experiment 10 times and compute the regression of y on x.

(b) Explain why you do (or don't) think $y = \frac{1}{2}x$ would be a good guess for the answer to (a).

6 The following data is from a student evaluation of faculty and courses. The first number after each professor is the percentage of students in a class of his who rated him as an above average professor and the second figure is the percentage of students who rated the course above average. Find the regression line of course rating on the rating of the professor.

Professor	His Rating	Course Rating
A	94	79
B	75	81
C	95	80
D	11	6
E	71	71
F	79	52
G	66	49
H	92	62
I	100	96
J	37	41

7 Suppose that the regression line of y on x for the points $(x_1, y_1), (x_2, y_2), \ldots, (x_n, y_n)$ is $y = 3x + 2$. Find the regression line of y on x for each of the following sets of points:

(a) $(x_1, y_1 + 5), (x_2, y_2 + 5), \ldots, (x_n, y_n + 5)$
(b) $(3x_1, y_1), (3x_2, y_2), \ldots, (3x_n, y_n)$
(c) $(x_1 - 2, y_1), (x_2 - 2, y_2), \ldots, (x_n - 2, y_n)$
(d) $(x_1 + 2, 2y_1), (x_2 + 2, 2y_2), \ldots, (x_n + 2, 2y_n)$

8 The results in Exercise 7 are special cases of the relationship between the regression line of y on x for points $(x_1, y_1), \ldots, (x_n, y_n)$ and the regression line of y on x for the points $(x'_1, y'_1), \ldots, (x'_n, y'_n)$ where (a) x'_i is obtained from x_i by adding a constant and/or multiplying by a constant for all i; and (b) y'_i is obtained from y_i by adding a constant and/or multiplying by a constant for all i.

Based on your results in Exercise 7, guess at the statement of this relationship and prove it.

2 CORRELATION

In the previous section we learned how to find relations between two or more variables. What we want to do now is learn how to decide how good these relationships are. We shall restrict our attention to the case where we found a linear relation between two variables.

Given the data points $(x_1, y_1), (x_2, y_2), \ldots, (x_n, y_n)$, we obtained the regression line of y on x in the form

$$y = mx + b$$

where

$$m = \frac{\sum_{i=1}^{n} x_i y_i - n\bar{x}\bar{y}}{nS_x^2}$$

and

$$b = \bar{y} - m\bar{x}$$

This was attained by defining a function of m and b

$$f(m, b) = \sum_{i=1}^{n} (mx_i + b - y_i)^2$$

and finding the values of m and b which made it a minimum.

One way to view the "goodness" of the regression line is to see how small this minimum value of f is. Letting T be the minimum, we have

$$\begin{aligned} T &= \sum_{i=1}^{n} (mx_i + b - y_i)^2 \\ &= nS_y^2 - m^2 nS_x^2 \qquad \text{(after some algebraic manipulation)} \end{aligned}$$

Some more algebraic chicanery will give us a better insight into the size of T. We have

$$\begin{aligned} T &= nS_y^2 - m^2 nS_x^2 \\ &= nS_y^2 - m^2 nS_x^2 \frac{S_y^2}{S_y^2} \\ &= nS_y^2 \left(1 - m^2 \frac{S_x^2}{S_y^2}\right) \end{aligned}$$

If the factor

$$\left(1 - m^2 \frac{S_x^2}{S_y^2}\right)$$

is close to 0, T is very small; whereas, when this factor is large, T is large.

Let us look at this expression more closely. Since T is defined as a sum of squares, we know $T \geq 0$. Similarly, $nS_y^2 \geq 0$ so

$$1 - \left(m\frac{S_x}{S_y}\right)^2 \geq 0$$

or

$$\left(m\frac{S_x}{S_y}\right)^2 \leq 1$$

and so

$$-1 \leq \left(m\frac{S_x}{S_y}\right) \leq 1$$

When $m(S_x/S_y)$ is close to 1 or -1, T is small and the equation

$$y = mx + b$$

is a good fit for the points. As $m(S_x/S_y)$ gets closer to 0, T gets larger and the equation is a poorer fit.

Definition $r = m(S_x/S_y)$ is called the (*Pearson Product-Moment*) *coefficient of correlation of* y *on* x or simply the *correlation coefficient*.

There is a second useful interpretation of the correlation coefficient. We have regressed the variable y on x. Suppose we choose

$$\sum_{i=1}^{n} (y_i - \bar{y})^2 = nS_y^2$$

as a measure of the variation in y. Compute

$$T = \sum (mx_i + b - y_i)^2$$

as before. We can think of this as the variation in y *not* caused by x. That is, the variation in y due to x is handled by the linear equation

$$y = mx + b$$

and when

$$mx_i + b - y_i$$

is not 0, this means other factors are at work. Therefore we can think of

$$\frac{T}{nS_y^2}$$

as the fraction of variation in y caused by factors other than x. Hence

$$1 - \frac{T}{nS_y^2}$$

is the fraction of variation in y due to x. Let us calculate this. We obtain

$$\begin{aligned} 1 - \frac{T}{nS_y^2} &= 1 - \frac{nS_y^2\left(1 - m^2\dfrac{S_x^2}{S_y^2}\right)}{nS_y^2} \\ &= 1 - (1 - r^2) \\ &= r^2 \end{aligned}$$

So, if r is the correlation coefficient, we may think of r^2 as that fraction of the variance of y due to the variance of x.

Once again, when r^2 is close to 1 (r is close to $+1$ or -1), most of the variation in y is due to x. When r^2 (and hence r) is close to 0, little of the variation in y is being caused by x.

One frequently calculates the correlation coefficient without finding the regression line in order to decide if two variables are related. To do this it is useful to have a formula for r in terms of the x_i's and y_i's. A little calculation reveals that

$$r = \frac{\sum_{i=1}^{n} (x_i - \bar{x})(y_i - \bar{y})}{nS_xS_y}$$

$$= \frac{\sum_{i=1}^{n} x_iy_i - \frac{1}{n}\left(\sum_{i=1}^{n} x_i\right)\left(\sum_{i=1}^{n} y_i\right)}{\sqrt{\left[\sum_{i=1}^{n} x_i^2 - \frac{1}{n}\left(\sum_{i=1}^{n} x_i\right)^2\right] \cdot \left[\sum_{i=1}^{n} y_i^2 - \frac{1}{n}\left(\sum_{i=1}^{n} y_i\right)^2\right]}}$$

By way of examples we shall calculate the correlation coefficients for the data in Examples, 1, 2, and 3 of the previous section.

EXAMPLE 1 Find the correlation coefficient of the points (1, 1), (2, 3), and (4, 3).

Using our formula, we obtain

$$r = \frac{(1 \cdot 1 + 2 \cdot 3 + 4 \cdot 3) - \frac{1}{3}(1 + 2 + 4)(1 + 3 + 3)}{\sqrt{[(1^2 + 2^2 + 4^2) - \frac{1}{3}(1 + 2 + 4)^2][(1^2 + 3^2 + 3^2) - \frac{1}{3}(1 + 3 + 3)^2]}}$$

$$= \frac{19 - \frac{1}{3} \cdot 7 \cdot 7}{\sqrt{[21 - \frac{1}{3} \cdot 49][19 - \frac{1}{3} \cdot 49]}}$$

$$= \frac{57 - 49}{\sqrt{(63 - 49)(57 - 49)}}$$

$$= \frac{8}{\sqrt{14 \cdot 8}} = \frac{8}{4\sqrt{7}} = \frac{2}{\sqrt{7}} \approx .76$$

EXAMPLE 2 Find the correlation coefficient of the points (1, 0), (2, 1), and (4, 3).

$$r = \frac{1 \cdot 0 + 2 \cdot 1 + 4 \cdot 3 - \frac{1}{3}(1 + 2 + 4)(0 + 1 + 3)}{\sqrt{[(1^2 + 2^2 + 4^2) - \frac{1}{3}(1 + 2 + 4)^2][(0^2 + 1^2 + 3^2) - \frac{1}{3}(0 + 1 + 3)^2]}}$$

$$= \frac{14 - \frac{1}{3}(7)(4)}{\sqrt{[21 - \frac{1}{3} \cdot 49][10 - \frac{1}{3} \cdot 16]}}$$

$$= \frac{42 - 28}{\sqrt{(63 - 49)(30 - 16)}} = \frac{14}{\sqrt{14 \cdot 14}} = 1$$

We had remarked earlier that when the points are collinear, the correlation coefficient is +1 or −1, and also pointed out in Section 1 that the three points in this example were all on the line $y = x - 1$.

EXAMPLE 3 The correlation coefficient for the regression problem done as Example 3 in Section 1 turns out to be .65801.

So far we have defined the (linear) correlation coefficient r, we have interpreted r^2 as that fraction of the variance of y explained by the variance of x, and we have said that $r = \pm 1$ when y is a linear function of x and $r = 0$ when

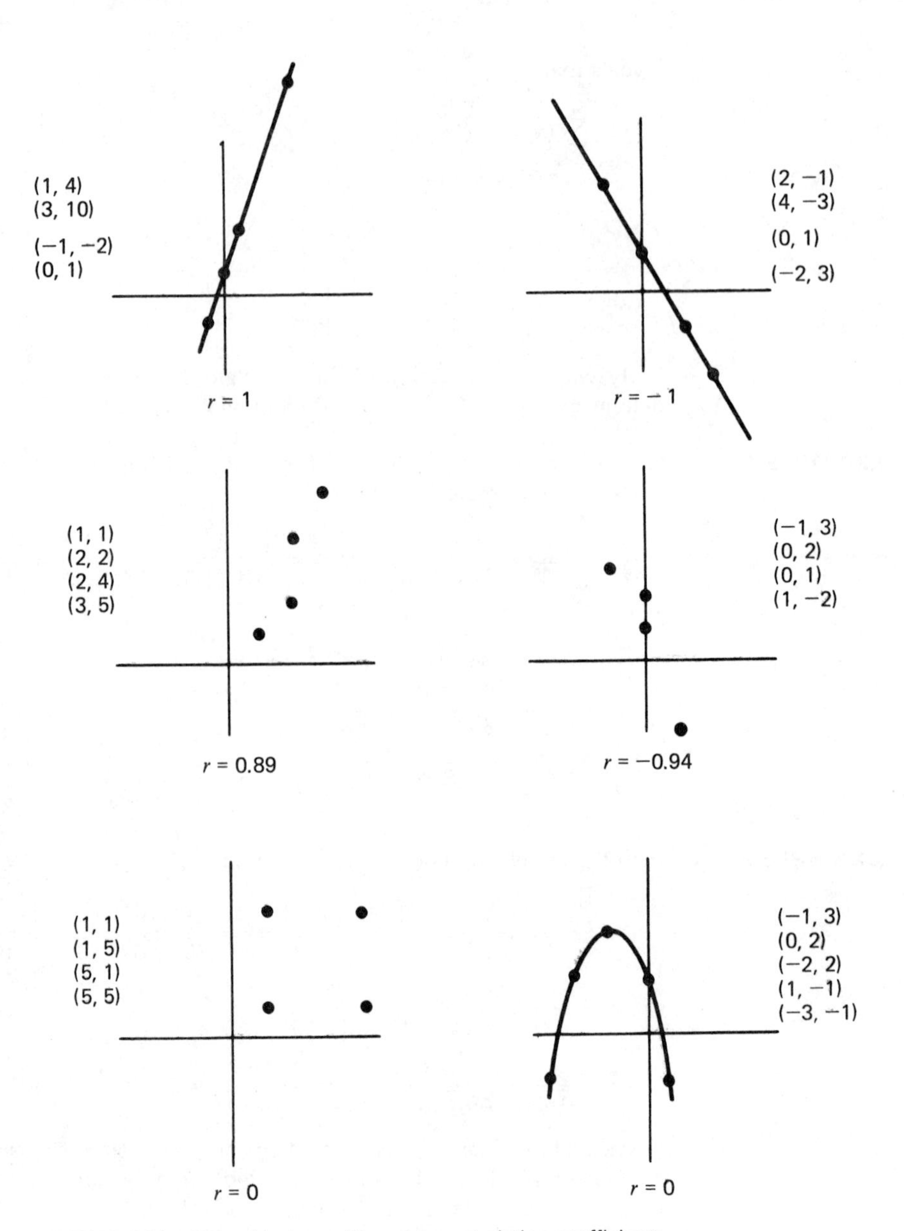

FIGURE 6.11 Sets of points with various correlation coefficients.

there is no linear relation. However, we have not said much about interpreting r when r is not ± 1 or 0. If we obtain a value such as $r = .63$ or $r = .37$, how do we interpret it? What do these (or other) numbers tell us about the value of the regression line? The answer to these questions lies in a theorem whose proof is well beyond the scope of this course. For the theorem we view our points (x_i, y_i) as a sample of size n drawn from two random variables X and Y on some sample space. There is a theoretical correlation coefficient, ρ, which is ± 1 if the two random variables are linearly related and 0 if there is no linear relationship between them. We view r as a statistic on the sample which estimates ρ.

Theorem If X and Y are normally distributed, then if $\rho = 0$, the statistic

$$\frac{r\sqrt{n-2}}{\sqrt{1-r^2}}$$

has a t-distribution with $n - 2$ degrees of freedom.

Computing this for the r of Example 1, we obtain

$$\frac{.76\sqrt{3-2}}{\sqrt{1-\frac{4}{7}}} = \frac{.76}{\sqrt{\frac{3}{7}}}$$
$$\approx 1.15$$

To view this as an hypothesis test we make

$$H_0 : \rho = 0$$

In order to reject this at a .10 significance level, the value of t with one degree of freedom should 3.078. Since our value, 1.15, is much less than this, we cannot reject H_0 and we accept that there is some degree of linear relation between the variables.

One word of warning is needed. Do not confuse high correlation with causal relation. As was pointed out early in the controversy over lung cancer and cigarettes, there is a high correlation between lung cancer and electricity. While some people thought that this exposed a weakness in the methods used by the Surgeon General's office, it did no such thing. The case against cigarettes did not rest wholly upon correlation. Evidence of a causal relation was also found. But no one established a causal relation between consumption of electricity and lung cancer. The correlation is undoubtedly the result of certain other factors (for example, perhaps affluence allows the purchase of more electricity as well as more cigarettes leading to lung cancer). In general, it is well to be cautious about the use of correlation. In particular, one should not infer causal relations from correlation.

Summary

We have defined the correlation coefficient

$$r = m\frac{S_x}{S_y}$$

where m is the slope of the regression line of y on x. We introduced the formula

$$r = \frac{\sum_{i=1}^{n} x_i y_i - \frac{1}{n}\left(\sum_{i=1}^{n} x_i\right)\left(\sum_{i=1}^{n} y_i\right)}{\sqrt{\left[\sum_{i=1}^{n} x_i^2 - \frac{1}{n}\left(\sum_{i=1}^{n} x_i\right)^2\right]\left[\sum_{i=1}^{n} y_i^2 - \frac{1}{n}\left(\sum_{i=1}^{n} y_i\right)^2\right]}}$$

for computing r directly from the data. It was observed that r^2 is that fraction of the variance of y which can be explained by x. Finally, we said that if we regard $(x_1, y_1), \ldots, (x_n, y_n)$ as a sample of size n from two non-linearly correlated populations, then the statistic

$$\frac{r\sqrt{n-2}}{\sqrt{1-r^2}}$$

is distributed as a t-variable with $n - 2$ degrees of freedom.

Exercises Section 2

1 Calculate the correlation coefficient for each pair of variables in Exercises 1 and 3 of the previous section.

2 For the data on 100 students calculate:
  (a) the correlation coefficient between math and verbal SAT's.
  (b) the correlation coefficient between freshmen and sophomore GPA's.
  (c) Is there a statistically significant correlation between the variables in (a) or in (b)?

3 The following data are the scores on two successive tests by ten students in a class. Is there a significant correlation between scores on the two tests at the .05 level?

Test 1	Test 2
56	61
64	84
66	76
74	94
72	90
79	95
75	88
64	77
67	72
51	63

If you were told that a student got a score of 60 on Test 1, what would you predict as his score on Test 2?

4 Suppose the points $(x_1, y_1), (x_2, y_2), \ldots, (x_n, y_n)$ have correlation coefficient .3. How large must n be before you can say at the .05 significance level that some linear relation exists?

5 Compute the correlation coefficient for the data of Exercise 6 in the previous section.

6 Compute the correlation coefficient for your data from Exercise 5 in the previous section.

7 Explain verbally why the correlation coefficient of two points should always be $+1$ or -1 and then, using the equation for calculating r, show it.

8 Assume that in Exercise 7 of the preceding section, the correlation coefficient is .4. Find the correlation coefficients of the sets of points in parts (a)–(d). Then redo Exercise 8 of that section with the words regression line replaced by correlation coefficient.

CHAPTER

7

Non-Parametric Techniques

0 INTRODUCTION

Thus far in our investigations, whether descriptive or inferential, we have had as a recurring objective the finding or the predicting of some parameter, say the mean or variance of a sample or a population. Once we knew that a certain random variable was binomially distributed, we were able to completely describe its distribution by specifying the values of p and n. Likewise, if the variable were normally distributed, we learned what we wanted to know by assigning values to μ and σ.

In many situations where we seek information about a random variable (or two or more variables), however, the exact nature of the distribution might be difficult to assess. In other situations we might be faced with the problem of making an educated guess about a population based on a relatively small sample. In cases such as these, there are a variety of statistical methods that do not rely on knowledge of the type of distribution, but rather provide a sometimes crude analysis of the samples at hand that could lead to further and more refined examinations. These methods, because they do not fit data to distributions that would involve selecting parameters, are called *non-parametric* or sometimes *distribution-free* methods. In this chapter we shall examine a few of these techniques. While they are in general easy to apply, the reader should be forewarned that their derivations are frequently complicated. More importantly, we should bear in mind that, as a class, non-parametric methods are crude in the sense that they rarely provide the kind of detailed information possible when the actual distribution is known.

In fairness to the methods we are about to discuss, it should be mentioned that in some situations the available data is of such a nature that parametric techniques seem to be of little value, whereas an appropriate non-parametric approach could yield usable information. Here is a case in point.

Suppose a professional golfer plays a certain course 20 times over a 4-year period, getting the following scores (arranged in chronological order):

72, 74, 72, 76, 73, 64, 80, 80, 73, 63, 63, 69, 82, 67, 69, 85, 68, 70, 67, 70

How can we determine whether this golfer's scores are improving over the years?

It is by no means clear how any of the parametric techniques at our disposal could shed light on this problem. We could, of course, compute the mean and variance of this set of data and perhaps fit a normal curve to it; but then how do we proceed? As it happens the mean score over the 4-year period is 71.85. We observe that each of the last four scores is below this mean and none of the first five scores is. But is this sufficient evidence to conclude that the golfer has tended to improve? How do we explain the two lowest scores (63) about half-way through the sequence? How do we account for the two scores in the 80's during the last eight rounds? More to the point, how can we invoke probability theory to lend its support to any claims we make as we did so carefully when we discussed hypothesis testing? Answers to these questions, some more satisfying than others, will be given in this chapter.

1 THE RUNS TEST

The *Runs Test* is used to help determine whether a sequence (of numbers, letters, or other symbols) could reasonably be viewed as the result of random arrangement or whether the sequence exhibits trends or patterns not attributable to chance. For instance, if a coin is flipped 12 times, we would probably be surprised to find any of the following sequences of results

H, H, H, H, H, H, T, T, T, T, T, T

H, H, T, T, H, H, T, T, H, H, T, T

H, T, H, T, H, T, H, T, H, T, H, T

In each case the patterns are evident. Granted, *some* sequence must occur and all sequences are equally likely, but our intuition tells us that this sort of regularity is unlikely.

With sequences of numbers, the Runs Test can help distinguish these sequences that could be viewed as the result of random sampling from a fixed population from sequences where the terms tend to increase or decrease or oscillate (say from larger to smaller to larger to smaller, and so on). One device for employing the Runs Test on a numerical sequence can be illustrated with the golf scores given earlier. We list the scores again, this time affixing the letter "a" to those terms that lie above the median score (71) and the letter "b" to those terms lying below the median.

72, 74, 72, 76, 73, 64, 80, 80, 73, 63, 63, 69, 82, 67, 69, 85, 68, 70, 67, 70

a a a a a b a a a b b b a b b a b b b b

We then mark off the various *runs* in the sequence of a's and b's, each *run* consisting of a (sub) sequence of like letters flanked by unlike letters or no letter at all. Thus we find first a run of 5 a's, then a run of 1 b, then a run of 3 a's, and so on. There are eight runs in all.

We ask the following question:

If a sequence is formed from 10 a's and 10 b's, where the letters are positioned at random, are there likely to be as few as eight runs?

Now there are two extreme cases:

aaaaaaaaaabbbbbbbbbb (Only two runs)

abababababababababab (20 runs!)

Moreover, each of these extremes can occur in only two ways (the ones shown and the sequences obtained by interchanging the a's and the b's). Though it may not be obvious, there are 18 sequences that contain three runs. (Leave the a's together and put some b's at each end; this gives nine sequences. Then get nine more by reversing the roles of a's and b's.) There are many more sequences yielding four runs and so on. Indeed there are $C(20, 10)$ sequences in all, since any sequence is determined by placing the 10 a's in any of the 20 positions, then filling in the rest with b's. Our problem can be made more explicit:

> What is the probability that a randomly selected sequence of 10 a's and 10 b's contain as few as eight runs?

We can complete the details of such probability questions in cases involving only a few a's and b's. For purposes of making our test more versatile we allow the number of a's to be different from the number of b's. Of course, this would not happen when "a" and "b" represent "above" and "below" the median and there are an even number altogether, but the abstract question of how many runs still makes sense when there are fewer or more a's than b's. Here are some simple cases that might shed some light on the general theory.

Case 1: Sequences of 2 a's and 2 b's.

Sequence	Number of Runs
aabb	2
abba	3
bbaa	2
abab	4
baba	4
baab	3

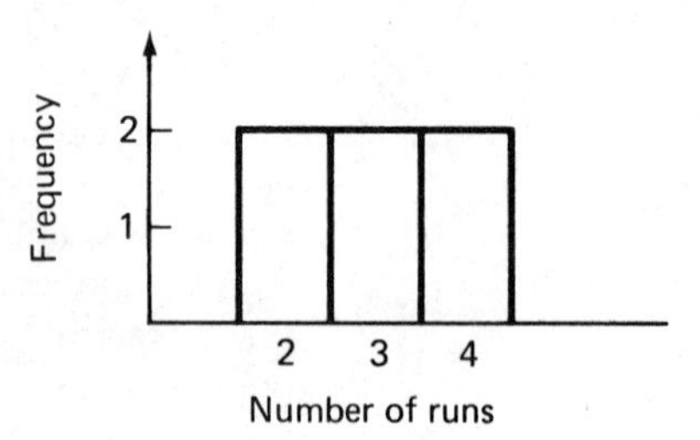

Case 2: Sequences of 3 a's and 2 b's.

Sequence	Number of Runs
aaabb	2
aabba	3
abbaa	3
bbaaa	2
aabab	4
ababa	5
abaab	4
baaba	4
baaab	3
babaa	4

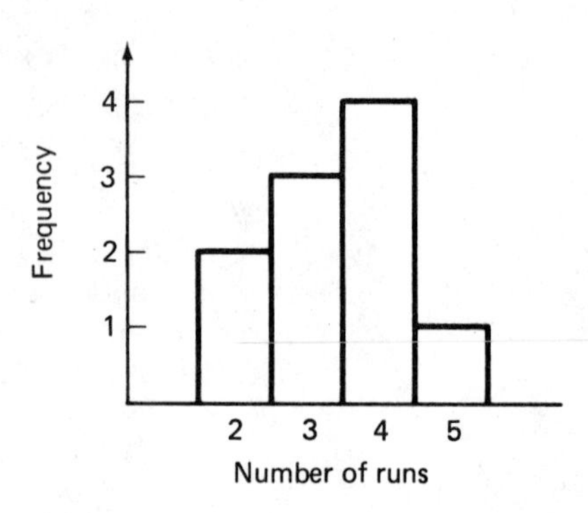

Case 3: Sequences of 3 a's and 3 b's.

Sequence	Number of Runs	Sequence	Number of Runs
aaabbb	2	bbbaaa	2
aababb	4	bbabaa	4
aabbab	4	bbaaba	4
aabbba	3	bbaaab	3
abaabb	4	babbaa	4
ababba	5	babaab	5
abbaba	5	baabab	5
abbbaa	3	baaabb	3
abbaab	4	baabba	4
ababab	6	bababa	6

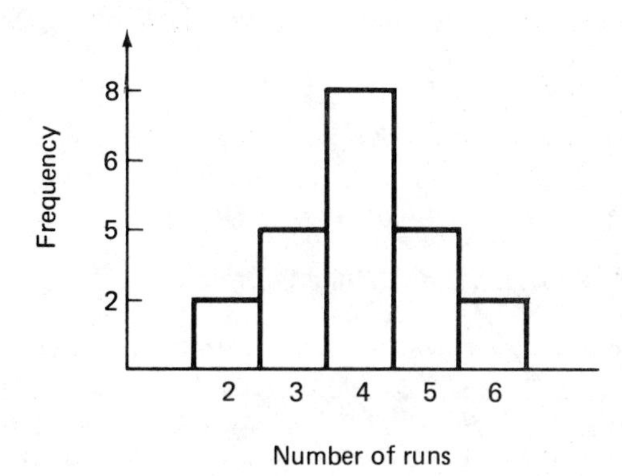

For example, then, we could argue that the probability of getting as few as three runs (i.e., either two runs or three runs) in a randomly selected sequence of 3 a's and 3 b's is

$$\frac{2}{20} + \frac{4}{20} = \frac{2+4}{20} = .3$$

In general, we take advantage of specially designed tables such as Table H in the Appendix. We are usually interested in one of two kinds of questions. In effect we are testing the null hypothesis:

H_0: The sequence at hand represents a random assignment of a's and b's. That is, the n_1 a's are equally likely to fall into *any* n_1 positions of the sequence.

The two types of questions amount to the suspicions that:

1. There are too few (or too many) runs (one-tailed test).
2. There are *either* too few or too many runs, we don't know which (two-tailed test).

Interpretation of Table

Here is a portion of the table corresponding to values $n_1 = 10$ and $n_2 = 10$:

	α	n_1 (10)
	⋮	
	.10	7, 15
n_2 (10)	.05	6, 16
	.01	5, 17

The α represents the probability for the left-handed (or right-handed) critical region. The row corresponding to $\alpha = .10$, for instance, indicates that the left critical region *ends in 7* while the right critical region *begins with 15*. That is,

$$P(\text{Number of runs} \leq 7) \leq .10$$

$$P(\text{Number of runs} \geq 15) \leq .10$$

Similarly the other two categories in the table for $n_1 = n_2 = 10$ inform us that

$$P(\text{Number of runs} \leq 5) = P(\text{Number of runs} \geq 17) \leq .01$$

For the golfer problem, we have eight runs. The logical test to apply is a one-tailed test: We suspect that due to the golfer's improvement over the years, there should be a tendency for *fewer* runs (i.e., there should be more b's clustered early in the sequence, and more a's clustered late in the sequence). But we find that we *cannot* reject the hypothesis of randomness of the sequence even at a 10% level, for that would require seven or fewer runs.

In general then, to test the randomness hypothesis at a 10% level we look for $\alpha = .10$ for a one-tailed test and $\alpha = .05$ for a two-tailed test; the table provides us with the "endpoints" of the critical regions.

EXAMPLE A classic example of the Runs Test involves the conjecture that people arriving at a lunch counter tend to seat themselves apart from other customers. In other words, if a person entered a diner and observed the following seating arrangement

O	O	E	E	E	O	E
1	2	3	4	5	6	7

where O stands for "occupied" and E for "empty," then he would be likely to select seat number 4 placing himself between two empty seats, rather than sit immediately adjacent to another customer. This conjecture could be tested by replacing a's and b's with E's and O's, viewing a seating arrangement at a lunch counter as a sequence of E's and O's. For the sake of argument let us suppose that a particular counter seats 24 people and has 10 customers. From our table we extract

		n_1 (10)
	.10	9, 17
n_2 (14)	.05	8, 17
	.01	6, 19

If our conjecture is correct, we should expect *more* runs than average. Thus we apply a one-tailed test seeking a right-hand critical region. If we find as many as 17 runs, we can reject the randomness hypothesis at a .05 significance level. While if we find as many as 19 runs, we can reject randomness at a .01 significance level.

Another useful application of the Runs Test is in connection with quality control in a factory. If items are sampled periodically as they come off a machine, then their measurements (weights, diameters, tensile strengths, etc.) can comprise a sequence of a's and b's (above and below their medians). Malfunctioning of the machine or carelessness of the operator could produce trends to increase or decrease the measurements, which in turn would lead to *too few runs*. Similarly if there are two shifts working on a production line, one sample from each shift taken each day could yield a sequence of a's and b's that could be used to detect whether one shift generally performed differently than the other. In this case there would be a tendency to alternate a's and b's, thus producing too many runs.

The Runs Test is among the most widely applicable of the non-parametric techniques. But as believers in the Fundamental Law of Economics (there is no such thing as a free lunch!) might suspect, the Runs Test is nearly as weak as it is versatile. A moment's thought would convince you that we give up a lot of information when we convert a number to an "a" (above the median) or a "b" (below the median) in the sense that we ignore the *extent* to which the number is above or below the median. That is, after the conversion, all "a's" look the same and some information is not taken into account. A good case in point is the golf score example where a more powerful test (in Section 3) will reveal that the golfer really has improved!

When we apply the Runs Test to numbers above and below the median, there is the possibility that *some numbers will equal the median*. In that event, we simply discard all such numbers (that equal the median) and apply the test to the remaining sequence of a's and b's. For instance, we would convert

3, 1, 6, 7, 2, 5, 4, 2, 3   (Median = 3)

to

b a a b a a b

Our table is good only for n_1 and n_2 as large as 20. When we want to detect a lack of randomness for larger sets of data, the following theorem can be invoked.

Let X be the number of runs in a sequence containing n_1 a's and n_2 b's. Then for $n_1 \geq 20$ and $n_2 \geq 20$, X is approximately normally distributed with

$$\mu = \frac{n_1 + 2n_1n_2 + n_2}{n_1 + n_2}$$

and

$$\sigma^2 = \frac{2n_1n_2(2n_1n_2 - n_1 - n_2)}{(n_1 + n_2)^2(n_1 + n_2 - 1)}$$

Actually, the approximation is fairly good for values of n_1 and n_2 as small as 10 (see Exercise 8).

Exercises
Section 1

1 What are the maximum and minimum number of runs possible (you should support your answers by listing suitable sequences) in a sequence having
 (a) 4 a's and 4 b's?
 (b) 4 a's and 5 b's?
 (c) 6 a's and 9 b's?
 (d) n_1 a's and n_2 b's, where $n_1 \leq n_2$?

2 A popular restaurant has a waiting line of 22 people, 12 men and 10 women, at their lunch hour. Explain how you would test the conjecture that nearly all the women are accompanied by a man.

3 Without consulting the Runs Test Table, write out a sequence of 12 a's and 13 b's that you think would be randomly arranged. Then determine by the table whether it is.

4 The owner of a furniture factory suspects that his day shift is producing more trestle tables than his night shift. His records for a 2-week period show the following numbers of tables were produced. Use the Runs Test to advise the owner.

Day	Night	Day	Night	...							
26	25	27	30	26	33	24	24	26	31	26	32
		21	31	31	20	28	31	29	32		

5 In a classroom in which you can find three (consecutive) rows with at least six students per row, write down the left-to-right sequence of M's (males) and F's (females) and apply the Runs Test to decide whether there is a tendency for students to sit next to members of the same sex or of the opposite sex.

6 Look in the library for records of average temperature for a particular city during a particular month over a 60-year period. Write down two sequences: the temperatures for the 30 even-numbered years, then the temperatures for the odd-numbered years. Apply the Runs Test to each sequence to decide whether there is a tendency for temperatures to rise over the years.

7 Stand at the entrance to a large classroom and record the sex of students arriving for class to determine whether women tend to arrive earlier than

men. A sample of 25 or 30 should be sufficient, although it might be interesting to try as many as 50 or 100. *Note:* If your school is not coeducational, make up your own problem.

8 **(a)** Find the left-tail critical region corresponding to a .10 significance level for the Runs Test with $n_1 = 10$ and $n_2 = 10$ using the normal approximation and compare this to the value in the table.

(b) Repeat (a) with $n_1 = 20$ and $n_2 = 20$.

2 THE SIGN TEST

The *Sign Test* is used in testing whether paired observations

$$(x_1, y_1), (x_2, y_2), \ldots, (x_n, y_n)$$

are drawn from populations with *equal medians*. The assumptions on the underlying random variables X and Y are limited to their being continuous and symmetric. Continuity helps prevent ties in actual practice; symmetry forces the median to be the mean and guarantees that a randomly selected value of X (or Y) has as much chance being above the mean as it does being below the mean. Thus we shall generally be testing the hypothesis

$$H_0: E(X) = E(Y)$$

Thus we are presenting a non-parametric analogy to the "difference between means" method discussed in Chapter 5. Now, however, we need to know very little about the underlying distributions and nothing about their variances.

EXAMPLE 1 An independent research laboratory desires to compare two types of gasoline to determine whether Brand X gives better mileage than Brand Y. Twenty cars are selected and each car is driven twice over a test track at identical s speeds, once with Brand X and once with Brand Y. We let $\{x_1, \ldots, x_{20}\}$ and $\{y_1, \ldots, y_{20}\}$ be the respective miles per gallon of the 20 cars using Brands X and Y. That is, we let X be the random variable that measures miles per gallon using Brand X and Y the miles per gallon using Brand Y. We define a new random variable U by

$U =$ the number of times we have $x_i > y_i$

Our assumption that the variables X and Y are continuous prevents any *ties* (i.e., we never have $x_i = y_i$, and indeed if our measuring device is extremely precise, we should expect some discrepancy between x_i and y_i for each i). Now we test the hypothesis

$H_0: E(X) = E(Y)$. That is, on the average, we get the same miles per gallon with Brand X as we do with Brand Y.

If H_0 were true, then the variable U would be a discrete, binomially distributed variable with $p = \frac{1}{2}$. We have thus reduced the problem to an examination of the distribution

$$B(u; 20, \tfrac{1}{2})$$

Hence if U is too small or too large, we can reject H_0 at some appropriate significance level.

In practice we have a decision to make as usual about the type of critical region, or whether to use a one- or two-tailed test. If we suspect in advance that Brand X is superior to Brand Y, we would probably elect to use a one-tailed test, rejecting H_0 if U were sufficiently large.

In fact, sometimes it is useful to extend the above technique to test whether the mean of one random variable is greater than another by some constant c. For instance, suppose that prior experience has led to the belief that Brand X yields at least 5 miles per gallon more than Brand Y. We define a variable

$$U' = \text{the number of times } x_i > y_i + 5$$

and seek a right-handed critical region in $B(u'; 20, \frac{1}{2})$ in which we can, at a .10 or .05 significance level, confirm our suspicions.

It should be noted that to apply the Sign Test, the pairs (x_i, y_i) should be independent of one another, so that the variable U can be considered random. In the gas mileage test, for example, we would be at fault in using the *same* car 20 times for the experiment.

Just as we did earlier with the binomial distribution, we frequently use the normal approximation when seeking critical regions for the variable U, particularly when the number of matched pairs (x_i, y_i) is large. Since we have already listed values of $B(x; 20, \frac{1}{2})$ we would not use the normal approximation.

If we had 31 pairs instead, the normal approximation would probably be used.

EXAMPLE 2

In the introduction it was pointed out that some kinds of quantitative data could not be suitably analyzed by parametric methods. In this example we illustrate this comment. Suppose that the student body of a certain college publishes a critique of faculty instruction. In so doing, each student rates each of his instructors on a scale from 1 to 100, where 1 denotes "lousy" and 100 denotes "perfect." The Student Senate meanwhile has been considering two professors for recognition as an outstanding teacher and decides to avail themselves of the ratings in making their final judgment. Accordingly, they select 17 questionnaires in which both Professor X and Professor Y are rated. Letting x_i and y_i be the scores for Professors X and Y, respectively, in these forms, they use the Sign Test, putting

$$U = \text{the number of times } x_i > y_i$$

If U is sufficiently large, Professor X wins the award; if U is sufficiently small, Professor Y is the recipient.

Note that it might be foolish to use an alternate technique where the Senate simply compares the total scores

$$\sum_{i=1}^{17} x_i \quad \text{to} \quad \sum_{i=1}^{17} y_i$$

for if 16 of the 17 students were rather moderate in their praise and condemnation, restricting their ratings of both professors to a range between 60 and 90, then one student could distort the total scores by rating Professor X as "1" and Professor Y as "100." In effect, the Sign Test does away with the absolute scores and reduces each student's opinion to a mere preference between Professors X and Y.

Incidentally, the range of 1 to 100 in the ratings was used to help eliminate ties that would be more likely to occur if the range were 1 to 10 for example. Theoretically, the Sign Test requires that X and Y be continuous, but in situations like Example 2, that restriction can be mildly abused in practice. *When ties do occur,* all tied values are discarded and the sample size is reduced accordingly.

Finally, the Sign Test, like the Runs Test, is very simple to use since the computation involved is minimal. Again we must remark that the Sign Test is not terribly powerful and will not always reject hypotheses that might succumb to techniques that use more of the available information.

Exercises Section 2

1 How big must U be in Example 1 to conclude at a .05 significance level that Brand X gives better mileage than Brand Y? at a .01 significance level?

2 Use the Sign Test to determine at a .05 significance level whether the following data support the conjecture that husbands outweigh wives by at least 25 pounds. Can the significance level be changed to .01?

Husbands' weights (x_i)	Wives' weights (y_i)
120	102
135	107
140	181
140	107
150	120
150	170
155	116
165	124
165	119
170	141
170	131
180	131
180	102
185	125
200	131

3 Repeat Exercise 2, changing the constant from 25 to 35 pounds.

4 A tire manufacturer contends that his Brand X tires outlast his competitor's Brand Y tires by at least 4,000 miles while selling for the same price. Does the following data lend credence to his claim? at what level of significance?

Test Car Number	BRAND X (Mileage in thousands)	BRAND Y (Mileage in thousands)
1	24.6	18.0
2	25.1	22.0
3	37.2	39.0
4	19.1	14.7
5	19.8	15.2
6	23.7	17.5
7	19.0	13.2
8	20.1	15.6
9	21.0	16.2
10	27.2	26.7
11	26.4	21.0
12	30.1	26.0

5 Using the data in Section 10, Chapter 5 on students numbered 1 through 50, see whether the Sign Test can reject the hypothesis that
(a) Freshman GPA's and sophomore GPA's have the same means.
(b) Math SAT and verbal SAT scores have the same means.

6 An instructor argues that his second exam was easier than his first. Should he use the Sign Test to support his claim?

Student	1	2	3	4	5	6	7	8	9	10	11	12	13
Exam 1	61	94	72	71	72	57	64	33	63	56	71	46	46
Exam 2	63	93	84	69	87	68	61	19	78	61	96	53	64

3 THE RANK TEST

The *Rank Test*, like the Sign Test, is used to compare two samples $\{x_i\}$ and $\{y_i\}$, being values of continuous random variables X and Y. Once again the question is whether the two random variables have the same density function, and in particular if their density functions have the same means. Actually there is more than one "Rank Test," but they all have a common starting point, namely, as the name implies, combining the x_i's and the y_i's into one increasing sequence and assigning ranks to each term. Thereafter, one generally resorts to either of two tests: the Wilcoxon T-test (if the samples are small and of the same size) or the Mann-Whitney U-test (if the samples are both at least eight in number). We begin by considering an example using a small sample to get the flavor of the Rank Tests.

Suppose we are given two samples

$$\{x_1, x_2, x_3, x_4\} \quad \text{and} \quad \{y_1, y_2, y_3, y_4\}$$

We arrange these into one sequence whose terms increase in size; omitting subscripts, here are some possible outcomes:

Sequence	Sum of Ranks of y's (T)
$xxyxyyxy$	$3 + 5 + 6 + 8 = 22$
$xyxyxyxy$	$2 + 4 + 6 + 8 = 20$
$xxxxyyyy$	$5 + 6 + 7 + 8 = 26$
$yxyyyxxx$	$1 + 3 + 4 + 5 = 13$
$yxxyxxyy$	$1 + 4 + 7 + 8 = 20$

To each sequence we have assigned a number T, being the sum of the ranks of the y terms. The value of T can range from 10 (when $T = 1 + 2 + 3 + 4$) to 26 (when $T = 5 + 6 + 7 + 8$). As we might suspect, some values of T appear with greater frequency than others. With some effort we could list all possible sequences of 4 x's and 4 y's, compile a histogram of their frequencies, and use that to determine critical regions corresponding to very small or very large T-values. To get a small T-value, the y_i's would have to be, on the whole, smaller than the x_i's, and vice-versa for large T-values. Thus if we suspect that the random variable X has a smaller mean than Y, we could support our argument by taking samples and showing T to be relatively large; if it were large enough, we could reject the hypothesis that X and Y have identical means.

Of course, we do not wish to develop the appropriate histogram each time we care to use the test. There are tables available when the samples are fairly small. They are similar in design to the table for the Runs Test. However, once the samples both contain at least eight terms, we can appeal to the following theorem.

Rank Test Theorem

Let $\{x_1, x_2, \ldots, x_n\}$ and $\{y_1, y_2, \ldots, y_m\}$ be random samples of the variables X and Y. Combine the samples into an increasing sequence

$$z_1, z_2, \ldots, z_{m+n} \qquad (z_1 < z_2 < \cdots < z_{m+n})$$

Let T = sum of the ranks of the y_i's in this sequence and put

$$U = mn + \frac{m(m+1)}{2} - T$$

If X and Y have identical frequency functions, then U is approximately normally distributed with

$$E(U) = \frac{mn}{2} \quad \text{and} \quad \sigma_U^2 = \frac{mn(m+n+1)}{12}$$

Interpretation of Theorem

When T (i.e., the sum of the y-ranks) is relatively small, U will be relatively big and may fall into a right-hand critical region determined by $E(U)$ and σ_U. Likewise as T gets large, U becomes small and could lie in a left-hand critical region.

EXAMPLE 1 A college professor suspects that his 9 o'clock calculus class is superior to his 1 o'clock calculus class. Confident that he gives equally stimulating lectures to both groups and that he covers precisely the same material at both hours, he decides to test his theory by ranking their performances on a common final exam. He compiles the following data:

	Final Exam Scores
9 o'clock (x_i)	37 51 63 71 71 76 79 79 83 85 85 91 93 97 99 99
1 o'clock (y_i)	36 42 48 52 56 68 68 70 78 82 82 90 100

We proceed to rank the combined classes, assigning "1" to the lowest score and "29" to the highest. Then we let T be the sum of the y_i-ranks. A convenient method of doing this is to list the sequence of grades in increasing order, underscoring the y_i's and indicating their ranks:

$$\underline{36},\ 37,\ \underline{42},\ \underline{48},\ 51,\ \underline{52},\ \underline{56},\ 63,\ \underline{68},\ \underline{68},\ \underline{70},\ 71,\ 71,\ 76,$$
(1) (3) (4) (6) (7) (9) (10) (11)

$$\underline{78},\ 79,\ 79,\ \underline{82},\ \underline{82},\ 83,\ 85,\ 85,\ \underline{90},\ 91,\ 93,\ 97,\ 99,\ 99,\ \underline{100}$$
(15) (18) (19) (23) (29)

Thus we get for a sum of y_i-ranks,

$$T = 1 + 3 + 4 + 6 + 7 + 9 + 10 + 11 + 15 + 18 + 19 + 23 + 29$$
$$= 155$$

Now referring to our theorem, we have

Number of x_i's $= n = 16$

Number of y_i's $= m = 13$

$$U = mn + \frac{m(m+1)}{2} - T$$
$$= (16)(13) + \frac{13(14)}{2} - 155$$
$$= 208 + 91 - 155$$
$$= 144$$

We test the hypothesis that both groups of scores are samples of random variables with the same mean by computing

$$E(U) = \frac{mn}{2} = \frac{16 \cdot 13}{2} = 104$$

$$\sigma_U^2 = \frac{mn(m+n+1)}{12} = \frac{16(13)(30)}{12} = 520$$

Thus

$$\sigma = \sqrt{520} \approx 22.8$$

But we now see that

$$P(U \geq 144) = \int_{144}^{\infty} N(x; 104, 22.8)\,dx$$
$$= \int_{1.75}^{\infty} N(z; 0, 1)\,dz$$
$$\approx .0401$$

since our standard unit conversion gives

$$z = \frac{144 - 104}{22.8} \approx 1.75$$

We may then conclude that, since the computed value of U falls into a right-hand critical region of probability .0401, we can reject the hypothesis (that x_i's and y_i's are samples from populations with the same mean) at better than a .05 significance level; we accept the alternative hypothesis that the mean of X is greater than the mean of Y. In short, we conclude that the 9 o'clock class is better.

EXAMPLE 2 Golf Scores Revisited

With a little bit of juggling, we can apply the Rank Test to the problem posed in the section on the Runs Test concerning golf scores. Recall that we were given a sequence of 20 scores recorded in chronological order. We formed a corresponding sequence of a's and b's that looked like

a, a, a, a, a, b, a, a, a, b, a, b, a, b, b, a, b, b, b, b

where a denoted "above median" and b "below median." We form two sets $\{x_i\}$ and $\{y_i\}$ as follows:

x_i = position of ith b

y_i = position of ith a

That is,

y_1 = position of first a = 1

y_2 = position of second a = 2

$y_3 = 3$

$y_4 = 4$

$y_5 = 5$

$y_6 = 7$ (since the sixth a appears in the seventh position)

Now to say that the a's and b's are more or less randomly distributed in the a,b-sequence is tantamount to saying that these two samples $\{x_i\}$ and $\{y_i\}$ could be viewed as values of random variables with identical means. But in fact we have

x_i's: $\{6, 10, 11, 12, 14, 15, 17, 18, 19, 20\}$

y_i's: $\{1, 2, 3, 4, 5, 6, 8, 9, 13, 16\}$

We apply the Rank Test to these samples. We have $m = n = 10$:

$$\begin{aligned} T &= 1 + 2 + 3 + 4 + 5 + 7 + 8 + 9 + 13 + 16 \\ &= 68 \end{aligned}$$

$$\begin{aligned} U &= mn + \frac{m(m+1)}{2} - T \\ &= 100 + 55 - 68 \\ &= 87 \end{aligned}$$

$$E(U) = \frac{mn}{2} = 50$$

$$\sigma_U^2 = \frac{mn(m+n+1)}{12} = \frac{10(10)(21)}{12} = 175$$

Thus

$$\sigma = \sqrt{175} \approx 13.23$$

Converting to standard units, we find

$$z = \frac{87 - 50}{13.23} = \frac{37}{13.23} \approx 2.87$$

and

$$P(U \geq 87) = \int_{2.87}^{\infty} N(z; 0, 1)\, dz \approx .0021$$

We therefore conclude, at more than a .01 significance level, that the $\{x_i\}$ and $\{y_i\}$ come from random variables X and Y with mean of Y less than mean of X. Going back to the a's and b's, we conclude that the positions of the a's are significantly *earlier* than the positions of the b's. In short, the sequence of a's and b's is far from random.

The reader is apt to recall that the Runs Test did not reject the randomness hypothesis even at a .10 significance level. Rather than feeling disturbed by this drastic discrepancy between the results of two different tests, the reader should take heart in the fact that generally the Rank Tests are more likely to reject a hypothesis than the Runs Test in those situations where both tests are applicable.

When ties occur, each of the tied observations is assigned the mean of the ranks which they jointly occupy. Thus for example if we have

X-values = $\{13, 14, 15, 15, 16\}$

Y-values = $\{14, 15, 17, 18\}$

we would convert to

$$\begin{array}{ccccccccc} 13, & 14, & \underline{14,} & 15, & \underline{15,} & 15, & 16, & \underline{17,} & \underline{18} \\ & & (2.5) & & (5) & & & (8) & (9) \end{array}$$

Notice that it wouldn't matter whether we view the first or the second "14" to be the Y-value since both are assigned the rank of 2.5. Similarly all three of the "15's" are assigned the rank of 5. When there are numerous ties, "correction factors" are available in the literature.

Summary

In these brief sections we have introduced three types of non-parametric tests. Their common features are simplicity of computation and generality of applicability. For small samples, one must resort to tables such as the Runs Test Table in the Appendix in the case of the Runs Test or the Rank Test. The Sign Test relies only on the binomial distribution for all sample sizes.

As a general rule when the underlying distribution (or pair of distributions) is known, one should attempt to use techniques such as estimation of parameters, hypothesis testing and the χ^2 test for goodness-of-fit because the results are apt to be more conclusive. Yet these non-parametric techniques, along with several others that can be found in the literature, can be extremely useful as an initial step in the analysis of data or as a suitable alternative when little is known of the underlying distributions.

Exercises Section 3

1 Redo Example 1 but modify the scores by trading the "36" in the afternoon class for the "97" in the morning class.

2 Refer to the data of Exercise 2 of the preceding section to get two sequences: the x_i's as they stand, and a sequence of y_i's obtained by adding 25 to each of the y_i's. Now use the Rank Test to determine whether husbands as a group weigh an average of at least 25 pounds more than wives.

3 Repeat Exercise 2 using 35 pounds instead of 25 pounds.

4 Use the Rank Test on Exercise 4 of the preceding section.

5 The following well-known formula can be used to help examine the variables U and T in the Rank Test Theorem.

$$1 + 2 + \cdots + k = \frac{k(k+1)}{2}, \qquad \text{for any positive integer } k$$

(a) What is the smallest value that T can assume in the theorem?

(b) What is the value of U when T has this smallest value?

(c) What is the largest value that T can assume? [*Hint:* $(5 + 6 + 7) = (1 + 2 + 3 + 4 + 5 + 6 + 7) - (1 + 2 + 3 + 4)$.]

(d) What is the value of U when T has this largest value?

6 Redo Exercise 6 of the preceding section using the Rank Test.

7 Test the hypothesis that there is no significant difference between freshman GPA and sophomore GPA by using the grades of the students numbered 1 through 20 in the table in Chapter 5, Section 10.

4 RANK CORRELATION

In this section we present two non-parametric analogues to the correlation coefficient discussed in Chapter 6. The first method uses a test statistic called the *Spearman* ρ ("rho") which evidently was used by Sir Francis Galton. The second statistic is called the *Kendall* τ ("tau"). In many respects the two statistics yield comparable results, yet ρ is somewhat easier to compute whereas τ has a slight edge in certain problems of statistical inference. Once the sample size is as large as 10, both variables can be converted to easily manageable variables: ρ converts to a t-variable and τ converts to a normal variable.

The context in each case can best be explained by reexamination of a problem we discussed in Chapter 6.

EXAMPLE 1 High School Rank-in-Class Revisited

Table 7.1 contains the class sizes (X-values) and the class standings (Y-values) for 20 college freshmen. In addition, we have ordered the class sizes from largest to smallest and assigned ranks from 1 to 20. We have also listed the ranks (in decreasing order) of the 20 class standings.

Table 7.1

Class Size	Rank	Class Standing	Rank
914	1	112	6
736	2	141	2
641	3	81	11
591	4	154	1
537	5	137	3
336	6	83	10
330	7	114	5
264	8	33	16
222	9	130	4
221	10	31	17
205	11	84	9
195	12	24	19
180	13	42	13
176	14	58	12
174	15	34	15
162	16	40	14
161	17	110	7
160	18	85	8
154	19	30	18
54	20	10	20

Computation of the Spearman ρ:

> ρ is simply the usual correlation coefficient computed on the *ranks* rather than on the original values.

However, since the ranks will always take on values

$1, 2, 3, \ldots, n$

it turns out that the formula for the correlation coefficient simplifies to

$$\rho = 1 - \frac{6 \Sigma (x_i' - y_i')^2}{n(n^2 - 1)}$$

where (x_i', y_i') are the *ranks* of the original data pairs (x_i, y_i). Taking the differences of the values in the second and fourth columns of Table 7.1, squaring them, and adding yields

$$\begin{aligned}\sum_{i=1}^{20} (x_i' - y_i')^2 &= (1-6)^2 + (2-2)^2 + (3-11)^2 + (4-1)^2 + (5-3)^2 \\ &\quad + (6-10)^2 + (7-5)^2 + (8-16)^2 + (9-4)^2 \\ &\quad + (10-17)^2 + (11-9)^2 + (12-19)^2 + (13-13)^2 \\ &\quad + (14-12)^2 + (15-15)^2 + (16-14)^2 + (17-7)^2 \\ &\quad + (18-8)^2 + (19-18)^2 + (20-20)^2 \\ &= 25 + 0 + 64 + 9 + 4 + 16 + 4 + 64 + 25 + 49 + 4 \\ &\quad + 49 + 0 + 4 + 0 + 4 + 100 + 100 + 1 + 0 \\ &= 522\end{aligned}$$

Thus

$$\begin{aligned}\rho &= 1 - \frac{6(522)}{20(400-1)} \\ &\approx 1 - .393 \\ &= .617\end{aligned}$$

Interpretation of the Spearman ρ:

When $n \geq 10$, the variable $t = [\rho/\sqrt{(1-\rho^2)/(n-2)}]$ is approximately distributed as a t-variable with $(n-2)$ degrees of freedom *if there is no relationship* between the original variables X and Y.

In other words, we can test the hypothesis

H_0: X and Y are not related

by computing the value

$$t = \frac{.617}{\sqrt{(1 - .617)^2/(20 - 2)}}$$
$$\approx \frac{.617}{.0344}$$
$$\approx 18.0$$

With 18 degrees of freedom the value $t = 18.0$ is significant at better than a .0005 level. We conclude that the original variables X (class size) and Y (class standing) *are* related, i.e., there is some (positive) correlation between them.

Remark 1 Once again it is common to find the correlation coefficient squared and the result used as a measure of relationship between X and Y. In the above example, we would say that since

$$\rho^2 = (.617)^2 \approx .38$$

it follows that 38% of the variation of Y is explained by its relationship with X (or vice versa). Likewise, when one is interested in comparing two or more relationships, say between X and Y as well as between X and Z, the squares of the respective correlation coefficients are generally compared rather than the coefficients themselves.

Remark 2 When $n < 10$, the approximation to the t-variable should not be used. Instead one must appeal to tables that can be found in books specializing in non-parametric methods.

Remark 3 (Ties) Quite often when assigning ranks to data there will be ties. When this occurs, we simply assign the *average rank* to each tied value and proceed as before. For instance we might have

X	Rank	Y	Rank
13	1.5	23	4
13	1.5	28	1
12	3	23	4
11	4	26	2
10	5	23	4

When there are only a relatively few number of ties, no problem arises from this averaging procedure. There are "correction" techniques available for situations that call for more precision. Remember, however, that the whole spirit of non-parametric methods is such that extensive refinement could defeat one advantage (ease and simplicity of computation) of these procedures.

Let us turn to another correlation statistic, the Kendall τ. Again we shall use Table 7.1, but now we compute an intermediate value S from the sequence of Y-ranks

6, 2, 11, 1, 3, 10, 5, 16, 4, 17, 9, 19, 13, 12, 15, 14, 7, 8, 18, 20

$$S = \sum_{k=1}^{20} \left[\begin{Bmatrix} \text{the number of} \\ \text{terms after } k \\ \text{that exceed } k \end{Bmatrix} - \begin{Bmatrix} \text{the number of terms} \\ \text{after } k \text{ that are less} \\ \text{than } k \end{Bmatrix} \right]$$

{There are 14 terms after 6 that exceed 6 and 5 terms after 6 that are less than 6.}

$$= (14 - 5) + (17 - 1) + (9 - 8) + (16 - 0) + (15 - 0) + (9 - 5) + (12 - 1) + (4 - 8) + (11 - 0) + (3 - 7) + (7 - 2) + (1 - 7) + (4 - 3) + (4 - 2) + (2 - 3) + (2 - 2) + (3 - 0) + (2 - 0) + (1 - 0) + (0 - 0)$$

$$= 9 + 16 + 1 + 16 + 15 + 4 + 11 - 4 + 11 - 4 + 5 - 6 + 1 + 2 - 1 + 0 + 3 + 2 + 1 + 0$$

$$= 97 - 15$$

$$= 82$$

Now just what does this value of S measure? Well, if the Y-ranks were in perfect correlation with the X-ranks, the sequence would be

1, 2, 3, 4, . . . , 20

and S would be

$$S = (19 - 0) + (18 - 0) + (17 - 0) + \cdots + (0 - 0)$$

and this would be the maximum possible value of S. Incidentally, there is a useful formula (see the exercises in Section 1, Chapter 4) that says

$$1 + 2 + \cdots + n = \frac{n(n + 1)}{2}$$

for any positive integer n. It follows that the maximum possible value for S is

$$\text{Maximum } S = 1 + 2 + \cdots + 19 = \frac{19(20)}{2} = 190$$

Similarly if the Y-ranks were in perfect (inverse) correlation with the X-ranks, then the sequence would be

20, 19, 18, 17, . . . , 1

and we would get

$$S = (0 - 19) + (0 - 18) + (0 - 17) + \cdots + (0 - 0)$$

$$= -(19 + 18 + 17 + \cdots + 1)$$

$$= -190 \text{ (the minimum possible value of } S)$$

Notice that if we divide S by 190, we always get a value between -1 and $+1$, and that these extreme values correspond to perfect inverse and perfect direct correlation, respectively. In essence, then, S simply measures the extent of disarray of the Y-rank sequence, and we form the statistic

$$\tau = \frac{S}{\frac{1}{2}n(n-1)}$$

which is called the *Kendall τ correlation coefficient.* As with the Spearman ρ, we have a convenient conversion:

When $n \geq 10$, the variable τ itself is approximately normally distributed with

$$\mu_\tau = 0$$

$$\sigma_\tau = \sqrt{\frac{2(2n+5)}{9n(n-1)}}$$

if there is no relationship between the original values of X and Y. Let us complete our analysis of Table 1. Since $n = 20$, we have (assuming X and Y are independent)

$$\sigma_\tau = \sqrt{\frac{2(40+5)}{180(19)}}$$

$$= \sqrt{\frac{1}{38}}$$

$$\approx .162$$

Our value of τ is

$$\tau = \frac{S}{\frac{1}{2}(n)(n-1)} = \frac{82}{190} \approx .432$$

Converting to standard units, we can check that

$$P(|\tau| \geq .432) = P(|z| \geq \tfrac{.432}{.162})$$

$$\approx .0070$$

And again we can conclude at better than a .01 significance level that X and Y are related.

Remark 4 *Ties* are dealt with as before by assigning average ranks to tied scores. Once again, when more precision is desired, there are "correction factor" methods available when ties occur. Also for small samples ($n < 10$) there are tables that should be used rather than the normal approximation.

As you may have observed, the value of τ, .432, is considerably smaller than the value of ρ, .617, for the same data. Moreover, their respective squares

$$\tau^2 \approx .19 \quad \text{and} \quad \rho^2 \approx .38$$

are different enough to cause us to question the usual interpretation of explained variance of Y due to its relationship to X. Recall however that the Pearson product-moment correlation in this same problem was

$r = .658$ (See Exercise 3, Section 2, Chapter 6.)

and we get

$$r^2 \approx .43$$

This value is respectively close to ρ^2, and in general this will be the case: When there is fairly conclusive correlation between X and Y, all three tests (using r, ρ, and τ) will be decisive. Moreover, in general, the r^2-value will be the largest, followed closely by ρ^2 and then by τ^2. Remember that all three statistics (r, ρ, and τ) allowed us to reject the hypothesis of no correlation at fairly extreme significance levels (.01 or smaller) and, in that sense, the three tests lead to the same conclusion.

Summary

The Spearman ρ and the Kendall τ are non-parametric analogues of the Pearson product-moment correlation coefficient. Both statistics are computed by first ranking the X-values and the Y-values from a sample

$$(x_1, y_1), (x_2, y_2), \ldots, (x_n, y_n)$$

of n paired observations. In general, ρ is slightly easier to compute, although hand computations are manageable for both τ and ρ when n is between 10 and 25. When n is less than 10, suitable tables must be consulted; otherwise both τ and ρ can be approximated by more familiar distributions (the normal distribution for τ and the t-distribution for ρ).

Incidentally, whereas the Pearson r is used to detect only a *linear* relationship between X and Y, both ρ and τ will detect a more general kind of functional relationship. Indeed, if Y is any *monotonic function* of X (i.e., either Y increases as X increases or Y decreases as X increases), then the respective ranks of X-samples and Y-samples will reflect this relationship just as well as if Y were a linear function of X. Thus, for example, if $Y \approx X^2$, $Y \approx \text{Log } X$ or $Y \approx \exp(-2X)$ for positive sample values, the ρ and τ values would indicate a significant correlation, whereas r would probably not.

Exercises Section 4

1 Find the values of ρ and τ for each of the following pairs of variables for the first 20 students in the list in Section 1 of Chapter 6.

(a) X = math SAT, Y = freshman GPA
(b) X = freshman GPA, Y = sophomore GPA
(c) X = class size, Y = math SAT
(d) X = class rank, Y = math SAT

2 **(Critics' choice.)** Two movie critics were asked to rank 12 movies in order of preference. Here are the results:

	Movie											
	1	2	3	4	5	6	7	8	9	10	11	12
Critic 1	3	9	12	8	1	4	7	10	11	2	6	5
Critic 2	3	11	12	7	2	1	4	5	9	6	10	8

(a) Compute both ρ and τ to see whether the critics tend to agree at all or whether they have entirely different standards of comparison.
(b) Draw a scatter diagram of this data letting x be the rank of critic 1.

3 A class of 15 students got the following scores on the first two hour exams and a 5-minute quiz in a mathematics course. Do the scores indicate a high degree of relative consistency in performance
(a) between the exams?
(b) between exam 1 and the quiz?

Exam 1	33	11	63	93	84	69	87	68	96	53	64	85	61	78	61
Exam 2	19	3	61	94	72	71	72	57	71	46	46	63	56	63	64
Quiz	0	3	3	3	2	3	4	5	5	4	3	0	5	5	1

4 According to the 1969 Statistical Handbook of Japan, the following table summarizes 12 categories of books published in 1950 and 1966. Use both ρ and τ to determine whether there was a significant shift in reading interest during that period

Subject	1950 Publications	1966 Publications
Official publications	3,804	3,528
Generalities	125	435
Philosophy	682	879
History	450	1,200
Social Science	1,788	3,424
Natural Science	1,035	1,248
Technology	715	1,768
Business	607	1,201
Arts	571	966
Psychology	677	574
Literature	2,118	2,708
Juvenile	2,433	559

5 A student course evaluation provides us with three random variables on the sample space of all courses surveyed.

X = percentage of students who feel their instructor is "one of the best"
Y = percentage of students who feel their instructor is either "one of the best" or "better than most"
Z = percentage of students who would "recommend this course to a friend"

In a sample of courses, we have the following:

Course	X	Y	Z
1	0	50	77
2	0	66	100
3	11	56	93
4	11	55	87
5	88	99	96
6	93	99	100
7	0	16	42
8	25	63	74
9	0	3	31
10	76	95	87
11	71	97	93
12	59	96	71
13	15	41	32
14	50	88	94
15	0	40	89
16	0	11	11
17	83	100	100
18	83	100	100
19	43	76	83
20	8	32	44
21	0	15	38
22	46	88	77
23	28	84	86
24	8	91	83
25	31	73	90

Which of the following pairs of variables are most highly correlated, and is there significant correlation?

(a) X and Z

(b) Y and Z

Answers to Selected Exercises

Chapter 1
Section 1

1

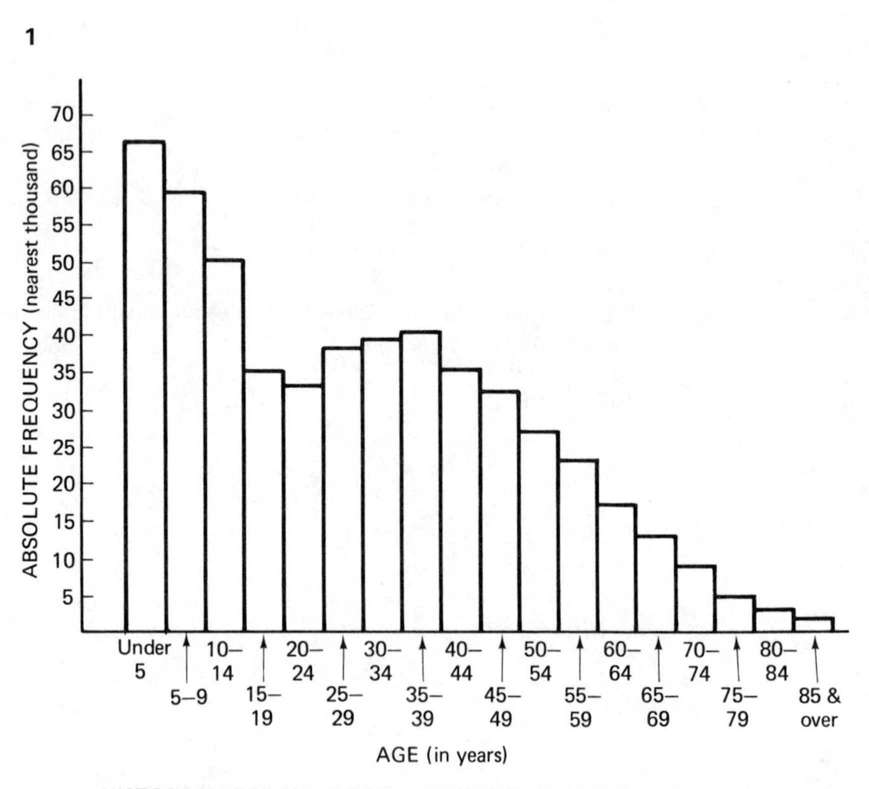

HISTOGRAM FOR MALE POPULATION IN DALLAS, TEXAS IN 1959

3 **(a)** Not able to answer
(b) 18.8%
5 **(a)** .8%
(b) 53.6%
7

Hours of Credit	Number of Students (in thousands) 4-year Schools	2-year Schools
1–3	17	11
4–6	18	8
7–9	11	4
10–12	23	6
13–15	61	10
16–18	57	7
19 or more	8	6
TOTAL	195	52
Rounded off	193	47

9 **(a)**

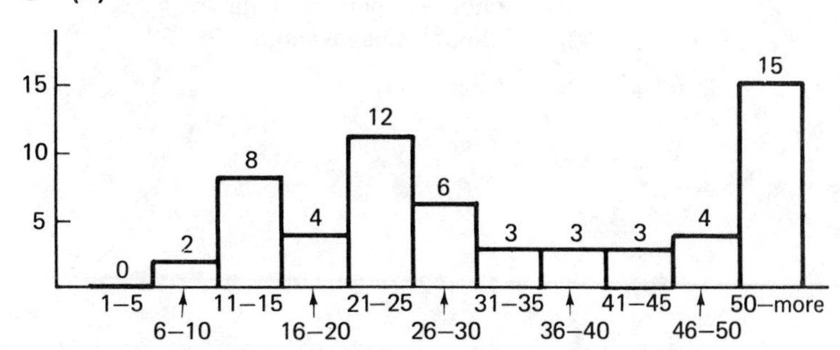

(b)

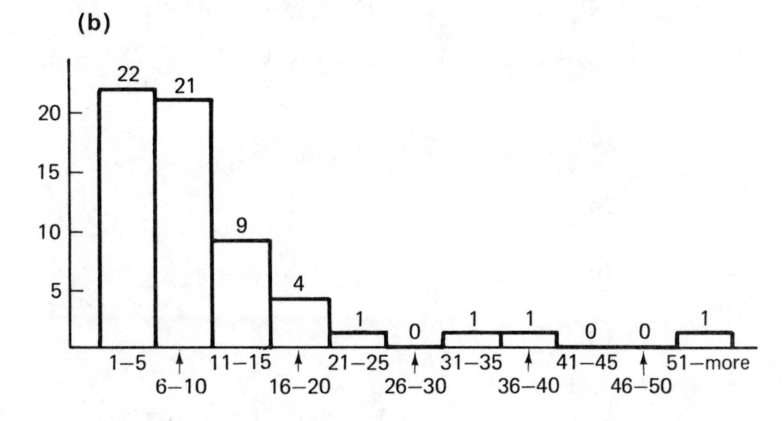

Chapter 1
Section 2

1 **(a)** Mean = 2, mode = any value (1, 2, or 3), median = 2, var = $\frac{2}{3}$, st dev $\approx$ .816
(c) Mean = 0, mode = any value (-1, 0, or 1), median = 0, var = $\frac{2}{3}$, st dev $\approx$ .816

(e) Mean = 0, mode = any value (−200, 0, or 200), median = 0, var = 26666.7, st dev ≈ 163.0

(g) Mean = −1, mode = −1, median = −1, var = 0, st dev = 0

3 $x = 1$, $y = 2$ work, as do infinitely many other choices, such as (0, 3), (−1, 4), (−2, 5), etc.

5 **(a)** $\{1, 1\}$ *only* possible set

(c) $\{3, -1\}$ *only* possible set

9 Mean = 3.55

11 **(a)** Mean $= \dfrac{x + y}{2}$, var $= \dfrac{(x - y)^2}{4}$

(c) Mean = 0, var $= \dfrac{(x - y)^2}{4}$

13 **(a)** Pro winners' average = 116.4
Pro losers' average = 102.7
College winners' average = 89.67
College losers' average = 72.13
High School winners' average = 74.33
High School losers' average = 63.53

Chapter 1
Section 3

1 **(a)** $\sum_{i=1}^{3} i^3 = 1^3 + 2^3 + 3^3 = 1 + 8 + 27 = 36$

(c) $\sum_{k=1}^{8} k^3 - \sum_{i=1}^{7} i^3 = 8^3 = 512$

(e) $\sum_{i=1}^{4} (-1)^i \frac{1}{i} = -1 + \frac{1}{2} - \frac{1}{3} + \frac{1}{4} = -\frac{7}{12}$

(g) $\sum_{k=2}^{5} \frac{1}{k} = \frac{1}{2} + \frac{1}{3} + \frac{1}{4} + \frac{1}{5} = \frac{77}{60}$

2 **(a)** $\sum_{k=1}^{6} 2k$

(b) $\sum_{k=1}^{5} \sqrt{k}$

(d) $\sum_{i=1}^{5} (-1)^{i+1} i$

3 $x_1 + x_2$

7
$$\sum_{i=1}^{2} (x_i - \bar{x}) = \left(x_1 - \frac{x_1 + x_2}{2}\right) + \left(x_2 - \frac{x_1 + x_2}{2}\right)$$
$$= \frac{2x_1 - x_1 - x_2}{2} + \frac{2x_2 - x_1 - x_2}{2}$$
$$= 0$$

Chapter 1
Section 4

1 (a)

Step	Box	P	L	N	A
1	Start				
2	1	1	1		
3	2	1	1	4	
4	3	1	1	4	1
5	4	1	1	4	1
6	5	– – True – –			
7	7	1	2	4	1
8	3	1	2	4	2
9	4	2	2	4	2
10	5	– – True – –			
11	7	2	3	4	2
12	3	2	3	4	3
13	4	6	3	4	3
14	5	– – True – –			
15	7	6	4	4	3
16	3	6	4	4	4
17	4	24	4	4	4
18	5	– – False – –			
19	6	Computer writes 24			
20	Stop				

(c)

Step	Box	P	L	N	A
1	Start				
2	1	1	1		
3	2	1	1	3	
4	3	1	1	3	5
5	4	5	1	3	5
6	5	– – True – –			
7	7	5	2	3	5
8	3	5	2	3	1
9	4	5	2	3	1
10	5	– – True – –			
11	7	5	3	3	1
12	3	5	3	3	0
13	4	0	3	3	0
14	5	– – False – –			
15	6	Computer writes 0			
16	Stop				

3 If $A = 17$, then $B = 71$.
If $A = 34$, then $B = 43$.
If $A = 20$, then $B = 2$ (=02).
The flow chart **reverses the digits** of a two-digit number.

5

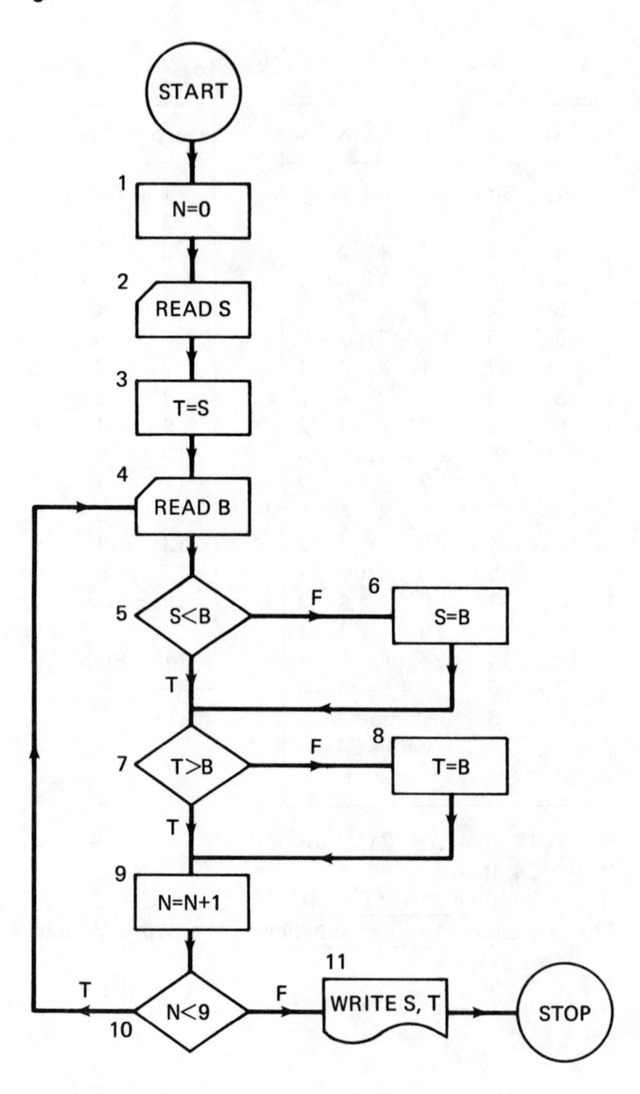
START
1
N=0
2
READ S
3
T=S
4
READ B
5
S<B
F
6
S=B
T
7
T>B
F
8
T=B
T
9
N=N+1
T
N<9
10
F
11
WRITE S, T
STOP

7

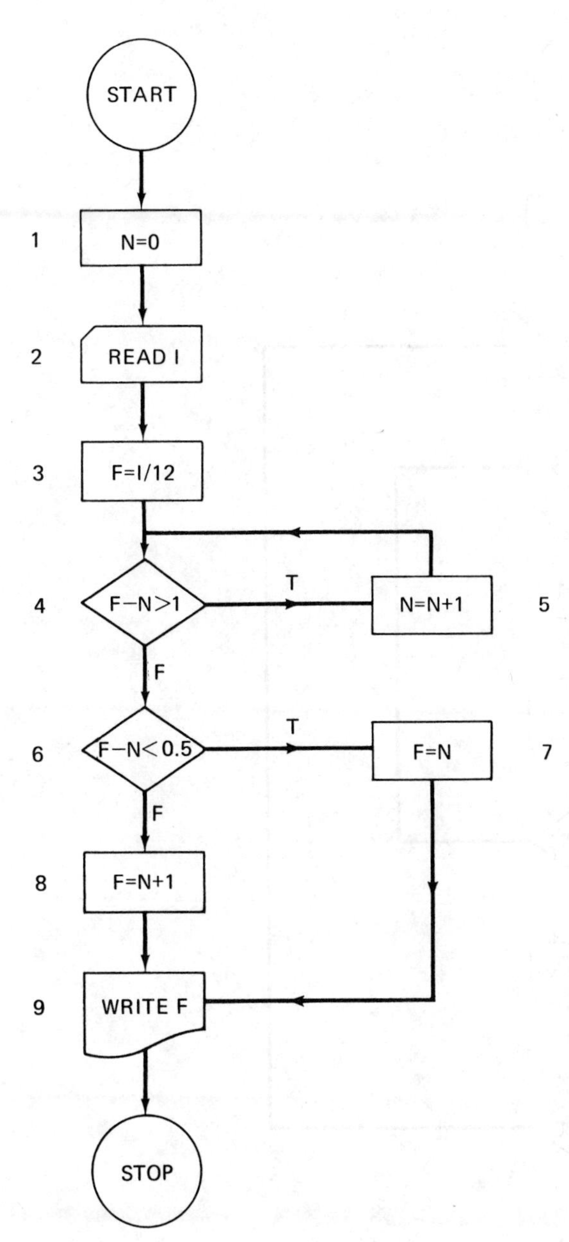
START
1
N=0
2
READ I
3
F=I/12
4
F−N>1
T
N=N+1
5
F
6
F−N<0.5
T
F=N
7
F
8
F=N+1
9
WRITE F
STOP

9

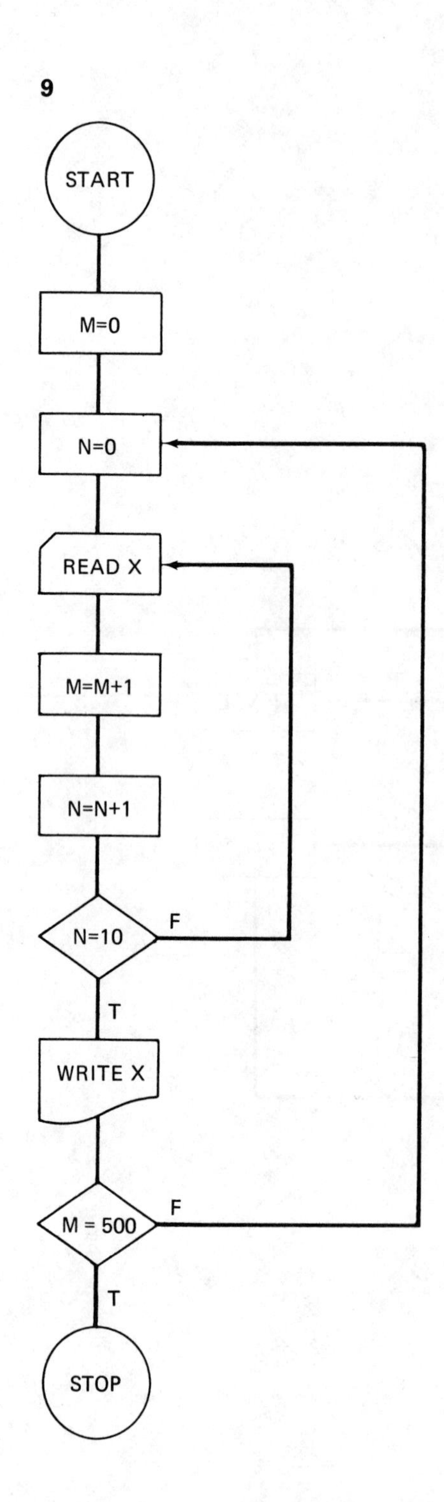

Chapter 2
Section 1

1

Sum of Both Faces	2	3	4	5	6	7
Relative Frequency	.0208	.0407	.0834	.1390	.1390	.1806

Sum of Both Faces	8	9	10	11	12
Relative Frequency	.1251	.1041	.0834	.0486	.0347

3 **(a)** $2^5 = 32$, 2^n
(b)

Number of Heads	5	4	3		2		1	0
Outcomes	HHHHH	HHHHT	HHHTT	HTHTH	HHTTT	TTHHT	HTTTT	TTTTT
		HHHTH	HHTHT	THHTH	HTHTT	HTTTH	THTTT	
		HHTHH	HTHHT	HTTHH	THHTT	THTTH	TTHTT	
		HTHHH	THHHT	THTHH	HTTHT	TTHTH	TTTHT	
		THHHH	HHTTH	TTHHH	THTHT	TTTHH	TTTTH	

(c)

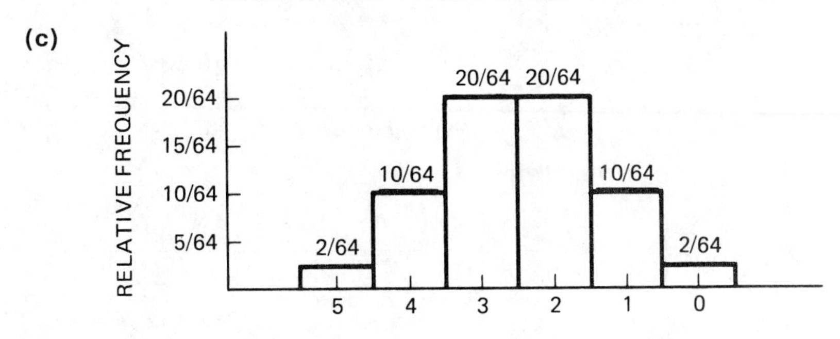

5 **(a)**

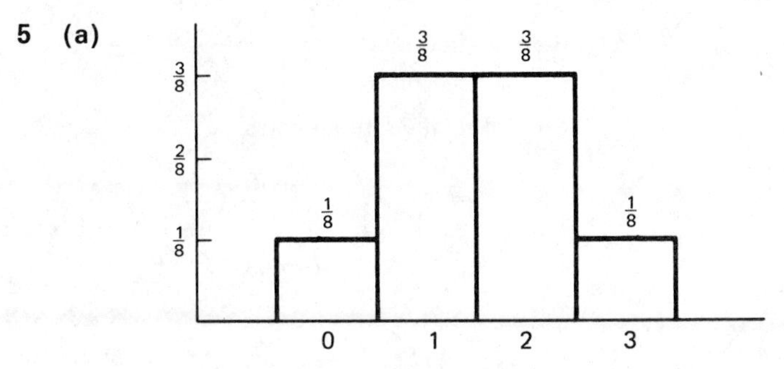

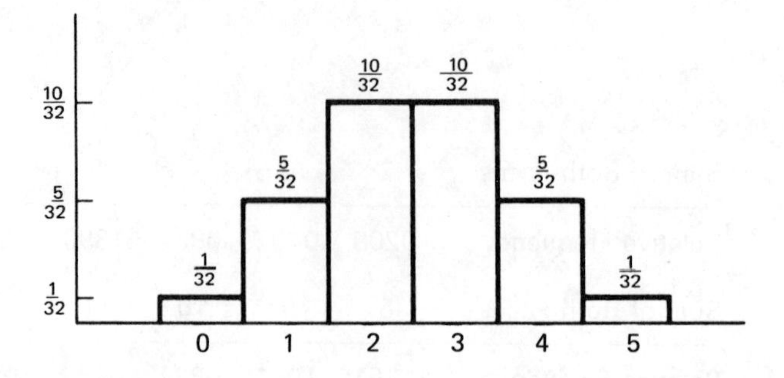

7 (a) 3.625

(b) 3.500

9 (a) $4^3 = 64$

(b) Loss $= 2.40 - 3.20 = -80$ cents in 64 plays

11 (a)

X	2	3	4	5	6	7	8	9
Relative Frequency	$\frac{1}{20}$	$\frac{2}{20}$	$\frac{3}{20}$	$\frac{4}{20}$	$\frac{4}{20}$	$\frac{3}{20}$	$\frac{2}{20}$	$\frac{1}{20}$

(c)

X	2	4	6	8	10	12	14	16
Relative Frequency	$\frac{1}{20}$	$\frac{2}{20}$	$\frac{3}{20}$	$\frac{4}{20}$	$\frac{4}{20}$	$\frac{3}{20}$	$\frac{2}{20}$	$\frac{1}{20}$

13 (a)

Data Point	X_1	X_2	X_3	. . .	X_n
Relative Frequency	$\frac{1}{n}$	$\frac{1}{n}$	$\frac{1}{n}$		$\frac{1}{n}$

(b) Relative frequency of $X_i = \dfrac{f_i}{f_1 + f_2 + \cdots + f_n}$

Sum of relative frequencies

$$= \frac{f_1}{f_1 + \cdots + f_n} + \frac{f_2}{f_1 + \cdots + f_n} + \cdots + \frac{f_n}{f_1 + \cdots + f_n}$$

$$= \frac{f_1 + f_2 + \cdots + f_n}{f_1 + f_2 + \cdots + f_n}$$

15 (a) Mean $= \dfrac{X_1 \cdot f_1 + X_2 \cdot f_2 + \cdots + X_n \cdot f_n}{f_1 + f_2 + \cdots + f_n}$

(b) Mean $= X_1 P_1 + X_2 P_2 + \cdots + X_n P_n$

Chapter 2
Section 2

1 (a) Continuous, 50 to 120

(c) Discrete, 0 to 4 million

(e) Discrete, 32 to 240

(g) Discrete, 0 to 10

(i) Discrete (VHF), 2 to 13

2 (a) Not equally likely
(b) Not equally likely
(c) Not equally likely

Chapter 2
Section 3

1 (a) $S = \{$HHH; HHT, HTH, THH; HTT, THT, TTH; TTT$\}$
(c) $A = \{$HHH, HHT, HTH, HTT$\}$
(e) $A \cap B = \{$HHH, HHT, HTH$\}$
$A \cup B = \{$HHH, HHT, HTH, HTT, THH$\}$
$A - B = \{$HTT$\}$
$B - A = \{$THH$\}$
$A^c = \{$THH, TTH, THT, TTT$\}$

3 (a) $P(A) = \frac{1}{6}$
(c) $\frac{1}{18}$
(e) $\frac{10}{36}$
(g) $\frac{5}{6}$

5

		Minimum	Maximum
(a)	$P(A \cap B)$	0	$\frac{1}{4}$
(c)	$P(S - (A \cup B))$	$\frac{5}{12}$	$\frac{2}{3}$

7 (a) $\frac{31}{365}$
(c) $\frac{11}{365}$
(e) $\frac{62}{365}$

9 (a) $A \cap B$
(c) $A \cup B$

Chapter 2
Section 4

1 $P(B) = \frac{3}{8}$, $P(A \cap B) = \frac{1}{8}$

3 (a) (1) $P(Y_2|A)$ means the probability that an A student is a sophomore.
(3) $P(M|\mathrm{D})$ means the probability that a D student is a male.
(5) $P(W \cup Y_3)$ means the probability of a student being either a woman or a junior.

b) (1) $P(Y_3|C)$
(3) $P(Y_4|M)$
(5) $P(Y_3|M \cap B)$

5

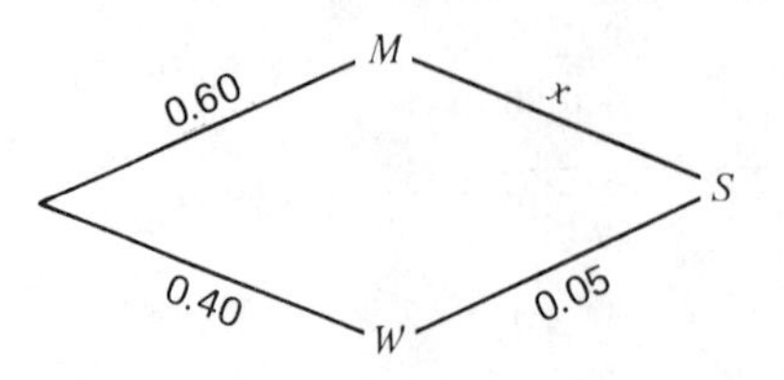

Let $x = P(\text{sleep}|\text{man})$
$P(S) = .10$, but
$P(S) = (.6)(x) + (.4)(.05) = .6x + .02$
$.10 = .6x + .02$
$x = \frac{8}{6}$

$$P(M|S) = \frac{(.6)(\frac{8}{6})}{.10} = .8$$

7

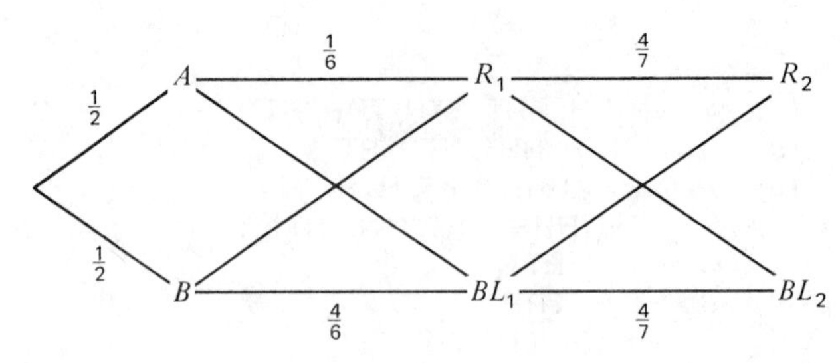

$P(A|R_2) = \frac{19}{39}$ $P(BL_2) = \frac{15}{28}$

9 $\frac{1}{3}$
11 $\frac{14}{17}$
13 (a) 1.9%
(c) 17%

Chapter 2
Section 5

3 (a) $P(B_2) = \frac{1}{6}$, $P(S_6) = \frac{5}{36}$, $P(B_2 \cap S_6) = \frac{1}{36}$; dependent
(c) $P(R_2 \cup B_2) = \frac{11}{36}$, $P(S_7) = \frac{6}{36}$, $P((R_2 \cup B_2) \cap S_7) = \frac{2}{36}$; dependent
(e) $P(R_1 \cap B_1) = \frac{1}{36}$, $P(S_2) = \frac{1}{36}$, $P((R_1 \cap B_1) \cap S_2) = \frac{1}{36}$; dependent
5 P (none of them gets an A) = .07
P (one A) = .38
P (two A's) = .43
P (three A's) = .12
7 Adding one dot outside $A \cup B$ will work; deleting one dot from $B - A$ will work.
9 $P(B|A) = \frac{P(B)P(A|B)}{P(A)} = \frac{P(B)P(A)}{P(A)} = P(B)$, if $P(A) \neq 0$

Chapter 2
Section 6

1 **(a)** 210
(c) 35
(e) 20
(g) 28
3 **(a)** .09
(b) .096

5 $P\text{ (no 2's)} = \dfrac{C(48, 13)}{C(52, 13)} = \dfrac{39 \cdot 38 \cdot 37 \cdot 36}{52 \cdot 51 \cdot 50 \cdot 49} = \dfrac{12,654}{41,650} \approx .304$

$P\text{ (no 3's)} \approx .304$

$P\text{ (no 2's and no 3's)} = \dfrac{C(44, 13)}{C(52, 13)} = \dfrac{15,366}{189,175} \approx .081$

Events are *dependent*.

7 $1 - \dfrac{C(5, 5)}{C(6, 5)} = \dfrac{5}{6}$

9 The following flow chart, when properly translated (e.g., don't divide in integer mode in FORTRAN) will work.

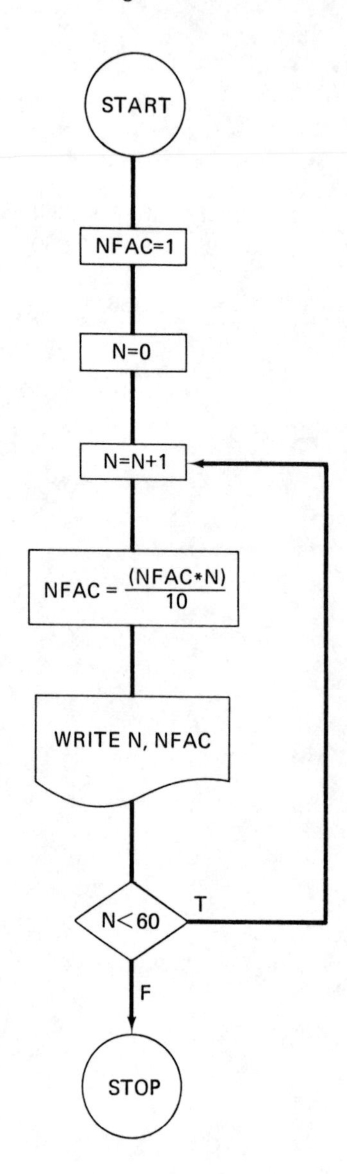

11 **(a)** $P\text{ (a pair)} = \dfrac{C(13,1)C(4,2)C(12,3)4^3}{C(52,5)} \approx .42$

(c) $P\text{ (3-of-a-kind)} = \dfrac{C(13,1)C(4,3)C(12,2)4^2}{C(52,5)} \approx .084$

(e) $P\text{ (flush)} = \dfrac{C(4, 1)C(13, 5)}{C(52, 5)} \approx .002$

(g) $P\text{ (4-of-a-kind)} = \dfrac{C(13, 1)C(12, 1)4}{C(52, 5)} \approx .00024$

13 **(a)** $2^5 = 32$

(c) 2

Chapter 3
Section 1

3 Same as for rolling a pair of fair dice

5 **(a)** $S = \{\text{HHH; HHT, HTH, THH; HTT, THT, TTH; TTT}\}$

(b) (Corresponding values of X): 3; 2, 2, 2; 1, 1, 1; 0

(c)

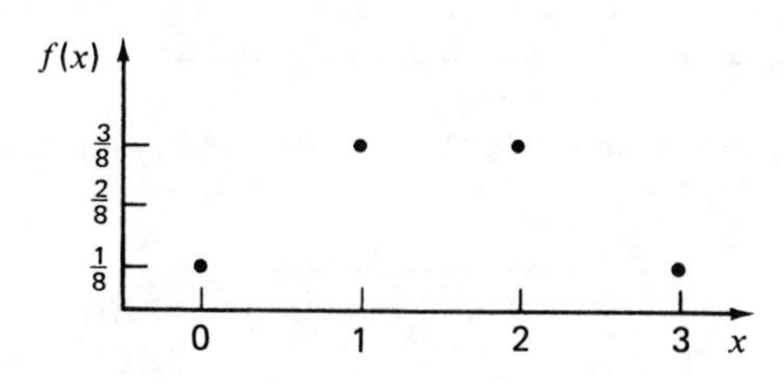

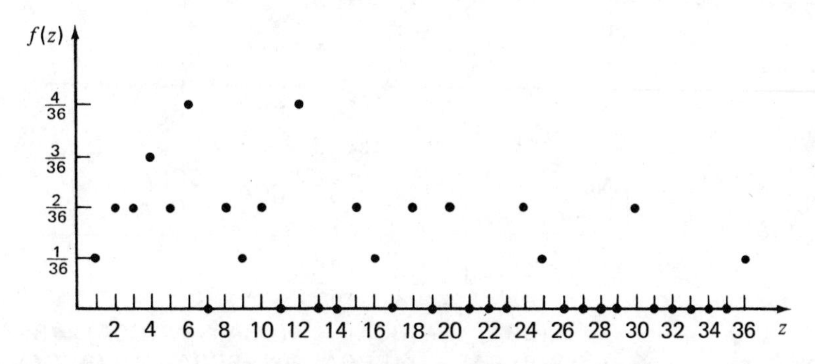

FREQUENCY FUNCTION FOR DISCRETE RANDOM VARIABLE Z, THE PRODUCT OF 2 FACES ON 1 TOSS OF A PAIR OF FAIR DICE

9 **(a)**

·A	·CDA
·BA	·DCA
·CA	·BCDA
·DA	·BDCA
·BCA	·CBDA
·CBA	·CDBA
·BDA	·DBCA
·DBA	·DCBA

9 (b) $X(\text{A}) = 1$
$X(\text{BA}) = X(\text{CA}) = X(\text{DA}) = 2$
$X(\text{BCA}) = X(\text{CBA}) = X(\text{BDA}) = X(\text{DBA}) = X(\text{CDA})$
$= X(\text{DCA}) = 3$
$X(\text{BCDA}) = X(\text{BDCA}) = X(\text{CBDA}) = X(\text{CDBA})$
$= X(\text{DBCA}) = X(\text{DCBA}) = 4$

9 (c)

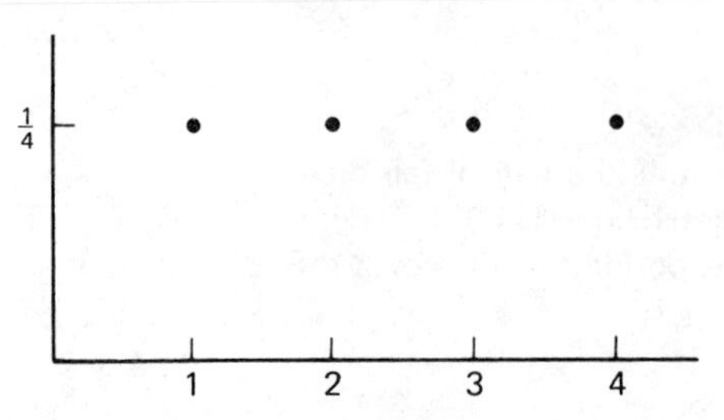

Frequency function for X, number of 4 keys drawn until correct one (a) is tried

11

·A
BA
·BBA
·BBBA . . .
·BBCA . . .
·BCA
·BCBA . . .
·BCCA . . .
CA
·CBA
·CBBA . . .
·CBCA . . .
·CCA
·CCBA . . .
·CCCA . . .

Not finite

$X(\text{A}) = 1$
$X(\text{BA}) = X(\text{CA}) = 2$
$X(\text{BBA}) = X(\text{BCA}) = X(\text{CBA}) = X(\text{CCA}) = 3$
$X(\text{BBBA}) = X(\text{BBCA}) = X(\text{BCBA}) = X(\text{BCCA}) = X(\text{CBBA})$
$= X(\text{CBCA}) = X(\text{CCBA}) = X(\text{CCCA}) = 4$

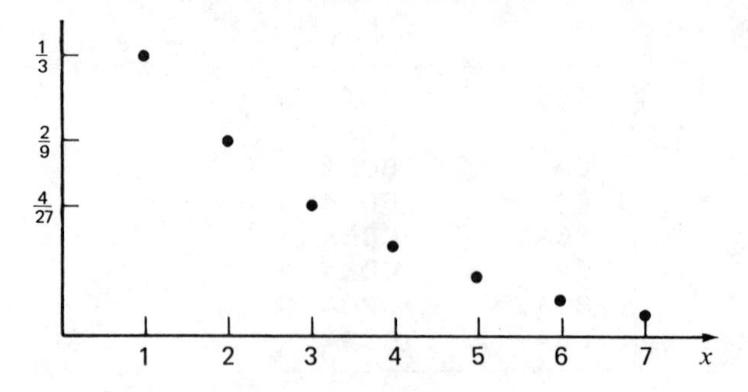

14 (a)

X	0	1	2	3	4
$f(x)$	.66	.30	.0033	.0003	.000019

(c)

Z	0	1	2	3	4	5
$f(z)$	.253	.422	.251	.066	.008	.0003

15

X	2	3
$f(x)$	$\frac{1}{2}$	$\frac{1}{2}$

17

X	7	11	12	15	16	20	27	31	35	36	40
$f(x)$	$\frac{3}{35}$	$\frac{6}{35}$	$\frac{1}{35}$	$\frac{1}{35}$	$\frac{6}{35}$	$\frac{3}{35}$	$\frac{1}{35}$	$\frac{6}{35}$	$\frac{3}{35}$	$\frac{2}{35}$	$\frac{3}{35}$

Chapter 3 Section 2

1 (a)

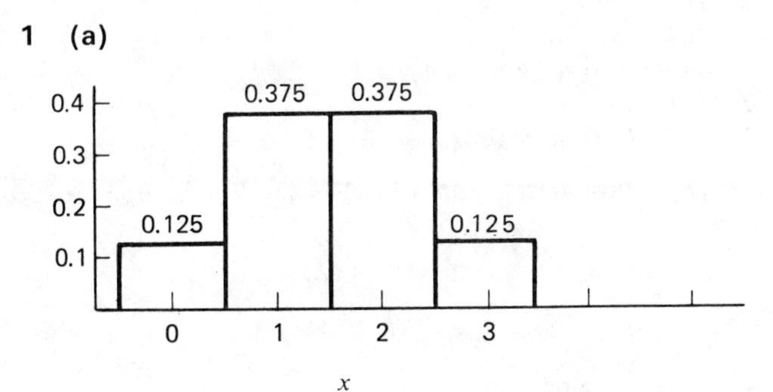

FREQUENCY FUNCTION GRAPH AND HISTOGRAM OF $B(x; 3, 0.5)$

(c)

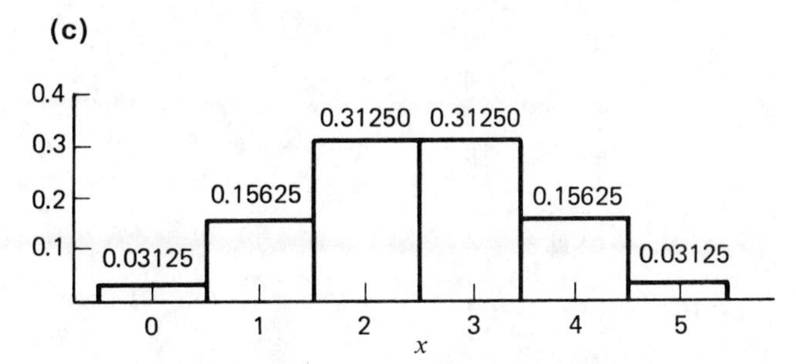

FREQUENCY FUNCTION GRAPH AND HISTOGRAM OF $B(x; 5, 0.5)$

(e)

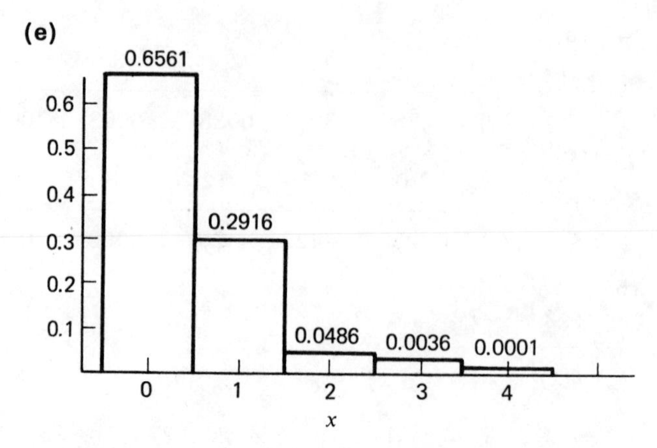

FREQUENCY FUNCTION GRAPH AND HISTOGRAM FOR $B(x; 4, 0.1)$

3 **(a)** $P(\text{at least 1 head}) = B(1; 3, .5) + B(2; 3, .5) + B(3; 3, .5)$
$= .375 + .375 + .125 = .875$

(c) .96875

4 **(a)** .5

(c) .31250

5 When $n = 4$, probability $= 1 - \frac{625}{1296}$

7 **(a)** $\frac{9}{28}$

9 **(a)** $P(\text{3 red marbles}) = B(3; 6, .4) = .2765$

(c) $P(\text{at least 3 non-white}) = \sum_{x=3}^{6} B(x; 6, .6) = .8208$

Chapter 3
Section 3

1 Mean = 2.83
Variance = 1.62

3 **(a)** $p = \frac{1}{2}$

n	μ	σ
4	2	1
8	4	$\sqrt{2}$
16	8	2
32	16	$2\sqrt{2}$
64	32	4

5 **(a)** $E(X_1) = 2.2$
$\sigma^2_{X_1} = .96$

(c) $E(X_3) = 2.2$
$\sigma^2_{X_3} = .32$

7 **(a)** $\mu = np = 6, \sigma^2 = npq = 2.4$

(c) $P(|x - \mu| > \sigma) = 1 - \sum_{x=5}^{7} B(x; 10, .6) = .3335$

(e) $P(|x - \mu| > 3\sigma) = B(0; 10, .6) + B(1; 10, .6) = .0017$

9 $\sigma' = 1$

11 (a) $f(x) = B(x; 3, \frac{1}{6})$

(c) Average loss = .79 cents (less than 1 penny) per play

Chapter 3
Section 4

1

	Chebychev Estimate	Actual Probability
(a)	$\frac{1}{4}$	.0586
(b)	$(\frac{1}{2})\frac{1}{16} = \frac{1}{32}$	.00004
(c)	$\frac{1}{2}(\frac{1}{36}) = \frac{1}{72}$	.0000000005

3 $|x - \mu| \geq k\sigma$ and $|x - \mu| < k\sigma$ are complementary events.

5 $n = 8400$

7 $E(Z) = E\left(\frac{X - \mu}{\sigma}\right) = \frac{1}{\sigma}E(X - \mu) = \frac{1}{\sigma}[E(X) - \mu] = 0$

$$E(Z^2) = E\left(\left(\frac{X - \mu}{\sigma}\right)^2\right) = E\left(\frac{(x - \mu)^2}{\sigma^2}\right) = \frac{1}{\sigma^2}(E(X - \mu)^2) = \frac{1}{\sigma^2}\cdot\sigma^2 = 1$$

Chapter 3
Section 5

1 (a) $B(x; 100, \frac{1}{2})$
$H_0: p = \frac{1}{2}$
Two-tailed test

(c) $B(x; 500, .9)$
$H_0: p < .9$
One-tailed test

3 Two-tailed Critical Regions

	$\alpha = .10$	$\alpha = .05$	$\alpha = .01$
(a)	0, 1, 2, 10, 11, 12	0, 1, 2, 10, 11, 12	0, 1, 11, 12
(c)	0, 1, 2, 3, 4, 5, 6, 7, 8, 9, 10, 20, 21, 22, 23, 24, 25, 26, 27, 28, 39, 30	0, 1, 2, 3, 4, 5, 6, 7, 8, 9, 21, 22, 23, 24, 25, 26, 27, 28, 29, 30	0, 1, 2, 3, 4, 5, 6, 7, , 23, 24, 25, 26, 27, 28, 29, 30
(e)	0, 1, 2, 3, 4, 5, 15, 16, 17, . . . , 28, 29, 30	0, 1, 2, 3, 4, 16, 17, 18, . . . , 28, 29, 30	0, 1, 2, 3, 18, 19, 20, . . . , 28, 29, 30

5 H_0 can be rejected at .05 level (not at .01 level).

7 More convincing that the drug is *not effective*

Chapter 3 Section 6

1 Since $\mathbf{P}(0; 3) = .0498$, we would expect $(.0498)(365) \approx 18$ days per year.

5 $B(0; 400, .01) \approx \mathbf{P}(0; 4) = .0183$

7

Number of Misprints	Expected Number of Poses
0	14.94
1	44.82
2	67.20

11 **(a)** .0821, **(c)** .133

Chapter 3 Section 7

1 **(a)** A_1: 2% cancellation; A_2: 5% cancellation; A_3 10% cancellation; A_4: 20% cancellation

Posterior probabilities
$P(A_1|X = 2) \approx .05$
$P(A_2|X = 2) \approx .29$
$P(A_3|X = 2) \approx .64$
$P(A_4|X = 2) \approx .014$

(c) Posterior probabilities
$P(A_1|X = 6) \approx .000$
$P(A_2|X = 6) \approx .020$
$P(A_3|X = 6) \approx .705$
$P(A_4|X = 6) \approx .279$

3 **(a)** $A_1: p = \frac{3}{4}$; $A_2: p = \frac{1}{2}$; $A_3: p = \frac{1}{4}$

Posterior probabilities
$P(A_1|X = 19) \approx .98$
$P(A_2|X = 19) \approx .02$
$P(A_3|X = 19) \approx .00$

(c) Posterior probabilities

$$P(A_1|X = 10) = \frac{.00001}{.04314} \approx .0002$$

$$P(A_2|X = 10) = \frac{.03896}{.04314} \approx .904$$

$$P(A_3|X = 10) = \frac{.00417}{.04314} \approx .096$$

Chapter 3 Section 8

3 $\frac{1}{4}, \frac{1}{4}, \frac{1}{4}, \frac{1}{4}$

5 $n = 10$

7 $\frac{16}{729}$

9 Let A be the event $X > m + n$ and B be the event $X > m$. Then $A \subset B$ so that $A \cap B = A$. Thus

$$P(A|B) = \frac{P(A)}{P(B)}$$

Now

$$P(A) = \sum_{k=m+n+1} q^{k-1}p = \frac{pq^{m+n}}{1-q} = q^{m+n}$$

Similarly, $P(B) = q^m$ and $P(A|B) = q^n = P(X > n)$.

Chapter 3
Section 9

1 (a)

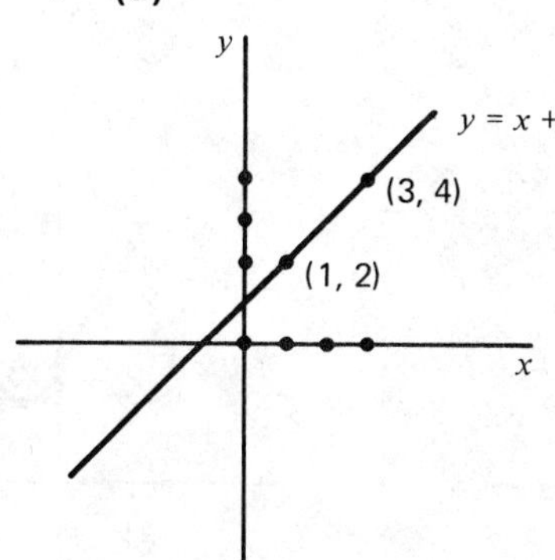

(c)

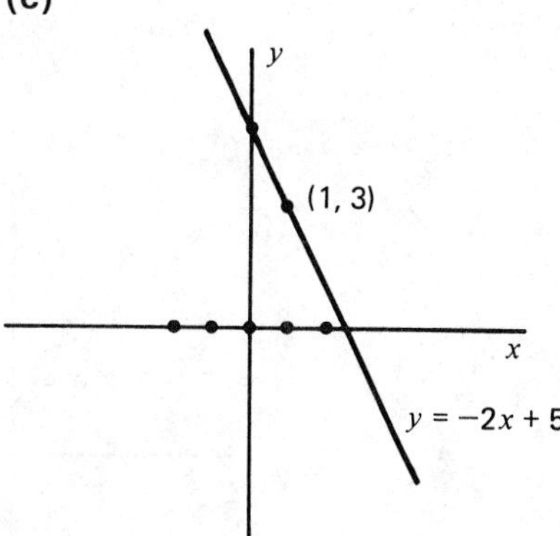

(e)

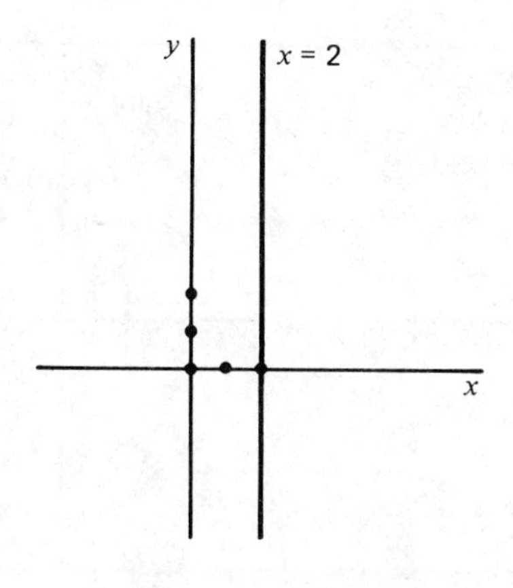

2 (a)

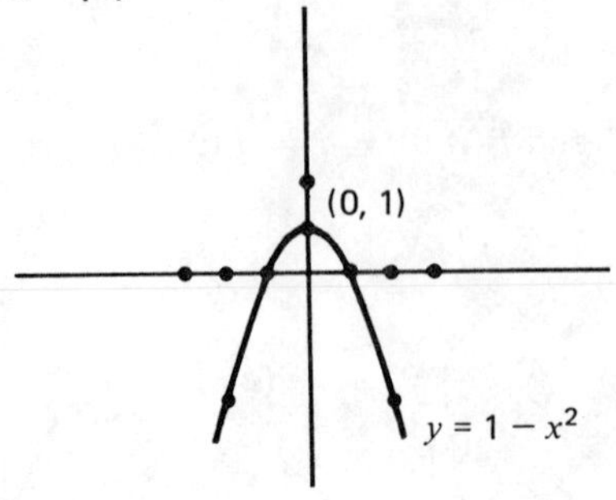

2 (c)

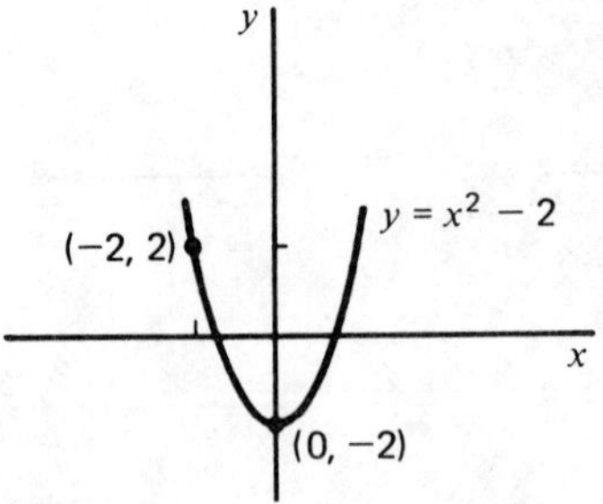

(e)

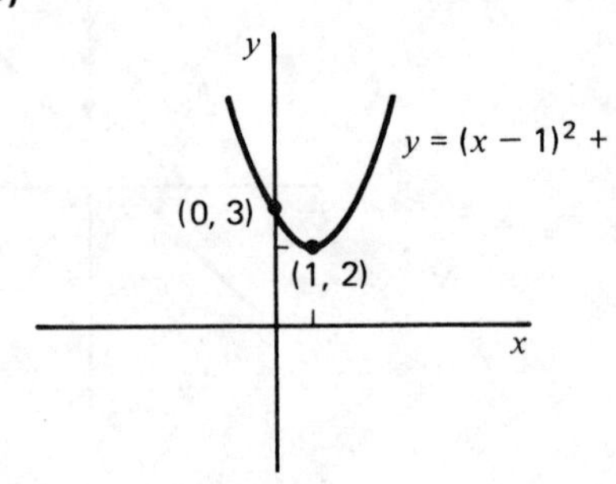

3 (a)

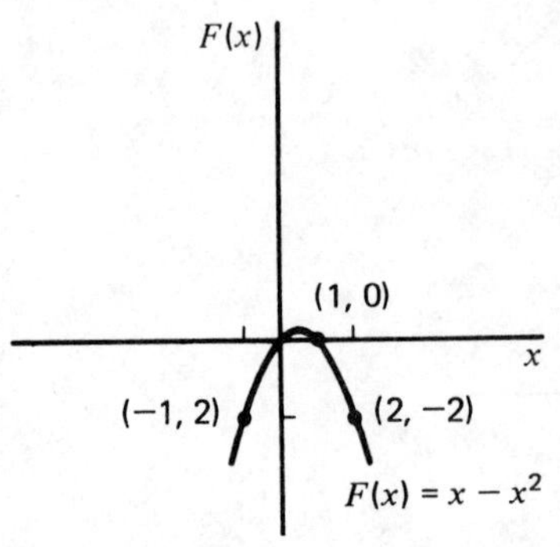

(d)

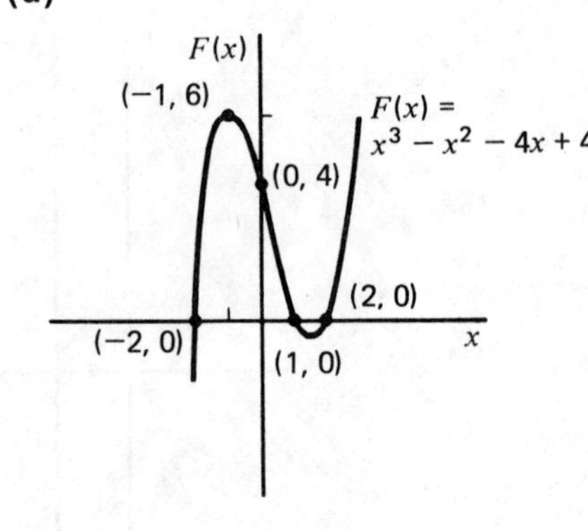

(e)

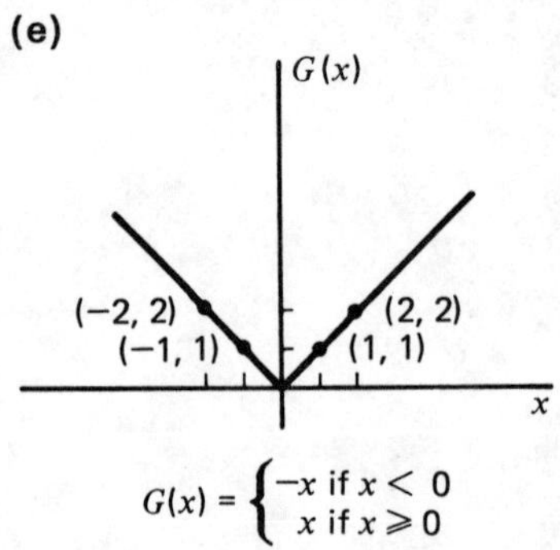

$$G(x) = \begin{cases} -x \text{ if } x < 0 \\ x \text{ if } x \geqslant 0 \end{cases}$$

5

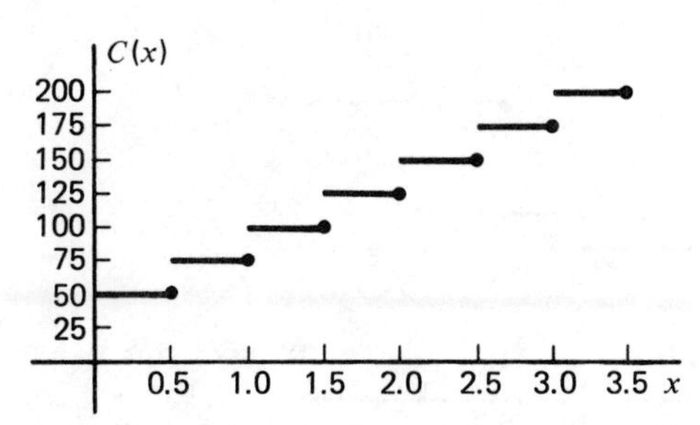

7 (a)

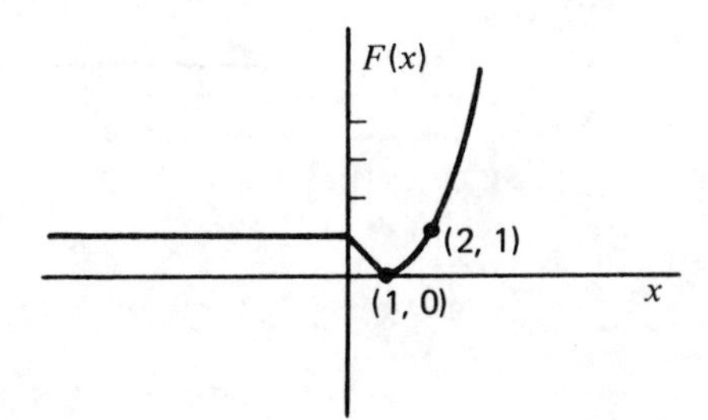

Chapter 3 Section 10

1 (a)

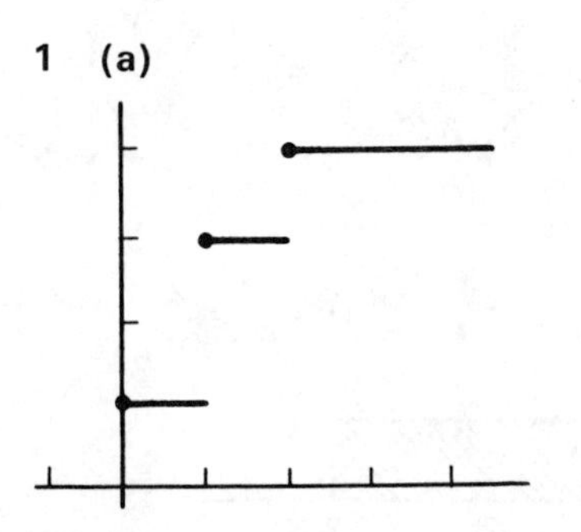

(c)

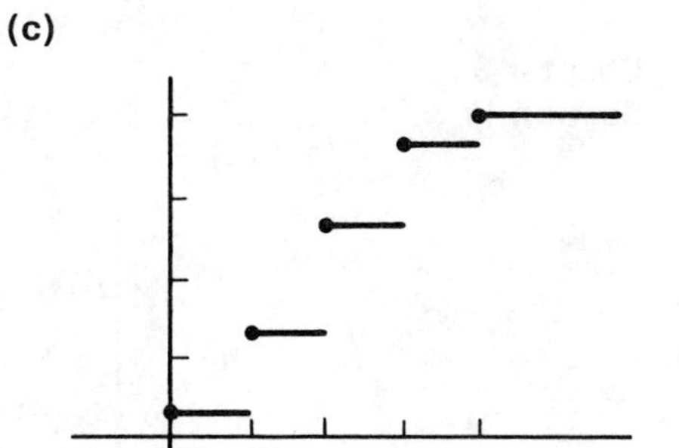

2 (a)

$$F(x) = \begin{cases} 0, & \text{if } x < 0 \\ .25, & \text{if } 0 \le x < 1 \\ .75, & \text{if } 1 \le x < 2 \\ 1.00, & \text{if } 2 \le x \end{cases}$$

(c)

$$F(x) = \begin{cases} 0, & \text{if } x < 0 \\ .0625, & \text{if } 0 \le x < 1 \\ .3125, & \text{if } 1 \le x < 2 \\ .6875, & \text{if } 2 \le x < 3 \\ .9375, & \text{if } 3 \le x < 4 \\ 1.0000, & \text{if } 4 \le x \end{cases}$$

3 (a)

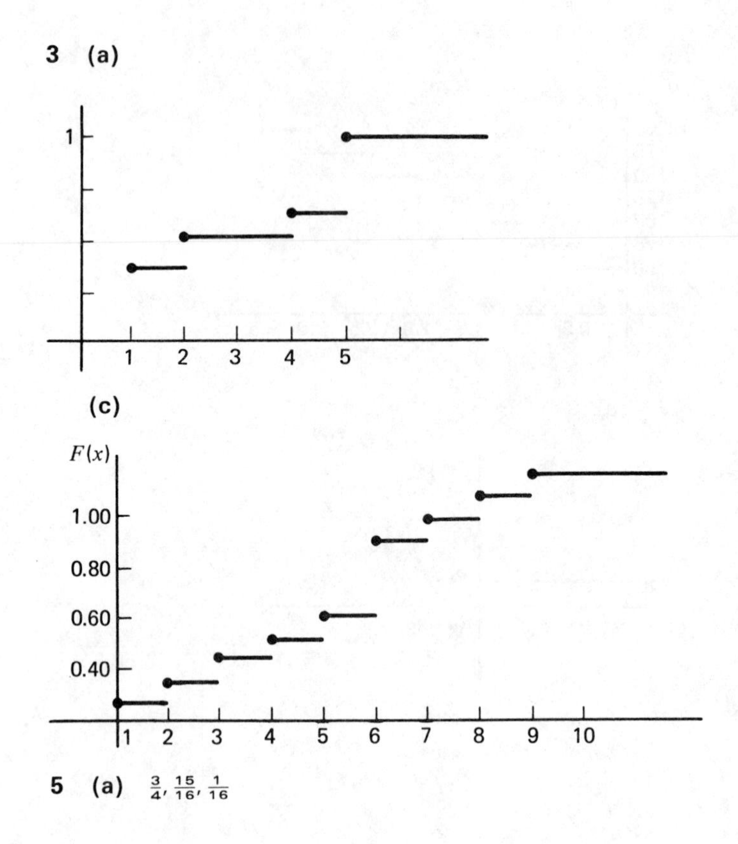

5 (a) $\frac{3}{4}, \frac{15}{16}, \frac{1}{16}$

Chapter 3
Section 11

1 (a)

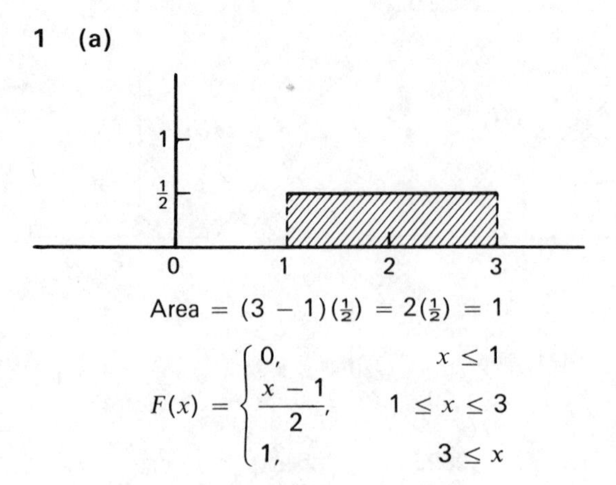

Area $= (3 - 1)(\frac{1}{2}) = 2(\frac{1}{2}) = 1$

$$F(x) = \begin{cases} 0, & x \leq 1 \\ \dfrac{x-1}{2}, & 1 \leq x \leq 3 \\ 1, & 3 \leq x \end{cases}$$

(c)

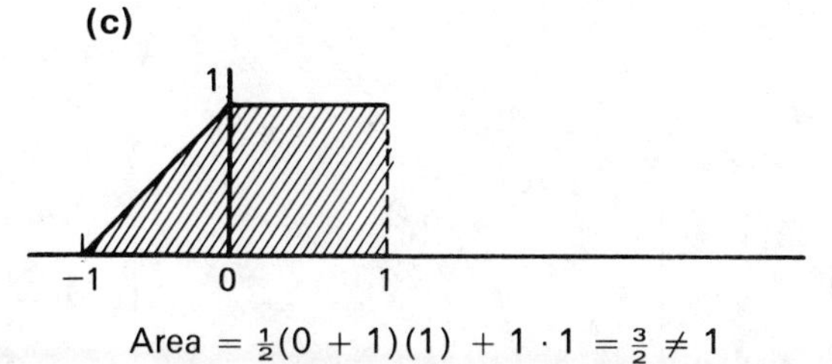

$$\text{Area} = \tfrac{1}{2}(0 + 1)(1) + 1 \cdot 1 = \tfrac{3}{2} \neq 1$$

(e)

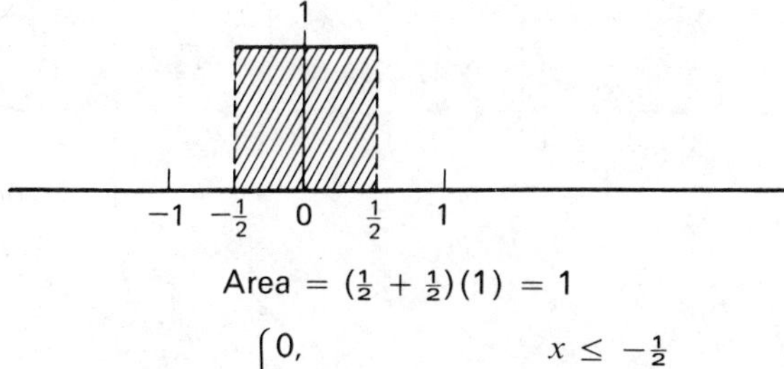

$$\text{Area} = (\tfrac{1}{2} + \tfrac{1}{2})(1) = 1$$

$$F(x) = \begin{cases} 0, & x \le -\frac{1}{2} \\ x + \frac{1}{2} & -\frac{1}{2} \le x \le \frac{1}{2} \\ 1, & \frac{1}{2} \le x \end{cases}$$

2 (a)

$$f(x) = \begin{cases} \dfrac{x}{9}, & 0 \le x \le 3 \\ 0, & \text{otherwise} \end{cases} \qquad \text{Area} = \tfrac{1}{2}(3 - 0)(\tfrac{1}{3}) = \tfrac{1}{2} \neq 1$$

not a density function

(c)

$$f(x) = \begin{cases} \dfrac{x}{8}, & 0 \le x \le 4 \\ 0, & \text{otherwise} \end{cases} \qquad \text{Area} = \tfrac{1}{2}(4 - 0)(\tfrac{1}{2}) = 1$$

(e)

$$f(x) = \begin{cases} +\sqrt{1 - x^2}, & -1 \le x \le 1 \\ 0, & \text{otherwise} \end{cases} \qquad \text{Area} = \frac{\pi}{2}(1)^2 \neq 1$$

not a density function

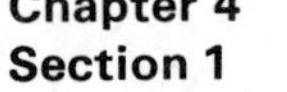

Chapter 4
Section 1

1 (a)

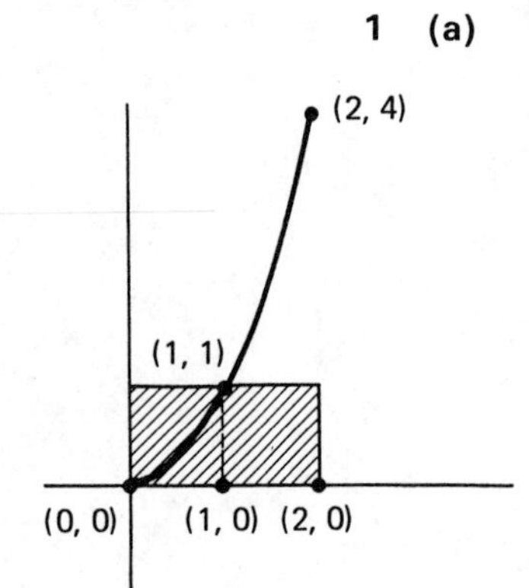

$R_1 = 2$

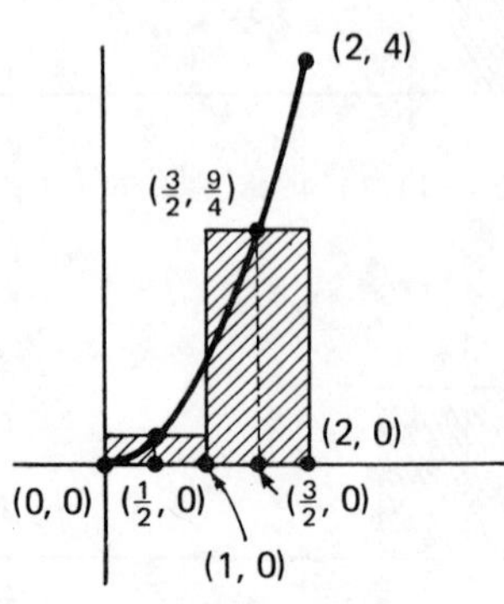

$R_2 = \frac{5}{2}$

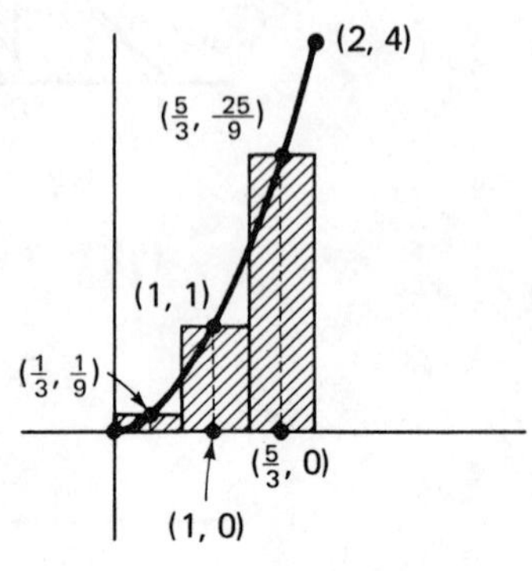

$R_3 = \frac{70}{27}$

(c)

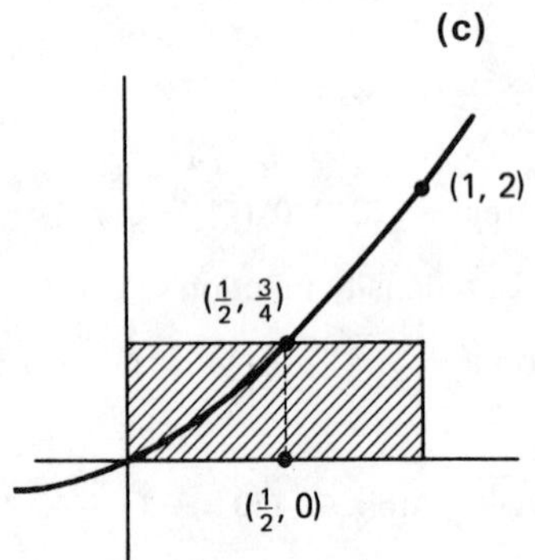

$R_1 = \frac{3}{4}$

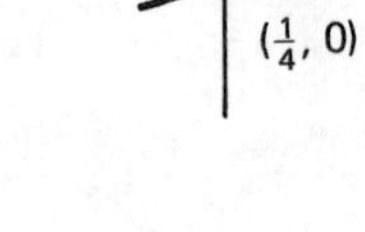

$R_2 = \frac{13}{16}$

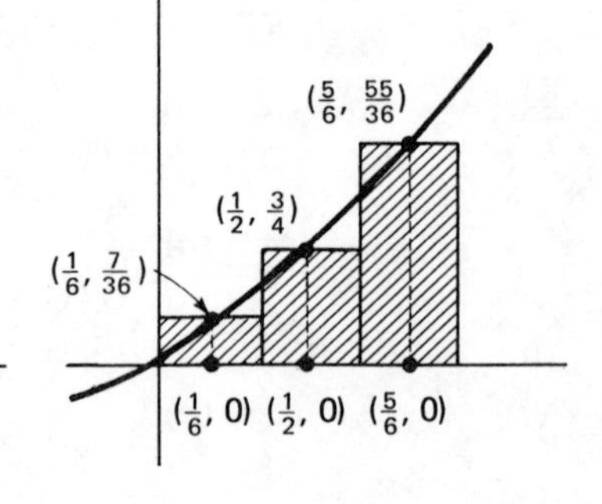

$R_3 = \frac{89}{108}$

5 $R_n = 8$

$$L_n = 4\left(2 - \frac{1}{n}\right)$$

$$U_n = 4\left(2 + \frac{1}{n}\right)$$

Chapter 4
Section 2

1 (a) $R_1 = \frac{9}{4}$; $R_2 = \frac{19}{8}$; $R_3 = \frac{251}{108}$

(c) $R_1 = \frac{15}{4}$; $R_2 = \frac{31}{8}$; $R_3 = \frac{413}{108}$

3 (a) $R_1 = \frac{1}{4}$; $R_2 = \frac{5}{16}$; $R_3 = \frac{35}{108}$

(c) $R_2 = \frac{5}{2}$; $R_4 = \frac{43}{16}$; $R_6 = \frac{286}{108}$

5 **(a)** is; **(b)**, **(c)**, **(d)**, and **(e)** aren't.
7 $R_1 = R_2 = R_3 = R_4 = -3$
9 $R_1 = \frac{1}{2}, R_2 = 0, R_3 = \frac{1}{6}, R_4 = 0$

Chapter 4 Section 3

1 **(a)** $g'(x) = 1$
(c) $g'(x) = 0$
(e) $g'(x) = 2x$
(g) $g'(x) = 2x + 1$
(i) $g'(x) = -\dfrac{1}{x^2}$
(k) $g'(x) = -3$
3 **(a)**

Δx	1	$-\frac{1}{2}$	$\frac{1}{4}$	$-\frac{1}{8}$	$\frac{1}{16}$
$\dfrac{(2 + \Delta x)^2 - 2^2}{\Delta x}$	5	$3\frac{1}{2}$	$4\frac{1}{4}$	$3\frac{7}{8}$	$4\frac{1}{16}$

Chapter 4 Section 4

1 1
3 $\frac{1}{2}$
5 2
7 1
9 $\frac{1}{3}$
11 $\frac{5}{2}$
13 **(a)** 5.26%
(b) 9.1%

Chapter 4 Section 5

1 $\frac{5}{2}$
3 6
5 0
7 1
9 **(a)** $g'(x) = \dfrac{1}{2\sqrt{x + 1}}$
(c) $g'(x) = -\dfrac{1}{x^2}$
(e) $g'(x) = 2 \exp(2x)$
11 12
14 **(a)** 0
(c) 303
(e) 2

Chapter 4 Section 6

3 (a) $G(x) = x^2 + 3x$

(c) $G(x) = \dfrac{1}{9}x^3 + 4$

5 $f'(x) = 1 + x + \dfrac{x^2}{2} + \dfrac{x^3}{3!} + \dfrac{x^4}{4!} + \dfrac{x^5}{5!}$

6 (a)

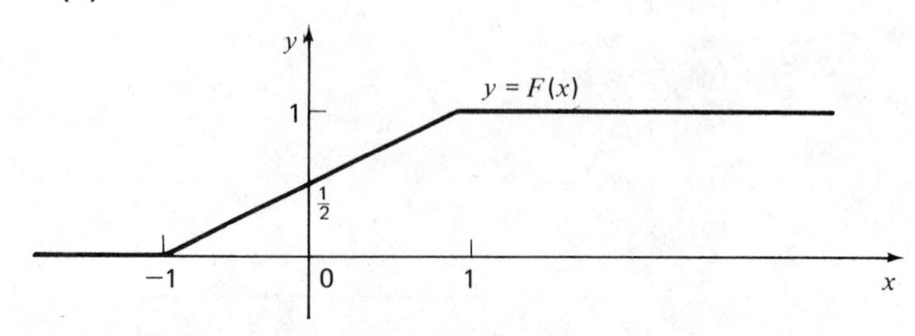

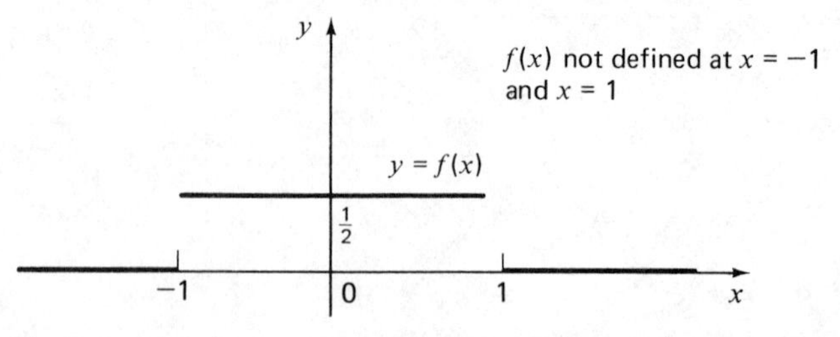

7 (a) $D_x(F \cdot G \cdot H) = F \cdot G \cdot D_xH + F \cdot H \cdot D_xG + G \cdot H \cdot D_xF$

(b) $D_xf(x) = (x + 3)(x^2 - 2)(3x^2 - 2) + (x + 3)(x^3 - 2x)(2x) + (x^2 - 2)(x^3 - 2x)(1)$

8 (a) Continuous and differentiable at $x = 0$

(c) Continuous but not differentiable at $x = 0$

(e) Continuous but not differentiable at $x = 0$

9 (a) $f'(x) = 5x^4, f''(x) = 20x^3, f'''(x) = 60x^2$

(c) $g^{(4)}(x) = 0$

(e) $h'(t) = 1 + t + \dfrac{t^2}{2} + \dfrac{t^3}{3!} + \dfrac{t^4}{4!}$

$h''(t) = 1 + t + \dfrac{t^2}{2} + \dfrac{t^3}{3!}$

$h'''(t) = 1 + t + \dfrac{t^2}{2}$

Chapter 4 Section 7

1 (a) $\frac{1}{2}$

(c) 0

(e) $e - 1$

(g) $\frac{1}{2}$

3 $\int_1^2 (x + 1)\, dx = \frac{1}{2}x^2 + x - 12\big|_1^2$

$$= (\tfrac{1}{2}(2)^2 + 2 - 12) - (\tfrac{1}{2}(1)^2 + 1 - 12) = \tfrac{5}{2}$$

5

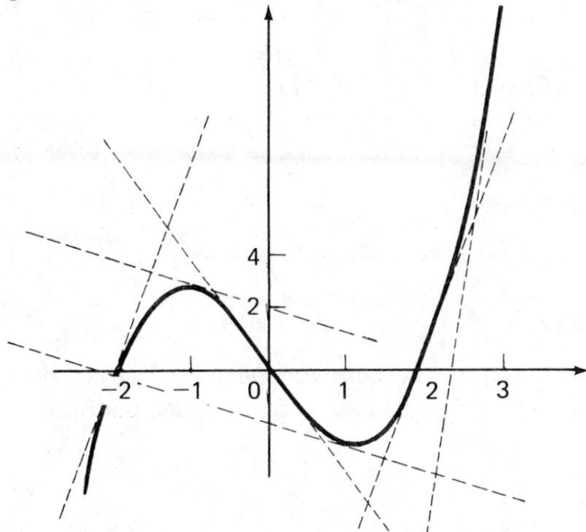

$f(x) = x^3 - 4x$
$f'(x) = 3x^2 - 4$
$f'(-2) = 8$
$f'(-1) = -1$
$f'(0) = -4$
$f'(1) = -1$
$f'(2) = 8$
$f'(3) = 23$

7

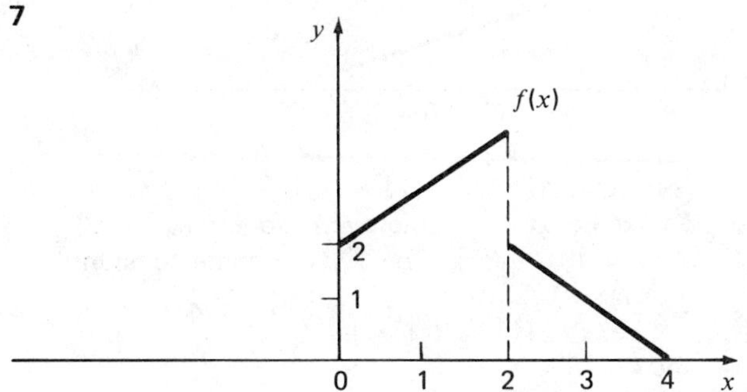

$$\int_0^4 f(x)\, dx = \int_0^2 f(x)\, dx + \int_2^4 f(x)\, dx$$
$$= \int_0^2 (2 + x)\, dx + \int_2^4 (4 - x)\, dx$$
$$= 6 + 2 = 8$$

9 Slope of chord $= \dfrac{f(b) - f(a)}{b - a} = \dfrac{(cb^2 + db + e) - (ca^2 + da + e)}{b - a}$

$$= c(b + a) + d$$

Slope of tangent at $x = f'(x) = 2cx + d$

Slope of tangent at $\left(m = \dfrac{a + b}{2}\right) = 2c\left(\dfrac{a + b}{2}\right) + d = c(a + b) + d$

Chapter 4
Section 8

1

	Maximum	Minimum
(a)	$x = 4$ $f(4) = 14$	$x = 1$ $f(1) = 5$
(c)	$x = 1$ $f(1) = e$	$x = -1$ $f(-1) = -\frac{1}{e}$
(e)	$x = 2$ $f(2) = 6$	$x = 0$ $f(0) = 4$

3 Let $x = \frac{4}{3}$ to get maximum volume ≈ 34.2 cubic inches.

5 Optimal speed $= \sqrt{3{,}200} \approx 56.56$ miles per hour.

7 **(a)**

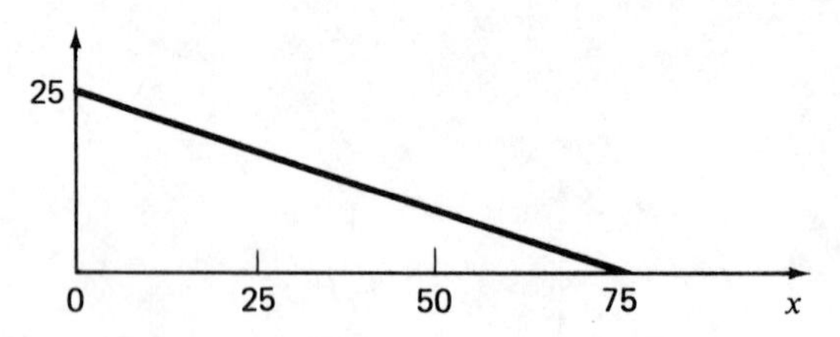

(c) $x = 30$

9 As fixed cost (including overhead, storage, etc.)

11 $C(x) = 45x - 3x^2 + k$ (k = some constant)

If $C(1) = 125$, then

$$125 = 45(1) - 3(1)^2 + k$$

and $k = 83$.

13 **(a)**

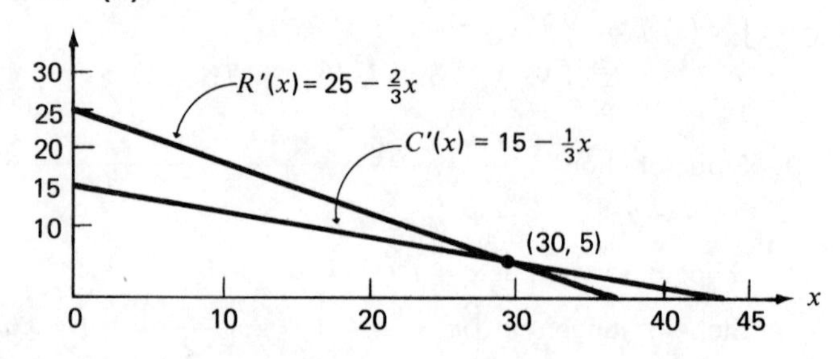

13 **(c)** Must have

$$R''(a) \leq C''(a)$$

which says that the rate of change (i.e., increase) of marginal revenue must be less than or equal to the rate of change (i.e., increase) of marginal cost.

Chapter 4
Section 9

1 (a) $5(3x^2 + 1)(x^3 + x)^4$
(c) $4(\frac{3}{2}x^{1/2})(x^{3/2} + 1)^3$
(e) $7(2x + 1)(x^2 + x)^6 \exp((x^2 + x)^7)$
(g) $(-x) \exp\left(\frac{-x^2}{2}\right)$
(i) $4(2x + 2) \exp^4 (x^2 + 2x)$

3 (a) $f'(x) = (x^n + nx^{n-1}) \exp(x)$
$f''(x) = (x^n + nx^{n-1} + n(n-1)x^{n-2}) \exp(x)$
(c) $f'(x) = \exp(\exp(x) + x)$
$f''(x) = (\exp(x) + 1) \exp(\exp(x) + x)$

5 (a) $c(x^2 + 1) \cdot 2x$
(c) $2(s(x^2)) \cdot c(x^2) \cdot (2x)$
(e) $\exp(c(x)) \cdot (-s(x))$
(g) $\frac{1}{2}(s(x))^2|_0^1 = \frac{1}{2}(s(1))^2 - \frac{1}{2}(s(0))^2$

Chapter 4
Section 10

1 (a) $f'(t) = \dfrac{3t^2 + 2t}{t^3 + t^2 + 1}$
(c) $h'(t) = \dfrac{1}{t^2(t + 3)} - \dfrac{2}{t^3} \text{Log}(t + 3)$
(e) $G'(x) = \left(1 + \dfrac{1}{x}\right) e^x e^{\text{Log}(x)} = (x + 1) e^x$

2 (a) Log 2
(c) 2 Log 3
(e) $\frac{29}{6}$

3 4,346 years ago

5 $D_x(a^x) = a^x \text{Log}\, a$

7 Since $a^x = e^{x \text{Log}(a)}$, we must have $a > 0$ to define Log (a).

9 (a) $x = \left(\dfrac{1}{y}\right) \text{Log}(\text{Log}(z))$
(c) $x = \left(\dfrac{1}{yz}\right) \exp(\exp(t))$

Chapter 4
Section 11

1 .4405

3 Want: $\mathbf{P}(0; 100, \lambda) = e^{-100\lambda} = .99$ or

$$\lambda = \frac{\text{Log}\ .99}{-100} \approx .0001$$

$$\mathbf{P}(0; 1, \lambda) = \mathbf{P}(0; \lambda) = e^{-\lambda} = e^{-.0001} \approx \frac{1}{1.00001005}$$

$$\approx .99999$$

5 6 hours

7 $e^{-5} \approx .0067$

Chapter 4
Section 12

1 **(a)**

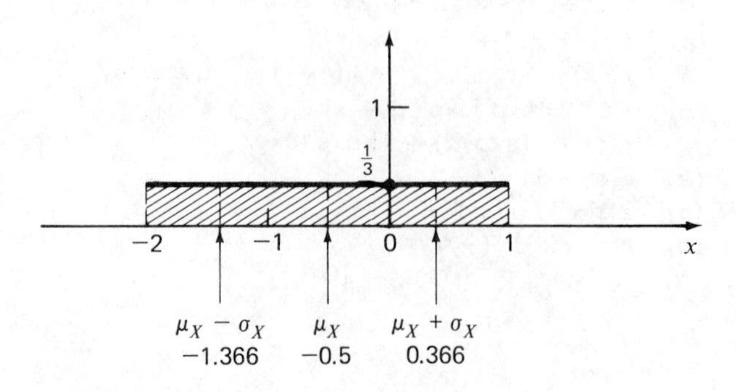

(c)

1

−2 −1 0 1 2 x

μ_X

$\mu_X - \sigma_X$ $\mu_X + \sigma_X$

−0.4083 0.4083

(e)

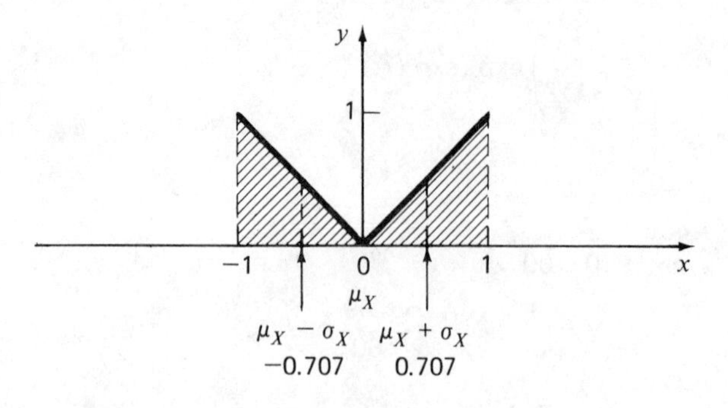

3 Chebychev's Theorem says

$$P(X \in [0, 3]) \geq .75$$

since $\mu = \sigma = 1$. But the actual area is $1 - e^{-3} \approx .95$.

7 **(a)**

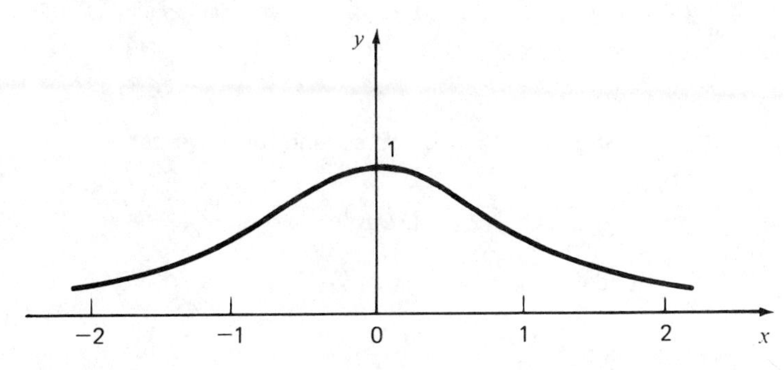

(c) Symmetry about $x = 0$ would suggest $\mu = 0$.

9 $$\mu_Y = \int_{-\infty}^{\infty} \frac{x-\mu}{\sigma} f(x)\,dx = \frac{1}{\sigma}\left\{\int_{-\infty}^{\infty} xf(x)\,dx - \int_{-\infty}^{\infty} \mu f(x)\,dx\right\}$$
$$= \frac{1}{\sigma}(\mu - \mu) = 0$$

And thus,

$$\sigma_Y^2 = \int_{-\infty}^{\infty} \left(\frac{x-\mu}{\sigma}\right)^2 f(x)\,dx = \frac{1}{\sigma^2}\int_{-\infty}^{\infty} (x-\mu)^2 f(x)\,dx = \frac{1}{\sigma^2}\cdot\sigma^2 = 1$$

Chapter 4
Section 13

1 **(a)** $F_x(x, y) = 2x$; $F_y(x, y) = -10y$

(c) $$H_x(x, y, z) = \frac{1}{x + 2y} - 5z^3(x + y)^4$$
$$H_y(x, y, z) = \frac{2}{x + 2y} - 5z^3(x + y)^4$$
$$H_z(x, y, z) = -3z^2(x + y)^5$$

(e) $E_m(m, c) = c^2$; $E_c(m, c) = 2mc$

(g) $A_n(n, r, P) = Pr$
$A_r(n, r, P) = Pn$
$A_p(n, r, p) = 1 + nr$

3 $x = 1, y = 1$

5 Optimal sales: $x = 45$ washers, $y = 10$ dryers
Net profit = 5,750 dollars

Chapter 4
Section 14

1 (a) $x = \frac{15}{11}, y = \frac{5}{11}$ (yields minimum value)

3 f has no minimal or maximal values: as $x \to \infty$, $y \to -\infty$, and $f(x, y) \to \infty$; as $x \to -\infty$, $y \to \infty$, and $f(x, y) \to -\infty$.

5 $F(L, K, \lambda) = pX - wL - rK - \lambda(X - AL^{\alpha}K^{\beta})$
$F_L = -w + \lambda A^{\alpha}L^{\alpha-1}K^{\beta}$
$F_K = -r + \lambda A\beta L^{\alpha}K^{\beta-1}$
$F_{\lambda} = -(X - AL^{\alpha}K^{\beta})$
Setting $F_L = F_K = 0$ and solving for λ, we get

$$\lambda = \frac{w}{A\alpha L^{\alpha-1}K^{\beta}} = \frac{r}{A\beta L^{\alpha}K^{\beta-1}}$$

Hence,

$$\frac{L}{K} = \frac{\alpha}{\beta}\frac{r}{w}$$

7 We have

$$\tfrac{1}{2}P_1L_1^{-1/2}C_1^{1/2} = \lambda_1 = \tfrac{1}{3}P_2L_2^{-2/3}C_2^{2/3}$$

and

$$\tfrac{1}{2}P_1L_1^{1/2}C_1^{-1/2} = \lambda_2 = \tfrac{2}{3}P_2L_2^{1/3}C_2^{-1/3}$$

from which we get

$$(*) \quad \left(\frac{C_1}{L_1}\right)^{1/2} = \frac{2}{3}\left(\frac{C_2}{L_2}\right)^{2/3}\frac{P_2}{P_1}$$

$$= \frac{2}{3}\left(\frac{C_2}{L_2}\right)^{2/3} \cdot \frac{3}{4}\left(\frac{L_1}{C_1}\right)^{1/2}\left(\frac{C_2}{L_2}\right)^{1/3}$$

$$= \frac{1}{2}\frac{C_2}{L_2}\left(\frac{L_1}{C_1}\right)^{1/2}$$

so that

$$(**) \quad \frac{C_1}{L_1} = \frac{1}{2}\frac{C_2}{L_2}$$

From (*) and (**) we get

$$\frac{C_2}{L_2} = \frac{1}{8}\left(\frac{3}{2}\right)^6\left(\frac{P_1}{P_2}\right)^6$$

which, together with the budget constraints, give us the solution.

Chapter 4
Section 15

1 (a) $R = 1$
(c) $R = 3$
(e) $R = 2$

3 $f(x) \approx 1 + \frac{1}{2}x - \frac{1}{8}x^2 + \frac{1}{16}x^3 - \frac{15}{384}x^4$
Thus,

$$\int_0^1 f(x)\,dx \approx 1 + \tfrac{1}{2}\cdot\tfrac{1}{2} - \tfrac{1}{8}\cdot\tfrac{1}{3} + \tfrac{1}{16}\cdot\tfrac{1}{4} - \tfrac{15}{384}\cdot\tfrac{1}{5}$$
$$\approx 1.216$$
$$\int_0^1 \sqrt{1+x}\,dx = \tfrac{2}{3}(1+x)^{3/2}\Big|_0^1 = \tfrac{2}{3}(2^{3/2} - 1) \approx 1.219$$

5 $f(x) = \sum_{n=0}^{\infty} (n+1)x^n \qquad (R = 1)$

$$f'(x) = \sum_{n=1}^{\infty} n(n+1)x^{n-1} = 1(1-x)^{-3}$$

Thus

$$(1-x)^{-3} = \sum_{n=1}^{\infty} \frac{n(n+1)}{2}x^{n-1}$$

7 (a)
$$\frac{e^x - 1}{x} = \frac{\left(1 + x + \dfrac{x^2}{2} + \dfrac{x^3}{6} + \cdots\right) - 1}{x}$$
$$= \frac{x\left(1 + \dfrac{x}{2} + \dfrac{x^2}{6} + \cdots\right)}{x}$$
$$= 1 + \frac{x}{2} + \frac{x^2}{6} + \cdots$$

$$\frac{e^x - 1}{x} \to 1 \quad \text{as} \quad x \to 0$$

Chapter 5
Section 1

1 (a) $\mu = 10, \sigma = \sqrt{5}$
(c) $\mu = \frac{100}{3}, \sigma = \frac{10}{3}\sqrt{2}$

3 (a) $\int_{39.5}^{\infty} N\left(x; 25, \dfrac{5}{\sqrt{2}}\right) dx$

(c) $\int_{-\infty}^{50.5} N(x; 48, 4\sqrt{2})\,dx$

5 (a) $q = \dfrac{\sigma^2}{\mu}, p = 1 - \dfrac{\sigma^2}{\mu}, n = \dfrac{\mu}{1 - (\sigma^2/\mu)} = \dfrac{\mu^2}{\mu - \sigma^2}$

(c) No, since n, p, q are uniquely determined by μ and σ

Chapter 5
Section 2

1 (a) $\int_{-\infty}^{0} N(z; 0, 1)\, dz = .5$

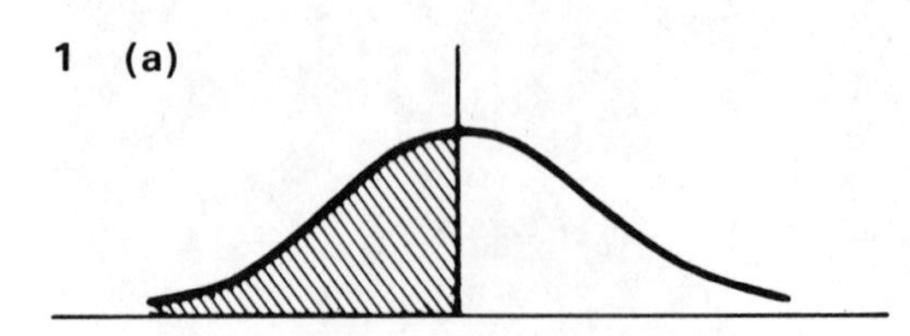

(c) 1 − .8412 = .1588
(e) .9772 − .0228 = .9444
(g) .9772 − .0014 = .9758
(i) .6915 − .3085 = .3820

3 (a) .8968
(c) .0228

5 (a) Exact .3438, approximate .2090
(c) Exact .1190, approximate .0793

7 (a) .4483
(c) .4761

Chapter 5
Section 3

1 (a) .0401
(c) .0062
(e) .6824

7 (a) 0
(c) .0035

9 (a) 15.6

Chapter 5
Section 4

1

	2-tailed Regions		
	.10	.05	.01
(a)	{0, 1, . . . , 41, 59, 60, . . . , 100}	{0, 1, . . . , 39, 61, 62, . . . , 100}	{0, . . . , 35, 65, . . . , 100}
(c)	{0, . . . , 240, 760, . . . , 1,000}	{0, . . . , 189, 811, . . . , 1,000}	{0, . . . , 55, 945, . . . , 1,000}
(e)	{0, . . . , 148, 185, . . . , 500}	{0, . . . , 145, 188, . . . , 500}	{0, . . . , 136, 197, . . . , 500}

3 Not effective at .05 level

5 Cannot conclude it was better at .05 level

Chapter 5
Section 5

1 $p = \dfrac{\sum x_i}{kn}$

3 $\theta = \dfrac{n}{\sum x_i}$

5

	a	b	c
Estimate	24	27	42
Count	25	38	43

7 $c =$ largest of $x_1, \ldots, x_n$

Chapter 5
Section 6

1 90% interval (\$39,746.16; \$40,253.84)
95% interval (\$39,694.24; \$40,305.76)
99% interval (\$39,597.52; \$40,402.48)

3 **(a_1)** 90% interval (\$39,770; \$40,230)
95% interval (\$39,726; \$40,274)
99% interval (\$39,639; \$40,361)

(a_2) 90% interval (30,656; 31,344)
95% interval (30,588; 31,412)
99% interval (30,458; 31,542)

(c_1) 90% interval (\$39,923; \$40,077)
95% interval (\$39,909; \$40,091)
99% interval (\$39,870; \$40,130)

(c_2) 90% interval (30,885; 31,115)
95% interval (30,863; 31,137)
99% interval (30,819; 31,181)

5 $(-\infty, 33.2)$

Chapter 5
Section 7

1 (121.44, 139.96)

3 **(a)** None
(b) 16,200

5 Claim is outside 90% confidence interval

Chapter 5
Section 8

1 At .01 significance, $n = 13{,}573$
At .1 significance, $n = 4{,}097$

5 .99 significance

7 Yes
9 No
11 There is a difference at the .05 significance level.
13 10%

Chapter 5
Section 9

1 Cannot reject
3 Not at .05 significance level

Chapter 5
Section 10

1 **(a)** $\chi^2 = 3.14$ with 3 degrees of freedom. The hypothesis that the distribution is normal cannot be rejected at the .10 significance level.
(b) $\chi^2 = 17.33$ with 5 degrees of freedom. The hypothesis that the distribution is uniform can be rejected at the .01 level of significance.
3 $\chi^2 = 2.59$; cannot reject at the .01 level of significance.

Chapter 6
Section 1

1 **(a)** $y = \frac{3}{4}x - \frac{1}{4}$
(b) $y = \frac{1}{2}$
3 $C = 59.6$, $a = 19.97$
7 **(a)** $y = 3x - 3$
(c) $y = 3x - 4$

1 **(a)** .866
(b) 0
3 $r = .634$

$$\frac{r\sqrt{10 - 2}}{\sqrt{1 - r^2}} = 2.82 > 2.306$$

So there is a significant correlation at the .05 level.
Predict a 73.95.
5 $r = .898$

1 **(a)** 8; abababab or babababa (only two possible sequences)
(b) 9; babababab (only possible sequence)
(c) 13; bbbababababababab (several others possible)

(d) Two cases: If $n_1 < n_2$, then $2n_1 + 1$ is maximum number.
If $n_1 = n_2$, then $2n_1$ is maximum number.

Chapter 7
Section 2

1 15; 16

3 Letting $U =$ number of times $x_i > y_i + 35$, we get

$$U = 0+0+0+0+0+0+1+1+1+0+1+1+1+1+1$$
$$= 8$$

and we *reserve judgment*!

5 **(a)** Reserve judgment (accept H_0).

(b) H_0 can be rejected at better than a .05 significance level (in favor of a two-sided alternative).

Chapter 7
Section 3

1 $U = 118$; the corresponding z-value is

$$\frac{118 - 104}{22.8} = \frac{16}{22.8} \approx .61$$

Conclusion: Cannot reject hypothesis that two classes are of equal caliber.

3 $U = 112.5$

$$z = \frac{112.5 - 112.5}{194} = 0$$

Conclusion: *Accept* H_0: $X = Y = +35$.

5 **(a)** $1 + 2 + \cdots + m = \dfrac{m(m+1)}{2}$

(b) $U = mn + \dfrac{m(m+1)}{2} - \dfrac{m(m+1)}{2} = mn$

(c) $(n+1) + (n+2) + \cdots + (n+m)$

$$= \sum_{k=1}^{n+m} k - \sum_{k=1}^{n} k$$
$$= \frac{(m+n)(m+n+1)}{2} - \frac{n(n+1)}{2}$$
$$= \frac{m^2 + 2mn + n^2 + m + n - n^2 - n}{2}$$
$$\frac{m^2 + 2mn + m}{2}$$
$$= mn + \frac{m(m+1)}{2}$$

(d) $U = 0$

7 $U = 192.5$

$$z = \frac{192.5 - 200}{37} \approx -.2$$

Accept H_0: sophomore GPA = freshman GPA

Chapter 7
Section 4

1 **(a)** $\rho = 1 - \dfrac{6(1173.50)}{20(399)} \approx 1 - .86 = .14$

$\rho \approx .14$, $\tau = .10$

(c) $\rho \approx -.34$, $\tau \approx -.25$

3 **(a)** $\rho = 1 - \dfrac{6(88.50)}{15(224)} \approx .84$, indicating *significant* correlation between exam 1 and exam 2

5 **(a)** $\rho = 1 - \dfrac{6(1057)}{25(624)} \approx .59$. But $t = 3.47 > t_{.005}$. So we can conclude that X and Z are significantly related variables.

Appendix

Table A Square Root and Exponential Values

Suggestions for using this table.

1. To compute $\sqrt{131}$, we would write

$$\sqrt{131} = \sqrt{100(1.31)} = 10\sqrt{1.31} = 10(1.1446) = 11.446$$

But to compute $\sqrt{13.1}$ we would write

$$\sqrt{13.1} = \sqrt{10(1.31)} = 3.6194 \text{ (using the column headed } \sqrt{10x}\text{)}$$

2. To find e^x and e^{-x} for values of x larger than those that appear in this table, we can use the facts

$$e^{a+b} = e^a \cdot e^b \text{ and } e^{ab} = (e^a)^b$$

For example,

$$e^{-6.2} = (e^{-3.1})^2 = (.045)^2$$

and

$$e^7 = e^4 \cdot e^3 = (54.598)(20.086)$$

x	$\sqrt{x}$	$\sqrt{10x}$	e^x	e^{-x}
0.01	0.1000	0.3162	1.0101	0.9900
0.02	0.1414	0.4472	1.0202	0.9802
0.03	0.1732	0.5477	1.0305	0.9704
0.04	0.2000	0.6325	1.0408	0.9608
0.05	0.2236	0.7071	1.0513	0.9512
0.06	0.2449	0.7746	1.0618	0.9418
0.07	0.2646	0.8367	1.0725	0.9324
0.08	0.2828	0.8944	1.0833	0.9231
0.09	0.3000	0.9487	1.0942	0.9139
0.10	0.3162	1.0000	1.1052	0.9048
0.11	0.3317	1.0488	1.1163	0.8958
0.12	0.3464	1.0954	1.1275	0.8869
0.13	0.3606	1.1402	1.1388	0.8781
0.14	0.3742	1.1832	1.1503	0.8694
0.15	0.3873	1.2247	1.1618	0.8607
0.16	0.4000	1.2649	1.1735	0.8521
0.17	0.4123	1.3038	1.1853	0.8437
0.18	0.4243	1.3416	1.1972	0.8353
0.19	0.4359	1.3784	1.2092	0.8270
0.20	0.4472	1.4142	1.2214	0.8187
0.21	0.4583	1.4491	1.2337	0.8106
0.22	0.4690	1.4832	1.2461	0.8025
0.23	0.4796	1.5166	1.2586	0.7945
0.24	0.4899	1.5492	1.2712	0.7866
0.25	0.5000	1.5811	1.2840	0.7788
0.26	0.5099	1.6125	1.2969	0.7711
0.27	0.5196	1.6432	1.3100	0.7634
0.28	0.5292	1.6733	1.3231	0.7558
0.29	0.5385	1.7029	1.3364	0.7483

x	$\sqrt{x}$	$\sqrt{10x}$	e^x	e^{-x}
0.30	0.5477	1.7321	1.3499	0.7408
0.31	0.5568	1.7607	1.3634	0.7334
0.32	0.5657	1.7889	1.3771	0.7261
0.33	0.5745	1.8166	1.3910	0.7189
0.34	0.5831	1.8439	1.4049	0.7118
0.35	0.5916	1.8708	1.4191	0.7047
0.36	0.6000	1.8974	1.4333	0.6977
0.37	0.6083	1.9235	1.4477	0.6907
0.38	0.6164	1.9494	1.4623	0.6839
0.39	0.6245	1.9748	1.4770	0.6771
0.40	0.6325	2.0000	1.4918	0.6703
0.41	0.6403	2.0248	1.5068	0.6637
0.42	0.6481	2.0494	1.5220	0.6570
0.43	0.6557	2.0736	1.5373	0.6505
0.44	0.6633	2.0976	1.5527	0.6440
0.45	0.6708	2.1213	1.5683	0.6376
0.46	0.6782	2.1448	1.5841	0.6313
0.47	0.6856	2.1679	1.6000	0.6250
0.48	0.6928	2.1909	1.6161	0.6188
0.49	0.7000	2.2136	1.6323	0.6126
0.50	0.7071	2.2361	1.6487	0.6065
0.51	0.7141	2.2583	1.6653	0.6005
0.52	0.7211	2.2804	1.6820	0.5945
0.53	0.7280	2.3022	1.6989	0.5886
0.54	0.7348	2.3238	1.7160	0.5827
0.55	0.7416	2.3452	1.7333	0.5769
0.56	0.7483	2.3664	1.7507	0.5712
0.57	0.7550	2.3875	1.7683	0.5655
0.58	0.7616	2.4083	1.7860	0.5599
0.59	0.7681	2.4290	1.8040	0.5543
0.60	0.7746	2.4495	1.8221	0.5488
0.61	0.7810	2.4698	1.8404	0.5434
0.62	0.7874	2.4900	1.8589	0.5379
0.63	0.7937	2.5100	1.8776	0.5326
0.64	0.8000	2.5298	1.8965	0.5273
0.65	0.8062	2.5495	1.9155	0.5220
0.66	0.8124	2.5690	1.9348	0.5169
0.67	0.8185	2.5884	1.9542	0.5117
0.68	0.8246	2.6077	1.9739	0.5066
0.69	0.8307	2.6268	1.9937	0.5016
0.70	0.8367	2.6458	2.0138	0.4966
0.71	0.8426	2.6646	2.0340	0.4916
0.72	0.8485	2.6833	2.0544	0.4868
0.73	0.8544	2.7019	2.0751	0.4819
0.74	0.8602	2.7203	2.0959	0.4771
0.75	0.8660	2.7386	2.1170	0.4724
0.76	0.8718	2.7568	2.1383	0.4677
0.77	0.8775	2.7749	2.1598	0.4630
0.78	0.8832	2.7928	2.1815	0.4584
0.79	0.8888	2.8107	2.2034	0.4538
0.80	0.8944	2.8284	2.2255	0.4493
0.81	0.9000	2.8460	2.2479	0.4449

x	$\sqrt{x}$	$\sqrt{10x}$	e^x	e^{-x}
0.82	0.9055	2.8636	2.2705	0.4404
0.83	0.9110	2.8810	2.2933	0.4360
0.84	0.9165	2.8983	2.3164	0.4317
0.85	0.9220	2.9155	2.3396	0.4274
0.86	0.9274	2.9326	2.3632	0.4232
0.87	0.9327	2.9496	2.3869	0.4190
0.88	0.9381	2.9665	2.4109	0.4148
0.89	0.9434	2.9833	2.4351	0.4107
0.90	0.9487	3.0000	2.4596	0.4066
0.91	0.9539	3.0166	2.4843	0.4025
0.92	0.9592	3.0332	2.5093	0.3985
0.93	0.9644	3.0496	2.5345	0.3946
0.94	0.9695	3.0659	2.5600	0.3906
0.95	0.9747	3.0822	2.5857	0.3867
0.96	0.9798	3.0984	2.6117	0.3829
0.97	0.9849	3.1145	2.6379	0.3791
0.98	0.9899	3.1305	2.6645	0.3753
0.99	0.9950	3.1464	2.6912	0.3716
1.00	1.0000	3.1623	2.7183	0.3679
1.01	1.0050	3.1780	2.7456	0.3642
1.02	1.0100	3.1937	2.7732	0.3606
1.03	1.0149	3.2094	2.8011	0.3570
1.04	1.0198	3.2249	2.8292	0.3535
1.05	1.0247	3.2404	2.8577	0.3499
1.06	1.0296	3.2558	2.8864	0.3465
1.07	1.0344	3.2711	2.9154	0.3430
1.08	1.0392	3.2863	2.9447	0.3396
1.09	1.0440	3.3015	2.9743	0.3362
1.10	1.0488	3.3166	3.0042	0.3329
1.11	1.0536	3.3317	3.0344	0.3296
1.12	1.0583	3.3466	3.0649	0.3263
1.13	1.0630	3.3615	3.0957	0.3230
1.14	1.0677	3.3764	3.1268	0.3198
1.15	1.0724	3.3912	3.1582	0.3166
1.16	1.0770	3.4059	3.1899	0.3135
1.17	1.0817	3.4205	3.2220	0.3104
1.18	1.0863	3.4351	3.2544	0.3073
1.19	1.0909	3.4496	3.2871	0.3042
1.20	1.0954	3.4641	3.3201	0.3012
1.21	1.1000	3.4785	3.3535	0.2982
1.22	1.1045	3.4928	3.3872	0.2952
1.23	1.1091	3.5071	3.4212	0.2923
1.24	1.1136	3.5214	3.4556	0.2894
1.25	1.1180	3.5355	3.4903	0.2865
1.26	1.1225	3.5496	3.5254	0.2837
1.27	1.1269	3.5637	3.5609	0.2808
1.28	1.1314	3.5777	3.5966	0.2780
1.29	1.1358	3.5917	3.6328	0.2753
1.30	1.1402	3.6056	3.6693	0.2725
1.31	1.1446	3.6194	3.7062	0.2698
1.32	1.1489	3.6332	3.7434	0.2671
1.33	1.1533	3.6469	3.7810	0.2645

x	$\sqrt{x}$	$\sqrt{10x}$	e^x	e^{-x}
1.34	1.1576	3.6606	3.8190	0.2618
1.35	1.1619	3.6742	4.8574	0.2592
1.36	1.1662	3.6878	3.8962	0.2567
1.37	1.1705	3.7014	3.9354	0.2541
1.38	1.1747	3.7148	3.9749	0.2516
1.39	1.1790	3.7283	4.0149	0.2491
1.40	1.1832	3.7417	4.0552	0.2466
1.41	1.1874	3.7550	4.0960	0.2441
1.42	1.1916	3.7683	4.1371	0.2417
1.43	1.1958	3.7815	4.1787	0.2393
1.44	1.2000	3.7947	4.2207	0.2369
1.45	1.2042	3.8079	4.2631	0.2346
1.46	1.2083	3.8210	4.3060	0.2322
1.47	1.2124	3.8341	4.3492	0.2299
1.48	1.2166	3.8471	4.3929	0.2276
1.49	1.2207	3.8601	4.4371	0.2254
1.50	1.2247	3.8730	4.4817	0.2231
1.51	1.2288	3.8859	4.5267	0.2209
1.52	1.2329	3.8987	4.5722	0.2187
1.53	1.2369	3.9115	4.6182	0.2165
1.54	1.2410	3.9243	4.6646	0.2144
1.55	1.2450	3.9370	4.7115	0.2122
1.56	1.2490	3.9497	4.7588	0.2101
1.57	1.2530	3.9623	4.8066	0.2080
1.58	1.2570	3.9749	4.8550	0.2060
1.59	1.2610	3.9875	4.9037	0.2039
1.60	1.2649	4.0000	4.9530	0.2019
1.61	1.2689	4.0125	5.0028	0.1999
1.62	1.2728	4.0249	5.0531	0.1979
1.63	1.2767	4.0373	5.1039	0.1959
1.64	1.2806	4.0497	5.1552	0.1940
1.65	1.2845	4.0620	5.2070	0.1920
1.66	1.2884	4.0743	5.2593	0.1901
1.67	1.2923	4.0866	5.3122	0.1882
1.68	1.2961	4.0988	5.3656	0.1864
1.69	1.3000	4.1110	5.4195	0.1845
1.70	1.3038	4.1231	5.4739	0.1827
1.71	1.3077	4.1352	5.5290	0.1809
1.72	1.3115	4.1473	5.5845	0.1791
1.73	1.3153	4.1593	5.6407	0.1773
1.74	1.3191	4.1713	5.6973	0.1755
1.75	1.3229	4.1833	5.7546	0.1738
1.76	1.3266	4.1952	5.8124	0.1720
1.77	1.3304	4.2071	5.8709	0.1703
1.78	1.3342	4.2190	5.9299	0.1686
1.79	1.3379	4.2308	5.9895	0.1670
1.80	1.3416	4.2426	6.0496	0.1653
1.81	1.3454	4.2544	6.1104	0.1637
1.82	1.3491	4.2661	6.1719	0.1620
1.83	1.3528	4.2778	6.2339	0.1604
1.84	1.3565	4.2895	6.2965	0.1588
1.85	1.3601	4.3012	6.3598	0.1572
1.86	1.3638	4.3128	6.4237	0.1557

x	$\sqrt{x}$	$\sqrt{10x}$	e^x	e^{-x}
1.87	1.3675	4.3243	6.4883	0.1541
1.88	1.3711	4.3359	6.5535	0.1526
1.89	1.3748	4.3474	6.6194	0.1511
1.90	1.3784	4.3589	6.6859	0.1496
1.91	1.3820	4.3704	6.7531	0.1481
1.92	1.3856	4.3818	6.8210	0.1466
1.93	1.3892	4.3932	6.8895	0.1451
1.94	1.3928	4.4045	6.9588	0.1437
1.95	1.3964	4.4159	7.0287	0.1423
1.96	1.4000	4.4272	7.0993	0.1409
1.97	1.4036	4.4385	7.1707	0.1395
1.98	1.4071	4.4497	7.2427	0.1381
1.99	1.4107	4.4609	7.3155	0.1367
2.00	1.4142	4.4721	7.3891	0.1353
2.01	1.4177	4.4833	7.4633	0.1340
2.02	1.4213	4.4944	7.5383	0.1327
2.03	1.4248	4.5056	7.6141	0.1313
2.04	1.4283	4.5166	7.6906	0.1300
2.05	1.4318	4.5277	7.7679	0.1287
2.06	1.4353	4.5387	7.8460	0.1275
2.07	1.4387	4.5497	7.9248	0.1262
2.08	1.4422	4.5607	8.0045	0.1249
2.09	1.4457	4.5717	8.0849	0.1237
2.10	1.4491	4.5826	8.1662	0.1225
2.11	1.4526	4.5935	8.2482	0.1212
2.12	1.4560	4.6043	8.3311	0.1200
2.13	1.4595	4.6152	8.4149	0.1188
2.14	1.4629	4.6260	8.4994	0.1177
2.15	1.4663	4.6368	8.5849	0.1165
2.16	1.4697	4.6476	8.6711	0.1153
2.17	1.4731	4.6583	8.7583	0.1142
2.18	1.4765	4.6690	8.8463	0.1130
2.19	1.4799	4.6797	8.9352	0.1119
2.20	1.4832	4.6904	9.0250	0.1108
2.21	1.4866	4.7011	9.1157	0.1097
2.22	1.4900	4.7117	9.2073	0.1086
2.23	1.4933	4.7223	9.2999	0.1075
2.24	1.4967	4.7329	9.3933	0.1065
2.25	1.5000	4.7434	9.4877	0.1054
2.26	1.5033	4.7539	9.5831	0.1044
2.27	1.5067	4.7645	9.6794	0.1033
2.28	1.5100	4.7749	9.7767	0.1023
2.29	1.5133	4.7854	9.8749	0.1013
2.30	1.5166	4.7958	9.9742	0.1003
2.31	1.5199	4.8062	10.0744	0.0993
2.32	1.5232	4.8166	10.1757	0.0983
2.33	1.5264	4.8270	10.2779	0.0973
2.34	1.5297	4.8374	10.3812	0.0963
2.35	1.5330	4.8477	10.4856	0.0954
2.36	1.5362	4.8580	10.5909	0.0944
2.37	1.5395	4.8683	10.6974	0.0935
2.38	1.5427	4.8785	10.8049	0.0926
2.39	1.5460	4.8888	10.9135	0.0916

x	$\sqrt{x}$	$\sqrt{10x}$	e^x	e^{-x}
2.40	1.5492	4.8990	11.0232	0.0907
2.41	1.5524	4.9092	11.1340	0.0898
2.42	1.5556	4.9193	11.2459	0.0889
2.43	1.5588	4.9295	11.3589	0.0880
2.44	1.5620	4.9396	11.4730	0.0872
2.45	1.5652	4.9497	11.5883	0.0863
2.46	1.5684	4.9598	11.7048	0.0854
2.47	1.5716	4.9699	11.8224	0.0846
2.48	1.5748	4.9800	11.9413	0.0837
2.49	1.5780	4.9900	12.0613	0.0829
2.50	1.5811	5.0000	12.1825	0.0821
2.51	1.5843	5.0100	12.3049	0.0813
2.52	1.5875	5.0200	12.4286	0.0805
2.53	1.5906	5.0299	12.5535	0.0797
2.54	1.5937	5.0398	12.6797	0.0789
2.55	1.5969	5.0498	12.8071	0.0781
2.56	1.6000	5.0596	12.9358	0.0773
2.57	1.6031	5.0695	13.0658	0.0765
2.58	1.6062	5.0794	13.1971	0.0758
2.59	1.6093	5.0892	13.3298	0.0750
2.60	1.6125	5.0990	13.4637	0.0743
2.61	1.6155	5.1088	13.5990	0.0735
2.62	1.6186	5.1186	13.7357	0.0728
2.63	1.6217	5.1284	13.8738	0.0721
2.64	1.6248	5.1381	14.0132	0.0714
2.65	1.6279	5.1478	14.1540	0.0707
2.66	1.6310	5.1575	14.2963	0.0699
2.67	1.6340	5.1672	14.4400	0.0693
2.68	1.6371	5.1769	14.5851	0.0686
2.69	1.6401	5.1865	14.7317	0.0679
2.70	1.6432	5.1962	14.8797	0.0672
2.71	1.6462	5.2058	15.0293	0.0665
2.72	1.6492	5.2154	15.1803	0.0659
2.73	1.6523	5.2249	15.3329	0.0652
2.74	1.6553	5.2345	15.4870	0.0646
2.75	1.6583	5.2440	15.6426	0.0639
2.76	1.6613	5.2536	15.7998	0.0633
2.77	1.6643	5.2631	15.9586	0.0627
2.78	1.6673	5.2726	16.1190	0.0620
2.79	1.6703	5.2820	16.2810	0.0614
2.80	1.6733	5.2915	16.4446	0.0608
2.81	1.6763	5.3009	16.6099	0.0602
2.82	1.6793	5.3104	16.7768	0.0596
2.83	1.6823	5.3198	16.9455	0.0590
2.84	1.6852	5.3292	17.1158	0.0584
2.85	1.6882	5.3385	17.2878	0.0578
2.86	1.6912	5.3479	17.4615	0.0573
2.87	1.6941	5.3572	17.6370	0.0567
2.88	1.6971	5.3666	17.8143	0.0561
2.89	1.7000	5.3759	17.9933	0.0556
2.90	1.7029	5.3852	18.1741	0.0550
2.91	1.7059	5.3944	18.3568	0.0545
2.92	1.7088	5.4037	18.5413	0.0539

x	$\sqrt{x}$	$\sqrt{10x}$	e^x	e^{-x}
2.93	1.7117	5.4129	18.7276	0.0534
2.94	1.7146	5.4222	18.9158	0.0529
2.95	1.7176	5.4314	19.1060	0.0523
2.96	1.7205	5.4406	19.2980	0.0518
2.97	1.7234	5.4498	19.4919	0.0513
2.98	1.7263	5.4589	19.6878	0.0503
2.99	1.7292	5.4681	19.8857	0.0508
3.00	1.7321	5.4772	20.0855	0.0498
3.01	1.7349	5.4863	20.2874	0.0493
3.02	1.7378	5.4955	20.4913	0.0488
3.03	1.7407	5.5045	20.6972	0.0483
3.04	1.7436	5.5136	20.9052	0.0478
3.05	1.7464	5.5227	21.1153	0.0474
3.06	1.7493	5.5317	21.3275	0.0469
3.07	1.7521	5.5408	21.5419	0.0464
3.08	1.7550	5.5498	21.7584	0.0460
3.09	1.7578	5.5588	21.9771	0.0455
3.10	1.7607	5.5678	22.1980	0.0450
3.11	1.7635	5.5767	22.4210	0.0446
3.12	1.7664	5.5857	22.6464	0.0442
3.13	1.7692	5.5946	22.8740	0.0437
3.14	1.7720	5.6036	23.1039	0.0433
3.15	1.7748	5.6125	23.3361	0.0429
3.16	1.7776	5.6214	23.5706	0.0424
3.17	1.7804	5.6303	23.8075	0.0420
3.18	1.7833	5.6391	24.0468	0.0416
3.19	1.7861	5.6480	24.2884	0.0412
3.20	1.7889	5.6569	24.5325	0.0408
3.21	1.7916	5.6657	24.7791	0.0404
3.22	1.7944	5.6745	25,0281	0.0400
3.23	1.7972	5.6833	25.2796	0.0396
3.24	1.8000	5.6921	25.5337	0.0392
3.25	1.8028	5.7009	25.7903	0.0388
3.26	1.8055	5.7096	26.0495	0.0384
3.27	1.8083	5.7184	26.3113	0.0380
3.28	1.8111	5.7271	26.5758	0.0376
3.29	1.8138	5.7359	26.8429	0.0373
3.30	1.8166	5.7446	27.1126	0.0369
3.31	1.8193	5.7533	27.3851	0.0365
3.32	1.8221	5.7619	27.6603	0.0362
3.33	1.8248	5.7706	27.9383	0.0358
3.34	1.8276	5.7793	28.2191	0.0354
3.35	1.8303	5.7879	28.5027	0.0351
3.36	1.8330	5.7966	28.7892	0.0347
3.37	1.8358	5.8052	29.0785	0.0344
3.38	1.8385	5.8138	29.3708	0.0340
3.39	1.8412	5.8224	29.6660	0.0337
3.40	1.8439	5.8310	29.9641	0.0334
3.41	1.8466	5.8395	30.2652	0.0330
3.42	1.8493	5.8481	30.5694	0.0327
3.43	1.8520	5.8566	30.8766	0.0324
3.44	1.8547	5.8652	31.1869	0.0321
3.45	1.8574	5.8737	31.5004	0.0317

x	$\sqrt{x}$	$\sqrt{10x}$	e^x	e^{-x}
3.46	1.8601	5.8822	31.8170	0.0314
3.47	1.8628	5.8907	32.1367	0.0311
3.48	1.8655	5.8992	32.4597	0.0308
3.49	1.8682	5.9076	32.7859	0.0305
3.50	1.8708	5.9161	33.1154	0.0302
3.51	1.8735	5.9245	33.4483	0.0299
3.52	1.8762	5.9330	33.7844	0.0296
3.53	1.8788	5.9414	34.1240	0.0293
3.54	1.8815	5.9498	34.4669	0.0290
3.55	1.8841	5.9582	34.8133	0.0287
3.56	1.8868	5.9666	35.1632	0.0284
3.57	1.8894	5.9749	35.5166	0.0282
3.58	1.8921	5.9833	35.8735	0.0279
3.59	1.8947	5.9917	36.2341	0.0276
3.60	1.8974	6.0000	36.5982	0.0273
3.61	1.9000	6.0083	36.9660	0.0271
3.62	1.9026	6.0166	37.3376	0.0268
3.63	1.9053	6.0249	37.7128	0.0265
3.64	1.9079	6.0332	38.0918	0.0263
3.65	1.9105	6.0415	38.4747	0.0260
3.66	1.9131	6.0498	38.8613	0.0257
3.67	1.9157	6.0581	39.2519	0.0255
3.68	1.9183	6.0663	39.6464	0.0252
3.69	1,9209	6.0745	40.0448	0.0250
3.70	1.9235	6.0828	40.4473	0.0247
3.71	1.9261	6.0910	40.8538	0.0245
3.72	1.9287	6.0992	41.2644	0.0242
3.73	1.9313	6.1074	41.6791	0.0240
3.74	1.9339	6.1156	42.0980	0.0238
3.75	1.9365	6.1237	42.5211	0.0235
3.76	1.9391	6.1319	42.9484	0.0233
3.77	1.9416	6.1400	43.3800	0.0231
3.78	1.9442	6.1482	43.8160	0.0228
3.79	1.9468	6.1563	44.2564	0.0226
3.80	1.9494	6.1644	44.7012	0.0224
3.81	1.9519	6.1725	45.1504	0.0221
3.82	1.9545	6.1806	45.6042	0.0219
3.83	1.9570	6.1887	46.0625	0.0217
3.84	1.9596	6.1968	46.5255	0.0215
3.85	1.9621	6.2048	46.9931	0.0213
3.86	1.9647	6.2129	47.4653	0.0211
3.87	1.9672	6.2209	47.9424	0.0209
3.88	1.9698	6.2290	48.4242	0.0207
3.89	1.9723	6.2370	48.9109	0.0204
3.90	1.9748	6.2450	49.4024	0.0202
3.91	1.9774	6.2530	49.8989	0.0200
3.92	1.9799	6.2610	50.4004	0.0198
3.93	1.9824	6.2690	50.9070	0.0196
3.94	1.9849	6.2769	51.4186	0.0194
3.95	1.9875	6.2849	51.9354	0.0193
3.96	1.9900	6.2929	52.4573	0.0191
3.97	1.9925	6.3008	52.9845	0.0189

x	$\sqrt{x}$	$\sqrt{10x}$	e^x	e^{-x}
3.98	1.9950	6.3087	53.5170	0.0187
3.99	1.9975	6.3166	54.0549	0.0185
4.00	2.0000	6.3246	54.5981	0.0183
4.01	2.0025	6.3325	55.1468	0.0181
4.02	2.0050	6.3403	55.7011	0.0180
4.03	2.0075	6.3482	56.2609	0.0178
4.04	2.0100	6.3561	56.8263	0.0176
4.05	2.0125	6.3640	57.3974	0.0174
4.06	2.0149	6.3718	57.9743	0.0172
4.07	2.0174	6.3797	58.5569	0.0171
4.08	2.0199	6.3875	59.1454	0.0169
4.09	2.0224	6.3953	59.7398	0.0167
4.10	2.0248	6.4031	60.3403	0.0166
4.11	2.0273	6.4109	60.9467	0.0164
4.12	2.0298	6.4187	61.5592	0.0162
4.13	2.0322	6.4265	62.1779	0.0161
4.14	2.0347	6.4343	62.8028	0.0159
4.15	2.0372	6.4420	63.4340	0.0158
4.16	2.0396	6.4498	64.0715	0.0156
4.17	2.0421	6.4576	64.7154	0.0155
4.18	2.0445	6.4653	65.3658	0.0153
4.19	2.0469	6.4730	66.0228	0.0151
4.20	2.0494	6.4807	66.6863	0.0150
4.21	2.0518	6.4885	67.3565	0.0148
4.22	2.0543	6.4962	68.0334	0.0147
4.23	2.0567	6.5038	68.7172	0.0146
4.24	2.0591	6.5115	69.4078	0.0144
4.25	2.0616	6.5192	70.1054	0.0143
4.26	2.0640	6.5269	70.8099	0.0141
4.27	2.0664	6.5345	71.5216	0.0140
4.28	2.0688	6.5422	72.2404	0.0138
4.29	2.0712	6.5498	72.9664	0.0137
4.30	2.0736	6.5574	73.6997	0.0136
4.31	2.0761	6.5651	74.4405	0.0134
4.32	2.0785	6.5727	75.1886	0.0133
4.33	2.0809	6.5803	75.9442	0.0132
4.34	2.0833	6.5879	76.7075	0.0130
4.35	2.0857	6.5955	77.4784	0.0129
4.36	2.0881	6.6030	78.2571	0.0128
4.37	2.0905	6.6106	79.0436	0.0127
4.38	2.0928	6,6182	79.8380	0.0125
4.39	2.0952	6.6257	80.6404	0.0124
4.40	2.0976	6.6332	81.4508	0.0123
4.41	2.1000	6.6408	82.2695	0.0122
4.42	2.1024	6.6483	83.0962	0.0120
4.43	2.1048	6.6558	83.9314	0.0119
4.44	2.1071	6.6633	84.7749	0.0118
4.45	2.1095	6.6708	85.6269	0.0117
4.46	2.1119	6.6783	86.4874	0.0116
4.47	2.1142	6.6858	87.3567	0.0114
4.48	2.1166	6.6933	88.2346	0.0113
4.49	2.1190	6.7007	89.1214	0.0112

x	$\sqrt{x}$	$\sqrt{10x}$	e^x	e^{-x}
4.50	2.1213	6.7082	90.0171	0.0111
4.51	2.1237	6.7157	90.9218	0.0110
4.52	2.1260	6.7231	91.8356	0.0109
4.53	2.1284	6.7305	92.7585	0.0108
4.54	2.1307	6.7380	93.6907	0.0107
4.55	2.1331	6.7454	94.6323	0.0106
4.56	2.1354	6.7528	95.5834	0.0105
4.57	2.1378	6.7602	96.5441	0.0104
4.58	2.1401	6.7676	97.5143	0.0103
4.59	2.1424	6.7750	98.4944	0.0102
4.60	2.1448	6.7823	99.4843	0.0101
4.61	2.1471	6.7897	100.4841	0.0100
4.62	2.1494	6.7971	101.4940	0.0099
4.63	2.1517	6.8044	102.5140	0.0098
4.64	2.1541	6.8118	103.5443	0.0097
4.65	2.1564	6.8191	104.5850	0.0096
4.66	2.1587	6.8264	105.6361	0.0095
4.67	2.1610	6.8337	106.6977	0.0094
4.68	2.1633	6.8411	107.7700	0.0093
4.69	2.1656	6.8484	108.8531	0.0092
4.70	2.1679	6.8557	109.9472	0.0091
4.71	2.1703	6.8629	111.0521	0.0090
4.72	2.1726	6.8702	112.1682	0.0089
4.73	2.1749	6.8775	113.2955	0.0088
4.74	2.1772	6.8848	114.4342	0.0087
4.75	2.1794	6.8920	115.5842	0.0087
4.76	2.1817	6.8993	116.7458	0.0086
4.77	2.1840	6.9065	117.9192	0.0085
4.78	2.1863	6.9138	119.1043	0.0084
4.79	2.1886	6.9210	120.3013	0.0083
4.80	2.1909	6.9282	121.5103	0.0082
4.81	2.1932	6.9354	122.7316	0.0081
4.82	2.1954	6.9426	123.9651	0.0081
4.83	2.1977	6.9498	125.2108	0.0080
4.84	2.2000	6.9570	126.4692	0.0079
4.85	2.2023	6.9642	127.7403	0.0078
4.86	2.2045	6.9714	129.0242	0.0078
4.87	2.2068	6.9785	130.3209	0.0077
4.88	2.2091	6.9857	131.6306	0.0076
4.89	2.2113	6.9929	132.9535	0.0075
4.90	2.2136	7.0000	134.2897	0.0074
4.91	2.2159	7.0071	135.6394	0.0074
4.92	2.2181	7.0143	137.0025	0.0073
4.93	2.2204	7.0214	138.3794	0.0072
4.94	2.2226	7.0285	139.7702	0.0072
4.95	2.2249	7.0356	141.1749	0.0071
4.96	2.2271	7.0427	142.5937	0.0070
4.97	2.2293	7.0498	144.0268	0.0069
4.98	2.2316	7.0569	145.4743	0.0069
4.99	2.2338	7.0640	146.9364	0.0068
5.00	2.2361	7.0711	148.4130	0.0067

Table B Binomial Probabilities

Use of Table: For various choices of x, n, and p this table gives

$$B(x; n, p) = C(n, x) p^x q^{n-x}$$

Note that p goes only to .50, but the fact that

$$B(x; n, p) = B(n - x; n, q)$$

allows us, for example, to find

$$B(4; 9, .75) = B(5; 9, .25) = .0389$$

	x	p = .10	p = .20	p = .25	p = .30	p = 1/3	p = .40	p = .50	$n - x$
	0	0.5905	0.3277	0.2373	0.1681	0.1317	0.0778	0.0313	5
	1	0.3280	0.4096	0.3955	0.3601	0.3292	0.2592	0.1563	4
	2	0.0729	0.2048	0.2637	0.3087	0.3292	0.3446	0.3125	3
n = 5	3	0.0081	0.0512	0.0879	0.1323	0.1646	0.2304	0.3125	2
	4	0.0004	0.0064	0.0146	0.0283	0.0412	0.0768	0.1563	1
	5	0.0000	0.0003	0.0010	0.0024	0.0041	0.0102	0.0313	0
	0	0.5314	0.2621	0.1780	0.1176	0.0878	0.0467	0.0156	6
	1	0.3543	0.3932	0.3560	0.3025	0.2634	0.1866	0.0938	5
	2	0.0984	0.2458	0.2966	0.3241	0.3292	0.3110	0.2344	4
n = 6	3	0.0146	0.0819	0.1318	0.1852	0.2195	0.2765	0.3125	3
	4	0.0012	0.0154	0.0330	0.0595	0.0823	0.1382	0.2344	2
	5	0.0001	0.0015	0.0044	0.0102	0.0165	0.0369	0.0937	1
	6	0.0000	0.0001	0.0002	0.0007	0.0014	0.0041	0.0156	0
	0	0.4783	0.2097	0.1335	0.0824	0.0585	0.0280	0.0078	7
	1	0.3720	0.3670	0.3115	0.2471	0.2048	0.1306	0.0547	6
	2	0.1240	0.2753	0.3115	0.3177	0.3073	0.2613	0.1641	5
	3	0.0230	0.1147	0.1730	0.2269	0.2561	0.2903	0.2734	4
n = 7	4	0.0026	0.0287	0.0577	0.0972	0.1280	0.1935	0.2734	3
	5	0.0002	0.0043	0.0115	0.0250	0.0384	0.0774	0.1641	2
	6	0.0000	0.0004	0.0013	0.0036	0.0064	0.0172	0.0547	1
	7	0.0000	0.0000	0.0001	0.0002	0.0005	0.0016	0.0078	0
	0	0.4305	0.1678	0.1001	0.0576	0.0390	0.0168	0.0039	8
	1	0.3826	0.3355	0.2670	0.1977	0.1561	0.0896	0.0313	7
	2	0.1488	0.2936	0.3115	0.2965	0.2731	0.2090	0.1094	6
	3	0.0331	0.1468	0.2076	0.2541	0.2731	0.2787	0.2188	5
n = 8	4	0.0046	0.0459	0.0865	0.1363	0.1707	0.2322	0.2734	4
	5	0.0004	0.0092	0.0231	0.0467	0.0683	0.1239	0.2187	3
	6	0.0000	0.0011	0.0038	0.0100	0.0171	0.0413	0.1094	2
	7	0.0000	0.0001	0.0004	0.0012	0.0024	0.0079	0.0312	1
	8	0.0000	0.0000	0.0000	0.0001	0.0002	0.0007	0.0039	0

	x	p = .10	p = .20	p = .25	p = .30	p = 1/3	p = .40	p = .50	$n-x$
	0	0.3874	0.1342	0.0751	0.0404	0.0260	0.0101	0.0020	9
	1	0.3874	0.3020	0.2253	0.1556	0.1171	0.0605	0.0176	8
	2	0.1722	0.3020	0.3003	0.2668	0.2341	0.1612	0.0703	7
	3	0.0446	0.1762	0.2336	0.2668	0.2731	0.2508	0.1641	6
	4	0.0074	0.0661	0.1168	0.1715	0.2048	0.2508	0.2461	5
n = 9	5	0.0008	0.0165	0.0389	0.0735	0.1024	0.1672	0.2461	4
	6	0.0001	0.0028	0.0087	0.0210	0.0341	0.0743	0.1641	3
	7	0.0000	0.0003	0.0012	0.0039	0.0073	0.0212	0.0703	2
	8	0.0000	0.0000	0.0001	0.0004	0.0009	0.0035	0.0176	1
	9	0.0000	0.0000	0.0000	0.0000	0.0001	0.0003	0.0020	0
	0	0.3487	0.1074	0.0563	0.0282	0.0173	0.0060	0.0010	10
	1	0.3874	0.2684	0.1877	0.1211	0.0867	0.0403	0.0098	9
	2	0.1937	0.3020	0.2816	0.2335	0.1951	0.1209	0.0439	8
	3	0.0574	0.2013	0.2503	0.2668	0.2601	0.2150	0.1172	7
	4	0.0112	0.0881	0.1460	0.2001	0.2276	0.2508	0.2051	6
n = 10	5	0.0015	0.0264	0.0584	0.1029	0.1366	0.2007	0.2461	5
	6	0.0001	0.0055	0.0162	0.0368	0.0569	0.1115	0.2051	4
	7	0.0000	0.0008	0.0031	0.0090	0.0163	0.0425	0.1172	3
	8	0.0000	0.0001	0.0004	0.0014	0.0030	0.0106	0.0439	2
	9	0.0000	0.0000	0.0000	0.0001	0.0003	0.0016	0.0098	1
	10	0.0000	0.0000	0.0000	0.0000	0.0000	0.0001	0.0010	0
	0	0.3138	0.0859	0.0422	0.0198	0.0116	0.0036	0.0005	11
	1	0.3835	0.2362	0.1549	0.0932	0.0636	0.0266	0.0054	10
	2	0.2131	0.2953	0.2581	0.1998	0.1590	0.0887	0.0269	9
	3	0.0710	0.2215	0.2581	0.2568	0.2384	0.1774	0.0806	8
	4	0.0158	0.1107	0.1721	0.2201	0.2384	0.2365	0.1611	7
n = 11	5	0.0025	0.0388	0.0803	0.1321	0.1669	0.2207	0.2256	6
	6	0.0003	0.0097	0.0268	0.0566	0.0835	0.1471	0.2256	5
	7	0.0000	0.0017	0.0064	0.0173	0.0298	0.0701	0.1611	4
	8	0.0000	0.0002	0.0011	0.0037	0.0075	0.0234	0.0806	3
	9	0.0000	0.0000	0.0001	0.0005	0.0012	0.0052	0.0269	2
	10	0.0000	0.0000	0.0000	0.0000	0.0001	0.0007	0.0054	1
	11	0.0000	0.0000	0.0000	0.0000	0.0000	0.0000	0.0005	0
	0	0.2824	0.0687	0.0317	0.0138	0.0077	0.0022	0.0002	12
	1	0.3766	0.2062	0.1267	0.0712	0.0462	0.0174	0.0029	11
	2	0.2301	0.2835	0.2323	0.1678	0.1272	0.0639	0.0161	10
	3	0.0852	0.2362	0.2581	0.2397	0.2120	0.1419	0.0537	9
	4	0.0213	0.1329	0.1936	0.2311	0.2384	0.2128	0.1208	8
	5	0.0038	0.0532	0.1032	0.1585	0.1908	0.2270	0.1934	7
n = 12	6	0.0005	0.0155	0.0401	0.0792	0.1113	0.1766	0.2256	6
	7	0.0000	0.0033	0.0115	0.0291	0.0477	0.1009	0.1934	5
	8	0.0000	0.0005	0.0024	0.0078	0.0149	0.0420	0.1208	4
	9	0.0000	0.0001	0.0004	0.0015	0.0033	0.0125	0.0537	3
	10	0.0000	0.0000	0.0000	0.0002	0.0005	0.0025	0.0161	2
	11	0.0000	0.0000	0.0000	0.0000	0.0000	0.0003	0.0029	1
	12	0.0000	0.0000	0.0000	0.0000	0.0000	0.0000	0.0002	0
	0	0.2542	0.0550	0.0238	0.0097	0.0051	0.0013	0.0001	13
	1	0.3672	0.1787	0.1029	0.0540	0.0334	0.0113	0.0016	12
n = 13	2	0.2448	0.2680	0.2059	0.1388	0.1002	0.0453	0.0095	11
	3	0.0997	0.2457	0.2517	0.2181	0.1837	0.1107	0.0349	10

	x	$p = .10$	$p = .20$	$p = .25$	$p = .30$	$p = 1/3$	$p = .40$	$p = .50$	$n - x$
	4	0.0277	0.1535	0.2097	0.2337	0.2296	0.1845	0.0873	9
	5	0.0055	0.0691	0.1258	0.1803	0.2067	0.2214	0.1571	8
	6	0.0008	0.0230	0.0559	0.1030	0.1378	0.1968	0.2095	7
	7	0.0001	0.0058	0.0186	0.0442	0.0689	0.1312	0.2095	6
$n = 13$	8	0.0000	0.0011	0.0047	0.0142	0.0258	0.0656	0.1571	5
(contd.)	9	0.0000	0.0001	0.0009	0.0034	0.0072	0.0243	0.0873	4
	10	0.0000	0.0000	0.0001	0.0006	0.0014	0.0065	0.0349	3
	11	0.0000	0.0000	0.0000	0.0001	0.0002	0.0012	0.0095	2
	12	0.0000	0.0000	0.0000	0.0000	0.0000	0.0001	0.0016	1
	13	0.0000	0.0000	0.0000	0.0000	0.0000	0.0000	0.0001	0
	0	0.2288	0.0440	0.0178	0.0068	0.0034	0.0008	0.0001	14
	1	0.3559	0.1539	0.0832	0.0407	0.0240	0.0073	0.0009	13
	2	0.2570	0.2501	0.1802	0.1134	0.0779	0.0317	0.0056	12
	3	0.1142	0.2501	0.2402	0.1943	0.1559	0.0845	0.0222	11
	4	0.0349	0.1720	0.2202	0.2290	0.2143	0.1549	0.1611	10
	5	0.0078	0.0860	0.1468	0.1963	0.2143	0.2066	0.1222	9
	6	0.0013	0.0322	0.0734	0.1262	0.1607	0.2066	0.1833	8
$n = 14$	7	0.0002	0.0092	0.0280	0.0618	0.0918	0.1574	0.2095	7
	8	0.0000	0.0020	0.0082	0.0232	0.0402	0.0918	0.1833	6
	9	0.0000	0.0003	0.0018	0.0066	0.0134	0.0408	0.1222	5
	10	0.0000	0.0000	0.0003	0.0014	0.0033	0.0136	0.0611	4
	11	0.0000	0.0000	0.0000	0.0002	0.0006	0.0033	0.0222	3
	12	0.0000	0.0000	0.0000	0.0000	0.0001	0.0005	0.0056	2
	13	0.0000	0.0000	0.0000	0.0000	0.0000	0.0001	0.0009	1
	14	0.0000	0.0000	0.0000	0.0000	0.0000	0.0000	0.0001	0
	0	0.2059	0.0352	0.0134	0.0047	0.0023	0.0005	0.0000	15
	1	0.3432	0.1319	0.0668	0.0305	0.0171	0.0047	0.0005	14
	2	0.2669	0.2309	0.1559	0.0916	0.0599	0.0219	0.0032	13
	3	0.1285	0.2501	0.2252	0.1700	0.1299	0.0634	0.0139	12
	4	0.0428	0.1876	0.2252	0.2186	0.1948	0.1268	0.0417	11
	5	0.0105	0.1032	0.1651	0.2061	0.2143	0.1859	0.0916	10
	6	0.0019	0.0430	0.0917	0.1472	0.1786	0.2066	0.1527	9
	7	0.0003	0.0138	0.0393	0.0811	0.1148	0.1771	0.1964	8
$n = 15$	8	0.0000	0.0035	0.0131	0.0348	0.0574	0.1181	0.1964	7
	9	0.0000	0.0007	0.0034	0.0116	0.0223	0.0612	0.1527	6
	10	0.0000	0.0001	0.0007	0.0030	0.0067	0.0245	0.0916	5
	11	0.0000	0.0000	0.0001	0.0006	0.0015	0.0074	0.0417	4
	12	0.0000	0.0000	0.0000	0.0001	0.0003	0.0016	0.0139	3
	13	0.0000	0.0000	0.0000	0.0000	0.0000	0.0003	0.0032	2
	14	0.0000	0.0000	0.0000	0.0000	0.0000	0.0000	0.0005	1
	15	0.0000	0.0000	0.0000	0.0000	0.0000	0.0000	0.0000	0
	0	0.1853	0.0281	0.0100	0.0033	0.0015	0.0003	0.0000	16
	1	0.3294	0.1126	0.0535	0.0228	0.0122	0.0030	0.0002	15
	2	0.2745	0.2111	0.1336	0.0732	0.0457	0.0150	0.0018	14
	3	0.1423	0.2463	0.2079	0.1465	0.1066	0.0468	0.0085	13
$n = 16$	4	0.0514	0.2001	0.2252	0.2040	0.1732	0.1014	0.0278	12
	5	0.0137	0.1201	0.1802	0.2099	0.2078	0.1623	0.0667	11
	6	0.0028	0.0550	0.1101	0.1649	0.1905	0.1983	0.1222	10
	7	0.0004	0.0197	0.0524	0.1010	0.1361	0.1889	0.1746	9
	8	0.0001	0.0055	0.0197	0.0487	0.0765	0.1417	0.1964	8

	x	p = .10	p = .20	p = .25	p = .30	p = 1/3	p = .40	p = .50	$n - x$
	9	0.0000	0.0012	0.0058	0.0185	0.0340	0.0840	0.1746	7
	10	0.0000	0.0002	0.0014	0.0056	0.0119	0.0392	0.1222	6
	11	0.0000	0.0000	0.0002	0.0013	0.0032	0.0142	0.0667	5
n = 16	12	0.0000	0.0000	0.0000	0.0002	0.0007	0.0040	0.0278	4
(contd.)	13	0.0000	0.0000	0.0000	0.0000	0.0001	0.0008	0.0085	3
	14	0.0000	0.0000	0.0000	0.0000	0.0000	0.0001	0.0018	2
	15	0.0000	0.0000	0.0000	0.0000	0.0000	0.0000	0.0002	1
	16	0.0000	0.0000	0.0000	0.0000	0.0000	0.0000	0.0000	0
	0	0.1668	0.0225	0.0075	0.0023	0.0010	0.0002	0.0000	17
	1	0.3150	0.0957	0.0426	0.0169	0.0086	0.0019	0.0001	16
	2	0.2800	0.1914	0.1136	0.0581	0.0345	0.0102	0.0010	15
	3	0.1556	0.2393	0.1893	0.1245	0.0863	0.0341	0.0052	14
	4	0.0605	0.2093	0.2209	0.1868	0.1510	0.0796	0.0182	13
	5	0.0175	0.1361	0.1914	0.2081	0.1963	0.1379	0.0472	12
	6	0.0039	0.0680	0.1276	0.1784	0.1963	0.1839	0.0944	11
	7	0.0007	0.0267	0.0668	0.1201	0.1542	0.1927	0.1484	10
	8	0.0001	0.0084	0.0279	0.0644	0.0964	0.1606	0.1855	9
n = 17	9	0.0000	0.0021	0.0093	0.0276	0.0482	0.1070	0.1855	8
	10	0.0000	0.0004	0.0025	0.0095	0.0193	0.0571	0.1484	7
	11	0.0000	0.0001	0.0005	0.0026	0.0061	0.0242	0.0944	6
	12	0.0000	0.0000	0.0001	0.0006	0.0015	0.0081	0.0472	5
	13	0.0000	0.0000	0.0000	0.0001	0.0003	0.0021	0.0182	4
	14	0.0000	0.0000	0.0000	0.0000	0.0000	0.0004	0.0052	3
	15	0.0000	0.0000	0.0000	0.0000	0.0000	0.0001	0.0010	2
	16	0.0000	0.0000	0.0000	0.0000	0.0000	0.0000	0.0001	1
	17	0.0000	0.0000	0.0000	0.0000	0.0000	0.0000	0.0000	0
	0	0.1501	0.0180	0.0056	0.0016	0.0007	0.0001	0.0000	18
	1	0.3002	0.0811	0.0338	0.0126	0.0061	0.0012	0.0001	17
	2	0.2835	0.1723	0.0958	0.0458	0.0259	0.0069	0.0006	16
	3	0.1680	0.2297	0.1704	0.1046	0.0690	0.0246	0.0031	15
	4	0.0700	0.2153	0.2130	0.1681	0.1294	0.0614	0.0117	14
	5	0.0218	0.1507	0.1988	0.2017	0.1812	0.1146	0.0327	13
	6	0.0052	0.0816	0.1436	0.1873	0.1963	0.1655	0.0708	12
	7	0.0010	0.0350	0.0820	0.1376	0.1682	0.1892	0.1214	11
	8	0.0002	0.0120	0.0376	0.0811	0.1157	0.1734	0.1669	10
n = 18	9	0.0000	0.0033	0.0139	0.0386	0.0643	0.1284	0.1855	9
	10	0.0000	0.0008	0.0042	0.0149	0.0289	0.0771	0.1669	8
	11	0.0000	0.0001	0.0010	0.0046	0.0105	0.0374	0.1214	7
	12	0.0000	0.0000	0.0002	0.0012	0.0031	0.0145	0.0708	6
	13	0.0000	0.0000	0.0000	0.0002	0.0007	0.0045	0.0327	5
	14	0.0000	0.0000	0.0000	0.0000	0.0001	0.0011	0.0117	4
	15	0.0000	0.0000	0.0000	0.0000	0.0000	0.0002	0.0031	3
	16	0.0000	0.0000	0.0000	0.0000	0.0000	0.0000	0.0006	2
	17	0.0000	0.0000	0.0000	0.0000	0.0000	0.0000	0.0001	1
	18	0.0000	0.0000	0.0000	0.0000	0.0000	0.0000	0.0000	0
	0	0.1351	0.0144	0.0042	0.0011	0.0005	0.0001	0.0000	19
	1	0.2852	0.0685	0.0268	0.0093	0.0043	0.0008	0.0000	18
n = 19	2	0.2852	0.1540	0.0803	0.0358	0.0193	0.0046	0.0003	17
	3	0.1796	0.2182	0.1517	0.0869	0.0546	0.0175	0.0018	16
	4	0.0798	0.2182	0.2023	0.1491	0.1093	0.0467	0.0074	15

	x	$p = .10$	$p = .20$	$p = .25$	$p = .30$	$p = 1/3$	$p = .40$	$p = .50$	$n - x$
	5	0.0266	0.1636	0.2023	0.1916	0.1639	0.0933	0.0222	14
	6	0.0069	0.0955	0.1574	0.1916	0.1912	0.1451	0.0518	13
	7	0.0014	0.0443	0.0974	0.1525	0.1776	0.1797	0.0961	12
	8	0.0002	0.0166	0.0487	0.0981	0.1332	0.1797	0.1442	11
	9	0.0000	0.0051	0.0198	0.0514	0.0814	0.1464	0.1762	10
$n = 19$	10	0.0000	0.0013	0.0066	0.0220	0.0407	0.0976	0.1762	9
(contd.)	11	0.0000	0.0003	0.0018	0.0077	0.0166	0.0532	0.1442	8
	12	0.0000	0.0000	0.0004	0.0022	0.0055	0.0237	0.0961	7
	13	0.0000	0.0000	0.0001	0.0005	0.0015	0.0085	0.0518	6
	14	0.0000	0.0000	0.0000	0.0001	0.0003	0.0024	0.0222	5
	15	0.0000	0.0000	0.0000	0.0000	0.0001	0.0005	0.0074	4
	16	0.0000	0.0000	0.0000	0.0000	0.0000	0.0001	0.0018	3
	17	0.0000	0.0000	0.0000	0.0000	0.0000	0.0000	0.0003	2
	18	0.0000	0.0000	0.0000	0.0000	0.0000	0.0000	0.0000	1
	19	0.0000	0.0000	0.0000	0.0000	0.0000	0.0000	0.0000	0
	0	0.1216	0.0115	0.0032	0.0008	0.0003	0.0000	0.0000	20
	1	0.2702	0.0576	0.0211	0.0068	0.0030	0.0005	0.0000	19
	2	0.2852	0.1369	0.0669	0.0278	0.0143	0.0031	0.0002	18
	3	0.1901	0.2054	0.1339	0.0716	0.0429	0.0123	0.0011	17
	4	0.0898	0.2182	0.1897	0.1304	0.0911	0.0350	0.0046	16
	5	0.0319	0.1746	0.2023	0.1789	0.1457	0.0746	0.0148	15
	6	0.0089	0.1091	0.1686	0.1916	0.1821	0.1244	0.0370	14
	7	0.0020	0.0545	0.1124	0.1643	0.1821	0.1659	0.0739	13
	6	0.0004	0.0222	0.0609	0.1144	0.1480	0.1797	0.1201	12
	9	0.0001	0.0074	0.0271	0.0654	0.0987	0.1597	0.1602	11
$n = 20$	10	0.0000	0.0020	0.0099	0.0308	0.0543	0.1171	0.1762	10
	11	0.0000	0.0005	0.0030	0.0120	0.0247	0.0710	0.1602	9
	12	0.0000	0.0001	0.0008	0.0039	0.0092	0.0355	0.1201	8
	13	0.0000	0.0000	0.0002	0.0010	0.0028	0.0146	0.0739	7
	14	0.0000	0.0000	0.0000	0.0002	0.0007	0.0049	0.0370	6
	15	0.0000	0.0000	0.0000	0.0000	0.0001	0.0013	0.0148	5
	16	0.0000	0.0000	0.0000	0.0000	0.0000	0.0003	0.0046	4
	17	0.0000	0.0000	0.0000	0.0000	0.0000	0.0000	0.0011	3
	18	0.0000	0.0000	0.0000	0.0000	0.0000	0.0000	0.0002	2
	19	0.0000	0.0000	0.0000	0.0000	0.0000	0.0000	0.0000	1
	20	0.0000	0.0000	0.0000	0.0000	0.0000	0.0000	0.0000	0
	0	0.1094	0.0092	0.0024	0.0006	0.0002	0.0000	0.0000	21
	1	0.2553	0.0484	0.0166	0.0050	0.0021	0.0003	0.0000	20
	2	0.2837	0.1211	0.0555	0.0215	0.0105	0.0020	0.0001	19
	3	0.1996	0.1917	0.1172	0.0585	0.0333	0.0086	0.0006	18
	4	0.0998	0.2156	0.1757	0.1128	0.0750	0.0259	0.0029	17
	5	0.0377	0.1833	0.1992	0.1643	0.1275	0.0588	0.0097	16
	6	0.0112	0.1222	0.1770	0.1878	0.1700	0.1045	0.0259	15
$n = 21$	7	0.0027	0.0655	0.1265	0.1725	0.1821	0.1493	0.0554	14
	8	0.0005	0.0286	0.0738	0.1294	0.1594	0.1742	0.0970	13
	9	0.0001	0.0103	0.0355	0.0801	0.1151	0.1677	0.1402	12
	10	0.0000	0.0031	0.0142	0.0412	0.0691	0.1342	0.1682	11
	11	0.0000	0.0008	0.0047	0.0176	0.0345	0.0895	0.1682	10
	12	0.0000	0.0002	0.0013	0.0063	0.0144	0.0497	0.1402	9
	13	0.0000	0.0000	0.0003	0.0019	0.0050	0.0229	0.0970	8

	x	$p = .10$	$p = .20$	$p = .25$	$p = .30$	$p = 1/3$	$p = .40$	$p = .50$	$n - x$
	14	0.0000	0.0000	0.0001	0.0005	0.0014	0.0087	0.0554	7
	15	0.0000	0.0000	0.0000	0.0001	0.0003	0.0027	0.0259	6
	16	0.0000	0.0000	0.0000	0.0000	0.0001	0.0007	0.0097	5
$n = 21$	17	0.0000	0.0000	0.0000	0.0000	0.0000	0.0001	0.0029	4
(contd.)	18	0.0000	0.0000	0.0000	0.0000	0.0000	0.0000	0.0006	3
	19	0.0000	0.0000	0.0000	0.0000	0.0000	0.0000	0.0001	2
	20	0.0000	0.0000	0.0000	0.0000	0.0000	0.0000	0.0000	1
	21	0.0000	0.0000	0.0000	0.0000	0.0000	0.0000	0.0000	0
	0	0.0985	0.0074	0.0018	0.0004	0.0001	0.0000	0.0000	22
	1	0.2407	0.0406	0.0131	0.0037	0.0015	0.0002	0.0000	21
	2	0.2808	0.1065	0.0458	0.0166	0.0077	0.0014	0.0001	20
	3	0.2080	0.1775	0.1017	0.0474	0.0257	0.0060	0.0004	19
	4	0.1098	0.2108	0.1611	0.0965	0.0611	0.0190	0.0017	18
	5	0.0439	0.1898	0.1933	0.1489	0.1100	0.0456	0.0063	17
	6	0.0138	0.1344	0.1826	0.1808	0.1558	0.0862	0.0178	16
	7	0.0035	0.0768	0.1391	0.1771	0.1781	0.1314	0.0407	15
	8	0.0007	0.0360	0.0869	0.1423	0.1670	0.1642	0.0762	14
	9	0.0001	0.0140	0.0451	0.0949	0.1299	0.1703	0.1186	13
	10	0.0000	0.0046	0.0195	0.0529	0.0844	0.1476	0.1542	12
$n = 22$	11	0.0000	0.0012	0.0071	0.0247	0.0460	0.1073	0.1682	11
	12	0.0000	0.0003	0.0022	0.0097	0.0211	0.0656	0.1542	10
	13	0.0000	0.0001	0.0006	0.0032	0.0081	0.0336	0.1186	9
	14	0.0000	0.0000	0.0001	0.0009	0.0026	0.0144	0.0762	8
	15	0.0000	0.0000	0.0000	0.0002	0.0007	0.0051	0.0407	7
	16	0.0000	0.0000	0.0000	0.0000	0.0002	0.0015	0.0178	6
	17	0.0000	0.0000	0.0000	0.0000	0.0000	0.0004	0.0063	5
	18	0.0000	0.0000	0.0000	0.0000	0.0000	0.0001	0.0017	4
	19	0.0000	0.0000	0.0000	0.0000	0.0000	0.0000	0.0004	3
	20	0.0000	0.0000	0.0000	0.0000	0.0000	0.0000	0.0001	2
	21	0.0000	0.0000	0.0000	0.0000	0.0000	0.0000	0.0000	1
	22	0.0000	0.0000	0.0000	0.0000	0.0000	0.0000	0.0000	0
	0	0.0886	0.0059	0.0013	0.0003	0.0001	0.0000	0.0000	23
	1	0.2265	0.0339	0.0103	0.0027	0.0010	0.0001	0.0000	22
	2	0.2768	0.0933	0.0376	0.0127	0.0056	0.0009	0.0000	21
	3	0.2153	0.1633	0.0878	0.0382	0.0197	0.0041	0.0002	20
	4	0.1196	0.2042	0.1463	0.0818	0.0493	0.0138	0.0011	19
	5	0.0505	0.1940	0.1853	0.1332	0.0937	0.0350	0.0040	18
	6	0.0168	0.1455	0.1853	0.1712	0.1405	0.0700	0.0120	17
	7	0.0045	0.0883	0.1500	0.1782	0.1707	0.1133	0.0292	16
	8	0.0010	0.0442	0.1000	0.1527	0.1707	0.1511	0.0584	15
$n = 23$	9	0.0002	0.0184	0.0555	0.1091	0.1422	0.1679	0.0974	14
	10	0.0000	0.0064	0.0259	0.0655	0.0996	0.1567	0.1364	13
	11	0.0000	0.0019	0.0102	0.0332	0.0588	0.1234	0.1612	12
	12	0.0000	0.0005	0.0034	0.0142	0.0294	0.0823	0.1612	11
	13	0.0000	0.0001	0.0010	0.0052	0.0124	0.0464	0.1364	10
	14	0.0000	0.0000	0.0001	0.0016	0.0044	0.0221	0.0974	9
	15	0.0000	0.0000	0.0000	0.0004	0.0013	0.0088	0.0584	8
	16	0.0000	0.0000	0.0000	0.0001	0.0003	0.0029	0.0292	7
	17	0.0000	0.0000	0.0000	0.0000	0.0001	0.0008	0.0120	6
	18	0.0000	0.0000	0.0000	0.0000	0.0000	0.0002	0.0040	5

	x	p = .10	p = .20	p = .25	p = .30	p = 1/3	p = .40	p = .50	$n-x$
n = 23 (contd.)	19	0.0000	0.0000	0.0000	0.0000	0.0000	0.0000	0.0011	4
	20	0.0000	0.0000	0.0000	0.0000	0.0000	0.0000	0.0002	3
	21	0.0000	0.0000	0.0000	0.0000	0.0000	0.0000	0.0000	2
	22	0.0000	0.0000	0.0000	0.0000	0.0000	0.0000	0.0000	1
	23	0.0000	0.0000	0.0000	0.0000	0.0000	0.0000	0.0000	0
n = 24	0	0.0798	0.0047	0.0010	0.0002	0.0001	0.0000	0.0000	24
	1	0.2127	0.0283	0.0080	0.0020	0.0007	0.0001	0.0000	23
	2	0.2718	0.0815	0.0308	0.0097	0.0041	0.0006	0.0000	22
	3	0.2215	0.1493	0.0752	0.0305	0.0150	0.0028	0.0001	21
	4	0.1292	0.1960	0.1316	0.0687	0.0395	0.0099	0.0006	20
	5	0.0574	0.1960	0.1755	0.1177	0.0789	0.0265	0.0025	19
	6	0.0202	0.1552	0.1853	0.1598	0.1249	0.0560	0.0080	18
	7	0.0058	0.0998	0.1588	0.1761	0.1606	0.0960	0.0206	17
	8	0.0014	0.0530	0.1125	0.1604	0.1707	0.1360	0.0438	16
	9	0.0003	0.0236	0.0667	0.1222	0.1517	0.1612	0.0779	15
	10	0.0000	0.0088	0.0333	0.0785	0.1138	0.1612	0.1169	14
	11	0.0000	0.0028	0.0141	0.0428	0.0724	0.1367	0.1488	13
	12	0.0000	0.0008	0.0051	0.0199	0.0392	0.0988	0.1612	12
	13	0.0000	0.0002	0.0016	0.0079	0.0181	0.0608	0.1488	11
	14	0.0000	0.0000	0.0004	0.0026	0.0071	0.0318	0.1169	10
	15	0.0000	0.0000	0.0001	0.0008	0.0024	0.0141	0.0779	9
	16	0.0000	0.0000	0.0000	0.0002	0.0007	0.0053	0.0438	8
	17	0.0000	0.0000	0.0000	0.0000	0.0002	0.0017	0.0206	7
	18	0.0000	0.0000	0.0000	0.0000	0.0000	0.0004	0.0080	6
	19	0.0000	0.0000	0.0000	0.0000	0.0000	0.0001	0.0025	5
	20	0.0000	0.0000	0.0000	0.0000	0.0000	0.0000	0.0006	4
	21	0.0000	0.0000	0.0000	0.0000	0.0000	0.0000	0.0001	3
	22	0.0000	0.0000	0.0000	0.0000	0.0000	0.0000	0.0000	2
	23	0.0000	0.0000	0.0000	0.0000	0.0000	0.0000	0.0000	1
	24	0.0000	0.0000	0.0000	0.0000	0.0000	0.0000	0.0000	0
n = 25	0	0.0718	0.0038	0.0008	0.0001	0.0000	0.0000	0.0000	25
	1	0.1994	0.0236	0.0063	0.0014	0.0005	0.0000	0.0000	24
	2	0.2659	0.0708	0.0251	0.0074	0.0030	0.0004	0.0000	23
	3	0.2265	0.1358	0.0641	0.0243	0.0114	0.0019	0.0001	22
	4	0.1384	0.1867	0.1175	0.0572	0.0313	0.0071	0.0004	21
	5	0.0646	0.1960	0.1645	0.1030	0.0658	0.0199	0.0016	20
	6	0.0239	0.1633	0.1828	0.1472	0.1096	0.0442	0.0053	19
	7	0.0072	0.1108	0.1654	0.1712	0.1487	0.0800	0.0143	18
	8	0.0018	0.0623	0.1241	0.1651	0.1673	0.1200	0.0322	17
	9	0.0004	0.0294	0.0781	0.1336	0.1580	0.1511	0.0609	16
	10	0.0001	0.0118	0.0417	0.0916	0.1264	0.1612	0.0974	15
	11	0.0000	0.0040	0.0189	0.0536	0.0862	0.1465	0.1328	14
	12	0.0000	0.0012	0.0074	0.0268	0.0503	0.1139	0.1550	13
	13	0.0000	0.0003	0.0025	0.0115	0.0251	0.0760	0.1550	12
	14	0.0000	0.0001	0.0007	0.0042	0.0108	0.0434	0.1328	11
	15	0.0000	0.0000	0.0002	0.0013	0.0030	0.0212	0.0974	10
	16	0.0000	0.0000	0.0000	0.0004	0.0012	0.0088	0.0609	9
	17	0.0000	0.0000	0.0000	0.0001	0.0003	0.0031	0.0322	8
	18	0.0000	0.0000	0.0000	0.0000	0.0001	0.0009	0.0143	7

	x	p = .10	p = .20	p = .25	p = .30	p = 1/3	p = .40	p = .50	$n-x$
	19	0.0000	0.0000	0.0000	0.0000	0.0000	0.0002	0.0053	6
	20	0.0000	0.0000	0.0000	0.0000	0.0000	0.0000	0.0016	5
n = 25	21	0.0000	0.0000	0.0000	0.0000	0.0000	0.0000	0.0004	4
(contd.)	22	0.0000	0.0000	0.0000	0.0000	0.0000	0.0000	0.0001	3
	23	0.0000	0.0000	0.0000	0.0000	0.0000	0.0000	0.0000	2
	24	0.0000	0.0000	0.0000	0.0000	0.0000	0.0000	0.0000	1
	25	0.0000	0.0000	0.0000	0.0000	0.0000	0.0000	0.0000	0

x	$B(x;50,.1)$	x	$B(x;50,\frac{1}{3})$	x	$B(x;50,\frac{1}{2})$	x	$B(x;100,.1)$	x	$B(x;100,\frac{1}{3})$	x	$B(x;100,\frac{1}{2})$
0	0.0052	4	0.0000	11	0.0000	0	0.0000	15	0.0000	30	0.0000
1	0.0286	5	0.0001	12	0.0001	1	0.0003	16	0.0001	31	0.0001
2	0.0779	6	0.0004	13	0.0003	2	0.0016	17	0.0001	32	0.0001
3	0.1386	7	0.0012	14	0.0008	3	0.0059	18	0.0003	33	0.0002
4	0.1809	8	0.0033	15	0.0020	4	0.0159	19	0.0006	34	0.0005
5	0.1849	9	0.0077	16	0.0044	5	0.0339	20	0.0013	35	0.0009
6	0.1541	10	0.0157	17	0.0087	6	0.0596	21	0.0024	36	0.0016
7	0.1076	11	0.0286	18	0.0160	7	0.0889	22	0.0043	37	0.0027
8	0.0643	12	0.0465	19	0.0270	8	0.1148	23	0.0073	38	0.0045
9	0.0333	13	0.0679	20	0.0419	9	0.1304	24	0.0117	39	0.0071
10	0.0152	14	0.0898	21	0.0598	10	0.1319	25	0.0178	40	0.0108
11	0.0061	15	0.1077	22	0.0788	11	0.1199	26	0.0256	41	0.0159
12	0.0022	16	0.1178	23	0.0960	12	0.0988	27	0.0351	42	0.0223
13	0.0007	17	0.1178	24	0.1080	13	0.0743	28	0.0458	43	0.0301
14	0.0002	18	0.1080	25	0.1123	14	0.0513	29	0.0569	44	0.0390
15	0.0001	19	0.0910	26	0.1080	15	0.0327	30	0.0673	45	0.0485
16	0.0000	20	0.0705	27	0.0960	16	0.0193	31	0.0760	46	0.0580
		21	0.0503	28	0.0788	17	0.0106	32	0.0819	47	0.0666
		22	0.0332	29	0.0598	18	0.0054	33	0.0844	48	0.0735
		23	0.0303	30	0.0419	19	0.0026	34	0.0831	49	0.0780
		24	0.0114	31	0.0270	20	0.0012	35	0.0784	50	0.0796
		25	0.0059	32	0.0160	21	0.0005	36	0.0708	51	0.0780
		26	0.0028	33	0.0087	22	0.0002	37	0.0612	52	0.0735
		27	0.0013	34	0.0044	23	0.0001	38	0.0507	53	0.0666
		28	0.0005	35	0.0020	24	0.0000	39	0.0403	54	0.0580
		29	0.0002	36	0.0008			40	0.0307	55	0.0485
		30	0.0001	37	0.0003			41	0.0225	56	0.0390
		31	0.0000	38	0.0001			42	0.0158	57	0.0301
				39	0.0000			43	0.0107	58	0.0223
								44	0.0069	59	0.0159
								45	0.0043	60	0.0108
								46	0.0026	61	0.0071
								47	0.0015	62	0.0045
								48	0.0008	63	0.0027
								49	0.0004	64	0.0016
								50	0.0002	65	0.0009
								51	0.0001	66	0.0005
								52	0.0001	67	0.0002
								53	0.0000	68	0.0001
										69	0.0001
										70	0.0000

Table C Poisson Probabilities

$$P(x;\lambda) = \frac{\lambda^x e^{-\lambda}}{x!}$$

Examples: $P(3; 2) = .1804$
$P(0; .8) = .4493$

Note: For values of λ not in this table, one may appeal to Table A (for values of $e^{-\lambda}$) and use the formula directly.

$x \backslash \lambda$	0.1	0.2	0.3	0.4	0.5	0.6	0.7	0.8	0.9	1.0
0	0.9048	0.8187	0.7408	0.6703	0.6065	0.5488	0.4966	0.4493	0.4066	0.3679
1	0.0905	0.1637	0.2222	0.2681	0.3033	0.3293	0.3476	0.3595	0.3659	0.3679
2	0.0045	0.0164	0.0333	0.0536	0.0758	0.0988	0.1217	0.1438	0.1647	0.1839
3	0.0002	0.0011	0.0033	0.0072	0.0126	0.0198	0.0284	0.0383	0.0494	0.0613
4	0.0000	0.0001	0.0003	0.0007	0.0016	0.0030	0.0050	0.0077	0.0111	0.0153
5	0.0000	0.0000	0.0000	0.0001	0.0002	0.0004	0.0007	0.0012	0.0020	0.0031
6	0.0000	0.0000	0.0000	0.0000	0.0000	0.0000	0.0001	0.0002	0.0003	0.0005
7	0.0000	0.0000	0.0000	0.0000	0.0000	0.0000	0.0000	0.0000	0.0000	0.0001

$x \backslash \lambda$	1.1	1.2	1.3	1.4	1.5	1.6	1.7	1.8	.19	2.0
0	0.3329	0.3012	0.2725	0.2466	0.2231	0.2019	0.1827	0.1653	0.1496	0.1353
1	0.3662	0.3614	0.3543	0.3452	0.3347	0.3230	0.3106	0.2975	0.2842	0.2707
2	0.2014	0.2169	0.2303	0.2417	0.2510	0.2584	0.2640	0.2678	0.2700	0.2707
3	0.0738	0.0867	0.0998	0.1128	0.1255	0.1378	0.1496	0.1607	0.1710	0.1804
4	0.0203	0.0260	0.0324	0.0395	0.0471	0.0551	0.0636	0.0723	0.0812	0.0902
5	0.0045	0.0062	0.0084	0.0111	0.0141	0.0176	0.0216	0.0260	0.0309	0.0361
6	0.0008	0.0012	0.0018	0.0026	0.0035	0.0047	0.0061	0.0078	0.0098	0.0120
7	0.0001	0.0002	0.0003	0.0005	0.0008	0.0011	0.0015	0.0020	0.0027	0.0034
8	0.0000	0.0000	0.0001	0.0001	0.0001	0.0002	0.0003	0.0005	0.0006	0.0009
9	0.0000	0.0000	0.0000	0.0000	0.0000	0.0000	0.0001	0.0001	0.0001	0.0002

$x \backslash \lambda$	3.0	4.0	5.0	6.0	7.0	8.0	9.0	10.0
0	0.0498	0.0183	0.0067	0.0025	0.0009	0.0003	0.0001	0.0000
1	0.1494	0.0733	0.0337	0.0149	0.0064	0.0027	0.0011	0.0005
2	0.2240	0.1465	0.0842	0.0446	0.0223	0.0107	0.0050	0.0023
3	0.2240	0.1954	0.1404	0.0892	0.0521	0.0286	0.0150	0.0076
4	0.1680	0.1954	0.1755	0.1339	0.0912	0.0573	0.0337	0.0189
5	0.1008	0.1563	0.1755	0.1606	0.1277	0.0916	0.0607	0.0378
6	0.0504	0.1042	0.1462	0.1606	0.1490	0.1221	0.0911	0.0631
7	0.0216	0.0595	0.1044	0.1377	0.1490	0.1396	0.1171	0.0901
8	0.0081	0.0298	0.0653	0.1033	0.1304	0.1396	0.1318	0.1126
9	0.0027	0.0132	0.0363	0.0688	0.1014	0.1241	0.1318	0.1251
10	0.0008	0.0053	0.0181	0.0413	0.0710	0.0993	0.1186	0.1251
11	0.0002	0.0019	0.0082	0.0225	0.0452	0.0722	0.0970	0.1137
12	0.0001	0.0006	0.0034	0.0113	0.0263	0.0481	0.0728	0.0948
13	0.0000	0.0002	0.0013	0.0052	0.0142	0.0296	0.0504	0.0729
14	0.0000	0.0001	0.0005	0.0022	0.0071	0.0169	0.0324	0.0521
15	0.0000	0.0000	0.0002	0.0009	0.0033	0.0090	0.0194	0.0347

x \ λ	3.0	4.0	5.0	6.0	7.0	8.0	9.0	10.0
16	0.0000	0.0000	0.0000	0.0003	0.0014	0.0045	0.0109	0.0217
17	0.0000	0.0000	0.0000	0.0001	0.0006	0.0021	0.0058	0.0128
18	0.0000	0.0000	0.0000	0.0000	0.0002	0.0009	0.0029	0.0071
19	0.0000	0.0000	0.0000	0.0000	0.0001	0.0004	0.0014	0.0037
20	0.0000	0.0000	0.0000	0.0000	0.0000	0.0002	0.0006	0.0019
21	0.0000	0.0000	0.0000	0.0000	0.0000	0.0001	0.0003	0.0009
22	0.0000	0.0000	0.0000	0.0000	0.0000	0.0000	0.0001	0.0004
23	0.0000	0.0000	0.0000	0.0000	0.0000	0.0000	0.0000	0.0002
24	0.0000	0.0000	0.0000	0.0000	0.0000	0.0000	0.0000	0.0001

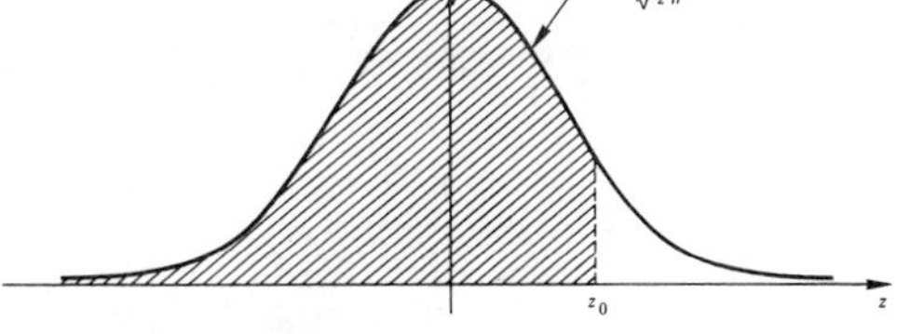

Table D The Standard Normal Distribution

Value in table represents the area of shaded region,

$$P(Z < z_0) = \int_{-\infty}^{z_0} N(z; 0, 1)\, dz$$

Example: To find $P(Z < 1.75)$, we look in row labeled "1.7" and column labeled ".05" to get .9599. An extended discussion of this table appears in Chapter 5, Section 2.

	Second decimal place of z_0									
z_0	.00	.01	.02	.03	.04	.05	.06	.07	.08	.09
0.0	0.5000	0.5040	0.5080	0.5120	0.5160	0.5199	0.5239	0.5279	0.5319	0.5359
0.1	0.5398	0.5438	0.5478	0.5517	0.5557	0.5596	0.5636	0.5675	0.5714	0.5753
0.2	0.5793	0.5832	0.5871	0.5910	0.5948	0.5987	0.6026	0.6064	0.6103	0.6141
0.3	0.6179	0.6217	0.6255	0.6293	0.6331	0.6368	0.6406	0.6443	0.6480	0.6517
0.4	0.6554	0.6591	0.6628	0.6664	0.6700	0.6736	0.6772	0.6808	0.6844	0.6879
0.5	0.6915	0.6950	0.6985	0.7019	0.7054	0.7088	0.7123	0.7157	0.7190	0.7224
0.6	0.7257	0.7291	0.7324	0.7357	0.7389	0.7422	0.7454	0.7486	0.7517	0.7549
0.7	0.7580	0.7611	0.7642	0.7673	0.7703	0.7734	0.7764	0.7793	0.7823	0.7852
0.8	0.7881	0.7910	0.7939	0.7967	0.7995	0.8023	0.8051	0.8078	0.8106	0.8133
0.9	0.8159	0.8186	0.8212	0.8238	0.8264	0.8289	0.8314	0.8340	0.8365	0.8389
1.0	0.8412	0.8438	0.8461	0.8484	0.8508	0.8531	0.8554	0.8577	0.8599	0.8621
1.1	0.8643	0.8665	0.8686	0.8708	0.8729	0.8749	0.8769	0.8790	0.8810	0.8829
1.2	0.8849	0.8869	0.8888	0.8906	0.8925	0.8943	0.8961	0.8980	0.8997	0.9015
1.3	0.9032	0.9049	0.9066	0.9082	0.9098	0.9114	0.9131	0.9147	0.9162	0.9177
1.4	0.9192	0.9207	0.9222	0.9235	0.9251	0.9265	0.9279	0.9292	0.9306	0.9319
1.5	0.9332	0.9344	0.9356	0.9370	0.9382	0.9394	0.9405	0.9418	0.9429	0.9441
1.6	0.9452	0.9463	0.9474	0.9484	0.9495	0.9505	0.9515	0.9525	0.9535	0.9545
1.7	0.9553	0.9564	0.9573	0.9581	0.9591	0.9599	0.9608	0.9616	0.9624	0.9633
1.8	0.9640	0.9649	0.9656	0.9664	0.9671	0.9678	0.9685	0.9693	0.9699	0.9706
1.9	0.9712	0.9718	0.9726	0.9732	0.9738	0.9744	0.9750	0.9756	0.9761	0.9766
2.0	0.9772	0.9777	0.9783	0.9788	0.9793	0.9798	0.9803	0.9808	0.9812	0.9817
2.1	0.9821	0.9826	0.9830	0.9833	0.9838	0.9842	0.9846	0.9850	0.9854	0.9857
2.2	0.9860	0.9864	0.9868	0.9871	0.9875	0.9877	0.9881	0.9884	0.9887	0.9890
2.3	0.9893	0.9896	0.9898	0.9901	0.9904	0.9906	0.9909	0.9910	0.9913	0.9915
2.4	0.9918	0.9920	0.9922	0.9925	0.9927	0.9929	0.9931	0.9931	0.9934	0.9936
2.5	0.9938	0.9940	0.9941	0.9942	0.9945	0.9946	0.9947	0.9949	0.9951	0.9952
2.6	0.9953	0.9955	0.9956	0.9957	0.9958	0.9960	0.9961	0.9962	0.9963	0.9964
2.7	0.9965	0.9966	0.9967	0.9968	0.9969	0.9969	0.9971	0.9972	0.9973	0.9974
2.8	0.9974	0.9975	0.9976	0.9977	0.9977	0.9978	0.9979	0.9979	0.9980	0.9980
2.9	0.9980	0.9981	0.9981	0.9983	0.9984	0.9984	0.9985	0.9985	0.9986	0.9986
3.0	0.9986	0.9987	0.9987	0.9988	0.9988	0.9989	0.9989	0.9989	0.9990	0.9990
3.1	0.9990	0.9991	0.9991	0.9991	0.9992	0.9992	0.9992	0.9992	0.9992	0.9992
3.2	0.9992	0.9992	0.9994	0.9994	0.9994	0.9994	0.9994	0.9995	0.9995	0.9995
3.3	0.9995	0.9995	0.9995	0.9996	0.9996	0.9996	0.9996	0.9996	0.9996	0.9997
3.4	0.9997	0.9997	0.9997	0.9997	0.9997	0.9997	0.9997	0.9997	0.9997	0.9998
3.5	0.9998	0.9998	0.9998	0.9998	0.9998	0.9998	0.9998	0.9998	0.9998	0.9998
3.6	0.9998	0.9998	0.9999	0.9999	0.9999	0.9999	0.9999	0.9999	0.9999	0.9999
3.7	0.9999	0.9999	0.9999	0.9999	0.9999	0.9999	0.9999	0.9999	0.9999	0.9999
3.8	0.9999	0.9999	0.9999	0.9999	0.9999	0.9999	0.9999	0.9999	0.9999	0.9999
3.9	1.0000	1.0000	1.0000	1.0000	1.0000	1.0000	1.0000	1.0000	1.0000	1.0000

Table E Percentiles for the t-Distribution

n	.8	.9	.95	.975	.99	.995
1	1.38	3.08	6.31	12.71	31.82	63.66
2	1.06	1.89	2.92	4.30	6.96	9.92
3	0.98	1.64	2.35	3.18	4.54	5.84
4	0.94	1.53	2.13	2.78	3.75	4.60
5	0.92	1.48	2.01	2.57	3.36	4.03
6	0.91	1.44	1.94	2.45	3.14	3.71
7	0.90	1.41	1.89	2.36	3.00	3.50
8	0.89	1.40	1.86	2.31	2.90	3.36
9	0.88	1.38	1.83	2.26	2.82	3.25
10	0.88	1.37	1.81	2.23	2.76	3.17
11	0.87	1.36	1.80	2.20	2.72	3.11
12	0.87	1.36	1.78	2.18	2.68	3.05
13	0.87	1.35	1.77	2.16	2.65	3.01
14	0.87	1.34	1.76	2.14	2.62	2.98
15	0.87	1.34	1.75	2.13	2.60	2.95
16	0.86	1.34	1.75	2.12	2.58	2.92
17	0.86	1.33	1.74	2.11	2.57	2.90
18	0.86	1.33	1.73	2.10	2.55	2.88
19	0.86	1.33	1.73	2.09	2.54	2.86
20	0.86	1.32	1.72	2.09	2.53	2.85
21	0.86	1.32	1.72	2.08	2.52	2.83
22	0.86	1.32	1.72	2.07	2.51	2.82
23	0.86	1.32	1.71	2.07	2.50	2.81
24	0.86	1.32	1.71	2.06	2.49	2.80
25	0.86	1.32	1.71	2.06	2.49	2.79
26	0.86	1.32	1.71	2.06	2.48	2.78
27	0.86	1.31	1.70	2.05	2.47	2.77
28	0.86	1.31	1.70	2.05	2.47	2.76
29	0.85	1.31	1.70	2.04	2.46	2.76
30	0.85	1.31	1.70	2.04	2.46	2.75
31	0.85	1.31	1.70	2.04	2.45	2.74
32	0.85	1.31	1.69	2.04	2.45	2.74
33	0.85	1.31	1.69	2.03	2.44	2.73
34	0.85	1.31	1.69	2.03	2.44	2.73
35	0.85	1.31	1.69	2.03	2.44	2.72
36	0.85	1.31	1.69	2.03	2.43	2.72
37	0.85	1.30	1.69	2.03	2.43	2.72
38	0.85	1.30	1.69	2.02	2.43	2.71
39	0.85	1.30	1.68	2.02	2.43	2.71
40	0.85	1.30	1.68	2.02	2.42	2.70
41	0.85	1.30	1.68	2.02	2.42	2.70
42	0.85	1.30	1.68	2.02	2.42	2.70
43	0.85	1.30	1.68	2.02	2.42	2.69
44	0.85	1.30	1.68	2.02	2.41	2.69
45	0.85	1.30	1.68	2.01	2.41	2.69
46	0.85	1.30	1.68	2.01	2.41	2.69
47	0.85	1.30	1.68	2.01	2.41	2.68
48	0.85	1.30	1.68	2.01	2.41	2.68
49	0.85	1.30	1.68	2.01	2.40	2.68
50	0.85	1.30	1.68	2.01	2.40	2.68

Table F Percentiles for the F-Distribution

Vertical degrees of freedom	Horizontal degrees of freedom													
	1	3	5	10	12	15	16	20	25	28	30	35	40	45
1	39.9	53.6	57.2	60.2	60.7	61.2	61.4	61.7	62.1	62.2	62.3	62.4	62.5	62.6
2	8.5	9.2	9.3	9.4	9.4	9.4	9.4	9.4	9.5	9.5	9.5	9.5	9.5	9.5
3	5.5	5.4	5.3	5.2	5.2	5.2	5.2	5.2	5.2	5.2	5.2	5.2	5.2	5.2
4	4.5	4.2	4.1	3.9	3.9	3.9	3.9	3.8	3.8	3.8	3.8	3.8	3.8	3.8
5	4.1	3.6	3.5	3.3	3.3	3.2	3.2	3.2	3.2	3.2	3.2	3.2	3.2	3.2
6	3.8	3.3	3.1	2.9	2.9	2.9	2.9	2.8	2.8	2.8	2.8	2.8	2.8	2.8
7	3.6	3.1	2.9	2.7	2.7	2.6	2.6	2.6	2.6	2.6	2.6	2.5	2.5	2.5
8	3.5	2.9	2.7	2.5	2.5	2.5	2.5	2.4	2.4	2.4	2.4	2.4	2.4	2.4
9	3.4	2.8	2.6	2.4	2.4	2.3	2.3	2.3	2.3	2.3	2.3	2.2	2.2	2.2
10	3.3	2.7	2.5	2.3	2.3	2.2	2.2	2.2	2.2	2.2	2.2	2.1	2.1	2.1
11	3.2	2.7	2.5	2.2	2.2	2.2	2.2	2.1	2.1	2.1	2.1	2.1	2.1	2.0
12	3.2	2.6	2.4	2.2	2.1	2.1	2.1	2.1	2.0	2.0	2.0	2.0	2.0	2.0
13	3.1	2.6	2.3	2.1	2.1	2.1	2.0	2.0	2.0	2.0	2.0	1.9	1.9	1.9
14	3.1	2.5	2.3	2.1	2.1	2.0	2.0	2.0	1.9	1.9	1.9	1.9	1.9	1.9
15	3.1	2.5	2.3	2.1	2.0	2.0	2.0	1.9	1.9	1.9	1.9	1.9	1.8	1.8
16	3.0	2.5	2.2	2.0	2.0	1.9	1.9	1.9	1.8	1.8	1.8	1.8	1.8	1.8
17	3.0	2.4	2.2	2.0	2.0	1.9	1.9	1.9	1.8	1.8	1.8	1.8	1.8	1.8
18	3.0	2.4	2.2	2.0	1.9	1.9	1.9	1.8	1.8	1.8	1.8	1.8	1.8	1.7
19	3.0	2.4	2.2	2.0	1.9	1.9	1.9	1.8	1.8	1.8	1.8	1.7	1.7	1.7
20	3.0	2.4	2.2	1.9	1.9	1.8	1.8	1.8	1.8	1.7	1.7	1.7	1.7	1.7
21	3.0	2.4	2.1	1.9	1.9	1.8	1.8	1.8	1.7	1.7	1.7	1.7	1.7	1.7
22	2.9	2.4	2.1	1.9	1.9	1.8	1.8	1.8	1.7	1.7	1.7	1.7	1.7	1.7
23	2.9	2.3	2.1	1.9	1.8	1.8	1.8	1.7	1.7	1.7	1.7	1.7	1.7	1.6
24	2.9	2.3	2.1	1.9	1.8	1.8	1.8	1.7	1.7	1.7	1.7	1.7	1.6	1.6
25	2.9	2.3	2.1	1.9	1.8	1.8	1.8	1.7	1.7	1.7	1.7	1.6	1.6	1.6
26	2.9	2.3	2.1	1.9	1.8	1.8	1.7	1.7	1.7	1.7	1.6	1.6	1.6	1.6
27	2.9	2.3	2.1	1.8	1.8	1.7	1.7	1.7	1.7	1.6	1.6	1.6	1.6	1.6
28	2.9	2.3	2.1	1.8	1.8	1.7	1.7	1.7	1.6	1.6	1.6	1.6	1.6	1.6
29	2.9	2.3	2.1	1.8	1.8	1.7	1.7	1.7	1.6	1.6	1.6	1.6	1.6	1.6
30	2.9	2.3	2.0	1.8	1.8	1.7	1.7	1.7	1.6	1.6	1.6	1.6	1.6	1.6
31	2.9	2.3	2.0	1.8	1.8	1.7	1.7	1.7	1.6	1.6	1.6	1.6	1.6	1.6
32	2.9	2.3	2.0	1.8	1.8	1.7	1.7	1.7	1.6	1.6	1.6	1.6	1.6	1.5
33	2.9	2.3	2.0	1.8	1.8	1.7	1.7	1.6	1.6	1.6	1.6	1.6	1.6	1.5
34	2.9	2.3	2.0	1.8	1.7	1.7	1.7	1.6	1.6	1.6	1.6	1.6	1.5	1.5
35	2.9	2.2	2.0	1.8	1.7	1.7	1.7	1.6	1.6	1.6	1.6	1.5	1.5	1.5
36	2.9	2.2	2.0	1.8	1.7	1.7	1.7	1.6	1.6	1.6	1.6	1.5	1.5	1.5
37	2.8	2.2	2.0	1.8	1.7	1.7	1.7	1.6	1.6	1.6	1.6	1.5	1.5	1.5
38	2.8	2.2	2.0	1.8	1.7	1.7	1.7	1.6	1.6	1.6	1.6	1.5	1.5	1.5
39	2.8	2.2	2.0	1.8	1.7	1.7	1.7	1.6	1.6	1.6	1.5	1.5	1.5	1.5
40	2.8	2.2	2.0	1.8	1.7	1.7	1.6	1.6	1.6	1.6	1.5	1.5	1.5	1.5
41	2.8	2.2	2.0	1.8	1.7	1.7	1.6	1.6	1.6	1.5	1.5	1.5	1.5	1.5
42	2.8	2.2	2.0	1.8	1.7	1.7	1.6	1.6	1.6	1.5	1.5	1.5	1.5	1.5
43	2.8	2.2	2.0	1.8	1.7	1.6	1.6	1.6	1.6	1.5	1.5	1.5	1.5	1.5
44	2.8	2.2	2.0	1.7	1.7	1.6	1.6	1.6	1.6	1.5	1.5	1.5	1.5	1.5
45	2.8	2.2	2.0	1.7	1.7	1.6	1.6	1.6	1.5	1.5	1.5	1.5	1.5	1.5
46	2.8	2.2	2.0	1.7	1.7	1.6	1.6	1.6	1.5	1.5	1.5	1.5	1.5	1.5
47	2.8	2.2	2.0	1.7	1.7	1.6	1.6	1.6	1.5	1.5	1.5	1.5	1.5	1.5
48	2.8	2.2	2.0	1.7	1.7	1.6	1.6	1.6	1.5	1.5	1.5	1.5	1.5	1.5
49	2.8	2.2	2.0	1.7	1.7	1.6	1.6	1.6	1.5	1.5	1.5	1.5	1.5	1.5
50	2.8	2.2	2.0	1.7	1.7	1.6	1.6	1.6	1.5	1.5	1.5	1.5	1.5	1.5

Table F (contd.)

Vertical degrees of freedom	Horizontal degrees of freedom													
	1	3	5	10	12	15	16	20	25	28	30	35	40	45
1	161.5	215.7	230.2	241.9	243.9	246.0	246.5	248.0	249.3	249.8	250.1	250.7	251.1	251.5
2	18.5	19.2	19.3	19.4	19.4	1.94	19.4	19.4	19.5	19.5	19.5	19.5	19.5	19.5
3	10.1	9.3	9.0	8.8	8.7	8.7	8.7	8.7	8.6	8.6	8.6	8.6	8.6	8.6
4	7.7	6.6	6.3	6.0	5.9	5.9	5.8	5.8	5.8	5.8	5.7	5.7	5.7	5.7
5	6.6	5.4	5.0	4.7	4.7	4.6	4.6	4.6	4.5	4.5	4.5	4.5	4.5	4.5
6	6.0	4.8	4.4	4.1	4.0	3.9	3.9	3.9	3.8	3.8	3.8	3.8	3.8	3.8
7	5.6	4.3	4.0	3.6	3.6	3.5	3.5	3.4	3.4	3.4	3.4	3.4	3.3	3.3
8	5.3	4.1	3.7	3.3	3.3	3.2	3.2	3.1	3.1	3.1	3.1	3.1	3.0	3.0
9	5.1	3.9	3.5	3.1	3.1	3.0	3.0	2.9	2.9	2.9	2.9	2.8	2.8	2.8
10	5.0	3.7	3.3	3.0	2.9	2.8	2.8	2.8	2.7	2.7	2.7	2.7	2.7	2.6
11	4.8	3.6	3.2	2.9	2.8	2.7	2.7	2.6	2.6	2.6	2.6	2.5	2.5	2.5
12	4.7	3.5	3.1	2.8	2.7	2.6	2.6	2.5	2.5	2.5	2.5	2.4	2.4	2.4
13	4.7	3.4	3.0	2.7	2.6	2.5	2.5	2.5	2.4	2.4	2.4	2.4	2.3	2.3
14	4.6	3.3	3.0	2.6	2.5	2.5	2.4	2.4	2.3	2.3	2.3	2.3	2.3	2.3
15	4.5	3.3	2.9	2.5	2.5	2.4	2.4	2.3	2.3	2.3	2.2	2.2	2.2	2.2
16	4.5	3.2	2.9	2.5	2.4	2.4	2.3	2.3	2.2	2.2	2.2	2.2	2.2	2.1
17	4.4	3.2	2.8	2.4	2.4	2.3	2.3	2.2	2.2	2.2	2.1	2.1	2.1	2.1
18	4.4	3.2	2.8	2.4	2.3	2.3	2.2	2.2	2.1	2.1	2.1	2.1	2.1	2.0
19	4.4	3.1	2.7	2.4	2.3	2.2	2.2	2.2	2.1	2.1	2.1	2.0	2.0	2.0
20	4.3	3.1	2.7	2.3	2.3	2.2	2.2	2.1	2.1	2.1	2.0	2.0	2.0	2.0
21	4.3	3.1	2.7	2.3	2.2	2.2	2.2	2.1	2.0	2.0	2.0	2.0	2.0	1.9
22	4.3	3.0	2.7	2.3	2.2	2.2	2.1	2.1	2.0	2.0	2.0	2.0	1.9	1.9
23	4.3	3.0	2.6	2.3	2.2	2.1	2.1	2.0	2.0	2.0	2.0	1.9	1.9	1.9
24	4.3	3.0	2.6	2.3	2.2	2.1	2.1	2.0	2.0	2.0	1.9	1.9	1.9	1.9
25	4.2	3.0	2.6	2.2	2.2	2.1	2.1	2.0	2.0	1.9	1.9	1.9	1.9	1.9
26	4.2	3.0	2.6	2.2	2.1	2.1	2.1	2.0	1.9	1.9	1.9	1.9	1.9	1.8
27	4.2	3.0	2.6	2.2	2.1	2.1	2.0	2.0	1.9	1.9	1.9	1.9	1.8	1.8
28	4.2	2.9	2.6	2.2	2.1	2.0	2.0	2.0	1.9	1.9	1.9	1.8	1.8	1.8
29	4.2	2.9	2.5	2.2	2.1	2.0	2.0	1.9	1.9	1.9	1.9	1.8	1.8	1.8
30	4.2	2.9	2.5	2.2	2.1	2.0	2.0	1.9	1.9	1.9	1.8	1.8	1.8	1.8
31	4.2	2.9	2.5	2.2	2.1	2.0	2.0	1.9	1.9	1.8	1.8	1.8	1.8	1.8
32	4.1	2.9	2.5	2.1	2.1	2.0	2.0	1.9	1.9	1.8	1.8	1.8	1.8	1.7
33	4.1	2.9	2.5	2.1	2.1	2.0	2.0	1.9	1.8	1.8	1.8	1.8	1.8	1.7
34	4.1	2.9	2.5	2.1	2.1	2.0	2.0	1.9	1.8	1.8	1.8	1.8	1.7	1.7
35	4.1	2.9	2.5	2.1	2.0	2.0	1.9	1.9	1.8	1.8	1.8	1.8	1.7	1.7
36	4.1	2.9	2.5	2.1	2.0	2.0	1.9	1.9	1.8	1.8	1.8	1.7	1.7	1.7
37	4.1	2.9	2.5	2.1	2.0	1.9	1.9	1.9	1.8	1.8	1.8	1.7	1.7	1.7
38	4.1	2.9	2.5	2.1	2.0	1.9	1.9	1.9	1.8	1.8	1.8	1.7	1.7	1.7
39	4.1	2.8	2.5	2.1	2.0	1.9	1.9	1.8	1.8	1.8	1.8	1.7	1.7	1.7
40	4.1	2.8	2.4	2.1	2.0	1.9	1.9	1.8	1.8	1.8	1.7	1.7	1.7	1.7
41	4.1	2.8	2.4	2.1	2.0	1.9	1.9	1.8	1.8	1.8	1.7	1.7	1.7	1.7
42	4.1	2.8	2.4	2.1	2.0	1.9	1.9	1.8	1.8	1.7	1.7	1.7	1.7	1.7
43	4.1	2.8	2.4	2.1	2.0	1.9	1.9	1.8	1.8	1.7	1.7	1.7	1.7	1.7
44	4.1	2.8	2.4	2.1	2.0	1.9	1.9	1.8	1.8	1.7	1.7	1.7	1.7	1.6
45	4.1	2.8	2.4	2.1	2.0	1.9	1.9	1.8	1.8	1.7	1.7	1.7	1.7	1.6
46	4.1	2.8	2.4	2.0	2.0	1.9	1.9	1.8	1.7	1.7	1.7	1.7	1.7	1.6
47	4.0	2.8	2.4	2.0	2.0	1.9	1.9	1.8	1.7	1.7	1.7	1.7	1.6	1.6
48	4.0	2.8	2.4	2.0	2.0	1.9	1.9	1.8	1.7	1.7	1.7	1.7	1.6	1.6
49	4.0	2.8	2.4	2.0	2.0	1.9	1.9	1.8	1.7	1.7	1.7	1.7	1.6	1.6
50	4.0	2.8	2.4	2.0	2.0	1.9	1.9	1.8	1.7	1.7	1.7	1.7	1.6	1.6

Vertical degrees of freedom	Horizontal degrees of freedom													
	1	3	5	10	12	15	16	20	25	28	30	35	40	45
1	4051.4	5403.7	5764.0	6055.9	6106.3	6157.7	6170.1	6208.7	6240.2	6253.2	6260.6	6275.9	6286.8	6295.8
2	98.5	99.2	99.3	99.4	99.4	99.4	99.4	99.4	99.5	99.5	99.5	99.5	99.5	99.5
3	34.1	29.5	28.2	27.2	27.1	26.9	26.8	26.7	26.6	26.5	26.5	26.5	26.4	26.4
4	21.2	16.7	15.5	14.5	14.4	14.2	14.2	14.0	13.9	13.9	13.9	13.8	13.7	13.7
5	16.3	12.1	11.0	10.0	9.9	9.7	9.7	9.6	9.4	9.4	9.4	9.3	9.3	9.3
6	13.7	9.8	8.7	7.9	7.7	7.6	7.5	7.4	7.3	7.3	7.2	7.2	7.1	7.1
7	12.2	8.5	7.5	6.6	6.5	6.3	6.3	6.2	6.1	6.0	6.0	5.9	5.9	5.9
8	11.3	7.6	6.6	5.8	5.7	5.5	5.5	5.4	5.3	5.2	5.2	5.2	5.1	5.1
9	10.6	7.0	6.1	5.3	5.1	5.0	4.9	4.8	4.7	4.7	4.6	4.6	4.6	4.5
10	10.0	6.6	5.6	4.8	4.7	4.6	4.5	4.4	4.3	4.3	4.2	4.2	4.2	4.1
11	9.6	6.2	5.3	4.5	4.4	4.3	4.2	4.1	4.0	4.0	3.9	3.9	3.9	3.8
12	9.3	6.0	5.1	4.3	4.2	4.0	4.0	3.9	3.8	3.7	3.7	3.7	3.6	3.6
13	9.1	5.7	4.9	4.1	4.0	3.8	3.8	3.7	3.6	3.5	3.5	3.5	3.4	3.4
14	8.9	5.6	4.7	3.9	3.8	3.7	3.6	3.5	3.4	3.4	3.3	3.3	3.3	3.2
15	8.7	5.4	4.6	3.8	3.7	3.5	3.5	3.4	3.3	3.2	3.2	3.2	3.1	3.1
16	8.5	5.3	4.4	3.7	3.6	3.4	3.4	3.3	3.2	3.1	3.1	3.1	3.0	3.0
17	8.4	5.2	4.3	3.6	3.5	3.3	3.3	3.2	3.1	3.0	3.0	3.0	2.9	2.9
18	8.3	5.1	4.2	3.5	3.4	3.2	3.2	3.1	3.0	2.9	2.9	2.9	2.8	2.8
19	8.2	5.0	4.2	3.4	3.3	3.2	3.1	3.0	2.9	2.9	2.8	2.8	2.8	2.7
20	8.1	4.9	4.1	3.4	3.2	3.1	3.1	2.9	2.8	2.8	2.8	2.7	2.7	2.7
21	8.0	4.9	4.0	3.3	3.2	3.0	3.0	2.9	2.8	2.7	2.7	2.7	2.6	2.6
22	7.9	4.8	4.0	3.3	3.1	3.0	2.9	2.8	2.7	2.7	2.7	2.6	2.6	2.6
23	7.9	4.8	3.9	3.2	3.1	2.9	2.9	2.8	2.7	2.6	2.6	2.6	2.5	2.5
24	7.8	4.7	3.9	3.2	3.0	2.9	2.9	2.7	2.6	2.6	2.6	2.5	2.5	2.5
25	7.8	4.7	3.9	3.1	3.0	2.9	2.8	2.7	2.6	2.6	2.5	2.5	2.5	2.4
26	7.7	4.6	3.8	3.1	3.0	2.8	2.8	2.7	2.6	2.5	2.5	2.5	2.4	2.4
27	7.7	4.6	3.8	3.1	2.9	2.8	2.7	2.6	2.5	2.5	2.5	2.4	2.4	2.4
28	7.6	4.6	3.8	3.0	2.9	2.8	2.7	2.6	2.5	2.5	2.4	2.4	2.4	2.3
29	7.6	4.5	3.7	3.0	2.9	2.7	2.7	2.6	2.5	2.4	2.4	2.4	2.3	2.3
30	7.6	4.5	3.7	3.0	2.8	2.7	2.7	2.5	2.5	2.4	2.4	2.3	2.3	2.3
31	7.5	4.5	3.7	3.0	2.8	2.7	2.6	2.5	2.4	2.4	2.4	2.3	2.3	2.2
32	7.5	4.5	3.7	2.9	2.8	2.7	2.6	2.5	2.4	2.4	2.3	2.3	2.3	2.2
33	7.5	4.4	3.6	2.9	2.8	2.6	2.6	2.5	2.4	2.3	2.3	2.3	2.2	2.2
34	7.4	4.4	3.6	2.9	2.8	2.6	2.6	2.5	2.4	2.3	2.3	2.2	2.2	2.2
35	7.4	4.4	3.6	2.9	2.7	2.6	2.6	2.4	2.3	2.3	2.3	2.2	2.2	2.2
36	7.4	4.4	3.6	2.9	2.7	2.6	2.5	2.4	2.3	2.3	2.3	2.2	2.2	2.1
37	7.4	4.4	3.6	2.8	2.7	2.6	2.5	2.4	2.3	2.3	2.2	2.2	2.2	2.1
38	7.4	4.3	3.5	2.8	2.7	2.5	2.5	2.4	2.3	2.3	2.2	2.2	2.1	2.1
39	7.3	4.3	3.5	2.8	2.7	2.5	2.5	2.4	2.3	2.2	2.2	2.2	2.1	2.1
40	7.3	4.3	3.5	2.8	2.7	2.5	2.5	2.4	2.3	2.2	2.2	2.2	2.1	2.1
41	7.3	4.3	3.5	2.8	2.7	2.5	2.5	2.4	2.3	2.2	2.2	2.1	2.1	2.1
42	7.3	4.3	3.5	2.8	2.6	2.5	2.5	2.3	2.2	2.2	2.2	2.1	2.1	2.1
43	7.3	4.3	3.5	2.8	2.6	2.5	2.4	2.3	2.2	2.2	2.2	2.1	2.1	2.0
44	7.2	4.3	3.5	2.8	2.6	2.5	2.4	2.3	2.2	2.2	2.2	2.1	2.1	2.0
45	7.2	4.3	3.5	2.7	2.6	2.5	2.4	2.3	2.2	2.2	2.1	2.1	2.1	2.0
46	7.2	4.2	3.4	2.7	2.6	2.5	2.4	2.3	2.2	2.2	2.1	2.1	2.0	2.0
47	7.2	4.2	3.4	2.7	2.6	2.4	2.4	2.3	2.2	2.1	2.1	2.1	2.0	2.0
48	7.2	4.2	3.4	2.7	2.6	2.4	2.4	2.3	2.2	2.1	2.1	2.1	2.0	2.0
49	7.2	4.2	3.4	2.7	2.6	2.4	2.4	2.3	2.2	2.1	2.1	2.1	2.0	2.0
50	7.2	4.2	3.4	2.7	2.6	2.4	2.4	2.3	2.2	2.1	2.1	2.0	2.0	2.0

Table G Percentiles for the X^2 - Distribution

n	.50	.75	.90	.95	.99
1	0.455	1.323	2.706	3.842	6.638
2	1.386	2.773	4.605	5.991	9.210
3	2.366	4.109	6.252	7.815	1.346
4	3.357	5.385	7.779	9.488	13.277
5	4.352	6.626	9.237	11.071	15.087
6	5.348	7.841	10.645	12.592	16.812
7	6.346	9.037	12.017	14.067	18.475
8	7.344	10.219	13.362	15.507	20.090
9	8.343	11.389	14.684	16.919	21.666
10	9.342	12.549	15.987	18.307	23.209
11	10.341	13.701	17.275	19.675	24.725
12	11.340	14.845	18.549	21.026	26.217
13	12.340	15.984	19.812	22.362	27.688
14	13.339	17.117	21.064	23.685	29.141
15	14.339	18.245	22.307	24.996	30.578
16	15.338	19.369	23.542	26.296	32.000
17	16.338	20.489	24.769	27.587	33.409
18	17.338	21.605	25.989	28.869	34.806
19	18.338	22.718	27.204	30.144	36.191
20	19.337	23.828	28.412	31.410	37.567
21	20.337	24.935	29.615	32.671	38.932
22	21.337	26.039	30.813	33.924	40.290
23	22.337	27.141	32.007	35.173	41.639
24	23.337	28.241	33.196	36.415	42.980
25	24.337	29.339	34.382	37.653	44.315
26	25.336	30.435	35.563	38.885	45.642
27	26.336	31.528	36.741	40.113	46.963
28	27.336	32.621	37.916	41.337	48.278
29	28.336	33.711	39.088	42.557	49.588
30	29.336	34.800	40.256	43.773	50.893
31	30.336	35.887	41.422	44.985	52.192
32	31.336	37.973	42.585	46.194	53.486
33	32.336	38.058	43.745	47.400	54.776
34	33.336	39.141	44.903	48.603	56.061
35	34.336	40.223	46.059	49.802	57.343
36	35.336	41.304	47.212	50.999	58.620
37	36.336	42.383	48.364	52.192	59.893
38	37.335	43.462	49.513	53.384	61.163
39	38.335	44.539	50.660	54.572	62.428
40	39.335	45.616	51.805	55.759	63.691
41	40.335	46.692	52.949	56.942	64.951
42	41.335	47.766	54.090	58.124	66.207
43	42.335	48.840	55.230	59.304	67.460
44	43.335	49.913	56.369	60.481	68.710
45	44.335	50.985	57.505	61.656	69.957
46	45.335	52.056	58.641	62.830	71.202
47	46.335	53.127	59.774	64.001	72.444
48	47.335	54.196	60.907	65.171	73.683
49	48.335	55.265	62.038	66.339	74.920
50	49.335	56.334	63.167	67.505	76.154

Table H Critical Values for the Runs Test

Use of the table: Values in the table represent critical values for *1-tailed* critical regions at the given α levels. For 2-sided tests, the pair of integers will determine a critical region corresponding to a 2α level of significance. Further discussion of the table appears in Chapter 7, Section 1.

n_1	n_2	2	3	4	5	6	7	8	9	10	11
3	0.20	2, 5	2, 6								
	0.10	0, 5	2, 6								
	0.05	0, 6	0, 7								
	0.01	0, 6	0, 7								
				(4)							
4	0.20	2, 5	3, 7	3, 7							
	0.10	0, 6	2, 7	2, 8							
	0.05	0, 6	0, 7	2, 8							
	0.01	0, 6	0, 8	0, 9							
					(5)						
5	0.20	2, 6	3, 7	3, 8	4, 8						
	0.10	2, 6	2, 7	3, 8	3, 9						
	0.05	0, 6	2, 8	2, 9	3, 9						
	0.01	0, 6	0, 8	0, 9	2,10						
						(6)					
6	0.20	2, 6	3, 7	4, 8	4, 9	5, 9					
	0.10	2, 6	2, 8	3, 9	3, 9	4,10					
	0.05	0, 6	2, 8	3, 9	3,10	3,11					
	0.01	0, 6	0, 8	2,10	2,11	2,12					
							(7)				
7	0.20	2, 6	3, 7	4, 8	5, 9	5,10	5,11				
	0.10	2, 6	3, 8	3, 9	4,10	4,11	5,11				
	0.05	0, 6	2, 8	3, 9	3,10	4,11	4,12				
	0.01	0, 6	0, 8	2,10	2,11	3,12	3,13				
								(8)			
8	0.20	2, 6	3, 8	4, 9	5,10	5,10	6,11	6,12			
	0.10	2, 6	3, 8	3, 9	4,10	5,11	5,12	5,13			
	0.05	2, 6	2, 8	3,10	3,11	4,12	4,13	5,13			
	0.01	0, 6	0, 8	2,10	2,12	3,13	3,14	4,14			
									(9)		
9	0.20	3, 6	4, 8	4, 9	5,10	6,11	6,11	7,12	7,13		
	0.10	2, 6	3, 8	4, 9	4,10	5,11	5,12	6,13	6,14		
	0.05	2, 6	2, 8	3,10	4,11	4,12	5,13	5,14	6,14		
	0.01	0, 6	2, 8	2,10	3,12	3,13	4,14	4,15	4,16		
										(10)	
10	0.20	3, 6	4, 8	4, 9	5,10	6,11	7,12	7,13	8,13	8,14	
	0.10	2, 6	3, 8	4,10	5,11	5,12	6,13	6,13	7,14	7,15	
	0.05	2, 6	3, 8	3,10	4,11	5,12	5,13	6,14	6,15	6,16	
	0.01	0, 6	2, 8	2,10	3,12	3,14	4,15	4,15	5,16	5,17	
											(11)
11	0.20	3, 6	4, 8	5, 9	5,10	6,11	7,12	8,13	8,14	9,14	9,15
	0.10	2, 6	3, 8	4,10	5,11	5,12	6,13	7,14	7,15	8,15	8,16
	0.05	2, 6	3, 8	3,10	4,12	5,13	5,14	6,15	6,15	7,16	7,17
	0.01	0, 6	2, 8	2,10	3,12	4,14	4,15	5,16	5,17	5,18	6,18

Table H (contd.)

n_1	n_2	2	3	4	5	6	7	8	9	10
12	0.20	3, 6	4, 8	5, 9	6,10	6,12	7,12	8,13	8,14	9,15
	0.10	2, 6	3, 8	4,10	5,11	6,12	6,13	7,14	7,15	8,16
	0.05	2, 6	3, 8	4,10	4,12	5,13	6,14	6,15	7,16	7,17
	0.01	0, 6	2, 8	3,10	3,12	4,14	4,15	5,16	5,17	6,18
13	0.20	3, 6	4, 8	5,10	6,11	7,12	7,13	8,14	9,15	9,15
	0.10	2, 6	3, 8	4,10	5,11	6,12	7,14	7,15	8,15	8,16
	0.05	2, 6	3, 8	4,10	4,12	5,13	6,14	6,15	7,16	8,17
	0.01	0, 6	2, 8	3,10	3,12	4,14	5,16	5,17	6,18	6,19
14	0.20	3, 6	4, 8	5,10	6,11	7,12	8,13	8,14	9,15	10,16
	0.10	2, 6	3, 8	4,10	5,12	6,13	7,14	7,15	8,16	9,17
	0.05	2, 6	3, 8	4,10	5,12	5,13	6,14	7,16	7,17	8,17
	0.01	0, 6	2, 8	3,10	3,12	4,14	5,16	5,17	6,18	6,19
15	0.20	3, 6	4, 8	5,10	6,11	7,12	8,13	9,14	9,15	10,16
	0.10	2, 6	4, 8	4,10	5,12	6,13	7,14	8,15	8,16	9,17
	0.05	2, 6	3, 8	4,10	5,12	6,14	6,15	7,16	8,17	8,18
	0.01	0, 6	2, 8	3,10	4,12	4,14	5,16	5,17	6,18	7,19
16	0.20	3, 6	4, 8	5,10	6,11	7,12	8,14	9,14	10,15	10,16
	0.10	2, 6	4, 8	5,10	6,12	6,13	7,14	8,15	9,16	9,17
	0.05	2, 6	3, 8	4,10	5,12	6,14	6,15	7,16	8,17	8,18
	0.01	0, 6	2, 8	3,10	4,12	4,14	5,16	6,17	6,18	7,20
17	0.20	3, 6	4, 8	5,10	6,11	7,12	8,14	9,15	10,16	11,17
	0.10	2, 6	4, 8	5,10	6,12	6,13	7,14	8,16	9,17	10,18
	0.05	2, 6	3, 8	4,10	5,12	6,14	7,15	7,16	8,17	9,18
	0.01	0, 6	2, 8	3,10	4,12	5,14	5,16	6,18	7,19	7,20
18	0.20	3, 6	4, 8	6,10	6,11	8,12	8,14	9,15	10,16	11,17
	0.10	2, 6	4, 8	5,10	6,12	7,13	8,14	8,16	9,17	10,18
	0.05	2, 6	3, 8	4,10	5,12	6,14	7,15	8,16	8,18	9,19
	0.01	0, 6	2, 8	3,10	4,12	5,14	5,16	6,18	7,19	7,20
19	0.20	3, 6	4, 8	6,10	7,12	8,13	9,14	10,15	10,16	11,17
	0.10	3, 6	4, 8	5,10	6,12	7,14	8,15	8,16	9,17	10,18
	0.05	2, 6	3, 8	4,10	5,12	6,14	7,15	8,16	8,18	9,19
	0.01	2, 6	2, 8	3,10	4,12	5,14	6,16	6,18	7,19	8,20
20	0.20	3, 6	4, 8	6,10	7,12	8,13	9,14	10,15	10,16	11,17
	0.10	3, 6	4, 8	5,10	6,12	7,14	8,15	9,16	10,17	10,18
	0.05	2, 6	3, 8	4,10	5,12	6,14	7,16	8,17	9,18	9,19
	0.01	2, 6	2, 8	3,10	4,12	5,14	6,16	6,18	7,19	8,20

	11	12	13	14	15	16	17	18	19	20
12	9,15	10,16								
	9,16	9,17								
	8,17	8,18								
	6,19	7,19								
			(13)							
13	10,16	10,17	11,17							
	9,17	9,18	10,18							
	8,18	9,18	9,19							
	6,19	7,20	7,21							
				(14)						
14	10,16	11,17	11,18	12,18						
	9,17	10,18	10,19	11,19						
	8,18	9,19	9,20	10,20						
	7,20	7,21	8,21	8,22						
					(15)					
15	11,17	11,17	12,18	12,19	13,19					
	10,18	10,19	11,19	11,20	12,20					
	9,19	9,19	10,20	10,21	22,21					
	7,20	8,21	8,22	8,23	9,23					
						(16)				
16	11,17	12,18	12,19	13,19	13,20	14,20				
	10,18	10,19	11,20	11,20	12,21	12,22				
	9,19	10,20	10,21	11,21	11,22	11,23				
	7,20	8,22	8,22	9,23	9,24	10,24				
							(17)			
17	11,17	12,18	13,19	13,20	14,20	14,21	15,21			
	10,18	11,19	11,20	12,21	12,21	13,22	13,23			
	9,19	10,20	10,21	11,22	11,22	12,23	12,24			
	8,21	8,22	9,23	9,24	10,24	10,25	10,26			
								(18)		
18	12,18	12,19	13,19	13,20	14,21	15,21	15,22	15,23		
	10,19	11,20	12,20	12,21	13,22	13,23	14,23	14,24		
	10,20	10,21	11,21	11,22	12,23	12,24	13,24	13,25		
	8,21	8,22	9,23	9,24	10,25	10,26	11,26	11,27		
									(19)	
19	12,18	13,19	13,20	14,20	14,21	15,22	15,22	16,23	16,24	
	11,19	11,20	12,21	13,22	13,22	14,23	14,24	15,24	15,25	
	10,20	10,21	11,22	12,23	12,23	13,24	13,25	14,25	14,26	
	8,22	9,23	9,24	10,24	10,25	11,26	11,27	12,27	12,28	
										(20)
20	12,18	13,19	13,20	14,21	15,22	15,22	16,23	16,23	17,24	17,25
	11,19	12,20	12,21	13,22	13,23	14,24	15,24	15,25	16,25	16,26
	10,20	11,21	11,22	12,23	12,24	13,25	13,25	14,26	14,27	15,27
	8,22	9,23	10,24	10,25	11,26	11,26	11,27	12,20	12,29	13,29

Index